U.S. Customary Units and Their SI Equivalents

Quantity	U.S. Customary Unit	SI Equivalent
Acceleration	ft/s^2	0.3048 m/s^2
	in./s^2	0.0254 m/s^2
Area	ft^2	0.0929 m^2
	in^2	645.2 mm^2
Energy	ft $\cdot$ lb	1.356 J
Force	kip	4.448 kN
	lb	4.448 N
	oz	0.2780 N
Impulse	lb $\cdot$ s	4.448 N $\cdot$ s
Length	ft	0.3048 m
	in.	25.40 mm
	mi	1.609 km
Mass	oz mass	28.35 g
	lb mass	0.4536 kg
	slug	14.59 kg
	ton	907.2 kg
Moment of a force	lb $\cdot$ ft	1.356 N $\cdot$ m
	lb $\cdot$ in.	0.1130 N $\cdot$ m
Moment of inertia		
Of an area	in^4	0.4162×10^6 mm^4
Of a mass	lb $\cdot$ ft $\cdot$ s^2	1.356 kg $\cdot$ m^2
Momentum	lb $\cdot$ s	4.448 kg $\cdot$ m/s
Power	ft $\cdot$ lb/s	1.356 W
	hp	745.7 W
Pressure or stress	lb/ft^2	47.88 Pa
	lb/in^2 (psi)	6.895 kPa
Velocity	ft/s	0.3048 m/s
	in./s	0.0254 m/s
	mi/h (mph)	0.4470 m/s
	mi/h (mph)	1.609 km/h
Volume, solids	ft^3	0.02832 m^3
	in^3	16.39 cm^3
Liquids	gal	3.785 l
	qt	0.9464 l
Work	ft $\cdot$ lb	1.356 J

Mechanics for Engineers
DYNAMICS

McGraw-Hill Book Company

New York
St. Louis
San Francisco
Auckland
Düsseldorf
Johannesburg
Kuala Lumpur
London
Mexico
Montreal
New Delhi
Panama
Paris
São Paulo
Singapore
Sydney
Tokyo
Toronto

Third Edition

Mechanics for Engineers

DYNAMICS

Ferdinand P. Beer

Professor and Chairman
Department of Mechanical Engineering and Mechanics
Lehigh University

E. Russell Johnston, Jr.

Professor and Head
Department of Civil Engineering
University of Connecticut

Mechanics for Engineers
DYNAMICS

Copyright © 1957, 1962, 1976 by McGraw-Hill, Inc. All rights reserved.
Printed in the United States of America. No part of this publication may be
reproduced, stored in a retrieval system, or transmitted, in any form or by any
means, electronic, mechanical, photocopying, recording, or otherwise,
without the prior written permission of the publisher.

0   DODO   8 3

This book was set in Laurel by York Graphic Services, Inc. The editors were
B. J. Clark and J. W. Maisel; the designer was Merrill Haber; the production
supervisor was Thomas J. LoPinto. The drawings were done by Felix
Cooper. Cover photographs were taken by Milton J. Heiberg.
R. R. Donnelley & Sons Company was printer and binder.

Library of Congress Cataloging in Publication Data

Beer, Ferdinand Pierre, date
    Mechanics for engineers.

    CONTENTS: [1] Statics.—[2] Dynamics.
    1. Mechanics. I. Title. I. Johnston, Elwood
Russell, date   joint author. II. Title.
QA807.B39   1976   531   75-34005
ISBN 0-07-004271-3 (Statics)
ISBN 0-07-004273-x (Dynamics)

CONTENTS

Preface   ix

List of Symbols   xiii

11

KINEMATICS OF PARTICLES   399

11.1   Introduction to Dynamics   399

RECTILINEAR MOTION OF PARTICLES   400

11.2   Position, Velocity, and Acceleration   400
11.3   Determination of the Motion of a Particle   404
11.4   Uniform Rectilinear Motion   412
11.5   Uniformly Accelerated Rectilinear Motion   412
11.6   Motion of Several Particles   414
*__11.7__   Graphical Solution of Rectilinear-Motion Problems   420
*__11.8__   Other Graphical Methods   422

CURVILINEAR MOTION OF PARTICLES   428

11.9   Position Vector, Velocity, and Acceleration   428
11.10 Component Motions   432
11.11 Relative Motion   433
11.12 Tangential and Normal Components   442
11.13 Radial and Transverse Components   444

12

KINETICS OF PARTICLES:
FORCE, MASS, AND ACCELERATION   454

12.1   Newton's Second Law of Motion   454
12.2   Systems of Units   456
12.3   Equations of Motion. Dynamic Equilibrium   459
12.4   Systems of Particles   460
12.5   Motion of the Mass Center of a System of Particles   462
12.6   Rectilinear Motion of a Particle   463

v

12.7 Curvilinear Motion of a Particle   472
12.8 Motion under a Central Force   473
12.9 Newton's Law of Gravitation   474
***12.10** Trajectory of a Particle under a Central Force   484
***12.11** Application to Space Mechanics   485
***12.12** Kepler's Laws of Planetary Motion   489

13
KINETICS OF PARTICLES: WORK AND ENERGY   496

13.1 Introduction   496
13.2 Work of a Force   496
13.3 Kinetic Energy of a Particle. Principle of Work and Energy   500
13.4 Applications of the Principle of Work and Energy   503
13.5 Systems of Particles   504
13.6 Potential Energy   513
13.7 Conservation of Energy   516
13.8 Power and Efficiency   525

14
KINETICS OF PARTICLES: IMPULSE AND MOMENTUM   532

14.1 Principle of Impulse and Momentum   532
14.2 Impulsive Motion   534
14.3 Systems of Particles   535
14.4 Conservation of Momentum   536
14.5 Impact   543
14.6 Direct Central Impact   543
14.7 Oblique Central Impact   546
14.8 Problems Involving Energy and Momentum   547
14.9 Angular Momentum   556
14.10 Conservation of Angular Momentum   556
14.11 Application to Space Mechanics   557
14.12 Principle of Impulse and Momentum for a System of Particles   558
***14.13** Variable Systems of Particles   566
***14.14** Steady Stream of Particles   566
***14.15** Systems Gaining or Losing Mass   570

15
KINEMATICS OF RIGID BODIES
583

15.1	Various Types of Plane Motion	583
15.2	Translation	585
15.3	Rotation	586
15.4	Linear and Angular Velocity, Linear and Angular Acceleration in Rotation	588
15.5	General Plane Motion	593
15.6	Absolute and Relative Velocity in Plane Motion	595
15.7	Instantaneous Center of Rotation in Plane Motion	603
15.8	Absolute and Relative Acceleration in Plane Motion	610
***15.9**	Analysis of Plane Motion in Terms of a Parameter	613
***15.10**	Particle Moving on a Slab in Translation	619
***15.11**	Particle Moving on a Rotating Slab. Coriolis Acceleration	621

16
KINETICS OF RIGID BODIES: FORCES AND ACCELERATIONS
631

16.1	Introduction	631
16.2	Plane Motion of a Rigid Body. D'Alembert's Principle	632
16.3	Solution of Problems Involving the Plane Motion of Rigid Bodies	635
16.4	Systems of Rigid Bodies	637
16.5	Constrained Plane Motion	653
16.6	Rotation of a Three-dimensional Body about a Fixed Axis	673

17
KINETICS OF RIGID BODIES: WORK AND ENERGY
685

17.1	Principle of Work and Energy for a Rigid Body	685
17.2	Work of Forces Acting on a Rigid Body	687
17.3	Kinetic Energy in Translation	688
17.4	Kinetic Energy in Rotation	688
17.5	Systems of Rigid Bodies	689
17.6	Kinetic Energy in Plane Motion	695
17.7	Conservation of Energy	696
17.8	Power	697

18
KINETICS OF RIGID BODIES: IMPULSE AND MOMENTUM 709

18.1	Principle of Impulse and Momentum for a Rigid Body	709
18.2	Momentum of a Rigid Body in Plane Motion	710
18.3	Application of the Principle of Impulse and Momentum to the Analysis of the Plane Motion of a Rigid Body	713
18.4	Systems of Rigid Bodies	715
18.5	Conservation of Angular Momentum	716
18.6	Impulsive Motion	725
18.7	Eccentric Impact	725
***18.8**	Gyroscopes	735

19
MECHANICAL VIBRATIONS 745

19.1	Introduction	745
VIBRATIONS WITHOUT DAMPING		746
19.2	Free Vibrations of Particles. Simple Harmonic Motion	746
19.3	Simple Pendulum (Approximate Solution)	750
***19.4**	Simple Pendulum (Exact Solution)	751
19.5	Free Vibrations of Rigid Bodies	757
19.6	Application of the Principle of Conservation of Energy	765
19.7	Forced Vibrations	771
DAMPED VIBRATIONS		777
***19.8**	Damped Free Vibrations	777
***19.9**	Damped Forced Vibrations	780
***19.10**	Electrical Analogues	781

Appendix	**Moments of Inertia of Masses**	791
Index		809
Answers to Even-numbered Problems		819

PREFACE

The main objective of a first course in mechanics should be to develop in the engineering student the ability to analyze any problem in a simple and logical manner and to apply to its solution a few, well-understood, basic principles. It is hoped that this text, as well as the preceding volume, *Mechanics for Engineers: Statics*, will help the instructor achieve this goal.†

In this third edition, the vectorial character of mechanics has again been emphasized. The concept of vectors and the laws governing the addition and the resolution of vectors were discussed at the beginning of the volume on statics. Throughout the text, forces, velocities, accelerations, and other vector quantities have been clearly distinguished from scalar quantities through the use of boldface type. Products and derivatives of vectors, however, are not used in this text.‡

One of the characteristics of the approach used in these volumes is that the mechanics of *particles* has been clearly separated from the mechanics of *rigid bodies*. This approach makes it possible to consider simple practical applications at an early stage and to postpone the introduction of more difficult concepts. In the volume on statics, the statics of particles was treated first, and the principle of equilibrium was immediately applied to practical situations involving only concurrent forces. The statics of rigid bodies was considered later, at which time the principle of transmissibility and the associated concept of moment of a force were introduced. In this volume, the same division is observed. The basic concepts of force, mass, and acceleration, of work and energy, and of impulse and momentum are introduced and first applied to problems involving only particles. Thus the student may familiarize himself with the three basic methods used in dynamics and learn their respective advantages before facing the difficulties associated with the motion of rigid bodies.

Although the authors strongly believe in the advantage of the approach they use, they do not wish to impose their choice on the instructor. Consequently, the various chapters have been written

---

† Both texts are also available in a single volume, *Mechanics for Engineers: Statics and Dynamics.*

‡ In a parallel text, *Vector Mechanics for Engineers: Dynamics,* vector analysis is used throughout the presentation of dynamics, making it possible to analyze more advanced problems in three-dimensional kinematics and kinetics.

so that they may be taught in alternate sequences. For example, Chap. 15 may be taken immediately after Chap. 11 if the instructor prefers to keep kinematics apart from kinetics, and Chaps. 13 and 14 may be postponed if he wishes to teach the method of work and energy (Chaps. 13 and 17) and the method of impulse and momentum (Chaps. 14 and 18) as single units. Whatever the sequence used, the very fact that distinct chapters are devoted to the dynamics of particles and to the dynamics of rigid bodies should help the student organize his own thoughts.

Since this text is designed for a first course in dynamics, new concepts have been presented in simple terms and every step explained in detail. On the other hand, by discussing the broader aspects of the problems considered and by stressing methods of general applicability, a definite maturity of approach has been achieved. For example, the concepts of mass and weight are carefully distinguished, and the limitations of the methods used to analyze plane motion are indicated, so that the student will not be tempted to apply them to cases in which they are not valid.

The fact that mechanics is essentially a *deductive* science based on a few fundamental principles has been stressed. Derivations have been presented in their logical sequence and with all the rigor warranted at this level. However, the learning process being largely *inductive*, simple applications have been considered first. Thus the dynamics of particles precedes the dynamics of rigid bodies, and, in the latter, the emphasis has been placed on the study of plane motion.

Free-body diagrams were introduced early in statics. They were used not only to solve equilibrium problems but also to express the equivalence of two systems of forces or, more generally, of two systems of vectors. The advantage of this approach becomes apparent in the study of the dynamics of rigid bodies. By placing the emphasis on "free-body-diagram equations" rather than on the standard algebraic equations of motion, a more intuitive and more complete understanding of the fundamental principles of dynamics may be achieved. This approach, which was first introduced in 1962 in the first edition of *Vector Mechanics for Engineers*, has now gained wide acceptance among mechanics teachers in this country. It is, therefore, being used in preference to the method of dynamic equilibrium in the solution of sample problems in this new edition of *Mechanics for Engineers*.

Color has been used in this edition to distinguish forces from other elements of the free-body diagrams. This makes it easier for the students to identify the forces acting on a given particle or rigid body and to follow the discussion of sample problems and other examples given in the text.

Because of the current trend among American engineers to

adopt the international system of units (SI metric units), the SI units most frequently used in mechanics were introduced in Chap. 1 of *Statics*. They are discussed again in Chap. 12 of this volume and used throughout the text. Half the sample problems and problems to be assigned have been stated in these units, while the other half retain U.S. customary units. The authors believe that this approach will best serve the needs of the students, who will be entering the engineering profession during the period of transition from one system of units to the other. It also should be recognized that the passage from one system to the other entails more than the use of conversion factors. Since the SI system of units is an absolute system based on the units of time, length, and mass, whereas the U.S. customary system is a gravitational system based on the units of time, length, and force, different approaches are required for the solution of many problems. For example, when SI units are used, a body is generally specified by its mass expressed in kilograms; in most problems of statics it was necessary to determine the weight of the body in newtons, and an additional calculation was required for this purpose. On the other hand, when U.S. customary units are used, a body is specified by its weight in pounds and, in dynamics problems, an additional calculation will be required to determine its mass in slugs (or $lb \cdot sec^2/ft$). The authors, therefore, believe that problems assignments should include both types of units. A sufficient number of problems, however, have been provided so that if so desired, two complete sets of assignments may be selected from problems stated in SI units only and two others from problems stated in U.S. customary units. Since the answers to all even-numbered problems stated in U.S. customary units have been given in both systems of units, teachers who wish to give special instructions to their students in the conversion of units may assign these problems and ask their students to use SI units in their solutions. This has been illustrated in two sample problems involving, respectively, the kinetics of particles (Sample Prob. 12.2) and the computation of mass moments of inertia (Sample Prob. 9.13 in the appendix).

A number of optional sections have been included. These sections are indicated by asterisks and may thus easily be distinguished from those which form the core of the basic dynamics course. They may be omitted without prejudice to the understanding of the rest of the text. The topics covered in these additional sections include graphical methods for the solution of rectilinear-motion problems, the trajectory of a particle under a central force, the deflection of fluid streams, problems involving jet and rocket propulsion, Coriolis acceleration, damped mechanical vibrations, and electrical analogues. These topics will be found of particular interest when dynamics is taught in the junior year.

The material presented in the text and most of the problems require no previous mathematical knowledge beyond algebra, trigonometry, and elementary calculus. However, special problems have been included, which make use of a more advanced knowledge of calculus, and certain sections, such as Secs. 19.8 and 19.9 on damped vibrations, should be assigned only if the students possess the proper mathematical background.

The text has been divided into units, each consisting of one or several theory sections, one or several sample problems, and a large number of problems to be assigned. Each unit corresponds to a well-defined topic and generally may be covered in one lesson. In a number of cases, however, the instructor will find it desirable to devote more than one lesson to a given topic. The sample problems have been set up in much the same form that a student will use in solving the assigned problems. They thus serve the double purpose of amplifying the text and demonstrating the type of neat and orderly work that the student should cultivate in his own solutions. Most of the problems to be assigned are of a practical nature and should appeal to the engineering student. They are primarily designed, however, to illustrate the material presented in the text and to help the student understand the basic principles of mechanics. The problems have been grouped according to the portions of material they illustrate and have been arranged in order of increasing difficulty. Problems requiring special attention have been indicated by asterisks. Answers to all even-numbered problems are given at the end of the book.

The authors wish to acknowledge gratefully the many helpful comments and suggestions offered by the users of the two previous editions of *Mechanics for Engineers* and of the first and second editions of *Vector Mechanics for Engineers*.

FERDINAND P. BEER
E. RUSSELL JOHNSTON, JR.

LIST OF SYMBOLS

$\mathbf{a}, a$	Acceleration
a	Constant; radius; distance; semimajor axis of ellipse
$\bar{\mathbf{a}}, \bar{a}$	Acceleration of mass center
$\mathbf{a}_{B/A}$	Relative acceleration of B with respect to A
$\mathbf{a}_c$	Coriolis acceleration
$\mathbf{A}, \mathbf{B}, \mathbf{C}, \ldots$	Reactions at supports and connections
$A, B, C, \ldots$	Points
A	Area
b	Width; distance; semiminor axis of ellipse
c	Constant; coefficient of viscous damping
C	Centroid; instantaneous center of rotation; capacitance
d	Distance
e	Coefficient of restitution; base of natural logarithms
E	Total mechanical energy; voltage
f	Frequency; function
$\mathbf{F}$	Force; friction force
g	Acceleration of gravity
G	Center of gravity; mass center; constant of gravitation
h	Angular momentum per unit mass
H_O	Angular momentum about point O
i	Current
$I, I_x, \ldots$	Moment of inertia
$\bar{I}$	Centroidal moment of inertia
J	Polar moment of inertia
k	Spring constant
k_x, k_y, k_o	Radius of gyration
$\bar{k}$	Centroidal radius of gyration
l	Length
L	Length; inductance
m	Mass
$\mathbf{M}$	Couple
$\mathbf{M}_O$	Moment about point O
M	Moment; mass of earth
n	Normal direction
N	Normal component of reaction

O	Origin of coordinates
p	Circular frequency
$\mathbf{P}$	Force; vector
$P_{xy}, \ldots$	Product of inertia
q	Electric charge
$\mathbf{Q}$	Force; vector
$\mathbf{r}$	Position vector
r	Radius; distance; polar coordinate
$\mathbf{R}$	Resultant force; resultant vector; reaction
R	Radius of earth; resistance
$\mathbf{s}$	Position vector
s	Distance; length of arc
t	Time; thickness; tangential direction
$\mathbf{T}$	Force
T	Tension; kinetic energy
$\mathbf{u}$	Velocity
u	Rectangular coordinate
U	Work
$\mathbf{v}, v$	Velocity
v	Speed; rectangular coordinate
$\overline{\mathbf{v}}, \overline{v}$	Velocity of mass center
$\mathbf{v}_{B/A}$	Relative velocity of B with respect to A
V	Volume; potential energy
w	Load per unit length
$\mathbf{W}, W$	Weight; load
x, y, z	Rectangular coordinates; distances
$\dot{x}, \dot{y}, \dot{z}$	Time derivatives of coordinates x, y, z
$\overline{x}, \overline{y}, \overline{z}$	Rectangular coordinates of centroid, center of gravity, or mass center
$\boldsymbol{\alpha}, \alpha$	Angular acceleration
α, β, γ	Angles
γ	Specific weight
δ	Elongation
ε	Eccentricity of conic section or of orbit
η	Efficiency
θ	Angular coordinate; angle; polar coordinate
μ	Coefficient of friction
ρ	Density; radius of curvature
τ	Period; periodic time
ϕ	Angle of friction; phase angle; angle
φ	Phase difference
$\boldsymbol{\omega}, \omega$	Angular velocity
ω	Circular frequency of forced vibration
Ω	Rate of precession

Mechanics for Engineers
DYNAMICS

Chapter
11

Kinematics of Particles

11.1. Introduction to Dynamics. Chapters 1 to 10 were devoted to *statics,* i.e., to the analysis of bodies at rest. We shall now begin the study of *dynamics,* which is the part of mechanics dealing with the analysis of bodies in motion.

While the study of statics goes back to the time of the Greek philosophers, the first significant contribution to dynamics was made by Galileo (1564–1642). His experiments on uniformly accelerated bodies led Newton (1642–1727) to formulate his fundamental laws of motion.

Dynamics is divided into two parts: (1) *Kinematics,* which is the study of the geometry of motion; kinematics is used to relate displacement, velocity, acceleration, and time, without reference to the cause of the motion. (2) *Kinetics,* which is the study of the relation existing between the forces acting on a body, the mass of the body, and the motion of the body; kinetics is used to predict the motion caused by given forces or to determine the forces required to produce a given motion.

Chapters 11 to 14 are devoted to the *dynamics of particles,* and Chap. 11 more particularly to the *kinematics of particles.* The use of the word particles does not imply that we shall restrict our study to that of small corpuscles; it rather indicates that in these first chapters we shall study the motion of bodies— possibly as large as cars, rockets, or airplanes—without regard to their size. By saying that the bodies are analyzed as particles, we mean that only their motion as an entire unit will be considered; any rotation about their own mass center will be neglected. There are cases, however, when such a rotation is not negligible; the bodies, then, may not be considered as particles. The analysis of such motions will be carried out in later chapters dealing with the *dynamics of rigid bodies*.

RECTILINEAR MOTION OF PARTICLES

11.2. Position, Velocity, and Acceleration. A particle moving along a straight line is said to be in *rectilinear motion*. At any given instant t, the particle will occupy a certain position on the straight line. To define the position P of the particle, we choose a fixed origin O on the straight line and a positive direction along the line. We measure the distance x from O to P and record it with a plus or minus sign, according to whether P is reached from O by moving along the line in the positive or the negative direction. The distance x, with the appropriate sign, completely defines the position of the particle; it is called the *position coordinate* of the particle considered. For example, the position coordinate corresponding to P in Fig. 11.1*a* is $x = +5$ m, while the coordinate corresponding to P' in Fig. 11.1*b* is $x' = -2$ m.

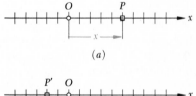

(*a*)

(*b*)   **Fig. 11.1**

When the position coordinate x of a particle is known for every value of time t, we say that the motion of the particle is known. The "timetable" of the motion may be given in the form of an equation in x and t, such as $x = 6t^2 - t^3$, or in the form of a graph of x vs. t as shown in Fig. 11.6. The units most generally used to measure the position coordinate x are the meter (m) in the SI system of units,† and the foot (ft) in the U.S. customary system of units. Time t will generally be measured in seconds (s).

† Cf. Sec. 1.3.

Consider the position P occupied by the particle at time t and the corresponding coordinate x (Fig. 11.2). Consider also the position P' occupied by the particle at a later time $t + \Delta t$; the position coordinate of P' may be obtained by adding to the coordinate x of P the small displacement Δx, which will be positive or negative according to whether P' is to the right or to the left of P. The *average velocity* of the particle over the time interval Δt is defined as the quotient of the displacement Δx and the time interval Δt,

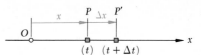

Fig. 11.2

$$\text{Average velocity} = \frac{\Delta x}{\Delta t}$$

If SI units are used, Δx is expressed in meters and Δt in seconds; the average velocity will thus be expressed in meters per second (m/s). If U.S. customary units are used, Δx is expressed in feet and Δt in seconds; the average velocity will then be expressed in feet per second (ft/s).

The *instantaneous velocity* v of the particle at the instant t is obtained from the average velocity by choosing shorter and shorter time intervals Δt and displacements Δx,

$$\text{Instantaneous velocity} = v = \lim_{\Delta t \to 0} \frac{\Delta x}{\Delta t}$$

The instantaneous velocity will also be expressed in m/s or ft/s. Observing that the limit of the quotient is equal, by definition, to the derivative of x with respect to t, we write

$$v = \frac{dx}{dt} \tag{11.1}$$

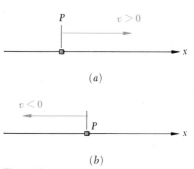

Fig. 11.3

The velocity v is represented by an algebraic number which may be positive or negative.† A positive value of v indicates that x increases, i.e., that the particle moves in the positive direction (Fig. 11.3a); a negative value of v indicates that x decreases, i.e., that the particle moves in the negative direction (Fig. 11.3b). The magnitude of v is known as the *speed* of the particle.

---

† As we shall see in Sec. 11.9, the velocity is actually a vector quantity. However, since we are considering here the rectilinear motion of a particle, where the velocity of the particle has a known and fixed direction, we need only specify the sense and magnitude of the velocity; this may be conveniently done by using a scalar quantity with a plus or minus sign. The same remark will apply to the acceleration of a particle in rectilinear motion.

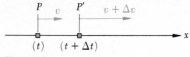

Fig. 11.4

Consider the velocity v of the particle at time t and also its velocity $v + \Delta v$ at a later time $t + \Delta t$ (Fig. 11.4). The *average acceleration* of the particle over the time interval Δt is defined as the quotient of Δv and Δt,

$$\text{Average acceleration} = \frac{\Delta v}{\Delta t}$$

If SI units are used, Δv is expressed in m/s and Δt in seconds; the average acceleration will thus be expressed in m/s². If U.S. customary units are used, Δv is expressed in ft/s and Δt in seconds; the average acceleration will then be expressed in ft/s².

The *instantaneous acceleration* a of the particle at the instant t is obtained from the average acceleration by choosing smaller and smaller values for Δt and Δv,

$$\text{Instantaneous acceleration} = a = \lim_{\Delta t \to 0} \frac{\Delta v}{\Delta t}$$

The instantaneous acceleration will also be expressed in m/s² or ft/s². The limit of the quotient is by definition the derivative of v with respect to t and measures the rate of change of the velocity. We write

$$a = \frac{dv}{dt} \tag{11.2}$$

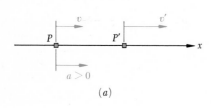

(a)

or, substituting for v from (11.1),

$$a = \frac{d^2x}{dt^2} \tag{11.3}$$

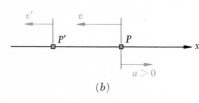

(b)

The acceleration a is represented by an algebraic number which may be positive or negative.† A positive value of a indicates that the velocity (i.e., the algebraic number v) increases. This may mean that the particle is moving faster in the positive direction (Fig. 11.5a) or that it is moving more slowly in the negative direction (Fig. 11.5b); in both cases, Δv is positive. A negative value of a indicates that the velocity decreases; either the particle is moving more slowly in the positive direction (Fig. 11.5c), or it is moving faster in the negative direction (Fig. 11.5d).

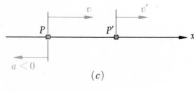

(c)

The term *deceleration* is sometimes used to refer to a when the speed of the particle (i.e., the magnitude of v) decreases; the particle is then moving more slowly. For example, the particle of Fig. 11.5 is decelerated in parts b and c, while it is truly accelerated (i.e., moves faster) in parts a and d.

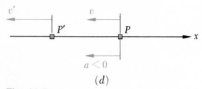

(d)

Fig. 11.5

Another expression may be obtained for the acceleration by eliminating the differential dt in Eqs. (11.1) and (11.2). Solving

†See footnote, page 401.

(11.1) for dt, we obtain $dt = dx/v$; carrying into (11.2), we write

$$a = v\frac{dv}{dx} \qquad (11.4)$$

Example. Consider a particle moving in a straight line, and assume that its position is defined by the equation

$$x = 6t^2 - t^3$$

where t is expressed in seconds and x in meters. The velocity v at any time t is obtained by differentiating x with respect to t,

$$v = \frac{dx}{dt} = 12t - 3t^2$$

The acceleration a is obtained by differentiating again with respect to t,

$$a = \frac{dv}{dt} = 12 - 6t$$

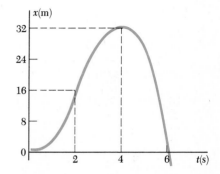

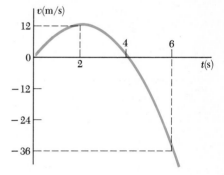

The position coordinate, the velocity, and the acceleration have been plotted against t in Fig. 11.6. The curves obtained are known as *motion curves*. It should be kept in mind, however, that the particle does not move along any of these curves; the particle moves in a straight line. Since the derivative of a function measures the slope of the corresponding curve, the slope of the x–t curve at any given time is equal to the value of v at that time and the slope of the v–t curve is equal to the value of a. Since $a = 0$ at $t = 2$ s, the slope of the v–t curve must be zero at $t = 2$ s; the velocity reaches a maximum at this instant. Also, since $v = 0$ at $t = 0$ and at $t = 4$ s, the tangent to the x–t curve must be horizontal for both of these values of t.

A study of the three motion curves of Fig. 11.6 shows that the motion of the particle from $t = 0$ to $t = \infty$ may be divided into four phases:

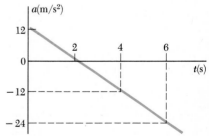

Fig. 11.6

1. The particle starts from the origin, $x = 0$, with no velocity but with a positive acceleration. Under this acceleration, the particle gains a positive velocity and moves in the positive direction. From $t = 0$ to $t = 2$ s, x, v, and a are all positive.
2. At $t = 2$ s, the acceleration is zero; the velocity has reached its maximum value. From $t = 2$ s to $t = 4$ s, v is positive, but a is negative; the particle still moves in the positive direction but more and more slowly; the particle is decelerated.
3. At $t = 4$ s, the velocity is zero; the position coordinate x has reached its maximum value. From then on, both v and a are negative; the particle is accelerated and moves in the negative direction with increasing speed.
4. At $t = 6$ s, the particle passes through the origin; its coordinate x is then zero, while the total distance traveled since the beginning of the motion is 64 m. For values of t larger than 6 s, x, v, and a will all be negative. The particle keeps moving in the negative direction, away from O, faster and faster.

11.3. Determination of the Motion of a Particle.

We saw in the preceding section that the motion of a particle is said to be known if the position of the particle is known for every value of the time t. In practice, however, a motion is seldom defined by a relation between x and t. More often, the conditions of the motion will be specified by the type of acceleration that the particle possesses. For example, a freely falling body will have a constant acceleration, directed downward and equal to 9.81 m/s^2 or 32.2 ft/s^2; a mass attached to a spring which has been stretched will have an acceleration proportional to the instantaneous elongation of the spring measured from the equilibrium position; etc. In general, the acceleration of the particle may be expressed as a function of one or more of the variables x, v, and t. In order to determine the position coordinate x in terms of t, it will thus be necessary to perform two successive integrations.

We shall consider three common classes of motion:

1. $a = f(t)$. *The Acceleration Is a Given Function of t.* Solving (11.2) for dv and substituting $f(t)$ for a, we write

$$dv = a \, dt$$
$$dv = f(t) \, dt$$

Integrating both members, we obtain the equation

$$\int dv = \int f(t) \, dt$$

which defines v in terms of t. It should be noted, however, that an arbitrary constant will be introduced as a result of the integration. This is due to the fact that there are many motions which correspond to the given acceleration $a = f(t)$. In order to uniquely define the motion of the particle, it is necessary to specify the *initial conditions* of the motion, i.e., the value v_0 of the velocity and the value x_0 of the position coordinate at $t = 0$. Replacing the indefinite integrals by *definite integrals* with lower limits corresponding to the initial conditions $t = 0$ and $v = v_0$ and upper limits corresponding to $t = t$ and $v = v$, we write

$$\int_{v_0}^{v} dv = \int_{0}^{t} f(t) \, dt$$

$$v - v_0 = \int_{0}^{t} f(t) \, dt$$

which yields v in terms of t.

We shall now solve (11.1) for dx,

$$dx = v \, dt$$

and substitute for v the expression just obtained. Both members are then integrated, the left-hand member with respect to x from $x = x_0$ to $x = x$, and the right-hand member with respect to t from $t = 0$ to $t = t$. The position coordinate x is thus obtained in terms of t; the motion is completely determined.

Two important particular cases will be studied in greater detail in Secs. 11.4 and 11.5: the case when $a = 0$, corresponding to a *uniform motion*, and the case when $a =$ constant, corresponding to a *uniformly accelerated motion*.

2. $a = f(x)$. *The Acceleration Is a Given Function of x.* Rearranging Eq. (11.4) and substituting $f(x)$ for a, we write

$$v \, dv = a \, dx$$
$$v \, dv = f(x) \, dx$$

Since each member contains only one variable, we may integrate the equation. Denoting again by v_0 and x_0, respectively, the initial values of the velocity and of the position coordinate, we obtain

$$\int_{v_0}^{v} v \, dv = \int_{x_0}^{x} f(x) \, dx$$
$$\tfrac{1}{2}v^2 - \tfrac{1}{2}v_0^2 = \int_{x_0}^{x} f(x) \, dx$$

which yields v in terms of x. We now solve (11.1) for dt,

$$dt = \frac{dx}{v}$$

and substitute for v the expression just obtained. Both members may be integrated, and the desired relation between x and t is obtained.

3. $a = f(v)$. *The Acceleration Is a Given Function of v.* We may then substitute $f(v)$ for a either in (11.2) or in (11.4) to obtain either of the following relations:

$$f(v) = \frac{dv}{dt} \qquad\qquad f(v) = v \frac{dv}{dx}$$

$$dt = \frac{dv}{f(v)} \qquad\qquad dx = \frac{v \, dv}{f(v)}$$

Integration of the first equation will yield a relation between v and t; integration of the second equation will yield a relation between v and x. Either of these relations may be used in conjunction with Eq. (11.1) to obtain the relation between x and t which characterizes the motion of the particle.

The position of a particle which moves along a straight line is defined by the relation $x = t^3 - 6t^2 - 15t + 40$, where x is expressed in feet and t in seconds. Determine (a) the time at which the velocity will be zero, (b) the position and distance traveled by the particle at that time, (c) the acceleration of the particle at that time, (d) the distance traveled by the particle from $t = 4\,\text{s}$ to $t = 6\,\text{s}$.

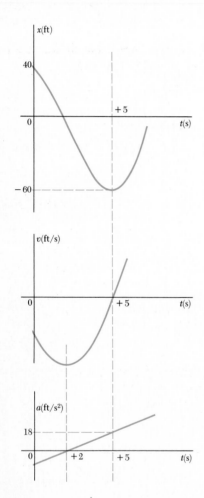

Solution. The equations of motion are

$$x = t^3 - 6t^2 - 15t + 40 \tag{1}$$

$$v = \frac{dx}{dt} = 3t^2 - 12t - 15 \tag{2}$$

$$a = \frac{dv}{dt} = 6t - 12 \tag{3}$$

a. Time at Which $v = 0$. We make $v = 0$ in (2),

$$3t^2 - 12t - 15 = 0 \qquad t = -1\,\text{s} \qquad \text{and} \qquad t = +5\,\text{s} \quad \blacktriangleleft$$

Only the root $t = +5\,\text{s}$ corresponds to a time after the motion has begun: for $t < 5\,\text{s}$, $v < 0$, the particle moves in the negative direction; for $t > 5\,\text{s}$, $v > 0$, the particle moves in the positive direction.

b. Position and Distance Traveled When $v = 0$. Carrying $t = +5\,\text{s}$ into (1), we have

$$x_5 = (5)^3 - 6(5)^2 - 15(5) + 40 \qquad x_5 = -60\,\text{ft} \quad \blacktriangleleft$$

The initial position at $t = 0$ was $x_0 = +40\,\text{ft}$. Since $v \neq 0$ during the interval $t = 0$ to $t = 5\,\text{s}$, we have

$$\text{Distance traveled} = x_5 - x_0 = -60\,\text{ft} - 40\,\text{ft} = -100\,\text{ft}$$

Distance traveled = 100 ft in the negative direction $\quad \blacktriangleleft$

c. Acceleration When $v = 0$. We carry $t = +5\,\text{s}$ into (3):

$$a_5 = 6(5) - 12 \qquad a_5 = +18\,\text{ft/s}^2 \quad \blacktriangleleft$$

d. Distance Traveled from $t = 4\,\text{s}$ to $t = 6\,\text{s}$. Since the particle moves in the negative direction from $t = 4\,\text{s}$ to $t = 5\,\text{s}$ and in the positive direction from $t = 5\,\text{s}$ to $t = 6\,\text{s}$, we shall compute separately the distance traveled during each of these time intervals.

From $t = 4\,\text{s}$ to $t = 5\,\text{s}$: $\qquad x_5 = -60\,\text{ft}$
$$x_4 = (4)^3 - 6(4)^2 - 15(4) + 40 = -52\,\text{ft}$$

$$\text{Distance traveled} = x_5 - x_4 = -60\,\text{ft} - (-52\,\text{ft}) = -8\,\text{ft}$$
$$= 8\,\text{ft in the negative direction}$$

From $t = 5\,\text{s}$ to $t = 6\,\text{s}$: $\qquad x_5 = -60\,\text{ft}$

$$x_6 = (6)^3 - 6(6)^2 - 15(6) + 40 = -50\,\text{ft}$$

$$\text{Distance traveled} = x_6 - x_5 = -50\,\text{ft} - (-60\,\text{ft}) = +10\,\text{ft}$$
$$= 10\,\text{ft in the positive direction}$$

Total distance traveled from $t = 4\,\text{s}$ to $t = 6\,\text{s}$ is

$$8\,\text{ft} + 10\,\text{ft} = 18\,\text{ft} \quad \blacktriangleleft$$

A ball is thrown from the top of a tower 18 m high, with a velocity of 12 m/s directed vertically upward. Knowing that the acceleration of the ball is constant and equal to 9.81 m/s² downward, determine (a) the velocity v and elevation y of the ball above the ground at any time t, (b) the highest elevation reached by the ball and the corresponding value of t, (c) the time when the ball will hit the ground and the corresponding velocity. Draw the v–t and y–t curves.

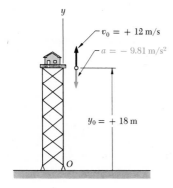

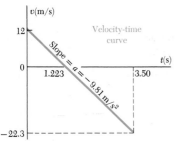

a. Velocity and Elevation. The y axis measuring the position coordinate (or elevation) is chosen with its origin O on the ground and its positive sense upward. The value of the acceleration and the initial values of v and y are as indicated. Substituting for a in $a = dv/dt$ and noting that, at $t = 0$, $v_0 = +12$ m/s, we have

$$\frac{dv}{dt} = a = -9.81 \text{ m/s}^2$$

$$\int_{v_0=12}^{v} dv = -\int_{0}^{t} 9.81 \, dt$$

$$[v]_{12}^{v} = -[9.81t]_0^t$$

$$v - 12 = -9.81t$$

$$v = 12 - 9.81t \quad (1) \quad \blacktriangleleft$$

Substituting for v in $v = dy/dt$ and noting that, at $t = 0$, $y_0 = 18$ m, we have

$$\frac{dy}{dt} = v = 12 - 9.81t$$

$$\int_{y_0=18}^{y} dy = \int_{0}^{t} (12 - 9.81t) \, dt$$

$$[y]_{18}^{y} = [12t - 4.90t^2]_0^t$$

$$y - 18 = 12t - 4.90t^2$$

$$y = 18 + 12t - 4.90t^2 \quad (2) \quad \blacktriangleleft$$

b. Highest Elevation. When the ball reaches its highest elevation, we have $v = 0$. Substituting into (1), we obtain

$$12 - 9.81t = 0 \qquad t = 1.223 \text{ s} \quad \blacktriangleleft$$

Carrying $t = 1.223$ s into (2), we have

$$y = 18 + 12(1.223) - 4.90(1.223)^2 \qquad y = 25.3 \text{ m} \quad \blacktriangleleft$$

c. Ball Hits the Ground. When the ball hits the ground, we have $y = 0$. Substituting into (2), we obtain

$$18 + 12t - 4.90t^2 = 0 \qquad t = -1.05 \text{ s} \qquad \text{and} \qquad t = +3.50 \text{ s} \quad \blacktriangleleft$$

Only the root $t = +3.50$ s corresponds to a time after the motion has begun. Carrying this value of t into (1), we have

$$v = 12 - 9.81(3.50) = -22.3 \text{ m/s} \qquad v = 22.3 \text{ m/s} \downarrow \quad \blacktriangleleft$$

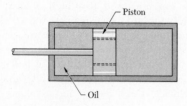

The brake mechanism used to reduce recoil in certain types of guns consists essentially of a piston which is attached to the barrel and may move in a fixed cylinder filled with oil. As the barrel recoils with an initial velocity v_0, the piston moves and oil is forced through orifices in the piston, causing the piston and the barrel to decelerate at a rate proportional to their velocity, i.e., $a = -kv$. Express (a) v in terms of t, (b) x in terms of t, (c) v in terms of x. Draw the corresponding motion curves.

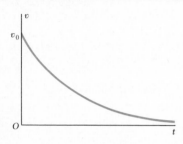

a. v in Terms of t. Substituting $-kv$ for a in the fundamental formula defining acceleration, $a = dv/dt$, we write

$$-kv = \frac{dv}{dt} \qquad \frac{dv}{v} = -k\,dt \qquad \int_{v_0}^{v} \frac{dv}{v} = -k \int_0^t dt$$

$$\ln \frac{v}{v_0} = -kt \qquad\qquad v = v_0 e^{-kt} \blacktriangleleft$$

b. x in Terms of t. Substituting the expression just obtained for v into $v = dx/dt$, we write

$$v_0 e^{-kt} = \frac{dx}{dt}$$

$$\int_0^x dx = v_0 \int_0^t e^{-kt}\,dt$$

$$x = -\frac{v_0}{k}[e^{-kt}]_0^t = -\frac{v_0}{k}(e^{-kt} - 1)$$

$$x = \frac{v_0}{k}(1 - e^{-kt}) \blacktriangleleft$$

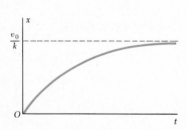

c. v in Terms of x. Substituting $-kv$ for a in $a = v\,dv/dx$, we write

$$-kv = v\frac{dv}{dx}$$

$$dv = -k\,dx$$

$$\int_{v_0}^{v} dv = -k \int_0^x dx$$

$$v - v_0 = -kx \qquad\qquad v = v_0 - kx \blacktriangleleft$$

Check. Part c could have been solved by eliminating t from the answers obtained for parts a and b. This alternate method may be used as a check. From part a we obtain $e^{-kt} = v/v_0$; substituting in the answer of part b, we obtain

$$x = \frac{v_0}{k}(1 - e^{-kt}) = \frac{v_0}{k}\left(1 - \frac{v}{v_0}\right) \qquad v = v_0 - kx \qquad \text{(checks)}$$

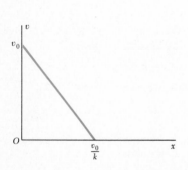

PROBLEMS

11.1 The motion of a particle is defined by the relation $x = t^3 - 9t^2 + 24t - 2$, where x is expressed in inches and t in seconds. Determine the position, velocity, and acceleration when $t = 5$ s.

11.2 The motion of a particle is defined by the relation $x = 2t^3 - 6t^2 + 15$, where x is expressed in inches and t in seconds. Determine the time, position, and acceleration when $v = 0$.

11.3 The motion of a particle is defined by the relation $x = t^2 - 10t + 30$, where x is expressed in meters and t in seconds. Determine (a) when the velocity is zero, (b) the position and the total distance traveled when $t = 8$ s.

11.4 The motion of a particle is defined by the relation $x = \frac{1}{3}t^3 - 3t^2 + 8t + 2$, where x is expressed in meters and t in seconds. Determine (a) when the velocity is zero, (b) the position and the total distance traveled when the acceleration is zero.

11.5 The acceleration of a particle is directly proportional to the time t. At $t = 0$, the velocity of the particle is $v = -16$ m/s. Knowing that both the velocity and the position coordinate are zero when $t = 4$ s, write the equations of motion for the particle.

11.6 The acceleration of a particle is defined by the relation $a = -2$ m/s². If $v = +8$ m/s and $x = 0$ when $t = 0$, determine the velocity, position, and total distance traveled when $t = 6$ s.

11.7 The acceleration of a particle is defined by the relation $a = 64 - 12t^2$. The particle starts at $t = 0$ with $v = 0$ and $x = 25$ in. Determine (a) the time when the velocity is again zero, (b) the position and velocity when $t = 5$ s, (c) the total distance traveled by the particle from $t = 0$ to $t = 5$ s.

11.8 The acceleration of a particle is defined by the relation $a = kt^2$. (a) Knowing that $v = -32$ ft/s when $t = 0$ and that $v = +32$ ft/s when $t = 4$ s, determine the constant k. (b) Write the equations of motion knowing also that $x = 0$ when $t = 4$ s.

11.9 The acceleration of a particle is defined by the relation $a = 21 - 12x^2$, where a is expressed in m/s² and x in meters. The particle starts with no initial velocity at the position $x = 0$. Determine (a) the velocity when $x = 1.5$ m, (b) the position where the velocity is again zero, (c) the position where the velocity is maximum.

11.10 The acceleration of an oscillating particle is defined by the relation $a = -kx$. Find the value of k such that $v = 10$ m/s when $x = 0$ and $x = 2$ m when $v = 0$.

11.11 The acceleration of a particle is defined by the relation $a = -kx^{-2}$. The particle starts with no initial velocity at $x = 8$ in., and it is observed that its velocity is 5 in./s when $x = 4$ in. Determine (*a*) the value of k, (*b*) the velocity of the particle when $x = 2$ in.

11.12 The acceleration of a particle moving in a straight line is directed toward a fixed point O and is inversely proportional to the distance of the particle from O. At $t = 0$, the particle is 2 ft to the right of O, has a velocity of 4 ft/s to the right, and has an acceleration of 3 ft/s² to the left. Determine (*a*) the velocity of the particle when it is 3 ft away from O, (*b*) the position of the particle at which its velocity is zero.

11.13 The acceleration of a particle is defined by the relation $a = -10v$, where a is expressed in m/s² and v in m/s. Knowing that at $t = 0$ the velocity is 30 m/s, determine (*a*) the distance the particle will travel before coming to rest, (*b*) the time required for the particle to come to rest, (*c*) the time required for the velocity of the particle to be reduced to 1 percent of its initial value.

11.14 The acceleration of a particle is defined by the relation $a = -0.0125v^2$, where a is the acceleration in m/s² and v is the velocity in m/s. If the particle is given an initial velocity v_0, find the distance it will travel (*a*) before its velocity drops to half the initial value, (*b*) before it comes to rest.

11.15 The acceleration of a particle falling through the atmosphere is defined by the relation $a = g(1 - k^2v^2)$. Knowing that the particle starts at $t = 0$ and $x = 0$ with no initial velocity, (*a*) show that the velocity at any time t is $v = (1/k) \tanh kgt$, (*b*) write an equation defining the velocity for any value of x. (*c*) Why is $v_t = 1/k$ called the terminal velocity?

11.16 It has been determined experimentally that the magnitude in ft/s² of the deceleration due to air resistance of a projectile is $0.001v^2$, where v is expressed in ft/s. If the projectile is released from rest and keeps pointing downward, determine its velocity after it has fallen 500 ft. (*Hint.* The total acceleration is $g - 0.001v^2$, where $g = 32.2$ ft/s².)

11.17 The acceleration of a particle is defined by the relation $a = k \sin (\pi t/T)$. Knowing that both the velocity and the position coordinate of the particle are zero when $t = 0$, determine (*a*) the equations of motion, (*b*) the maximum velocity, (*c*) the position at $t = 2T$, (*d*) the average velocity during the interval $t = 0$ to $t = 2T$.

11.18   The position of an oscillating particle is defined by the relation $x = A \sin(pt + \phi)$. Denoting the velocity and position coordinate when $t = 0$ by v_0 and x_0, respectively, show (a) that $\tan \phi = x_0 p / v_0$, (b) that the maximum value of the position coordinate is

$$A = \sqrt{x_0^2 + \left(\frac{v_0}{p}\right)^2}$$

11.19   A more accurate expression for the acceleration due to gravity is

$$g = \frac{32.2}{\left(1 + \dfrac{y}{20.9 \times 10^6}\right)^2}$$

where g is expressed in ft/s^2 and y is the altitude in feet. Using this value of g, compute the height reached by a bullet fired vertically upward with the following initial velocities: (a) 1000 ft/s, (b) 10,000 ft/s, (c) 36,700 ft/s.

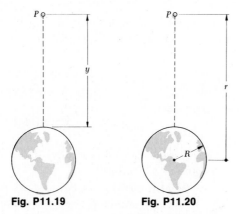

Fig. P11.19   **Fig. P11.20**

11.20   The acceleration due to gravity of a particle falling toward the earth is $a = -gR^2/r^2$, where r is the distance from the *center* of the earth to the particle, R is the radius of the earth, and g is the acceleration due to gravity at the surface of the earth. Derive an expression for the *escape velocity*, i.e., for the minimum velocity with which a particle should be projected vertically upward from the surface of the earth if it is not to return to the earth. (*Hint.* $v = 0$ for $r = \infty$.)

＊11.21   A particle is released from rest at a distance R above the surface of the earth. Using the expression for the acceleration due to gravity given in Prob. 11.20, determine (a) the velocity of the particle as it strikes the earth, (b) the time required for it to fall to the surface of the earth. (Radius of earth $= R = 6370$ km, $g = 9.81$ m/s^2.)

*11.22 When a package is dropped on a rigid surface, the acceleration of its cushioned contents may be defined by the relation $a = -k \tan (\pi x/2L)$, where L is the distance through which the cushioning material can be compressed. Denoting by v_0 the velocity when $x = 0$, show that $v^2 = v_0^2 + (4kL/\pi) \ln \cos (\pi x/2L)$. If $k = 300 \text{ m/s}^2$ and $L = 0.36 \text{ m}$, compute the initial velocity v_0 for which the maximum value of the position coordinate x is (a) 0.18 m, (b) 0.36 m.

11.4. Uniform Rectilinear Motion. This is a type of straight-line motion which is frequently encountered in practical applications. In this motion, the acceleration a of the particle is zero for every value of t. The velocity v is therefore constant, and Eq. (11.1) becomes

$$\frac{dx}{dt} = v = \text{constant}$$

The position coordinate x is obtained by integrating this equation. Denoting by x_0 the initial value of x, we write

$$\int_{x_0}^{x} dx = v \int_{0}^{t} dt$$

$$x - x_0 = vt$$

$$x = x_0 + vt \tag{11.5}$$

This equation may be used *only if the velocity of the particle is known to be constant.*

11.5. Uniformly Accelerated Rectilinear Motion. This is another common type of motion. In this motion, the acceleration a of the particle is constant, and Eq. (11.2) becomes

$$\frac{dv}{dt} = a = \text{constant}$$

The velocity v of the particle is obtained by integrating this equation,

$$\int_{v_0}^{v} dv = a \int_{0}^{t} dt$$

$$v - v_0 = at$$

$$v = v_0 + at \tag{11.6}$$

where v_0 is the initial velocity. Substituting for v into (11.1), we write

$$\frac{dx}{dt} = v_0 + at$$

Denoting by x_0 the initial value of x and integrating, we have

$$\int_{x_0}^{x} dx = \int_{0}^{t} (v_0 + at) \, dt$$

$$x - x_0 = v_0 t + \tfrac{1}{2} a t^2$$

$$x = x_0 + v_0 t + \tfrac{1}{2} a t^2 \tag{11.7}$$

We may also use Eq. (11.4) and write

$$v \frac{dv}{dx} = a = \text{constant}$$

$$v \, dv = a \, dx$$

Integrating both sides, we obtain

$$\int_{v_0}^{v} v \, dv = a \int_{x_0}^{x} dx$$

$$\tfrac{1}{2}(v^2 - v_0^2) = a(x - x_0)$$

$$v^2 = v_0^2 + 2a(x - x_0) \tag{11.8}$$

The three equations we have derived provide useful relations among position coordinate, velocity, and time in the case of a uniformly accelerated motion, as soon as appropriate values have been substituted for a, v_0, and x_0. The origin O of the x axis should first be defined and a positive direction chosen along the axis; this direction will be used to determine the signs of a, v_0, and x_0. Equation (11.6) relates v and t and should be used when the value of v corresponding to a given value of t is desired, or inversely. Equation (11.7) relates x and t; Eq. (11.8) relates v and x. An important application of uniformly accelerated motion is the motion of a *freely falling body*. The acceleration of a freely falling body (usually denoted by g) is equal to 9.81 m/s^2 or 32.2 ft/s^2.

It is important to keep in mind that the three equations above may be used *only when the acceleration of the particle is known to be constant*. If the acceleration of the particle is variable, its motion should be determined from the fundamental equations (11.1) to (11.4), according to the methods outlined in Sec. 11.3.

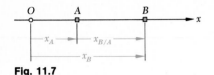

Fig. 11.7

11.6. Motion of Several Particles. When several particles move independently along the same line, independent equations of motion may be written for each particle. Whenever possible, time should be recorded from the same initial instant for all particles, and displacements should be measured from the same origin and in the same direction. In other words, a single clock and a single measuring tape should be used.

Relative Motion of Two Particles. Consider two particles A and B moving along the same straight line (Fig. 11.7). If the position coordinates x_A and x_B are measured from the same origin, the difference $x_B - x_A$ defines the *relative position coordinate of B with respect to A* and is denoted by $x_{B/A}$. We write

$$x_{B/A} = x_B - x_A \qquad \text{or} \qquad \boxed{x_B = x_A + x_{B/A}} \qquad (11.9)$$

A positive sign for $x_{B/A}$ means that B is to the right of A, a negative sign that B is to the left of A, regardless of the position of A and B with respect to the origin.

The rate of change of $x_{B/A}$ is known as the *relative velocity of B with respect to A* and is denoted by $v_{B/A}$. Differentiating (11.9), we write

$$v_{B/A} = v_B - v_A \qquad \text{or} \qquad \boxed{v_B = v_A + v_{B/A}} \qquad (11.10)$$

A positive sign for $v_{B/A}$ means that B is *observed from A* to move in the positive direction; a negative sign, that it is observed to move in the negative direction.

The rate of change of $v_{B/A}$ is known as the *relative acceleration of B with respect to A* and is denoted by $a_{B/A}$. Differentiating (11.10), we obtain

$$a_{B/A} = a_B - a_A \qquad \text{or} \qquad \boxed{a_B = a_A + a_{B/A}} \qquad (11.11)$$

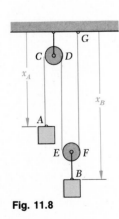

Fig. 11.8

Dependent Motions. Sometimes, the position of a particle will depend upon the position of another or of several other particles. The motions are then said to be dependent. For example, the position of block B in Fig. 11.8 depends upon the position of block A. Since the rope $ACDEFG$ is of constant length, and since the lengths of the portions of rope CD and EF wrapped around the pulleys remain constant, it follows that the sum of the lengths of the segments AC, DE, and FG is constant. Observing that the length of the segment AC differs from x_A only by a constant, and that, similarly, the lengths of the segments DE and FG differ from x_B only by a constant, we write

$$x_A + 2x_B = \text{constant}$$

Since only one of the two coordinates x_A and x_B may be chosen arbitrarily, we say that the system shown in Fig. 11.8 has *one degree of freedom.* From the relation between the position coordinates x_A and x_B, it follows that if x_A is given an increment Δx_A, i.e., if block A is lowered by an amount Δx_A, the coordinate x_B will receive an increment $\Delta x_B = -\frac{1}{2}\Delta x_A$, that is, block B will rise by half the same amount; this may easily be checked directly from Fig. 11.8.

In the case of the three blocks of Fig. 11.9, we may again observe that the length of the rope which passes over the pulleys is constant, and thus that the following relation must be satisfied by the position coordinates of the three blocks:

$$2x_A + 2x_B + x_C = \text{constant}$$

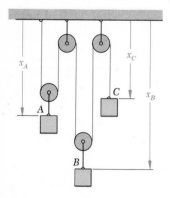

Fig. 11.9

Since two of the coordinates may be chosen arbitrarily, we say that the system shown in Fig. 11.9 has *two degrees of freedom.*

When the relation existing between the position coordinates of several particles is *linear*, a similar relation holds between the velocities and between the accelerations of the particles. In the case of the blocks of Fig. 11.9, for instance, we differentiate twice the equation obtained and write

$$2\frac{dx_A}{dt} + 2\frac{dx_B}{dt} + \frac{dx_C}{dt} = 0 \qquad \text{or} \qquad 2v_A + 2v_B + v_C = 0$$

$$2\frac{dv_A}{dt} + 2\frac{dv_B}{dt} + \frac{dv_C}{dt} = 0 \qquad \text{or} \qquad 2a_A + 2a_B + a_C = 0$$

A ball is thrown vertically upward from the 40-ft level in an elevator shaft, with an initial velocity of 50 ft/s. At the same instant an open-platform elevator passes the 10-ft level, moving upward with a constant velocity of 5 ft/s. Determine (a) when and where the ball will hit the elevator, (b) the relative velocity of the ball with respect to the elevator when the ball hits the elevator.

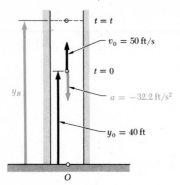

Motion of Ball. Since the ball has a constant acceleration, its motion is *uniformly accelerated*. Placing the origin O of the y axis at ground level and choosing its positive direction upward, we find that the initial position is $y_0 = +40$ ft, the initial velocity is $v_0 = +50$ ft/s, and the acceleration is $a = -32.2$ ft/s^2. Substituting these values in the equations for uniformly accelerated motion, we write

$$v_B = v_0 + at \qquad v_B = 50 - 32.2t \qquad (1)$$
$$y_B = y_0 + v_0 t + \tfrac{1}{2}at^2 \qquad y_B = 40 + 50t - 16.1t^2 \qquad (2)$$

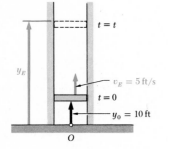

Motion of Elevator. Since the elevator has a constant velocity, its motion is *uniform*. Again placing the origin O at the ground level and choosing the positive direction upward, we note that $y_0 = +10$ ft and write

$$v_E = +5 \text{ ft/s} \qquad (3)$$
$$y_E = y_0 + v_E t \qquad y_E = 10 + 5t \qquad (4)$$

Ball Hits Elevator. We first note that the same time t and the same origin O were used in writing the equations of motion of both the ball and the elevator. We see from the figure that, when the ball hits the elevator,

$$y_E = y_B \qquad (5)$$

Substituting for y_E and y_B from (2) and (4) into (5), we have

$$10 + 5t = 40 + 50t - 16.1t^2$$
$$t = -0.56 \text{ s} \qquad \text{and} \qquad t = +3.35 \text{ s} \quad \blacktriangleleft$$

Only the root $t = 3.35$ s corresponds to a time after the motion has begun. Substituting this value into (4), we have

$$y_E = 10 + 5(3.35) = 26.7 \text{ ft}$$
$$\text{Elevation from ground} = 26.7 \text{ ft} \quad \blacktriangleleft$$

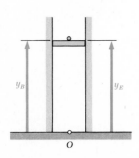

The relative velocity of the ball with respect to the elevator is

$$v_{B/E} = v_B - v_E = (50 - 32.2t) - 5 = 45 - 32.2t$$

When the ball hits the elevator at time $t = 3.35$ s, we have

$$v_{B/E} = 45 - 32.2(3.35) \qquad v_{B/E} = -62.9 \text{ ft/s} \quad \blacktriangleleft$$

The negative sign means that the ball is observed from the elevator to be moving in the negative sense (downward).

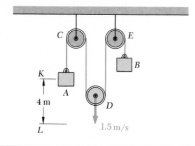

Two blocks A and B are connected by a cord passing over three pulleys C, D, and E as shown. Pulleys C and E are fixed, while D is pulled downward with a constant velocity of 1.5 m/s. At $t = 0$, block A starts moving downward from the position K with a constant acceleration and no initial velocity. Knowing that the velocity of block A is 6 m/s as it passes through point L, determine the change in elevation, the velocity, and the acceleration of block B when A passes through L.

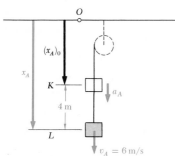

Motion of Block A. We place the origin O at the horizontal surface and choose the positive direction downward. We observe that when $t = 0$, block A is at position K and $(v_A)_0 = 0$. Since $v_A = 6$ m/s and $x_A - (x_A)_0 = 4$ m when the block passes through L, we write

$$v_A^2 = (v_A)_0^2 + 2a_A[x_A - (x_A)_0] \qquad (6)^2 = 0 + 2a_A(4)$$
$$a_A = 4.50 \text{ m/s}^2$$

The time at which block A reaches point L is obtained by writing

$$v_A = (v_A)_0 + a_A t \qquad 6 = 0 + 4.50t \qquad t = 1.333 \text{ s}$$

Motion of Pulley D. Recalling that the positive direction is downward, we write

$$a_D = 0 \qquad v_D = 1.5 \text{ m/s} \qquad x_D = (x_D)_0 + v_D t = (x_D)_0 + 1.5t$$

When block A reaches L, at $t = 1.333$ s, we have

$$x_D = (x_D)_0 + 1.5(1.333) = (x_D)_0 + 2$$

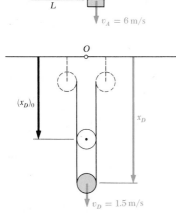

Thus,
$$x_D - (x_D)_0 = 2 \text{ m}$$

Motion of Block B. We note that the total length of cord $ACDEB$ differs from the quantity $(x_A + 2x_D + x_B)$ only by a constant. Since the cord length is constant during the motion, this quantity must also remain constant. Thus considering the times $t = 0$ and $t = 1.333$ s, we write

$$x_A + 2x_D + x_B = (x_A)_0 + 2(x_D)_0 + (x_B)_0 \qquad (1)$$
$$[x_A - (x_A)_0] + 2[x_D - (x_D)_0] + [x_B - (x_B)_0] = 0 \qquad (2)$$

But we know that $x_A - (x_A)_0 = 4$ m and $x_D - (x_D)_0 = 2$ m; substituting these values in (2), we find

$$4 + 2(2) + [x_B - (x_B)_0] = 0 \qquad x_B - (x_B)_0 = -8 \text{ m}$$

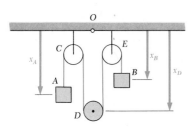

Thus: Change in elevation of $B = 8$ m ↑   ◄

Differentiating (1) twice, we obtain equations relating the velocities and the accelerations of A, B, and D. Substituting for the velocities and accelerations of A and D at $t = 1.333$ s, we have

$$v_A + 2v_D + v_B = 0: \qquad 6 + 2(1.5) + v_B = 0$$
$$v_B = -9 \text{ m/s} \qquad v_B = 9 \text{ m/s} ↑ \quad ◄$$
$$a_A + 2a_D + a_B = 0: \qquad 4.50 + 2(0) + a_B = 0$$
$$a_B = -4.50 \text{ m/s}^2 \qquad a_B = 4.50 \text{ m/s}^2 ↑ \quad ◄$$

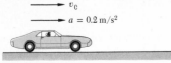

Fig. P11.23

PROBLEMS

11.23 An automobile travels 240 m in 30 s while being accelerated at a constant rate of 0.2 m/s^2. Determine (a) its initial velocity, (b) its final velocity, (c) the distance traveled during the first 10 s.

11.24 A stone is released from an elevator moving up at a speed of 5 m/s and reaches the bottom of the shaft in 3 s. (a) How high was the elevator when the stone was released? (b) With what speed does the stone strike the bottom of the shaft?

11.25 A man jumps from a 20-ft cliff with no initial velocity. (a) How long does it take him to reach the ground, and with what velocity does he hit the ground? (b) If this takes place on the moon, where g = 5.31 ft/s^2, what are the values obtained for the time and velocity? (c) If a motion picture is taken on the earth, but if the scene is supposed to take place on the moon, how many frames per second should be used so that the scene would appear realistic when projected at the standard speed of 24 frames per second?

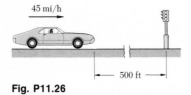

Fig. P11.26

11.26 A motorist is traveling at 45 mi/h when he observes that a traffic light 500 ft ahead of him turns red. The traffic light is timed to stay red for 10 s. If the motorist wishes to pass the light without stopping just as it turns green again, determine (a) the required uniform deceleration of the car, (b) the speed of the car as it passes the light.

11.27 Automobile A starts from O and accelerates at the constant rate of 4 ft/s^2. A short time later it is passed by truck B which is traveling in the opposite direction at a constant speed of 45 ft/s. Knowing that truck B passes point O, 25 s after automobile A started from there, determine when and where the vehicles passed each other.

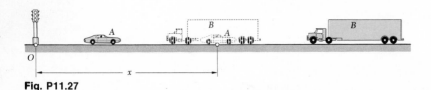

Fig. P11.27

11.28 An open-platform elevator is moving down a mine shaft at a constant velocity v_e when the elevator platform hits and dislodges a stone. Assuming that the stone starts falling with no initial velocity, (a) show that the stone will hit the platform with a relative velocity of magnitude v_e. (b) If v_e = 20 ft/s, determine when and where the stone will hit the elevator platform.

11.29 Two automobiles A and B are traveling in the same direction in adjacent highway lanes. Automobile B is stopped when it is passed by A, which travels at a constant speed of 36 km/h. Two seconds later automobile B starts and accelerates at a constant rate of 1.5 m/s^2. Determine (a) when and where B will overtake A, (b) the speed of B at that time.

11.30 Drops of water are observed to drip from a faucet at uniform intervals of time. As any drop B begins to fall freely, the preceding drop A has already fallen 0.3 m. Determine the distance drop A will have fallen by the time the distance between A and B will have increased to 0.9 m.

11.31 The elevator shown in the figure moves downward at the constant velocity of 15 ft/s. Determine (a) the velocity of the cable C, (b) the velocity of the counterweight W, (c) the relative velocity of the cable C with respect to the elevator, (d) the relative velocity of the counterweight W with respect to the elevator.

11.32 The elevator shown starts from rest and moves upward with a constant acceleration. If the counterweight W moves through 24 ft in 4 s, determine (a) the accelerations of the elevator and the cable C, (b) the velocity of the elevator after 4 s.

11.33 The slider block A moves to the left at a constant velocity of 300 mm/s. Determine (a) the velocity of block B, (b) the velocities of portions C and D of the cable, (c) the relative velocity of A with respect to B, (d) the relative velocity of portion C of the cable with respect to portion D.

11.34 The slider block B starts from rest and moves to the right with a constant acceleration. After 4 s the relative velocity of A with respect to B is 60 mm/s. Determine (a) the accelerations of A and B, (b) the velocity and position of B after 3 s.

11.35 Blocks A and C start from rest and move to the right with the following accelerations: $a_A = 12t$ ft/s^2 and $a_C = 3$ ft/s^2. Determine (a) the time at which the velocity of block B is zero, (b) the corresponding position of B.

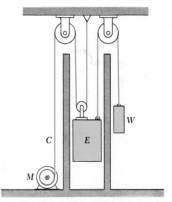

Fig. P11.31 and P11.32

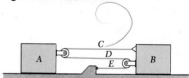

Fig. P11.33 and P11.34

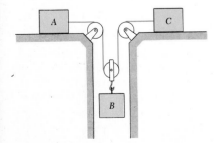

Fig. P11.35

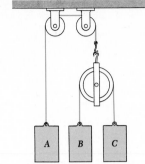

Fig. P11.36, P11.37, and P11.39

11.36 (a) Choosing the positive sense downward for each block, express the velocity of A in terms of the velocities of B and C. (b) Knowing that both blocks A and C start from rest and move downward with the respective accelerations $a_A = 2$ in./s^2 and $a_C = 5$ in./s^2, determine the position and velocity of B after 3 s.

11.37 The three blocks shown move with constant velocities. Find the velocity of each block, knowing that the relative velocity of A with respect to C is 200 mm/s upward and that the relative velocity of B with respect to A is 120 mm/s downward.

11.38 The three blocks of Fig. 11.9 move with constant velocities. Find the velocity of each block, knowing that C is observed from B to move downward with a relative velocity of 180 mm/s and A is observed from B to move downward with a relative velocity of 160 mm/s.

*11.39 The three blocks shown are equally spaced horizontally and move vertically with constant velocities. Knowing that initially they are at the same level and that the relative velocity of C with respect to B is 160 mm/s downward, determine the velocity of each block so that the three blocks will remain aligned during their motion.

*11.7. Graphical Solution of Rectilinear-Motion Problems. It was observed in Sec. 11.2 that the fundamental formulas

$$v = \frac{dx}{dt} \quad \text{and} \quad a = \frac{dv}{dt}$$

have a geometrical significance. The first formula expresses that the velocity at any instant is equal to the slope of the x–t curve at the same instant (Fig. 11.10). The second formula expresses that the acceleration is equal to the slope of the v–t curve. These

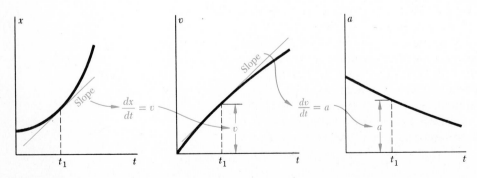

Fig. 11.10

two properties may be used to derive graphically the v–t and a–t curves of a motion when the x–t curve is known.

Integrating the two fundamental formulas from a time t_1 to a time t_2, we write

$$x_2 - x_1 = \int_{t_1}^{t_2} v \, dt \quad \text{and} \quad v_2 - v_1 = \int_{t_1}^{t_2} a \, dt \quad (11.12)$$

The first formula expresses that the area measured under the v–t curve from t_1 to t_2 is equal to the change in x during that time interval (Fig. 11.11). The second formula expresses similarly that the area measured under the a–t curve from t_1 to t_2 is equal to the change in v during that time interval. These two properties may be used to determine graphically the x–t curve of a motion when its v–t curve or its a–t curve is known (see Sample Prob. 11.6).

Graphical solutions are particularly useful when the motion considered is defined from experimental data and when x, v, and a are not analytical functions of t. They may also be used to advantage when the motion consists of distinct parts and when its analysis requires writing a different equation for each of its parts. When using a graphical solution, however, one should be careful to note (1) that the area under the v–t curve measures the *change in x*, not x itself, and, similarly, that the area under the a–t curve measures the change in v; (2) that, while an area above the t axis corresponds to an increase in x or v, an area located below the t axis measures a decrease in x or v.

It will be useful to remember, in drawing motion curves, that, if the velocity is constant, it will be represented by a horizontal straight line; the position coordinate x will then be a linear function of t and will be represented by an oblique straight line. If the acceleration is constant and different from zero, it will be represented by a horizontal straight line; v will then be a linear function of t, represented by an oblique straight line; and x will be expressed as a second-degree polynomial in t, represented by a parabola. If the acceleration is a linear function of t, the velocity and the position coordinate will be equal, respectively, to second-degree and third-degree polynomials; a is then represented by an oblique straight line, v by a parabola, and x by a cubic. In general, if the acceleration is a polynomial of degree n in t, the velocity will be a polynomial of degree $n + 1$ and the position coordinate a polynomial of degree $n + 2$; these polynomials are represented by motion curves of a corresponding degree.

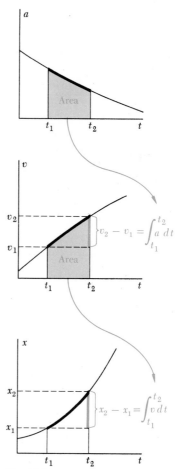

Fig. 11.11

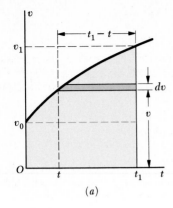

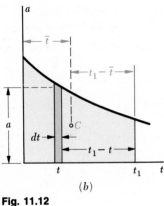

(a)

(b)

Fig. 11.12

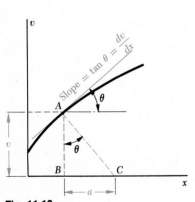

Fig. 11.13

*11.8. Other Graphical Methods. An alternate graphical solution may be used to determine directly from the a–t curve the position of a particle at a given instant. Denoting respectively by x_0 and v_0 the values of x and v at $t = 0$, by x_1 and v_1 their values at $t = t_1$, and observing that the area under the v–t curve may be divided into a rectangle of area $v_0 t_1$ and horizontal differential elements of area $(t_1 - t)\, dv$ (Fig. 11.12a), we write

$$x_1 - x_0 = \text{area under } v\text{–}t \text{ curve} = v_0 t_1 + \int_{v_0}^{v_1} (t_1 - t)\, dv$$

Substituting $dv = a\, dt$ in the integral, we obtain

$$x_1 - x_0 = v_0 t_1 + \int_{0}^{t_1} (t_1 - t) a\, dt$$

Referring to Fig. 11.12b, we note that the integral represents the first moment of the area under the a–t curve with respect to the line $t = t_1$ bounding the area on the right. This method of solution is known, therefore, as the *moment-area method*. If the abscissa $\bar{t}$ of the centroid C of the area is known, the position coordinate x_1 may be obtained by writing

$$x_1 = x_0 + v_0 t_1 + (\text{area under } a\text{–}t \text{ curve})(t_1 - \bar{t}) \quad (11.13)$$

If the area under the a–t curve is a composite area, the last term in (11.13) may be obtained by multiplying each component area by the distance from its centroid to the line $t = t_1$. Areas above the t axis should be considered as positive and areas below the t axis as negative.

Another type of motion curve, the v–x curve, is sometimes used. If such a curve has been plotted (Fig. 11.13), the acceleration a may be obtained at any time by drawing the normal to the curve and *measuring the subnormal BC*. Indeed, observing that the angle between AC and AB is equal to the angle θ between the horizontal and the tangent at A (the slope of which is $\tan \theta = dv/dx$), we write

$$BC = AB \tan \theta = v \frac{dv}{dx}$$

and thus, recalling formula (11.4),

$$BC = a$$

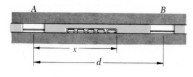

A subway train leaves station A; it gains speed at the rate of $4\ \text{ft/s}^2$ for 6 s, and then at the rate of $6\ \text{ft/s}^2$ until it has reached the speed of 48 ft/s. The train maintains the same speed until it approaches station B; brakes are then applied, giving the train a constant deceleration and bringing it to a stop in 6 s. The total running time from A to B is 40 s. Draw the a–t, v–t, and x–t curves, and determine the distance between stations A and B.

Acceleration-Time Curve. Since the acceleration is either constant or zero, the a–t curve is made of horizontal straight-line segments. The values of t_2 and a_4 are determined as follows:

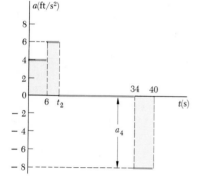

$0 < t < 6$:   Change in v = area under a–t curve
$$v_6 - 0 = (6\ \text{s})(4\ \text{ft/s}^2) = 24\ \text{ft/s}$$

$6 < t < t_2$:   Since the velocity increases from 24 to 48 ft/s,

Change in v = area under a–t curve
$$48 - 24 = (t_2 - 6)(6\ \text{ft/s}^2) \qquad t_2 = 10\ \text{s}$$

$t_2 < t < 34$:   Since the velocity is constant, the acceleration is zero.

$34 < t < 40$:   Change in v = area under a–t curve
$$0 - 48 = (6\ \text{s})a_4 \qquad a_4 = -8\ \text{ft/s}^2$$

The acceleration being negative, the corresponding area is below the t axis; this area represents a decrease in velocity.

Velocity-Time Curve. Since the acceleration is either constant or zero, the v–t curve is made of segments of straight line connecting the points determined above.

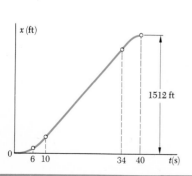

Change in x = area under v–t curve

$0 < t < 6$: $x_6 - 0 = \frac{1}{2}(6)(24) = 72\ \text{ft}$
$6 < t < 10$: $x_{10} - x_6 = \frac{1}{2}(4)(24 + 48) = 144\ \text{ft}$
$10 < t < 34$: $x_{34} - x_{10} = (24)(48) = 1152\ \text{ft}$
$34 < t < 40$: $x_{40} - x_{34} = \frac{1}{2}(6)(48) = 144\ \text{ft}$

Adding the changes in x, we obtain the distance from A to B:

$$d = x_{40} - 0 = 1512\ \text{ft}$$

$$d = 1512\ \text{ft} \quad \blacktriangleleft$$

Position-Time Curve. The points determined above should be joined by three arcs of parabola and one segment of straight line. The construction of the x–t curve will be performed more easily and more accurately if we keep in mind that for any value of t the slope of the tangent to the x–t curve is equal to the value of v at that instant.

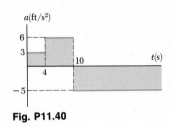

Fig. P11.40

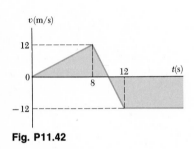

Fig. P11.42

PROBLEMS

11.40 A particle moves in a straight line with the acceleration shown in the figure. Knowing that it starts from the origin with $v_0 = -18$ ft/s, (a) plot the v–t and x–t curves for $0 < t < 20$ s, (b) determine its velocity, its position, and the total distance traveled after 12 s.

11.41 For the particle and motion of Prob. 11.40, plot the v–t and x–t curves for $0 < t < 20$ s and determine (a) the maximum value of the velocity of the particle, (b) the maximum value of its position coordinate.

11.42 A particle moves in a straight line with the velocity shown in the figure. Knowing that $x = -12$ m at $t = 0$, draw the a–t and x–t curves for $0 < t < 16$ s and determine (a) the total distance traveled by the particle after 12 s, (b) the two values of t for which the particle passes through the origin.

11.43 For the particle and motion of Prob. 11.42, plot the a–t and x–t curves for $0 < t < 16$ s and determine (a) the maximum value of the position coordinate of the particle, (b) the values of t for which the particle is at a distance of 15 m from the origin.

11.44 A series of city traffic signals is timed so that an automobile traveling at a constant speed of 30 mi/h will reach each signal just as it turns green. A motorist misses a signal and is stopped at signal A. Knowing that the next signal B is 800 ft ahead and that the maximum acceleration of his automobile is 5 ft/s², determine what the motorist should do to keep his maximum speed as small as possible, yet reach signal B just as it turns green. What is the maximum speed reached?

11.45 A bus starts from rest at point A and accelerates at the rate of 0.9 m/s² until it reaches a speed of 7.2 m/s. It then proceeds at 7.2 m/s until the brakes are applied; it comes to rest at point B, 18 m beyond the point where the brakes were applied. Assuming uniform deceleration and knowing that the distance between A and B is 90 m, determine the time required for the bus to travel from A to B.

11.46 The firing of a howitzer causes the barrel to recoil 800 mm before a braking mechanism brings it to rest. From a high-speed photographic record, it is found that the maximum value of the recoil velocity is 5.4 m/s and that this is reached 0.02 s after firing. Assuming that the recoil period consists of two phases during which the acceleration has, respectively, a constant positive value a_1 and a constant negative value a_2, determine (a) the values of a_1 and a_2, (b) the position of the barrel 0.02 s after firing, (c) the time at which the velocity of the barrel is zero.

11.47 A motorist is traveling at 60 mi/h when he observes that a traffic signal 1000 ft ahead of him turns red. He knows that the signal is timed to stay red for 20 s. What should he do to pass the signal at 60 mi/h just as it turns green again? Draw the v–t curve, selecting the solution which calls for the smallest possible deceleration and acceleration, and determine (a) the common value of the deceleration and acceleration in ft/s², (b) the minimum speed reached in mi/h.

11.48 A policeman on a motorcycle is escorting a motorcade which is traveling at 54 km/h. The policeman suddenly decides to take a new position in the motorcade, 70 m ahead. Assuming that he accelerates and decelerates at the rate of 2.5 m/s² and that he does not exceed at any time a speed of 72 km/h, draw the a–t and v–t curves for his motion and determine (a) the shortest time in which he can occupy his new position in the motorcade, (b) the distance he will travel in that time.

11.49 A freight elevator moving upward with a constant velocity of 5 m/s passes a passenger elevator which is stopped. Three seconds later, the passenger elevator starts upward with an acceleration of 1.25 m/s². When the passenger elevator has reached a velocity of 10 m/s, it proceeds at constant speed. Draw the v–t and y–t curves, and from them determine the time and distance required by the passenger elevator to overtake the freight elevator.

11.50 A car and a truck are both traveling at the constant speed of 45 mi/h; the car is 40 ft behind the truck. The driver of the car wants to pass the truck, i.e., he wishes to place his car at B, 40 ft in front of the truck, and then resume the speed of 45 mi/h. The maximum acceleration of the car is 5 ft/s² and the maximum deceleration obtained by applying the brakes is 20 ft/s². What is the shortest time in which the driver of the car can complete the passing operation if he does not at any time exceed a speed of 60 mi/h? Draw the v–t curve.

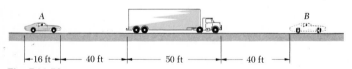

Fig. P11.50

16 ft — 40 ft — 50 ft — 40 ft

11.51 Car A is traveling at the constant speed v_A. It approaches car B which is traveling in the same direction at the constant speed of 40 mi/h. The driver of car B notices car A when it is still 150 ft behind him. He then accelerates at the constant rate of 2 ft/s² to avoid being passed by A. Knowing that the closest that A comes to B is 50 ft, determine the speed of car A.

11.52 A car and a truck are both traveling at the constant speed of 60 mi/h; the car is 30 ft behind the truck. The truck driver suddenly applies his brakes, causing the truck to decelerate at the constant rate of 9 ft/s². Two seconds later the driver of the car applies his brakes and just manages to avoid a rear-end collision. Determine the constant rate at which the car decelerated.

11.53 Two cars are traveling toward each other on a single-lane road at 16 and 12 m/s, respectively. When 120 m apart, both drivers realize the situation and apply their brakes. They succeed in stopping simultaneously, and just short of colliding. Assuming a constant deceleration for each car, determine (*a*) the time required for the cars to stop, (*b*) the deceleration of each car, and (*c*) the distance traveled by each car while slowing down.

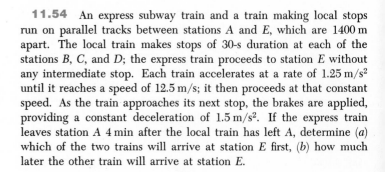

Fig. P11.53

11.54 An express subway train and a train making local stops run on parallel tracks between stations A and E, which are 1400 m apart. The local train makes stops of 30-s duration at each of the stations B, C, and D; the express train proceeds to station E without any intermediate stop. Each train accelerates at a rate of 1.25 m/s² until it reaches a speed of 12.5 m/s; it then proceeds at that constant speed. As the train approaches its next stop, the brakes are applied, providing a constant deceleration of 1.5 m/s². If the express train leaves station A 4 min after the local train has left A, determine (*a*) which of the two trains will arrive at station E first, (*b*) how much later the other train will arrive at station E.

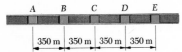

Fig. P11.54

11.55 An automobile at rest is passed by a truck traveling at a constant speed of 45 mi/h. The automobile starts and accelerates for 10 s at a constant rate until it reaches a speed of 60 mi/h. If the automobile then maintains a constant speed of 60 mi/h, determine when it will overtake the truck, assuming the automobile starts (*a*) just as the truck passes it, (*b*) 2 s after the truck has passed it.

11.56 The rate of change of acceleration is known as the *jerk*; large or abrupt rates of change of acceleration cause discomfort to elevator passengers. If the jerk, or rate of change of the acceleration, of an elevator is limited to ±0.5 m/s² per second, determine the shortest time required for an elevator, starting from rest, to rise 8 m and stop.

11.57   The acceleration record shown was obtained for a truck traveling on a straight highway. Knowing that the initial velocity of the truck was 18 km/h, determine (a) the velocity at $t = 6$ s, (b) the distance the truck travels during the 6-s test.

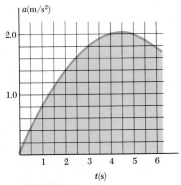

Fig. P11.57

11.58   A training airplane lands on an aircraft carrier and is brought to rest in 3.0 s by the arresting gear of the carrier. An accelerometer attached to the airplane provides the acceleration record shown. Determine by approximate means (a) the initial velocity of the airplane relative to the deck, (b) the distance the airplane travels along the deck before coming to rest.

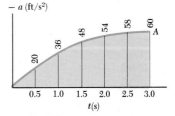

Fig. P11.58

11.59   The v–x curve shown was obtained experimentally during the motion of the bed of an industrial planer. Determine by approximate means the acceleration (a) when $x = 3$ in., (b) when $v = 40$ in./s.

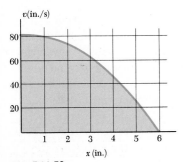

Fig. P11.59

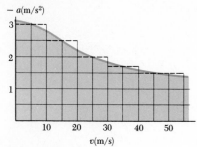

Fig. P11.60

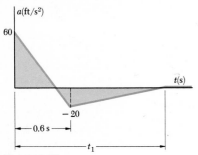

Fig. P11.63

11.60 The maximum possible acceleration of a passenger train under emergency conditions was determined experimentally; the results are shown (solid curve) in the figure. If the brakes are applied when the train is traveling at 90 km/h, determine by approximate means (a) the time required for the train to come to rest, (b) the distance traveled in that time.

11.61 Using the method of Sec. 11.8, derive the formula $x = x_0 + v_0 t + \frac{1}{2} a t^2$, for the position coordinate of a particle in uniformly accelerated rectilinear motion.

11.62 Using the method of Sec. 11.8, obtain an approximate solution for Prob. 11.58, assuming that the a–t curve is (a) a straight line from the origin to point A, (b) a parabola with vertex at A.

11.63 The acceleration of an object subjected to the pressure wave of a large explosion is defined approximately by the curve shown. The object is initially at rest and is again at rest at time t_1. Using the method of Sec. 11.8, determine (a) the time t_1, (b) the distance through which the object is moved by the pressure wave.

11.64 Using the method of Sec. 11.8, determine the position of the particle of Prob. 11.40 when $t = 14$ s.

11.65 For the particle of Prob. 11.42, draw the a–t curve and, using the method of Sec. 11.8, determine (a) the position of the particle when $t = 14$ s, (b) the maximum value of its position coordinate.

CURVILINEAR MOTION OF PARTICLES

11.9. Position Vector, Velocity, and Acceleration. When a particle moves along a curve other than a straight line, we say that the particle is in *curvilinear motion*. To define the position P occupied by the particle at a given time t, we select a fixed reference system, such as the x and y axes shown in Fig. 11.14a, and draw the vector **r** joining the origin O and point P. Since the vector **r** is characterized by its magnitude r and its direction with respect to the reference axes, it completely defines the position of the particle with respect to those axes; the vector **r** is referred to as the *position vector* of the particle at time t.

Consider now the vector **r′** defining the position *P′* occupied by the same particle at a later time $t + \Delta t$. The vector **Δr** joining *P* and *P′* represents the change in the position vector during the time interval Δt since, as we may easily check from Fig. 11.14*a*, the vector **r′** is obtained by adding the vectors **r** and **Δr** according to the triangle rule.† We note that **Δr** represents a change in *direction* as well as a change in *magnitude* of the position vector **r**. The *average velocity* of the particle over the time interval Δt is defined as the quotient of **Δr** and Δt. Since **Δr** is a vector and Δt a scalar, the quotient $\mathbf{\Delta r}/\Delta t$ is a vector attached at *P*, of the same direction as **Δr,** and of magnitude equal to the magnitude of **Δr** divided by Δt (Fig. 11.14*b*).

The *instantaneous velocity* of the particle at time *t* is obtained by choosing shorter and shorter time intervals Δt and, correspondingly, shorter and shorter vector increments **Δr**. The instantaneous velocity is thus represented by the vector

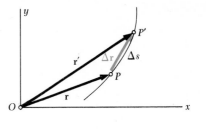

(*a*)

$$\mathbf{v} = \lim_{\Delta t \to 0} \frac{\mathbf{\Delta r}}{\Delta t} \qquad (11.14)$$

As Δt and **Δr** become shorter, the points *P* and *P′* get closer; the vector **v** obtained at the limit must therefore be tangent to the path of the particle (Fig. 11.14*c*). Its magnitude, denoted by *v* and called the *speed* of the particle, is obtained by substituting for the vector **Δr** in formula (11.14) its magnitude represented by the straight-line segment *PP′*. But the length of the segment *PP′* approaches the length Δs of the arc *PP′* as Δt decreases (Fig. 11.14*a*), and we may write

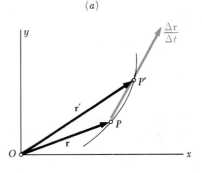

(*b*)

$$v = \lim_{\Delta t \to 0} \frac{PP'}{\Delta t} = \lim_{\Delta t \to 0} \frac{\Delta s}{\Delta t}$$

$$v = \frac{ds}{dt} \qquad (11.15)$$

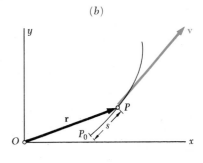

(*c*)

Fig. 11.14

The speed *v* may thus be obtained by differentiating with respect to *t* the length *s* of the arc described by the particle.

It is recalled that in many problems of statics it was found convenient to resolve a force **F** into component forces $\mathbf{F}_x$ and $\mathbf{F}_y$. Similarly, we shall often find it convenient in kinematics to resolve the vector **v** into rectangular components $\mathbf{v}_x$ and $\mathbf{v}_y$. To obtain these components, we resolve the vector **Δr** into com-

---

†Since the displacement from *O* to *P′* represented by **r′** is clearly equivalent to the two successive displacements represented by **r** and **Δr**, we verify that displacements satisfy the triangle rule and, therefore, the parallelogram law of addition. Thus, *displacements are truly vectors.*

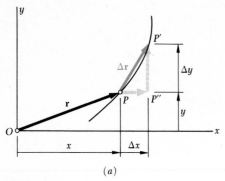

(a)

Fig. 11.15

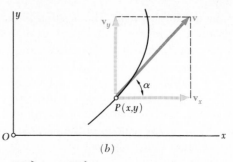

(b)

ponents $\overrightarrow{PP''}$ and $\overrightarrow{P''P'}$, respectively parallel to the x and y axes (Fig. 11.15a), and write

$$\Delta \mathbf{r} = \overrightarrow{PP''} + \overrightarrow{P''P'}$$

where the right-hand member represents a vector sum. Substituting in (11.14) the expression obtained for $\Delta \mathbf{r}$, we write

$$\mathbf{v} = \lim_{\Delta t \to 0} \frac{\overrightarrow{PP''}}{\Delta t} + \lim_{\Delta t \to 0} \frac{\overrightarrow{P''P'}}{\Delta t}$$

Since the first limit represents a vector parallel to the x axis, it must be equal to the component $\mathbf{v}_x$ of the velocity; similarly, the second limit must be equal to its component $\mathbf{v}_y$ (Fig. 11.15b). We write

$$\mathbf{v}_x = \lim_{\Delta t \to 0} \frac{\overrightarrow{PP''}}{\Delta t} \qquad \mathbf{v}_y = \lim_{\Delta t \to 0} \frac{\overrightarrow{P''P'}}{\Delta t}$$

The magnitudes v_x and v_y of the components of the velocity may be obtained by observing that the magnitudes of the vectors $\overrightarrow{PP''}$ and $\overrightarrow{P''P'}$ are respectively equal to the increments Δx and Δy of the coordinates x and y of the particle. We write

$$v_x = \lim_{\Delta t \to 0} \frac{\Delta x}{\Delta t} \qquad v_y = \lim_{\Delta t \to 0} \frac{\Delta y}{\Delta t}$$

$$v_x = \frac{dx}{dt} \qquad v_y = \frac{dy}{dt}$$

or, using dots to indicate time derivatives,

$$v_x = \dot{x} \qquad v_y = \dot{y} \qquad (11.16)$$

The expressions obtained are known as the *scalar components* of the velocity. A positive value for v_x indicates that the vector component $\mathbf{v}_x$ is directed to the right, a negative value that it is directed to the left; similarly, a positive value for v_y indicates that $\mathbf{v}_y$ is directed upward, a negative value that it is directed downward. We verify from Fig. 11.15b that the magnitude and direction of the velocity **v** may be obtained from its scalar components v_x and v_y as follows:

$$v^2 = v_x^2 + v_y^2 \qquad \tan \alpha = \frac{v_y}{v_x} \qquad (11.17)$$

Consider the velocity **v** of the particle at time t and also its velocity **v′** at a later time $t + \Delta t$ (Fig. 11.16a). Let us draw both vectors **v** and **v′** from the same origin O' (Fig. 11.16b). The vector Δ**v** joining Q and Q' represents the change in the velocity of the particle during the time interval Δt, since the vector **v′** may be obtained by adding the vectors **v** and Δ**v**. We should note that Δ**v** represents a change in the *direction* of the velocity as well as a change in *speed*. The *average acceleration* of the particle over the time interval Δt is defined as the quotient of Δ**v** and Δt. Since Δ**v** is a vector and Δt a scalar, the quotient Δ**v**$/\Delta t$ is a vector of the same direction as Δ**v**.

The *instantaneous acceleration* of the particle at time t is obtained by choosing smaller and smaller values for Δt and Δ**v**. The instantaneous acceleration is thus represented by the vector

$$\mathbf{a} = \lim_{\Delta t \to 0} \frac{\Delta \mathbf{v}}{\Delta t} \qquad (11.18)$$

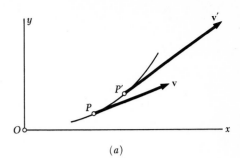

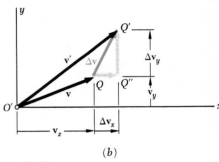

(a)

(b)

Fig. 11.16

It should be noted that, in general, *the acceleration is not tangent to the path of the particle;* also, in general, the magnitude of the acceleration *does not* represent the rate of change of the speed of the particle.

To obtain the rectangular components $\mathbf{a}_x$ and $\mathbf{a}_y$ of the acceleration, we resolve the vector Δ**v** into components $\overrightarrow{QQ''}$ and $\overrightarrow{Q''Q'}$, respectively parallel to the x and y axes (Fig. 11.16b), and write

$$\Delta \mathbf{v} = \overrightarrow{QQ''} + \overrightarrow{Q''Q'}$$

Substituting in (11.18) the expression obtained for Δ**v**, we write

$$\mathbf{a} = \lim_{\Delta t \to 0} \frac{\overrightarrow{QQ''}}{\Delta t} + \lim_{\Delta t \to 0} \frac{\overrightarrow{Q''Q'}}{\Delta t}$$

Since the first limit represents a vector parallel to the x axis, it must be equal to the component $\mathbf{a}_x$ of the acceleration; similarly, the second limit must be equal to its component $\mathbf{a}_y$ (Fig. 11.17). We write

$$\mathbf{a}_x = \lim_{\Delta t \to 0} \frac{\overrightarrow{QQ''}}{\Delta t} \qquad \mathbf{a}_y = \lim_{\Delta t \to 0} \frac{\overrightarrow{Q''Q'}}{\Delta t}$$

The magnitudes a_x and a_y of the components of the acceleration, which are known as the scalar components of the acceleration,

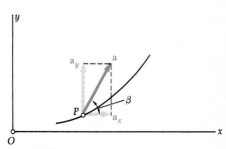

Fig. 11.17

may be obtained by observing that the magnitudes of the vectors $\overrightarrow{QQ''}$ and $\overrightarrow{Q''Q'}$ are respectively equal to the increments Δv_x and Δv_y of the scalar components v_x and v_y of the velocity of the particle. We write

$$a_x = \lim_{\Delta t \to 0} \frac{\Delta v_x}{\Delta t} \qquad a_y = \lim_{\Delta t \to 0} \frac{\Delta v_y}{\Delta t}$$

$$a_x = \frac{dv_x}{dt} \qquad a_y = \frac{dv_y}{dt}$$

$$a_x = \dot{v}_x \qquad a_y = \dot{v}_y \qquad\qquad (11.19)$$

or, substituting for v_x and v_y from (11.16),

$$a_x = \ddot{x} \qquad a_y = \ddot{y} \qquad\qquad (11.20)$$

The magnitude and direction of the acceleration **a** may be obtained from its scalar components a_x and a_y (Fig. 11.17) by formulas similar to formulas (11.17).

11.10. Component Motions. The use of rectangular components to describe the position, the velocity, and the acceleration of a particle is particularly effective when the component a_x of the acceleration is independent of the coordinate y and of the velocity component v_y and when the component a_y of the acceleration is independent of x and v_x. Equations (11.19) may then be integrated independently, and so may Eqs. (11.16). In other words, the motion of the particle in the x direction and its motion in the y direction may be considered separately.

In the case of the *motion of a projectile*, for example, it may be shown (Sec. 12.7), if the resistance of the air is neglected, that the components of the acceleration are

$$a_x = 0 \qquad a_y = -g$$

where g is 9.81 m/s^2 or 32.2 ft/s^2. Substituting into Eqs. (11.19), we write

$$\dot{v}_x = 0 \qquad\quad \dot{v}_y = -g$$
$$v_x = (v_x)_0 \qquad v_y = (v_y)_0 - gt$$

where $(v_x)_0$ and $(v_y)_0$ are the components of the initial velocity. Substituting into Eqs. (11.16), and assuming that the projectile is fired from the origin O, we write

$$\dot{x} = (v_x)_0 \qquad\quad \dot{y} = (v_y)_0 - gt$$
$$x = (v_x)_0 t \qquad y = (v_y)_0 t - \tfrac{1}{2} g t^2$$

These equations show that the motion of the projectile in the horizontal direction is uniform, while its motion in the vertical direction is uniformly accelerated. The motion of a projectile may thus be replaced by two independent rectilinear motions, which are easily visualized if we assume that the projectile is fired vertically with an initial velocity $(v_y)_0$ from a platform moving with a constant horizontal velocity $(v_x)_0$ (Fig. 11.18). The coordinate x of the projectile is equal at any instant to the distance traveled by the platform, while its coordinate y may be computed as if the projectile were moving along a vertical line.

It may be observed that the equations defining the coordinates x and y of a projectile at any instant are the parametric equations of a parabola. Thus, the trajectory of a projectile is *parabolic*. This result, however, ceases to be valid when the resistance of the air or the variation of g with altitude is taken into account.

11.11. Relative Motion.   Consider two particles A and B moving in the same plane (Fig. 11.19); the vectors $\mathbf{r}_A$ and $\mathbf{r}_B$ define their position at any given instant with respect to a fixed system of axes xy centered at O. Consider now a system of axes $x'y'$ centered at A and parallel to the x and y axes. While the origin of these axes moves, their orientation remains the same; such a system of axes is said to be in *translation*. The vector $\mathbf{r}_{B/A}$ joining A and B defines the position of the particle B with respect to the moving reference system $x'y'$; it is therefore called the position vector of B relative to the reference system centered at A or, for short, the *position vector of B relative to A*.

We note from Fig. 11.19 that the position vector $\mathbf{r}_B$ of particle B may be obtained by adding by the triangle rule the position vector $\mathbf{r}_A$ of particle A and the position vector $\mathbf{r}_{B/A}$ of B relative to A; we write

$$\mathbf{r}_B = \mathbf{r}_A + \mathbf{r}_{B/A} \qquad (11.21)$$

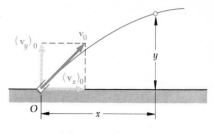

(*a*) Motion of a projectile

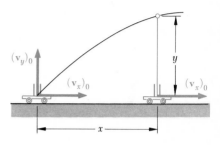

(*b*) Equivalent rectilinear motions

Fig. 11.18

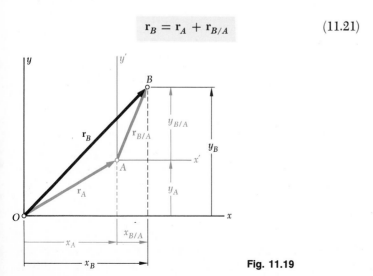

Fig. 11.19

In terms of x and y components, this relation reads

$$x_B = x_A + x_{B/A} \qquad y_B = y_A + y_{B/A} \qquad (11.22)$$

where x_B and y_B denote the coordinates of particle B with respect to the fixed x and y axes, while $x_{B/A}$ and $y_{B/A}$ represent the coordinates of B with respect to the moving x' and y' axes. Differentiating Eqs. (11.22) with respect to t and using dots to indicate time derivatives, we write

$$\dot{x}_B = \dot{x}_A + \dot{x}_{B/A} \qquad \dot{y}_B = \dot{y}_A + \dot{y}_{B/A} \qquad (11.23)$$

The derivatives $\dot{x}_A$ and $\dot{y}_A$ represent the x and y components of the velocity $\mathbf{v}_A$ of particle A, while the derivatives $\dot{x}_B$ and $\dot{y}_B$ represent the components of the velocity $\mathbf{v}_B$ of particle B. The vector $\mathbf{v}_{B/A}$ of components $\dot{x}_{B/A}$ and $\dot{y}_{B/A}$ is known as the *relative velocity of B with respect to A* or, more accurately, as the relative velocity of B with respect to a reference system in translation with A. Writing Eqs. (11.23) in vector form, we have

$$\mathbf{v}_B = \mathbf{v}_A + \mathbf{v}_{B/A} \qquad (11.24)$$

Differentiating Eqs. (11.23), we write

$$\ddot{x}_B = \ddot{x}_A + \ddot{x}_{B/A} \qquad \ddot{y}_B = \ddot{y}_A + \ddot{y}_{B/A} \qquad (11.25)$$

where the first two derivatives in each equation represent the components of the accelerations $\mathbf{a}_B$ and $\mathbf{a}_A$. The vector $\mathbf{a}_{B/A}$ of components $\ddot{x}_{B/A}$ and $\ddot{y}_{B/A}$ is known as the *relative acceleration of B with respect to A* or, more accurately, as the relative acceleration of B with respect to a reference system in translation with A. Writing Eqs. (11.25) in vector form, we obtain

$$\mathbf{a}_B = \mathbf{a}_A + \mathbf{a}_{B/A} \qquad (11.26)$$

The motion of B with respect to the fixed x and y axes is referred to as the *absolute motion of B*, while the motion of B with respect to the x' and y' axes attached at A is referred to as the *relative motion of B with respect to A*. The equations derived in this section show that *the absolute motion of B may be obtained by combining the motion of A and the relative motion of B with respect to A*. Equation (11.24), for example, expresses that the absolute velocity $\mathbf{v}_B$ of particle B may be obtained by adding vectorially the velocity of A and the velocity of B relative to A. Equation (11.26) expresses a similar property in terms of the accelerations. We should keep in mind, however, that the x' and y' axes are in translation, i.e., that, while they move with A, they maintain the same orientation. As we shall see later (Sec. 15.11), different relations must be used when the moving reference axes rotate.

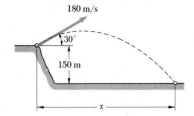

A projectile is fired from the edge of a 150-m cliff with an initial velocity of 180 m/s, at an angle of 30° with the horizontal. Neglecting air resistance, find (*a*) the horizontal distance from the gun to the point where the projectile strikes the ground, (*b*) the greatest elevation above the ground reached by the projectile.

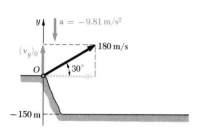

Solution. We shall consider separately the vertical and the horizontal motion.

Vertical Motion. Uniformly accelerated motion. Choosing the positive sense of the y axis upward and placing the origin O at the gun, we have

$$(v_y)_0 = (180 \text{ m/s}) \sin 30° = +90 \text{ m/s}$$
$$a = -9.81 \text{ m/s}^2$$

Substituting into the equations of uniformly accelerated motion, we have

$$\begin{array}{lll} v_y = (v_y)_0 + at & v_y = 90 - 9.81t & (1) \\ y = (v_y)_0 t + \tfrac{1}{2}at^2 & y = 90t - 4.90t^2 & (2) \\ v_y^2 = (v_y)_0^2 + 2ay & v_y^2 = 8100 - 19.62y & (3) \end{array}$$

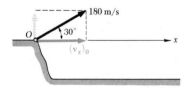

Horizontal Motion. Uniform motion. Choosing the positive sense of the x axis to the right, we have

$$(v_x)_0 = (180 \text{ m/s}) \cos 30° = +155.9 \text{ m/s}$$

Substituting into the equation of uniform motion, we obtain

$$x = (v_x)_0 t \qquad x = 155.9t \qquad (4)$$

a. Horizontal Distance. When the projectile strikes the ground, we have

$$y = -150 \text{ m}$$

Carrying this value into Eq. (2) for the vertical motion, we write

$$-150 = 90t - 4.90t^2 \qquad t^2 - 18.37t - 30.6 = 0 \qquad t = 19.91 \text{ s}$$

Carrying $t = 19.91$ s into Eq. (4) for the horizontal motion, we obtain

$$x = 155.9(19.91) \qquad x = 3100 \text{ m} \quad \blacktriangleleft$$

b. Greatest Elevation. When the projectile reaches its greatest elevation, we have $v_y = 0$; carrying this value into Eq. (3) for the vertical motion, we write

$$0 = 8100 - 19.62y \qquad y = 413 \text{ m}$$
$$\text{Greatest elevation above ground} = 150 \text{ m} + 413 \text{ m}$$
$$= 563 \text{ m} \quad \blacktriangleleft$$

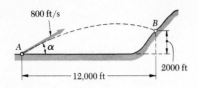

A projectile is fired with an initial velocity of 800 ft/s at a target B located 2000 ft above the gun A and at a horizontal distance of 12,000 ft. Neglecting air resistance, determine the value of the firing angle α.

Solution. We shall consider separately the horizontal and the vertical motion.

Horizontal Motion. Placing the origin of coordinates at the gun, we have

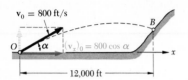

$$(v_x)_0 = 800 \cos \alpha$$

Substituting into the equation of uniform horizontal motion, we obtain

$$x = (v_x)_0 t \qquad x = (800 \cos \alpha)t$$

The time required for the projectile to move through a horizontal distance of 12,000 ft is obtained by making x equal to 12,000 ft.

$$12,000 = (800 \cos \alpha)t$$

$$t = \frac{12,000}{800 \cos \alpha} = \frac{15}{\cos \alpha}$$

Vertical Motion

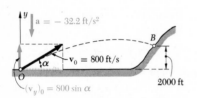

$$(v_y)_0 = 800 \sin \alpha \qquad a = -32.2 \text{ ft/s}^2$$

Substituting into the equation of uniformly accelerated vertical motion, we obtain

$$y = (v_y)_0 t + \tfrac{1}{2}at^2 \qquad y = (800 \sin \alpha)t - 16.1t^2$$

Projectile Hits Target. When $x = 12,000$ ft, we must have $y = 2000$ ft. Substituting for y and making t equal to the value found above, we write

$$2000 = 800 \sin \alpha \frac{15}{\cos \alpha} - 16.1\left(\frac{15}{\cos \alpha}\right)^2$$

Since $1/\cos^2 \alpha = \sec^2 \alpha = 1 + \tan^2 \alpha$, we have

$$2000 = 800(15) \tan \alpha - 16.1(15^2)(1 + \tan^2 \alpha)$$
$$3622 \tan^2 \alpha - 12,000 \tan \alpha + 5622 = 0$$

Solving this quadratic equation for $\tan \alpha$, we have

$$\tan \alpha = 0.565 \qquad \text{and} \qquad \tan \alpha = 2.75$$

$$\alpha = 29.5° \qquad \text{and} \qquad \alpha = 70.0° \quad \blacktriangleleft$$

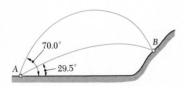

The target will be hit if either of these two firing angles is used (see figure).

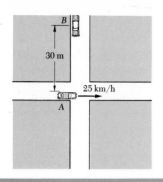

Automobile A is traveling east at the constant speed of 25 km/h. As automobile A crosses the intersection shown, automobile B starts from rest 30 m north of the intersection and moves south with a constant acceleration of 1.2 m/s². Determine the position, velocity, and acceleration of B relative to A five seconds after A crosses the intersection.

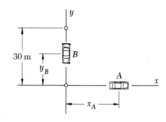

Solution. We choose x and y axes with origin at the intersection of the two streets and with positive senses directed respectively east and north.

Motion of Automobile A. First the speed is expressed in m/s:

$$25 \text{ km/h} = \frac{25 \text{ km}}{1 \text{ h}} = \frac{25\,000 \text{ m}}{3600 \text{ s}} = 6.94 \text{ m/s}$$

Noting that the motion of A is uniform, we write, for any time t,

$$a_A = 0$$
$$v_A = +6.94 \text{ m/s}$$
$$x_A = (x_A)_0 + v_A t = 0 + 6.94t$$

For $t = 5$ s, we have

$$a_A = 0 \qquad\qquad a_A = 0$$
$$v_A = +6.94 \text{ m/s} \qquad\qquad v_A = 6.94 \text{ m/s} \rightarrow$$
$$x_A = +(6.94 \text{ m/s})(5 \text{ s}) = +34.7 \text{ m} \qquad r_A = 34.7 \text{ m} \rightarrow$$

Motion of Automobile B. We note that the motion of B is uniformly accelerated, and write

$$a_B = -1.2 \text{ m/s}^2$$
$$v_B = (v_B)_0 + at = 0 - 1.2t$$
$$y_B = (y_B)_0 + (v_B)_0 t + \tfrac{1}{2}a_B t^2 = 30 + 0 - \tfrac{1}{2}(1.2)t^2$$

For $t = 5$ s, we have

$$a_B = -1.2 \text{ m/s}^2 \qquad\qquad a_B = 1.2 \text{ m/s}^2 \downarrow$$
$$v_B = -(1.2 \text{ m/s})(5 \text{ s}) = -6 \text{ m/s} \qquad\qquad v_B = 6 \text{ m/s} \downarrow$$
$$y_B = 30 - \tfrac{1}{2}(1.2 \text{ m/s})(5 \text{ s})^2 = +15 \text{ m} \qquad r_B = 15 \text{ m} \uparrow$$

Motion of B Relative to A. We draw the triangle corresponding to the vector equation $\mathbf{r}_B = \mathbf{r}_A + \mathbf{r}_{B/A}$ and obtain the magnitude and direction of the position vector of B relative to A.

$$r_{B/A} = 37.8 \text{ m} \qquad \alpha = 23.4° \qquad r_{B/A} = 37.8 \text{ m} \; \diagdown \; 23.4° \; \blacktriangleleft$$

Proceeding in a similar fashion, we find the velocity and acceleration of B relative to A.

$$\mathbf{v}_B = \mathbf{v}_A + \mathbf{v}_{B/A}$$
$$v_{B/A} = 9.17 \text{ m/s} \qquad \beta = 40.8° \qquad v_{B/A} = 9.17 \text{ m/s} \; \diagup \; 40.8° \; \blacktriangleleft$$
$$\mathbf{a}_B = \mathbf{a}_A + \mathbf{a}_{B/A} \qquad\qquad a_{B/A} = 1.2 \text{ m/s}^2 \downarrow \; \blacktriangleleft$$

PROBLEMS

Note. Neglect air resistance in problems concerning projectiles.

11.66   The motion of a particle is defined by the equations $x = \frac{1}{3}t^3 - 2t^2$ and $y = \frac{1}{2}t^2 - 2t$, where x and y are expressed in meters and t in seconds. Determine the velocity and acceleration when (*a*) $t = 1$ s, (*b*) $t = 3$ s.

11.67   In Prob. 11.66, determine (*a*) the time at which the value of the y coordinate is minimum, (*b*) the corresponding velocity and acceleration of the particle.

11.68   The motion of a particle is defined by the equations $x = (t + 1)^2$ and $y = (t + 1)^{-2}$, where x and y are expressed in feet and t in seconds. Show that the path of the particle is a rectangular hyperbola and determine the velocity and acceleration when (*a*) $t = 0$, (*b*) $t = 1$ s.

11.69   The motion of a particle is defined by the equations $x = 25(1 - e^{-t})$ and $y = -3t^2 + 15$, where x and y are expressed in feet and t in seconds. Determine the velocity and acceleration when $t = 2$ s.

11.70   The motion of a vibrating particle is defined by the equations $x = 100 \sin \pi t$ and $y = 25 \cos 2\pi t$, where x and y are expressed in millimeters and t in seconds. (*a*) Determine the velocity and acceleration when $t = 1$ s. (*b*) Show that the path of the particle is parabolic.

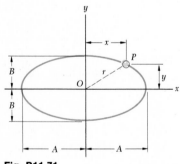

Fig. P11.71

11.71   A particle moves in an elliptic path according to the equations $x = A \cos pt$ and $y = B \sin pt$. Show that the acceleration (*a*) is directed toward the origin, (*b*) is proportional to the distance from the origin to the particle.

11.72   A particle moves in a circular path according to the equations $x = r \sin \omega t$ and $y = r \cos \omega t$. Show (*a*) that the magnitude v of the velocity is constant and equal to $r\omega$, (*b*) that the acceleration is of constant magnitude $r\omega^2$ (or v^2/r) and is directed toward the center of the circular path.

11.73   The motion of a vibrating particle is defined by the equations $x = A \sin pt$ and $y = B \sin (pt + \phi)$, where A, B, p, and ϕ are constants. If $A = B$, determine (*a*) the value of the constant ϕ for which the speed of the particle is constant, (*b*) the corresponding value of the speed, (*c*) the corresponding path of the particle.

11.74   A man standing on a bridge 20 m above the water throws a stone in a horizontal direction. Knowing that the stone hits the water 30 m from a point on the water directly below the man, determine (a) the initial velocity of the stone, (b) the distance at which the stone would hit the water if it were thrown with the same velocity from a bridge 5 m lower.

11.75   Water issues at A from a pressure tank with a horizontal velocity v_0. For what range of values v_0 will the water enter the opening BC?

11.76   A nozzle at A discharges water with an initial velocity of 40 ft/s at an angle of 60° with the horizontal. Determine where the stream of water strikes the roof. Check that the stream will clear the edge of the roof.

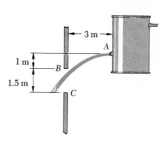

Fig. P11.75

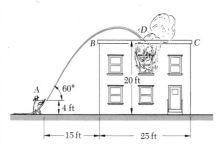

Fig. P11.76

11.77   In Prob. 11.76, determine the largest and smallest initial velocity for which the water will fall on the roof.

11.78   A ball is dropped vertically onto a 20° incline at A; the direction of rebound forms an angle of 40° with the vertical. Knowing that the ball next strikes the incline at B, determine (a) the velocity of rebound at A, (b) the time required for the ball to travel from A to B.

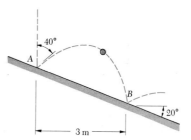

Fig. P11.78

11.79   Sand is discharged at A from a conveyor belt and falls onto a stockpile at B. Knowing that the conveyor belt forms an angle $\alpha = 20°$ with the horizontal and moves at a constant speed of 12 ft/s, determine the distance d to the top of the stockpile.

11.80   The conveyor belt moves at a constant speed of 12 ft/s. Knowing that $d = 12$ ft, determine the angle α for which the sand is deposited on the stockpile at B.

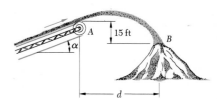

Fig. P11.79 and P11.80

11.81   A projectile is fired with an initial velocity of 210 m/s. Find the angle at which it should be fired if it is to hit a target located at a distance of 3600 m on the same level.

11.82 A boy can throw a baseball a maximum distance of 30 m in New York, where $g = 9.81$ m/s^2. How far could he throw the baseball (a) in Singapore, where $g = 9.78$ m/s^2? (b) On the moon, where $g = 1.618$ m/s^2?

11.83 A projectile is fired with an initial velocity $\mathbf{v}_0$ at an angle α with the horizontal. Determine (a) the maximum height h reached by the projectile, (b) the horizontal range R of the projectile, (c) the maximum horizontal range R and the corresponding firing angle α.

11.84 If the maximum horizontal range of a given gun is R, determine the firing angle which should be used to hit a target located at a distance $\frac{1}{2}R$ on the same level.

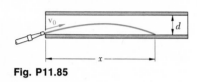
Fig. P11.85

11.85 A nozzle discharges a stream of water with an initial velocity $\mathbf{v}_0$ of 50 ft/s into the end of a horizontal pipe of inside diameter $d = 5$ ft. Determine the largest distance x that the stream can reach.

11.86 Show that the trajectory of a projectile is a parabola if air resistance is neglected.

11.87 Two airplanes A and B are each flying at a constant altitude of 3000 ft. Plane A is flying due east at a constant speed of 300 mi/h while plane B is flying southwest at a constant speed of 450 mi/h. Determine the change in position of plane B relative to plane A which takes place during a 2-min interval.

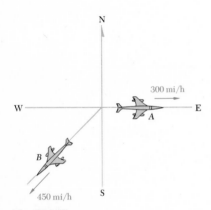
Fig. P11.87

11.88 Instruments in an airplane indicate that, with respect to the air, the plane is moving west at a speed of 250 mi/h. At the same time ground-based radar indicates the plane to be moving at a speed of 300 mi/h in a direction 12° south of west. Determine the magnitude and direction of the velocity of the air.

11.89 As he passes a pole, a man riding in a truck tries to hit the pole by throwing a stone with a horizontal velocity of 20 m/s relative to the truck. Knowing that the speed of the truck is 40 km/h, determine (a) the direction in which he must throw the stone, (b) the horizontal velocity of the stone with respect to the ground.

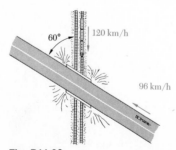
Fig. P11.90

11.90 An automobile and a train travel at the constant speeds shown. Three seconds after the train passes under the highway bridge the automobile crosses the bridge. Determine (a) the velocity of the train relative to the automobile, (b) the change in position of the train relative to the automobile during a 4-s interval, (c) the distance between the train and the automobile 5 s after the automobile has crossed the bridge.

11.91 During a rainstorm the paths of the raindrops appear to form an angle of 30° with the vertical when observed from a side window of a train moving at a speed of 15 km/h. A short time later, after the speed of the train has increased to 30 km/h, the angle between the vertical and the paths of the drops appears to be 45°. If the train were stopped, at what angle and with what velocity would the drops be observed to fall?

11.92 Ship A travels due south at 12 knots, while ship B travels toward the southeast at 10 knots. Determine the velocity (*a*) of ship B relative to ship A, (*b*) of ship A relative to ship B.

11.93 As observed from a ship moving due south at 10 mi/h, the wind appears to blow from the east. After the ship has changed course, and as it is moving due west at 10 mi/h, the wind appears to blow from the northeast. Assuming that the wind velocity is constant during the period of observation, determine the magnitude and direction of the true wind velocity.

11.94 An airplane is flying horizontally at an altitude of 2500 m and at a constant speed of 900 km/h on a path which passes directly over an antiaircraft gun. The gun fires a shell with a muzzle velocity of 500 m/s and hits the airplane. Knowing that the firing angle of the gun is 60°, determine the velocity and acceleration of the shell relative to the airplane at the time of impact.

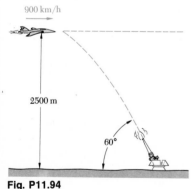

Fig. P11.94

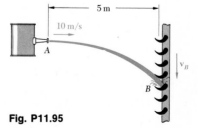

Fig. P11.95

11.95 Water is discharged at A with an initial velocity of 10 m/s and strikes a series of vanes at B. Knowing that the vanes move downward with a constant speed of 3 m/s, determine the velocity and acceleration of the water relative to the vane at B.

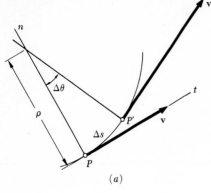

(a)

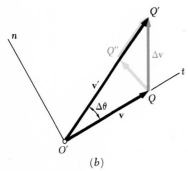

(b)

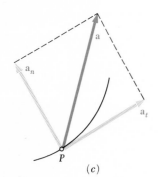

(c)

Fig. 11.20

11.12. Tangential and Normal Components. We saw in Sec. 11.9 that the velocity of a particle is a vector tangent to the path of the particle but that, in general, the acceleration is not tangent to the path. It is sometimes convenient to resolve the acceleration into components directed, respectively, along the tangent and the normal to the path of the particle. The positive sense along the tangent is chosen to coincide with the sense of motion of the particle, and the positive sense along the normal is chosen toward the inside of the path (Fig. 11.20).

We consider again, as we did in Sec. 11.9, the velocity $\mathbf{v}$ of the particle at a time t and its velocity $\mathbf{v}'$ at a time $t + \Delta t$ (Fig. 11.20a). Drawing both vectors from the same origin O', and joining the points Q and Q', we obtain the small vector $\Delta\mathbf{v}$ representing the difference between the vectors $\mathbf{v}$ and $\mathbf{v}'$ (Fig. 11.20b). As we saw earlier, the average acceleration is defined as the quotient of $\Delta\mathbf{v}$ and Δt, and the instantaneous acceleration $\mathbf{a}$ as the limit of this quotient. To obtain the tangential component $\mathbf{a}_t$ and the normal component $\mathbf{a}_n$ of the acceleration, we measure the distance $O'Q'' = O'Q$ along the line $O'Q'$ and resolve the vector $\Delta\mathbf{v}$ into its components $\overrightarrow{QQ''}$ and $\overrightarrow{Q''Q'}$ (Fig. 11.20b); we write

$$\Delta\mathbf{v} = \overrightarrow{QQ''} + \overrightarrow{Q''Q'}$$

The component $\overrightarrow{QQ''}$ represents a change in the *direction* of the velocity, while the component $\overrightarrow{Q''Q'}$ represents a change in the *magnitude* of the velocity, i.e., a change in *speed*. Substituting in (11.18) the expression obtained for $\Delta\mathbf{v}$, we write

$$\mathbf{a} = \lim_{\Delta t \to 0} \frac{\Delta\mathbf{v}}{\Delta t} = \lim_{\Delta t \to 0} \frac{\overrightarrow{QQ''}}{\Delta t} + \lim_{\Delta t \to 0} \frac{\overrightarrow{Q''Q'}}{\Delta t}$$

As Δt approaches zero, the vector $\overrightarrow{Q''Q'}$ becomes parallel to the tangent to the path at P and the vector $\overrightarrow{QQ''}$ becomes parallel to the normal. We must have, therefore,

$$\mathbf{a}_t = \lim_{\Delta t \to 0} \frac{\overrightarrow{Q''Q'}}{\Delta t} \qquad \mathbf{a}_n = \lim_{\Delta t \to 0} \frac{\overrightarrow{QQ''}}{\Delta t} \qquad (11.27)$$

Considering first the tangential component, we observe that the length of the vector $\overrightarrow{Q''Q'}$ measures the change in speed $v' - v = \Delta v$. The scalar component a_t of the acceleration is therefore

$$a_t = \lim_{\Delta t \to 0} \frac{\Delta v}{\Delta t}$$

$$a_t = \frac{dv}{dt} \qquad (11.28)$$

Formula (11.28) expresses that the *tangential component* a_t of the acceleration is equal to the *rate of change of the speed of the particle.*

Considering now the normal component of the acceleration, and denoting by $\Delta\theta$ the angle between the tangents to the path at P and P', we observe that, for small values of $\Delta\theta$, the length of the vector $\overrightarrow{QQ''}$ is equal to $v\,\Delta\theta$. Since the angle $\Delta\theta$ is equal to the angle between the normals at P and P' (Fig. 11.20a), we have $\Delta\theta = \Delta s/\rho$, where Δs is the length of the arc PP' and ρ the radius of curvature of the path at P. It thus follows from the second of the relations (11.27) that the scalar component a_n of the acceleration is

$$a_n = \lim_{\Delta t \to 0} \frac{v\,\Delta\theta}{\Delta t} = \lim_{\Delta t \to 0} \frac{v}{\rho}\frac{\Delta s}{\Delta t} = \frac{v}{\rho}\frac{ds}{dt}$$

Recalling from (11.15) that $ds/dt = v$, we have

$$a_n = \frac{v^2}{\rho} \qquad\qquad (11.29)$$

Formula (11.29) expresses that the *normal component* a_n of the acceleration at P is equal to the *square of the speed divided by the radius of curvature of the path at P.* It should be noted that the vector $\mathbf{a}_n$ is always directed *toward the center of curvature of the path* (Fig. 11.20c).

It appears from the above that the tangential component of the acceleration reflects a change in the speed of the particle, while its normal component reflects a change in the direction of motion of the particle. The acceleration of a particle will be zero only if both its components are zero. Thus, the acceleration of a particle moving with constant speed along a curve will not be zero, unless the particle happens to pass through a point of inflection of the curve (where the radius of curvature is infinite) or unless the curve is a straight line.

The fact that the normal component of the acceleration depends upon the radius of curvature of the path followed by the particle is taken into account in the design of structures or mechanisms as widely different as airplane wings, railroad tracks, and cams. In order to avoid sudden changes in the acceleration of the air particles flowing past a wing, wing profiles are designed without any sudden change in curvature. Similar care is taken in designing railroad curves, to avoid sudden changes in the acceleration of the cars (which would be hard on the equipment and unpleasant for the passengers). A straight section of track, for instance, is never directly followed by a circular section. Special transition sections are used, to help pass smoothly from the infinite radius of curvature of the straight section to the finite

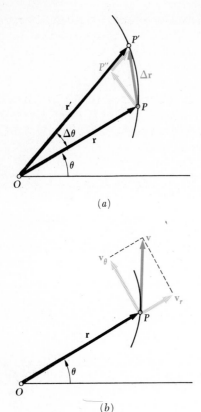

(a)

(b)

Fig. 11.21

radius of the circular track. Likewise, in the design of high-speed cams, abrupt changes in acceleration are avoided by using transition curves which produce a continuous change in acceleration.

11.13. Radial and Transverse Components. In certain problems, the position of a particle is defined by its polar coordinates r and θ. It is then convenient to resolve the velocity and acceleration of the particle into components parallel and perpendicular, respectively, to the position vector $\mathbf{r}$. These components are called *radial* and *transverse components*.

Consider the vectors $\mathbf{r}$ and $\mathbf{r}'$ defining the position of the particle respectively at times t and $t + \Delta t$ (Fig. 11.21a). We recall that the velocity of the particle at time t is defined as the limit of the quotient of $\Delta \mathbf{r}$ and Δt, where $\Delta \mathbf{r}$ is the vector joining P and P'. To obtain the radial component $\mathbf{v}_r$ and the transverse component $\mathbf{v}_\theta$ of the velocity, we measure a distance $OP'' = OP$ along the line OP' and resolve the vector $\Delta \mathbf{r}$ into its components $\overrightarrow{PP''}$ and $\overrightarrow{P''P'}$. The velocity $\mathbf{v}$ may thus be expressed as

$$\mathbf{v} = \lim_{\Delta t \to 0} \frac{\Delta \mathbf{r}}{\Delta t} = \lim_{\Delta t \to 0} \frac{\overrightarrow{PP''}}{\Delta t} + \lim_{\Delta t \to 0} \frac{\overrightarrow{P''P'}}{\Delta t}$$

As Δt approaches zero, the vectors $\overrightarrow{P''P'}$ and $\overrightarrow{PP''}$ become parallel and perpendicular, respectively, to $\mathbf{r}$. We must have, therefore,

$$\mathbf{v}_r = \lim_{\Delta t \to 0} \frac{\overrightarrow{P''P'}}{\Delta t} \qquad \mathbf{v}_\theta = \lim_{\Delta t \to 0} \frac{\overrightarrow{PP''}}{\Delta t}$$

We observe that the length of the vector $\overrightarrow{P''P'}$ measures the change in magnitude $r' - r = \Delta r$ of the radius vector. (Note that the change in magnitude Δr should not be confused with the magnitude of the vector $\Delta \mathbf{r}$, which represents a change in direction as well as in magnitude.) On the other hand, the length of the vector $\overrightarrow{PP''}$ is equal to $r \, \Delta \theta$ for small values of $\Delta \theta$. The scalar components of the velocity are therefore

$$v_r = \lim_{\Delta t \to 0} \frac{\Delta r}{\Delta t} \qquad v_\theta = \lim_{\Delta t \to 0} \frac{r \, \Delta \theta}{\Delta t}$$

$$v_r = \frac{dr}{dt} \qquad v_\theta = r \frac{d\theta}{dt}$$

or, using dots to indicate time derivatives,

$$v_r = \dot{r} \qquad v_\theta = r \dot{\theta} \tag{11.30}$$

Consider, now, the velocity $\mathbf{v}$ of P at a time t and its velocity $\mathbf{v}'$ at a time $t + \Delta t$ (Fig. 11.22a). Drawing the radial and transverse components of $\mathbf{v}$ and $\mathbf{v}'$ from the same origin O', we represent the change in velocity by the two vectors $\overrightarrow{RR'}$ and $\overrightarrow{TT'}$ (Fig. 11.22b). The vector $\overrightarrow{RR'}$ represents a change in magnitude and direction for the vector $\mathbf{v}_r$; similarly, the vector $\overrightarrow{TT'}$ represents a change in magnitude and direction for the vector $\mathbf{v}_\theta$. The sum of the vectors $\overrightarrow{RR'}$ and $\overrightarrow{TT'}$ represents the total change in velocity $\Delta\mathbf{v}$. As we saw earlier, the acceleration of the particle is defined as the limit of the quotient of $\Delta\mathbf{v}$ and Δt. To obtain the radial component $\mathbf{a}_r$ and the transverse component $\mathbf{a}_\theta$ of the acceleration, we resolve the vector $\overrightarrow{RR'}$ into its components $\overrightarrow{RR''}$ and $\overrightarrow{R''R'}$ (with $O'R'' = O'R$) and the vector $\overrightarrow{TT'}$ into its components $\overrightarrow{TT''}$ and $\overrightarrow{T''T'}$ (with $O'T'' = O'T$). The acceleration $\mathbf{a}$ of the particle may then be expressed as

$$\mathbf{a} = \lim_{\Delta t \to 0} \frac{\Delta\mathbf{v}}{\Delta t} = \lim_{\Delta t \to 0} \left(\frac{\overrightarrow{RR'}}{\Delta t} + \frac{\overrightarrow{TT'}}{\Delta t} \right)$$

$$= \lim_{\Delta t \to 0} \left(\frac{\overrightarrow{RR''}}{\Delta t} + \frac{\overrightarrow{R''R'}}{\Delta t} + \frac{\overrightarrow{TT''}}{\Delta t} + \frac{\overrightarrow{T''T'}}{\Delta t} \right)$$

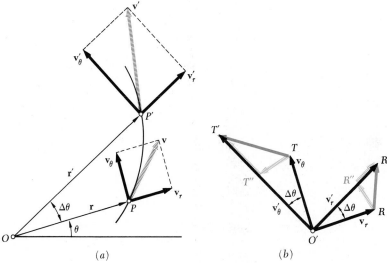

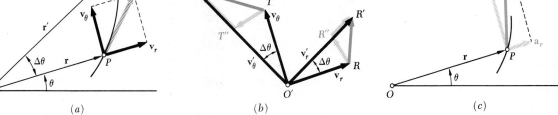

(a)   (b)   (c)

Fig. 11.22

As Δt approaches zero, the vectors $\overrightarrow{R''R'}$ and $\overrightarrow{TT''}$ become parallel to $\mathbf{r}$, while the vectors $\overrightarrow{RR''}$ and $\overrightarrow{T''T'}$ become perpendicular to $\mathbf{r}$. We must have, therefore,

$$\mathbf{a}_r = \lim_{\Delta t \to 0} \left(\frac{\overrightarrow{R''R'}}{\Delta t} + \frac{\overrightarrow{TT''}}{\Delta t} \right) \qquad \mathbf{a}_\theta = \lim_{\Delta t \to 0} \left(\frac{\overrightarrow{T''T'}}{\Delta t} + \frac{\overrightarrow{RR''}}{\Delta t} \right)$$

From Fig. 11.22b we observe that, for small values of Δt, the lengths of the various vectors involved are

$$R''R' = \Delta(v_r) \qquad T''T' = \Delta(v_\theta)$$
$$RR'' = v_r\,\Delta\theta \qquad TT'' = v_\theta\,\Delta\theta$$

where $\Delta(v_r)$ and $\Delta(v_\theta)$ represent the *change in magnitude* of the vectors $\mathbf{v}_r$ and $\mathbf{v}_\theta$, respectively. Noting also that the vector $\overrightarrow{TT''}$ has a sense opposite to that of $\mathbf{r}$, we obtain the following expressions for the scalar components of the acceleration:

$$a_r = \lim_{\Delta t \to 0} \left[\frac{\Delta(v_r)}{\Delta t} - \frac{v_\theta\,\Delta\theta}{\Delta t} \right] \qquad a_\theta = \lim_{\Delta t \to 0} \left[\frac{\Delta(v_\theta)}{\Delta t} + \frac{v_r\,\Delta\theta}{\Delta t} \right]$$

$$a_r = \frac{d(v_r)}{dt} - v_\theta \frac{d\theta}{dt} \qquad a_\theta = \frac{d(v_\theta)}{dt} + v_r \frac{d\theta}{dt}$$

or, using dots to indicate time derivatives,

$$a_r = \dot{v}_r - v_\theta \dot{\theta} \qquad a_\theta = \dot{v}_\theta + v_r \dot{\theta} \qquad (11.31)$$

We should note that, in general, a_r *is not equal to* $\dot{v}_r$ *and* a_θ *is not equal to* $\dot{v}_\theta$. Each of the relations (11.31) contains an extra term which represents the *change in direction* of the vectors $\mathbf{v}_r$ and $\mathbf{v}_\theta$. Differentiating the relations (11.30), we obtain

$$\dot{v}_r = \ddot{r} \qquad \dot{v}_\theta = \frac{d}{dt}(r\dot{\theta}) = r\ddot{\theta} + \dot{r}\dot{\theta} \qquad (11.32)$$

Substituting from (11.30) and (11.32) into (11.31), we write

$$a_r = \ddot{r} - r\dot{\theta}^2 \qquad a_\theta = r\ddot{\theta} + 2\dot{r}\dot{\theta} \qquad (11.33)$$

A positive value for a_r indicates that the vector $\mathbf{a}_r$ has the same sense as $\mathbf{r}$; a positive value for a_θ indicates that the vector $\mathbf{a}_\theta$ points toward increasing values of θ.

In the case of a particle moving along a circle of center O, we have $r = \text{constant}$, $\dot{r} = \ddot{r} = 0$, and the formulas (11.33) reduce to

$$a_r = -r\dot{\theta}^2 \qquad a_\theta = r\ddot{\theta} \qquad (11.34)$$

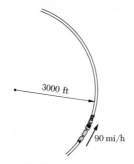

SAMPLE PROBLEM 11.10

A train is traveling on a curved section of track of radius 3000 ft at the speed of 90 mi/h. The brakes are suddenly applied, causing the train to slow down at a constant rate; after 6 s, the speed has been reduced to 60 mi/h. Determine the acceleration of a car immediately after the brakes have been applied.

Tangential Component of Acceleration. First the speeds are expressed in ft/s.

$$90 \text{ mi/h} = \left(90 \frac{\text{mi}}{\text{h}}\right)\left(\frac{5280 \text{ ft}}{1 \text{ mi}}\right)\left(\frac{1 \text{ h}}{3600 \text{ s}}\right) = 132 \text{ ft/s}$$

$$60 \text{ mi/h} = 88 \text{ ft/s}$$

Since the train slows down at a constant rate, we have

$$a_t = \text{average } a_t = \frac{\Delta v}{\Delta t} = \frac{88 \text{ ft/s} - 132 \text{ ft/s}}{6 \text{ s}} = -7.33 \text{ ft/s}^2$$

Normal Component of Acceleration. Immediately after the brakes have been applied, the speed is still 132 ft/s, and we have

$$a_n = \frac{v^2}{\rho} = \frac{(132 \text{ ft/s})^2}{3000 \text{ ft}} = 5.81 \text{ ft/s}^2$$

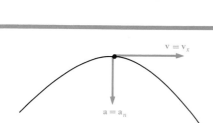

Magnitude and Direction of Acceleration. The magnitude and direction of the resultant **a** of the components **a**$_n$ and **a**$_t$ are

$$\tan \alpha = \frac{a_n}{a_t} = \frac{5.81 \text{ ft/s}^2}{7.33 \text{ ft/s}^2} \qquad \alpha = 38.4° \quad \blacktriangleleft$$

$$a = \frac{a_n}{\sin \alpha} = \frac{5.81 \text{ ft/s}^2}{\sin 38.4°} \qquad a = 9.35 \text{ ft/s}^2 \quad \blacktriangleleft$$

SAMPLE PROBLEM 11.11

Determine the minimum radius of curvature of the trajectory described by the projectile considered in Sample Prob. 11.7.

Solution. Since $a_n = v^2/\rho$, we have $\rho = v^2/a_n$. The radius will be small when v is small or when a_n is large. The speed v is minimum at the top of the trajectory since $v_y = 0$ at that point; a_n is maximum at that same point, since the direction of the vertical coincides with the direction of the normal. Therefore, the minimum radius of curvature occurs at the top of the trajectory. At this point, we have

$$v = v_x = 155.9 \text{ m/s} \qquad a_n = a = 9.81 \text{ m/s}^2$$

$$\rho = \frac{v^2}{a_n} = \frac{(155.9 \text{ m/s})^2}{9.81 \text{ m/s}^2} \qquad \rho = 2480 \text{ m} \quad \blacktriangleleft$$

447

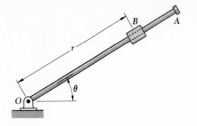

SAMPLE PROBLEM 11.12

The rotation of the 3-ft arm OA about O is defined by the relation $\theta = 0.15t^2$, where θ is expressed in radians and t in seconds. Block B slides along the arm in such a way that its distance from O is $r = 3 - 0.40t^2$, where r is expressed in feet and t in seconds. Determine the total velocity and the total acceleration of block B after the arm OA has rotated through $30°$.

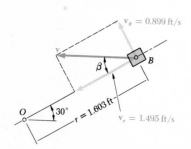

Solution. We first find the time t at which $\theta = 30°$. Substituting $\theta = 30° = 0.524$ rad into the expression for θ, we obtain

$$\theta = 0.15t^2 \qquad 0.524 = 0.15t^2 \qquad t = 1.869 \text{ s}$$

Equations of Motion. Substituting $t = 1.869$ s in the expressions for r, θ, and their first and second derivatives, we have

$$r = 3 - 0.40t^2 = 1.603 \text{ ft} \qquad \theta = 0.15t^2 = 0.524 \text{ rad}$$
$$\dot{r} = -0.80t = -1.495 \text{ ft/s} \qquad \dot{\theta} = 0.30t = 0.561 \text{ rad/s}$$
$$\ddot{r} = -0.80 = -0.800 \text{ ft/s}^2 \qquad \ddot{\theta} = 0.30 = 0.300 \text{ rad/s}^2$$

Velocity of B. Using Eqs. (11.30), we obtain the values of v_r and v_θ when $t = 1.869$ s.

$$v_r = \dot{r} = -1.495 \text{ ft/s}$$
$$v_\theta = r\dot{\theta} = 1.603(0.561) = 0.899 \text{ ft/s}$$

Solving the right triangle shown, we obtain the magnitude and direction of the velocity,

$$v = 1.744 \text{ ft/s} \qquad \beta = 31.0° \quad \blacktriangleleft$$

Acceleration of B. Using Eqs. (11.33), we obtain

$$a_r = \ddot{r} - r\dot{\theta}^2$$
$$= -0.800 - 1.603(0.561)^2 = -1.304 \text{ ft/s}^2$$
$$a_\theta = r\ddot{\theta} + 2\dot{r}\dot{\theta}$$
$$= 1.603(0.300) + 2(-1.495)(0.561) = -1.196 \text{ ft/s}^2$$
$$a = 1.770 \text{ ft/s} \qquad \gamma = 42.5° \quad \blacktriangleleft$$

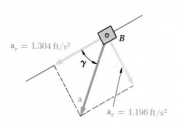

PROBLEMS

11.96 An automobile travels at a constant speed on a highway curve of 1000-m radius. If the normal component of the acceleration is not to exceed 1.2 m/s², determine the maximum allowable speed.

11.97 A car goes around a highway curve of 300-m radius at a speed of 90 km/h. (*a*) What is the normal component of its acceleration? (*b*) At what speed is the normal component of the acceleration one-half as large as that found in part *a*?

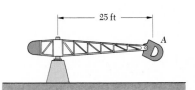

11.98 Determine the peripheral speed of the centrifuge test cab *A* for which the normal component of the acceleration is 10*g*.

Fig. P11.98

11.99 It is desired to have a normal component of acceleration of 8*g* during the testing of an airplane flying at 1200 mi/h. Determine the radius of the circular path along which the pilot should fly the airplane.

11.100 A motorist enters a curve of 500-ft radius at a speed of 30 mi/h. If he increases his speed at a constant rate of 3 ft/s², determine the magnitude of the total acceleration after he has traveled 400 ft along the curve.

11.101 A passenger train is traveling at 60 mi/h along a curve of 3000-ft radius. If the maximum total acceleration is not to exceed *g*/10, determine the maximum rate at which the speed may be decreased.

11.102 The speed of a racing car is increased at a constant rate from 90 km/h to 126 km/h over a distance of 150 m along a curve of 250-m radius. Determine the magnitude of the total acceleration of the car after it has traveled 100 m along the curve.

11.103 A monorail train is traveling at a speed of 144 km/h along a curve of 1000-m radius. Determine the maximum rate at which the speed may be decreased if the total acceleration of the train is not to exceed 2 m/s².

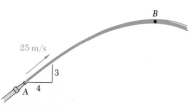

11.104 A nozzle discharges a stream of water in the direction shown with an initial velocity of 25 m/s. Determine the radius of curvature of the stream (*a*) as it leaves the nozzle, (*b*) at the maximum height of the stream.

Fig. P11.104

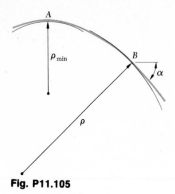

Fig. P11.105

11.105 (*a*) Show that the radius of curvature of the trajectory of a projectile reaches its minimum value at the highest point *A* of the trajectory. (*b*) Denoting by α the angle formed by the trajectory and the horizontal at a given point *B*, show that the radius of curvature of the trajectory at *B* is $\rho = \rho_{min}/\cos^3 \alpha$.

11.106 Determine the radius of curvature of the trajectory described by the projectile of Sample Prob. 11.7 as the projectile leaves the gun.

11.107 For each of the two firing angles obtained in Sample Prob. 11.8, determine the radius of curvature of the trajectory described by the projectile as it leaves the gun.

11.108 A satellite will travel indefinitely in a circular orbit around the earth if the normal component of its acceleration is equal to $g(R/r)^2$, where $g = 32.2 \text{ ft/s}^2$, $R =$ radius of the earth = 3960 mi, and $r =$ distance from the center of the earth to the satellite. Determine the height above the surface of the earth at which a satellite will travel indefinitely around the earth at a speed of 16,500 mi/h.

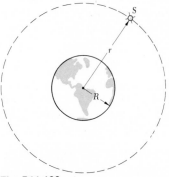

Fig. P11.108

11.109 Determine the speed of an earth satellite traveling in a circular orbit 1000 mi above the surface of the earth. (See information given in Prob. 11.108.)

11.110 Assuming the orbit of the moon to be a circle of radius 239,000 mi, determine the speed of the moon relative to the earth. (See information given in Prob. 11.108.)

11.111 Show that the speed of an earth satellite traveling in a circular orbit is inversely proportional to the square root of the radius of its orbit. Also, determine the minimum time in which a satellite can circle the earth. (See information given in Prob. 11.108.)

11.112 The two-dimensional motion of a particle is defined by the relations $r = 60t^2 - 20t^3$ and $\theta = 2t^2$, where r is expressed in millimeters, t in seconds, and θ in radians. Determine the velocity and acceleration of the particle when (a) $t = 0$, (b) $t = 1$ s.

11.113 The particle of Prob. 11.112 is at the origin at $t = 0$. Determine its velocity and acceleration as it returns to the origin.

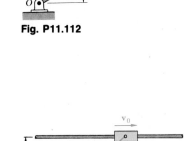

Fig. P11.112

11.114 The two-dimensional motion of a particle is defined by the relations $r = 2b \cos \omega t$ and $\theta = \omega t$, where b and ω are constants. Determine (a) the velocity and acceleration of the particle at any instant, (b) the radius of curvature of its path. What conclusion can you draw regarding the path of the particle?

11.115 The two-dimensional motion of a particle is defined by the relations $r = b \sin \pi t$ and $\theta = \frac{1}{2}\pi t$. Determine the velocity and acceleration of the particle when (a) $t = \frac{1}{2}$, (b) $t = 1$.

11.116 A wire OA connects the collar A and a reel located at O. Knowing that the collar moves to the right with a constant speed v_0, determine $d\theta/dt$ in terms of v_0, b, and θ.

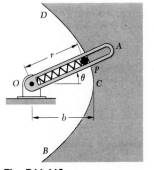

Fig. P11.116

11.117 A rocket is fired vertically from a launching pad at B. Its flight is tracked by radar from point A. Determine the velocity of the rocket in terms of d, θ, and $\dot{\theta}$.

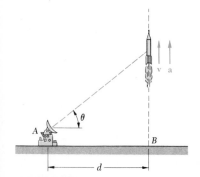

Fig. P11.117

11.118 Determine the acceleration of the rocket of Prob. 11.117 in terms of d, θ, $\dot{\theta}$, and $\ddot{\theta}$.

11.119 As the rod OA rotates, the pin P moves along the parabola BCD. Knowing that the equation of the parabola is $r = 2b/(1 + \cos \theta)$ and that $\theta = kt$, determine the velocity and acceleration of P when (a) $\theta = 0$, (b) $\theta = 90°$.

Fig. P11.119

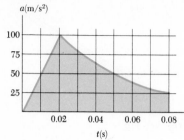

Fig. P11.120

11.120 The *a–t* curve shown was obtained during the motion of a test sled. Knowing that the sled started from rest at $t = 0$, determine the velocity and position of the sled at $t = 0.08$ s.

11.121 An experimental ion-propulsion engine is capable of giving a space vehicle a constant acceleration of 0.02 ft/s^2. If the engine is placed in operation when the speed of the vehicle is $18,000 \text{ mi/h}$, determine the time required to bring the speed of the vehicle to $19,000 \text{ mi/h}$. Assume that the vehicle is moving in a straight line, far from the sun or any planet.

11.122 As circle B rolls on the fixed circle A, point P describes a cardioid defined by the relations $r = 2b(1 + \cos \frac{1}{2}\pi t)$ and $\theta = \frac{1}{2}\pi t$. Determine the velocity and acceleration of P when (a) $t = 1$, (b) $t = 2$.

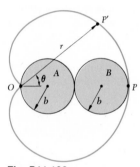

Fig. P11.122

11.123 The acceleration of a particle moving in a straight line is defined by the relation $a = -kv^{1.5}$. The particle starts at $t = 0$ and $x = 0$ with an initial velocity v_0. (a) Show that the velocity and position coordinate at any time t are related by the equation $x/t = \sqrt{v_0 v}$. (b) Determine the value of k, knowing that for $v_0 = 40 \text{ m/s}$ the particle comes to rest after traveling 2 m.

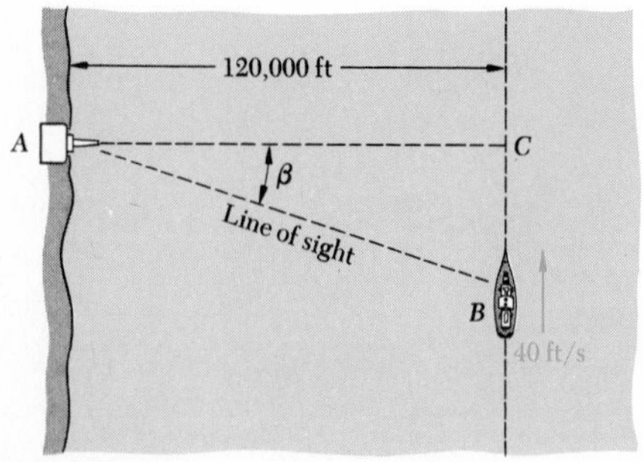

11.124 Ship B is proceeding due north on the course shown. The shore gun at A is trained due east and will be fired in an attempt to hit the ship at point C. Knowing that the muzzle velocity of the gun is 2400 ft/s, determine (a) the required firing angle α, (b) the required angle β between AC and the line of sight at the time of firing.

11.125 The magnitude in m/s^2 of the deceleration due to air resistance of the nose cone of a small experimental rocket is known to be $6 \times 10^{-4} v^2$, where v is expressed in m/s. If the nose cone is projected vertically from the ground with an initial velocity of 100 m/s, determine the maximum height that it will reach.

Fig. P11.124

11.126 Determine the velocity of the nose cone of Prob. 11.125 when it returns to the ground.

11.127 Standing on the side of a hill, an archer shoots an arrow with an initial velocity of 200 ft/s at an angle $\alpha = 15°$ with the horizontal. Determine the horizontal distance d traveled by the arrow before it strikes the ground at B.

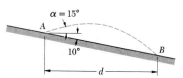

Fig. P11.127

11.128 In Prob. 11.127, determine the radius of curvature of the trajectory (*a*) immediately after the arrow has been shot, (*b*) as the arrow passes through its point of maximum elevation.

11.129 A train starts at a station and accelerates uniformly at a rate of 0.6 m/s^2 until it reaches a speed of 24 m/s; it then proceeds at the constant speed of 24 m/s. Determine the time and the distance traveled if its average velocity is (*a*) 16 m/s, (*b*) 22 m/s.

11.130. Knowing that block B moves downward with a constant velocity of 180 mm/s, determine (*a*) the velocity of block A, (*b*) the velocity of pulley D.

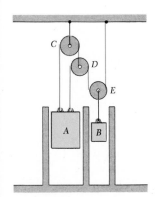

Fig. P11.130

11.131 Drops of water fall down a mine shaft at the uniform rate of one drop per second. A mine elevator moving up the shaft at 30 ft/s is struck by a drop of water when it is 300 ft below ground level. When and where will the next drop of water strike the elevator?

Chapter
12

Kinetics of Particles:
Force, Mass, and Acceleration

Fig. 12.1

12.1. Newton's Second Law of Motion. Newton's first and third laws of motion were used extensively in statics to study bodies at rest and the forces acting upon them. These two laws are also used in dynamics; in fact, they are sufficient for the study of the motion of bodies which have no acceleration. However, when bodies are accelerated, i.e., when the magnitude or the direction of their velocity changes, it is necessary to use the second law of motion in order to relate the motion of the body with the forces acting on it. This law may be stated as follows:

If the resultant force acting on a particle is not zero, the particle will have an acceleration proportional to the magnitude of the resultant and in the direction of this resultant force.

Newton's second law of motion may best be understood if we imagine the following experiment: A particle is subjected to a force $\mathbf{F}_1$ of constant direction and constant magnitude F_1. Under the action of that force, the particle will be observed to move in a straight line and *in the direction of the force* (Fig. 12.1a). By determining the position of the particle at various instants, we find that its acceleration has a constant magnitude a_1. If the experiment is repeated with forces $\mathbf{F}_2$, $\mathbf{F}_3$, etc., of different magnitude or direction (Fig. 12.1b and c), we find each time that the particle moves in the direction of the force acting on

454

it and that the magnitudes a_1, a_2, a_3, etc., of the accelerations are proportional to the magnitudes F_1, F_2, F_3, etc., of the corresponding forces,

$$\frac{F_1}{a_1} = \frac{F_2}{a_2} = \frac{F_3}{a_3} = \cdots = \text{constant}$$

The constant value obtained for the ratio of the magnitudes of the forces and accelerations is a characteristic of the particle under consideration. It is called the *mass* of the particle and is denoted by m. When a particle of mass m is acted upon by a force **F**, the force **F** and the acceleration **a** of the particle must therefore satisfy the relation

$$\mathbf{F} = m\mathbf{a} \qquad (12.1)$$

This relation provides a complete formulation of Newton's second law; it expresses not only that the magnitudes of **F** and **a** are proportional, but also (since m is a positive scalar) that the vectors **F** and **a** have the same direction (Fig. 12.2). We should note that Eq. (12.1) still holds when **F** is not constant but varies with t in magnitude or direction. The magnitudes of **F** and **a** remain proportional, and the two vectors have the same direction at any given instant. However, they will not, in general, be tangent to the path of the particle.

When a particle is subjected simultaneously to several forces, Eq. (12.1) should be replaced by

$$\Sigma\mathbf{F} = m\mathbf{a} \qquad (12.2)$$

Fig. 12.2

where $\Sigma\mathbf{F}$ represents the sum, or resultant, of all the forces acting on the particle.

It should be noted that the system of axes with respect to which the acceleration **a** is determined is not arbitrary. These axes must have a constant orientation with respect to the stars, and their origin must either be attached to the sun† or move with a constant velocity with respect to the sun. Such a system of axes is called a *newtonian frame of reference*.‡ A system of axes attached to the earth does *not* constitute a newtonian frame of reference, since the earth rotates with respect to the stars and is accelerated with respect to the sun. However, in most

---

† More accurately, to the mass center of the solar system.

‡ Since the stars are not actually fixed, a more rigorous definition of a newtonian frame of reference (also called *inertial system*) is *one with respect to which Eq. (12.2) holds*.

engineering applications, the acceleration **a** may be determined with respect to axes attached to the earth and Eqs. (12.1) and (12.2) used without any appreciable error. On the other hand, these equations do not hold if **a** represents a relative acceleration measured with respect to moving axes, such as axes attached to an accelerated car or to a rotating piece of machinery.

We may observe that, if the resultant $\Sigma\mathbf{F}$ of the forces acting on the particle is zero, it follows from Eq. (12.2) that the acceleration **a** of the particle is also zero. If the particle is initially at rest ($v_0 = 0$) with respect to the newtonian frame of reference used, it will thus remain at rest ($v = 0$). If originally moving with a velocity $\mathbf{v}_0$, the particle will maintain a constant velocity $\mathbf{v} = \mathbf{v}_0$; that is, it will move with the constant speed v_0 in a straight line. This, we recall, is the statement of Newton's first law (Sec. 2.9). Thus, Newton's first law is a particular case of Newton's second law and may be omitted from the fundamental principles of mechanics.

12.2. Systems of Units. In using the fundamental equation $\mathbf{F} = m\mathbf{a}$, the units of force, mass, length, and time cannot be chosen arbitrarily. If they are, the magnitude of the force **F** required to give an acceleration **a** to the mass m will *not* be numerically equal to the product ma; it will only be proportional to this product. Thus, we may choose three of the four units arbitrarily but must choose the fourth unit so that the equation $\mathbf{F} = m\mathbf{a}$ is satisfied. The units are then said to form a consistent system of kinetic units.

Two systems of consistent kinetic units are currently used by American engineers, the International System of Units (SI units†), and the U.S. customary units. Since both systems have been discussed in detail in Sec. 1.3, we shall describe them only briefly in this section.

International System of Units (SI Units). In this system, the base units are the units of length, mass, and time, and are called, respectively, the *meter* (m), the *kilogram* (kg), and the *second* (s). All three are arbitrarily defined (Sec. 1.3). The unit of force is a derived unit. It is called the *newton* (N) and is defined as the force which gives an acceleration of $1\ \text{m/s}^2$ to a mass of 1 kg (Fig. 12.3). From Eq. (12.1) we write

$$1\ \text{N} = (1\ \text{kg})(1\ \text{m/s}^2) = 1\ \text{kg} \cdot \text{m/s}^2$$

The SI units are said to form an *absolute* system of units. This means that the three base units chosen are independent of the location where measurements are made. The meter, the kilogram, and the second may be used anywhere on the earth; they

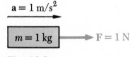

$a = 1\ \text{m/s}^2$

$m = 1\ \text{kg}$ $\rightarrow F = 1\ \text{N}$

Fig. 12.3

---

† SI stands for *Système International d'Unités* (French).

may even be used on another planet. They will always have the same significance.

Like any other force, the *weight* **W** of a body should be expressed in newtons. Since a body subjected to its own weight acquires an acceleration equal to the acceleration of gravity g, it follows from Newton's second law that the magnitude W of the weight of a body of mass m is

$$W = mg \qquad (12.3)$$

Recalling that $g = 9.81$ m/s², we find that the weight of a body of mass 1 kg (Fig. 12.4) is

$$W = (1\ kg)(9.81\ m/s^2) = 9.81\ N$$

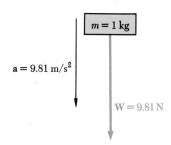

Fig. 12.4

Multiples and submultiples of the units of length, mass, and force are frequently used in engineering practice. They are, respectively, the *kilometer* (km) and the *millimeter* (mm); the *megagram*† (Mg) and the *gram* (g); and the *kilonewton* (kN). By definition

$$1\ km = 1000\ m \qquad 1\ mm = 0.001\ m$$
$$1\ Mg = 1000\ kg \qquad 1\ g = 0.001\ kg$$
$$1\ kN = 1000\ N$$

The conversion of these units to meters, kilograms, and newtons, respectively, can be effected by simply moving the decimal point three places to the right or to the left.

U.S. Customary Units. Most practicing American engineers still commonly use a system in which the base units are the units of length, force, and time. These units are, respectively, the *foot* (ft), the *pound* (lb), and the *second* (s). The second is the same as the corresponding SI unit. The foot is defined as 0.3048 m. The pound is defined as the *weight* of a platinum standard, called the *standard pound* and kept at the National Bureau of Standards in Washington, the mass of which is 0.453 592 43 kg. Since the weight of a body depends upon the gravitational attraction of the earth, which varies with location, it is specified that the standard pound should be placed at sea level and at the latitude of 45° to properly define a force of 1 lb. Clearly the U.S. customary units do not form an absolute system of units. Because of their dependence upon the gravitational attraction of the earth, they are said to form a *gravitational* system of units.

While the standard pound also serves as the unit of mass in commercial transactions in the United States, it cannot be so used in engineering computations since such a unit would not be consistent with the base units defined in the preceding para-

---

† Also known as a *metric ton*.

graph. Indeed, when acted upon by a force of 1 lb, that is, when subjected to its own weight, the standard pound receives the acceleration of gravity, $g = 32.2 \text{ ft/s}^2$ (Fig. 12.5), not the unit acceleration required by Eq. (12.1). The unit of mass consistent with the foot, the pound, and the second is the mass which receives an acceleration of 1 ft/s^2 when a force of 1 lb is applied to it (Fig. 12.6). This unit, sometimes called a *slug*, can be derived from the equation $F = ma$ after substituting 1 lb and 1 ft/s^2 for F and a, respectively. We write

$$F = ma \qquad 1 \text{ lb} = (1 \text{ slug})(1 \text{ ft/s}^2)$$

and obtain

$$1 \text{ slug} = \frac{1 \text{ lb}}{1 \text{ ft/s}^2} = 1 \text{ lb} \cdot \text{s}^2/\text{ft}$$

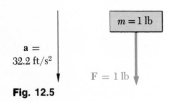

Fig. 12.5

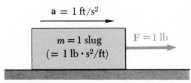

Fig. 12.6

Comparing Figs. 12.5 and 12.6, we conclude that the slug is a mass 32.2 times larger than the mass of the standard pound.

The fact that bodies are characterized in the U.S. customary system of units by their weight in pounds, rather than by their mass in slugs, was a convenience in the study of statics, where we were dealing constantly with weights and other forces and only seldom with masses. However, in the study of kinetics, where forces, masses, and accelerations are involved, we repeatedly shall have to express the mass m in slugs of a body, the weight W of which has been given in pounds. Recalling Eq. (12.3), we shall write

$$m = \frac{W}{g} \tag{12.4}$$

where g is the acceleration of gravity ($g = 32.2 \text{ ft/s}^2$).

The conversion from U.S. customary units to SI units, and vice versa, has been discussed in Sec. 1.4. We shall recall the conversion factors obtained respectively for the units of length, force, and mass:

Length: $1 \text{ ft} = 0.3048 \text{ m}$
Force: $1 \text{ lb} = 4.448 \text{ N}$
Mass: $1 \text{ slug} = 1 \text{ lb} \cdot \text{s}^2/\text{ft} = 14.59 \text{ kg}$

Although it cannot be used as a consistent unit of mass, we also recall that the mass of the standard pound is, by definition,

$$1 \text{ pound-mass} = 0.4536 \text{ kg}$$

This constant may be used to determine the *mass* in SI units (kilograms) of a body which has been characterized by its *weight* in U.S. customary units (pounds).

12.3. Equations of Motion. Dynamic Equilibrium. Consider a particle of mass m acted upon by several forces. Newton's second law of motion states that the resultant of these forces must be equal to the vector $m\mathbf{a}$ obtained by multiplying the acceleration $\mathbf{a}$ of the particle by its mass m. The given system of forces must therefore be equivalent to the vector $m\mathbf{a}$ (Fig. 12.7). Using rectangular components, we write the equations of motion,

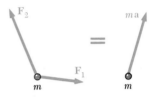

Fig. 12.7

$$\Sigma F_x = ma_x \qquad \Sigma F_y = ma_y \qquad \Sigma F_z = ma_z \qquad (12.5)$$

Newton's second law may also be expressed by considering a vector of magnitude ma, but of *sense opposite* to that of the acceleration. This vector is denoted by $(m\mathbf{a})_{\text{rev}}$, where the subscript indicates that the sense of the acceleration has been reversed, and is called an *inertia vector*. If the inertia vector $(m\mathbf{a})_{\text{rev}}$ is added to the forces acting on the particle, *we obtain a system of vectors equivalent to zero* (Fig. 12.8). The particle may thus be considered to be in equilibrium under the given forces and the inertia vector. The particle is said to be in *dynamic equilibrium*, and the problem under consideration may be solved by using the methods developed earlier in statics. We may, for instance, write that the sums of the components of the vectors shown in Fig. 12.8, *including the inertia vector*, are zero,

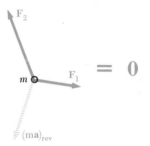

Fig. 12.8

$$\Sigma F_x = 0 \quad \Sigma F_y = 0 \quad \Sigma F_z = 0 \quad \textit{including inertia vector} \quad (12.6)$$

or if the forces are coplanar, we may draw all vectors tip to tail, including again the inertia vector, to form a closed vector polygon.

Because they measure the resistance that particles offer when we try to set them in motion or when we try to change the conditions of their motion, inertia vectors are often called *inertia forces*. The inertia forces, however, are not forces like the forces found in statics, which are either contact forces or gravitational forces (weights). Many people, therefore, object to the use of the word force when referring to the vector $(m\mathbf{a})_{\text{rev}}$ or even avoid altogether the concept of dynamic equilibrium. Others point out that inertia forces and actual forces such as gravitational forces affect our senses in the same way and cannot be distinguished by physical measurements. A man riding in an elevator which is accelerated upward will have the feeling that his weight has suddenly increased; and no measurement made within the elevator could establish whether the elevator is truly accelerated or whether the force of attraction exerted by the earth has suddenly increased.

Sample Problems have been solved in this text by the direct application of Newton's second law as illustrated in Fig. 12.7, rather than by the method of dynamic equilibrium.

12.4. Systems of Particles. When a problem involves the motion of several particles, the equations of motion, (12.5) or (12.6), may be written for each particle considered separately. However, in the case of a system involving a large number of particles, such as a rigid body, it will generally be more convenient to consider the system as a whole. In that case, we should note that the forces acting on a given particle of the system consist (1) of *external forces* exerted by bodies outside the system (such as the weight of the particle, which is exerted by the earth) and (2) of *internal forces* exerted by the other particles of the system. Figure 12.9 shows the particle P_1 of a system, the resultant $\mathbf{F}_1$ of the external forces exerted on P_1, and the internal forces $\mathbf{f}_{12}$, $\mathbf{f}_{13}$, etc., exerted by the other particles P_2, P_3, etc., of the system. According to Newton's second law, the resultant of the external and internal forces acting on P_1 is equal to the vector $m_1 \mathbf{a}_1$ obtained by multiplying the acceleration $\mathbf{a}_1$ of the particle by its mass m_1. This vector is generally known as the *effective force* of the particle P_1.

We now consider simultaneously all the particles of the system (Fig. 12.10). Since the relation we have established holds for any particle, it follows that the system of all the external and internal forces acting on the various particles is equivalent to the system of the effective forces of the particles (i.e., one system may be replaced by the other). Thus the sum of the components

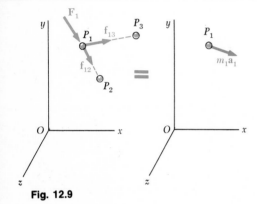

Fig. 12.9

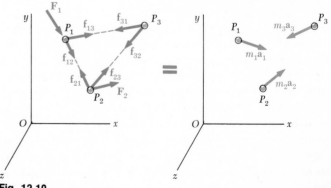

Fig. 12.10

in the x, y, or z direction of the external and internal forces must be equal to the sum of the components of the effective forces in the same direction. But the internal forces occur in pairs of *equal and opposite forces* (such as $\mathbf{f}_{12}$ and $\mathbf{f}_{21}$) and the sum of their components cancels out. We can therefore write the following relations between the components of the external and effective forces:

$$\Sigma(F_x)_{\text{ext}} = \Sigma ma_x \qquad \Sigma(F_y)_{\text{ext}} = \Sigma ma_y$$
$$\Sigma(F_z)_{\text{ext}} = \Sigma ma_z \qquad (12.7)$$

Consider now the moments about the x, y, or z axis of all the forces shown in Fig. 12.10. Since the internal forces occur in pairs of equal and opposite forces *directed along the same line of action*, the sum of their moments cancels out. Relations similar to Eqs. (12.7) therefore can be written between the moments of the external and effective forces.

Because the sums of their x, y, and z components and the sums of their moments about the x, y, and z axes are respectively equal, the system of the external forces and the system of the effective forces are said to be *equipollent*. This, however, does not mean that the two systems are *equivalent*. We note from Fig. 12.11 that the external force and the effective force acting on a given particle are *not* equal. The two systems of vectors, therefore, do not have the same effect on the individual particles; the two systems are *not* equivalent.

The distinction between equivalent and equipollent systems of vectors should be made in all cases involving a system of particles or a system of rigid bodies (as opposed to a single particle or a single rigid body). In all such cases we shall use a gray equals sign, as in Fig. 12.11, to indicate that the two systems of vectors are equipollent, i.e., have the same sums of

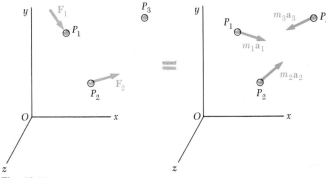

Fig. 12.11

components and the same sums of moments. We shall continue to use colored equals signs, as in Fig. 12.10, to indicate that two systems of vectors are truly equivalent.†

12.5. Motion of the Mass Center of a System of Particles. Equations (12.7) may be written in a modified form if the *mass center* of the system of particles is considered. The mass center of the system is the point G of coordinates $\bar{x}$, $\bar{y}$, $\bar{z}$ defined by the equations

$$(\Sigma m)\bar{x} = \Sigma mx \qquad (\Sigma m)\bar{y} = \Sigma my$$
$$(\Sigma m)\bar{z} = \Sigma mz \tag{12.8}$$

Since $W = mg$, we note that G is also the center of gravity of the system. However, in order to avoid any confusion, we shall call G the *mass center* of the system of particles when discussing properties of the system associated with the *mass* of the particles, and we shall refer to it as the *center of gravity* of the system when considering properties associated with the *weight* of the particles. Particles located outside the gravitational field of the earth, for example, have a mass but no weight. We may then properly refer to their mass center, but obviously not to their center of gravity.‡

We differentiate twice with respect to t the relations (12.8):

$$(\Sigma m)\ddot{\bar{x}} = \Sigma m\ddot{x} \qquad (\Sigma m)\ddot{\bar{y}} = \Sigma m\ddot{y}$$
$$(\Sigma m)\ddot{\bar{z}} = \Sigma m\ddot{z}$$

Recalling that the second derivatives of the coordinates of a point represent the components of the acceleration of that point, we have

$$(\Sigma m)\bar{a}_x = \Sigma ma_x \qquad (\Sigma m)\bar{a}_y = \Sigma ma_y$$
$$(\Sigma m)\bar{a}_z = \Sigma ma_z \tag{12.9}$$

where $\bar{a}_x$, $\bar{a}_y$, $\bar{a}_z$ represent the components of the acceleration $\bar{\mathbf{a}}$ of the mass center G of the system. Combining the relations (12.7) and (12.9), we write the equations

---

† The relation established in this section between the external and the effective forces of a system of particles is often referred to as *D'Alembert's principle*, after the French mathematician Jean le Rond d'Alembert (1717–1783). However, d'Alembert's original statement refers to the motion of a system of connected bodies, with $\mathbf{f}_{12}$, $\mathbf{f}_{21}$, etc., representing constraint forces which, if applied by themselves, will not cause the system to move. Since this is in general not the case for the internal forces acting on a system of free particles, we shall postpone the consideration of D'Alembert's principle until the study of the motion of rigid bodies (Chap. 16).

‡ It may also be pointed out that the mass center and the center of gravity of a system of particles do not exactly coincide, since the weights of the particles are directed toward the center of the earth and thus do not truly form a system of parallel forces.

$$\Sigma(F_x)_{\text{ext}} = (\Sigma m)\bar{a}_x \qquad \Sigma(F_y)_{\text{ext}} = (\Sigma m)\bar{a}_y$$
$$\Sigma(F_z)_{\text{ext}} = (\Sigma m)\bar{a}_z \qquad (12.10)$$

which define the motion of the mass center of the system. We note that Eqs. (12.10) are identical with the equations we would obtain for a particle of mass Σm, acted upon by all the external forces. We state therefore: *The mass center of a system of particles moves as if the entire mass of the system and all the external forces were concentrated at that point.*

This principle is best illustrated by the motion of an exploding shell. We know that, if the resistance of the air is neglected, a shell may be assumed to travel along a parabolic path. After the shell has exploded, the mass center G of the fragments of shell will continue to travel along the same path. Indeed, point G must move as if the mass and the weight of all fragments were concentrated at G; it must move, therefore, as if the shell had not exploded.

The principle we have established shows that we may replace a rigid body by a particle of the same mass if we are interested in the motion of the mass center of the body, and not in the motion of the body about its mass center. We should keep in mind, however, that Eqs. (12.9) and (12.10) relate only the components of the forces and vectors involved. To study the motion of a rigid body about its mass center, it will be necessary to also consider the relations between the moments of the external and effective forces (Chap. 16).

12.6. Rectilinear Motion of a Particle. Consider a particle of mass m moving in a straight line under the action of coplanar forces F_1, F_2, F_3, etc. The conditions of motion of the particle are expressed by Eq. (12.2). Since the particle moves in a straight line, its acceleration a must be directed along that line. Choosing the x axis in the same direction and the y axis in the plane of the forces (Fig. 12.12), we have $a_x = a$, $a_y = 0$ and write

$$\Sigma F_x = ma \qquad \Sigma F_y = 0 \qquad (12.11)$$

The equations obtained may be solved for two unknowns.

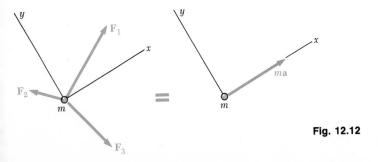

Fig. 12.12

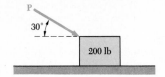

A 200-lb block rests on a horizontal plane. Find the magnitude of the force **P** required to give the block an acceleration of 10 ft/s² to the right. The coefficient of friction between the block and the plane is $\mu = 0.25$.

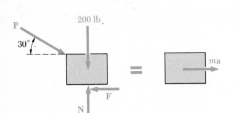

Solution. The mass of the block is

$$m = \frac{W}{g} = \frac{200 \text{ lb}}{32.2 \text{ ft/s}^2} = 6.21 \text{ lb} \cdot \text{s}^2/\text{ft}$$

We note that $F = \mu N = 0.25N$ and that $a = 10 \text{ ft/s}^2$. Expressing that the forces acting on the block are equivalent to the vector $m\mathbf{a}$, we write

$$\xrightarrow{+} \Sigma F_x = ma: \qquad P \cos 30° - 0.25N = (6.21 \text{ lb} \cdot \text{s}^2/\text{ft})(10 \text{ ft/s}^2)$$
$$P \cos 30° - 0.25N = 62.1 \text{ lb} \tag{1}$$

$$+\uparrow \Sigma F_y = 0: \qquad N - P \sin 30° - 200 \text{ lb} = 0, \tag{2}$$

Solving (2) for N and carrying the result into (1), we obtain

$$N = P \sin 30° + 200 \text{ lb}$$
$$P \cos 30° - 0.25(P \sin 30° + 200 \text{ lb}) = 62.1 \text{ lb} \qquad P = 151 \text{ lb} \quad \blacktriangleleft$$

Solve Sample Prob. 12.1 using SI units.

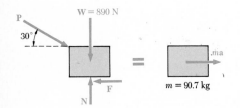

Solution. Using the conversion factors given in Sec. 12.2, we write

$$a = (10 \text{ ft/s}^2)(0.3048 \text{ m/ft}) = 3.05 \text{ m/s}^2$$
$$W = (200 \text{ lb})(4.448 \text{ N/lb}) = 890 \text{ N}$$

Recalling that, by definition, 1 lb is the weight of a mass of 0.4536 kg, we find that the mass of the 200-lb block is

$$m = 200(0.4536 \text{ kg}) = 90.7 \text{ kg}$$

Noting that $F = \mu N = 0.25N$ and expressing that the forces acting on the block are equivalent to the vector $m\mathbf{a}$, we write

$$\xrightarrow{+} \Sigma F_x = ma: \qquad P \cos 30° - 0.25N = (90.7 \text{ kg})(3.05 \text{ m/s}^2)$$
$$P \cos 30° - 0.25N = 277 \text{ N} \tag{1}$$

$$+\uparrow \Sigma F_y = 0: \qquad N - P \sin 30° - 890 \text{ N} = 0 \tag{2}$$

Solving (2) for N and carrying the result into (1), we obtain

$$N = P \sin 30° + 890 \text{ N}$$
$$P \cos 30° - 0.25(P \sin 30° + 890 \text{ N}) = 277 \text{ N} \qquad P = 674 \text{ N} \quad \blacktriangleleft$$

or, in U.S. customary units,

$$P = (674 \text{ N}) \div (4.448 \text{ N/lb}) \qquad P = 151 \text{ lb} \quad \blacktriangleleft$$

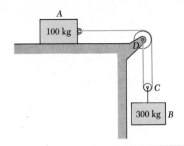

The two blocks shown start from rest. The horizontal plane and the pulley are frictionless, and the pulley is assumed to be of negligible mass. Determine the acceleration of each block and the tension in each cord.

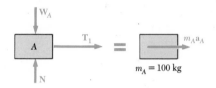

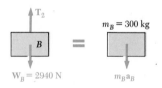

Solution. We denote by T_1 the tension in cord ACD and by T_2 the tension in cord BC. We note that if block A moves through s_A, block B moves through

$$s_B = \tfrac{1}{2}s_A$$

Differentiating twice with respect to t, we have

$$a_B = \tfrac{1}{2}a_A \qquad (1)$$

We shall apply Newton's second law successively to block A, block B, and pulley C.

Block A

$$\xrightarrow{+}\ \Sigma F_x = m_A a_A: \qquad\qquad T_1 = 100a_A \qquad (2)$$

Block B. Observing that the weight of block B is

$$W_B = m_B g = (300\,\text{kg})(9.81\,\text{m/s}^2) = 2940\,\text{N}$$

we write

$$+\downarrow \Sigma F_y = m_B a_B: \qquad 2940 - T_2 = 300a_B$$

or, substituting for a_B from (1),

$$2940 - T_2 = 300(\tfrac{1}{2}a_A)$$
$$T_2 = 2940 - 150a_A \qquad (3)$$

Pulley C. Since m_C is assumed to be zero, we have

$$+\downarrow \Sigma F_y = m_C a_C = 0: \qquad T_2 - 2T_1 = 0 \qquad (4)$$

Substituting for T_1 and T_2 from (2) and (3), respectively, into (4), we write

$$2940 - 150a_A - 2(100a_A) = 0$$
$$2940 - 350a_A = 0 \qquad a_A = 8.40\,\text{m/s}^2 \ \blacktriangleleft$$

Substituting the value obtained for a_A into (1) and (2), we have

$$a_B = \tfrac{1}{2}a_A = \tfrac{1}{2}(8.40\,\text{m/s}^2) \qquad\qquad a_B = 4.20\,\text{m/s}^2 \ \blacktriangleleft$$
$$T_1 = 100a_A = (100\,\text{kg})(8.40\,\text{m/s}^2) \qquad\qquad T_1 = 840\,\text{N} \ \blacktriangleleft$$

Recalling (4), we write

$$T_2 = 2T_1 \qquad T_2 = 2(840\,\text{N}) \qquad T_2 = 1680\,\text{N} \ \blacktriangleleft$$

We note that the value obtained for T_2 is *not* equal to the weight of block B.

PROBLEMS

12.1 The acceleration due to gravity on the moon is $5.31 \, \text{ft/s}^2$. Determine the weight in pounds, the mass in pounds, and the mass in $\text{lb} \cdot \text{s}^2/\text{ft}$, on the moon, of a silver bar whose mass is officially defined as 100.00 lb.

12.2 The value of g at any latitude ϕ may be obtained from the formula

$$g = 9.7807(1 + 0.0053 \sin^2 \phi) \quad \text{m/s}^2$$

Determine to four significant figures the weight in newtons and the mass in kilograms, at the latitudes of $0°$, $45°$, and $90°$, of a silver bar whose mass is officially defined as 10 kg.

12.3 Two boxes are weighed on the scales shown: scale a is a lever scale; scale b is a spring scale. The scales are attached to the roof of an elevator. When the elevator is at rest, each scale indicates a load of 20 lb. Determine the load which each scale will indicate when the elevator is accelerated upward at the rate of $5 \, \text{ft/s}^2$.

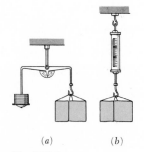

(a)   (b)

Fig. P12.3

12.4 If the spring scale of Prob. 12.3 indicates a load of 18 lb, determine the acceleration of the elevator and the load indicated by the lever scale.

12.5 (a) Knowing that the coefficient of friction between automobile tires and a wet concrete surface is 0.35, determine the shortest distance in which an automobile traveling at a speed of 75 mi/h can be stopped. (b) Solve part a assuming the surface is dry with a coefficient of friction of 0.80.

12.6 A motorist traveling at a speed of 45 mi/h suddenly applies his brakes and comes to a stop after skidding 150 ft. Determine (a) the time required for the car to stop, (b) the coefficient of friction between the tires and the pavement.

12.7 A 5-kg package is projected down the incline with an initial velocity of 4 m/s. Knowing that the coefficient of friction between the package and the incline is 0.35, determine (a) the velocity of the package after 3 m of motion, (b) the distance d at which the package comes to rest.

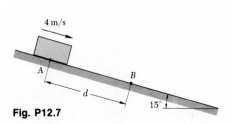

Fig. P12.7

12.8 A truck is proceeding up a long 3-percent grade at a constant speed of 60 km/h. If the driver does not change the setting of his throttle or shift gears, what will be the acceleration of the truck as it starts moving on the level section of the road?

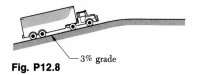

Fig. P12.8

3% grade

12.9 The 3-kg collar was moving down the rod with a velocity of 3 m/s when a force **P** was applied to the horizontal cable. Assuming negligible friction between the collar and the rod, determine the magnitude of the force **P** if the collar stopped after moving 1 m more down the rod.

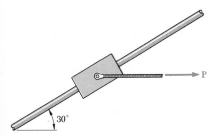

P

30°

Fig. P12.9

12.10 Solve Prob. 12.9, assuming a coefficient of friction of 0.20 between the collar and the rod.

12.11 Two packages are placed on a conveyor belt which is at rest. The coefficient of friction is 0.20 between the belt and package A, and 0.10 between the belt and package B. If the belt is suddenly started to the right and slipping occurs between the belt and the packages, determine (a) the acceleration of the packages, (b) the force exerted by package A on package B.

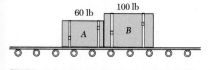

60 lb 100 lb

A B

Fig. P12.11

12.12 The subway train shown travels at a speed of 30 mi/h. Determine the force in each coupling when the brakes are applied, knowing that the braking force is 5000 lb on each car.

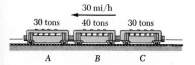

30 mi/h

30 tons 40 tons 30 tons

A B C

Fig. P12.12

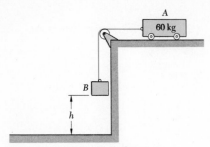

Fig. P12.13, P12.14, and P12.15

12.13 When the system shown is released from rest, the acceleration of block B is observed to be $3 \, \text{m/s}^2$ downward. Neglecting the effect of friction, determine (a) the tension in the cable, (b) the mass of block B.

12.14 The system shown is released from rest when $h = 1.4 \, \text{m}$. (a) Determine the mass of block B, knowing that it strikes the ground with a speed of $3 \, \text{m/s}$. (b) Attempt to solve part a, assuming the final speed to be $6 \, \text{m/s}$; explain the difficulty encountered.

12.15 The system shown is released from rest. Knowing that the mass of block B is 30 kg, determine how far the cart will move before it reaches a speed of $2.5 \, \text{m/s}$, (a) if the pulley may be considered as weightless and frictionless, (b) if the pulley "freezes" on its shaft and the cable must slip, with $\mu = 0.10$, over the pulley.

12.16 Each of the systems shown is initially at rest. Assuming the pulleys to be weightless and neglecting axle friction, determine for each system (a) the acceleration of block A, (b) the velocity of block A after 4 s, (c) the velocity of block A after it has moved 10 ft.

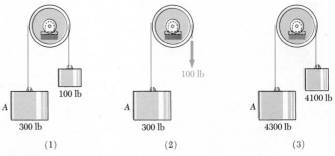

Fig. P12.16

12.17 The 100-kg block A is connected to a 25-kg counterweight B by the cable arrangement shown. If the system is released from rest, determine (a) the tension in the cable, (b) the velocity of B after 3 s, (c) the velocity of A after it has moved 1.2 m.

12.18 Block A is observed to move with an acceleration of $0.9 \, \text{m/s}^2$ directed upward. Determine (a) the mass of block B, (b) the corresponding tension in the cable.

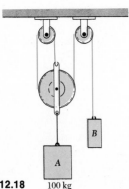

Fig. P12.17 and P12.18

12.19   A small ore car and its load weigh 8000 lb. The car is partially counterweighted by a 500-lb weight. If the car is initially at rest, determine the force **P** required if the velocity of the car is to be 10 ft/s after it has moved 20 ft up the incline.

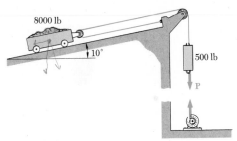

Fig. P12.19

12.20   Neglecting the effect of friction, determine (a) the acceleration of each block, (b) the tension in the cable.

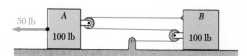

Fig. P12.20

12.21   It is assumed that friction exists between block A and the horizontal plane in Sample Prob. 12.3. If block B is observed to move 7.5 m in 2 s after starting from rest, determine the coefficient of friction between block A and the horizontal plane.

12.22   Solve Prob. 12.20, assuming the coefficient of friction between the blocks and the horizontal plane to be 0.20.

12.23   Two plates A and B, each of mass 50 kg, are placed as shown on a 15° incline. The coefficient of friction between A and B is 0.10; the coefficient of friction between A and the incline is 0.20. (a) If the plates are released from rest, determine the acceleration of each plate. (b) Solve part a assuming that plates A and B are welded together and act as a single rigid body.

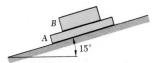

Fig. P12.23

12.24   The coefficient of friction between the load and the flat-bed trailer shown is 0.40. Knowing that the forward speed of the truck is 50 km/h, determine the shortest distance in which the truck can be brought to a stop if the load is not to shift.

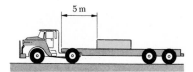

Fig. P12.24

12.25   The 400-lb plate A rests on small rollers which may be considered frictionless; the coefficient of friction between the two plates is $\mu = 0.20$. Determine the acceleration of plate A when a 160-lb force is suddenly applied to it, (a) for the system shown, (b) if the cable is cut.

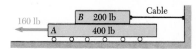

Fig. P12.25

*12.26 In Prob. 12.24, determine the shortest distance in which the truck can be stopped without having the load slide beyond the forward edge of the trailer. (*Note.* As the truck comes to a stop, the load is still moving relative to the trailer.)

12.27 A 60-lb crate rests on a 40-lb cart; the coefficient of static friction between the crate and the cart is 0.25. If the crate is not to slip with respect to the cart, determine (*a*) the maximum allowable magnitude of **P**, (*b*) the corresponding acceleration of the cart.

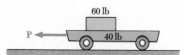

Fig. P12.27 and P12.28

12.28 The coefficients of friction between the 60-lb crate and the 40-lb cart are $\mu_s = 0.25$ and $\mu_k = 0.20$. If a force **P** of magnitude 30 lb is applied to the cart, determine the acceleration (*a*) of the cart, (*b*) of the crate, (*c*) of the crate with respect to the cart.

12.29 For the cart and crate of Prob. 12.28, determine the acceleration of the cart as a function of the magnitude of the force **P**.

12.30 A chain of length l and mass m per unit length rests over a small pulley of negligible mass. If the chain is released in the position shown ($h > \frac{1}{2}l$), determine the velocity of the chain when the end of the chain leaves the pulley. Also determine the maximum possible final velocity by letting h approach $\frac{1}{2}l$.

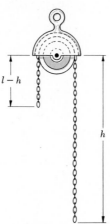

Fig. P12.30

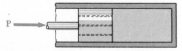

Fig. P12.31

12.31 A constant force **P** is applied to a piston and rod of total mass m in order to make them move in a cylinder filled with oil. As the piston moves, the oil is forced through orifices in the piston and exerts on the piston an additional force of magnitude kv, proportional to the speed v of the piston and in a direction opposite to its motion. Express the acceleration and velocity of the piston as a function of the time t, assuming that the piston starts from rest at time $t = 0$.

12.32 A ship of total mass m is anchored in the middle of a river which is flowing with a constant velocity $\mathbf{v}_0$. The horizontal component of the force exerted on the ship by the anchor chain is $\mathbf{T}_0$. If the anchor chain suddenly breaks, determine the time required for the ship to attain a velocity equal to $\frac{1}{2}\mathbf{v}_0$. Assume that the frictional resistance of the water is proportional to the velocity of the ship relative to the water.

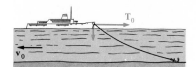

Fig. P12.32

12.33 A bomber flying horizontally in a straight line at a constant speed v_0 releases a powered missile which will fly ahead of the bomber in a parallel horizontal path. The missile engine, which is started at the time of release, provides a constant thrust T, while the atmosphere produces a drag equal to kv^2, where v is the speed of the missile. Neglecting any change in the mass of the missile, express the acceleration of the missile as a function of the distance traveled in free flight. It is assumed that $v_0 < \sqrt{T/k}$.

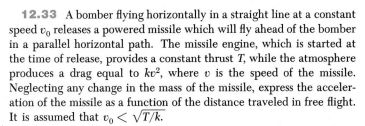

Bomber path

Missile path

Fig. P12.33

12.34 Determine the acceleration of each block when $W_A = W_B = W_C$.

12.35 Determine the acceleration of each block when $W_A = 5\,\text{lb}$, $W_B = 10\,\text{lb}$, and $W_C = 10\,\text{lb}$.

12.36 In the system shown, $W_A = 6\,\text{lb}$ and $W_C = 12\,\text{lb}$. Determine the required weight W_B if block B is not to move when the system is released from rest.

12.37 Determine the acceleration of each block when $m_A = 15\,\text{kg}$, $m_B = 10\,\text{kg}$, and $m_C = 5\,\text{kg}$.

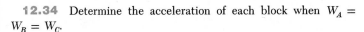

Fig. P12.34, P12.35, and P12.36

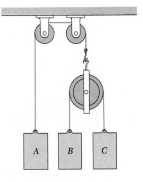

Fig. P12.37

12.38 Knowing that $\mu = 0.20$, determine the acceleration of each block when $m_A = 5\,\text{kg}$, $m_B = 10\,\text{kg}$, and $m_C = 10\,\text{kg}$.

12.39 Knowing that $\mu = 0.20$, determine the acceleration of each block when $m_A = 5\,\text{kg}$, $m_B = 10\,\text{kg}$, and $m_C = 20\,\text{kg}$.

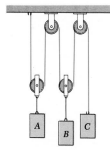

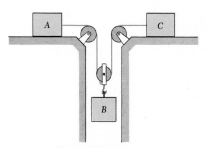

Fig. P12.38 and P12.39

12.7. Curvilinear Motion of a Particle. We saw in Chap. 11 that the acceleration of a particle moving in a plane along a curved path may be represented by its rectangular components $\mathbf{a}_x$ and $\mathbf{a}_y$, by its tangential and normal components $\mathbf{a}_t$ and $\mathbf{a}_n$, or by its radial and transverse components $\mathbf{a}_r$ and $\mathbf{a}_\theta$. We may therefore use any one of these methods of representation to express Newton's second law of motion.

Rectangular Components. Using the rectangular components $\mathbf{a}_x$ and $\mathbf{a}_y$ of the acceleration, and resolving the forces acting on the particle into their components $\mathbf{F}_x$ and $\mathbf{F}_y$, we write

$$\Sigma F_x = ma_x \qquad \Sigma F_y = ma_y \qquad (12.12)$$

These equations may be solved for two unknowns.

Consider, as an example, the *motion of a projectile.* If the resistance of the air is neglected, the only force acting on the projectile after it has been fired is its weight $\mathbf{W}$. Substituting $\Sigma F_x = 0$ and $\Sigma F_y = -W$ into Eqs. (12.12), we obtain

$$a_x = 0 \qquad a_y = -\frac{W}{m} = -g$$

where g is 9.81 m/s^2 or 32.2 ft/s^2. The equations obtained may be integrated independently, as was shown in Sec. 11.10, to obtain the velocity and displacement of the projectile at any instant.

Tangential and Normal Components. Using the tangential and normal components $\mathbf{a}_t$ and $\mathbf{a}_n$ of the acceleration, and resolving the forces acting on the particle into components $\mathbf{F}_t$ along the tangent to the path (in the direction of motion) and components $\mathbf{F}_n$ along the normal (toward the inside of the path), we write (Fig. 12.13)

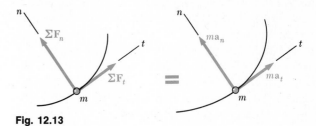

Fig. 12.13

$$\Sigma F_t = ma_t \qquad \Sigma F_n = ma_n \qquad (12.13)$$

Substituting for a_t and a_n from (11.28) and (11.29), we have

$$\Sigma F_t = m\frac{dv}{dt} \qquad \Sigma F_n = m\frac{v^2}{\rho} \qquad (12.14)$$

The equations obtained may be solved for two unknowns.

Radial and Transverse Components.   Using the radial and transverse components $\mathbf{a}_r$ and $\mathbf{a}_\theta$ of the acceleration, and resolving the forces acting on the particle into components $\mathbf{F}_r$ along the line *OP* joining the origin *O* to the particle *P*, and components $\mathbf{F}_\theta$ in the direction perpendicular to *OP* (Fig. 12.14), we write

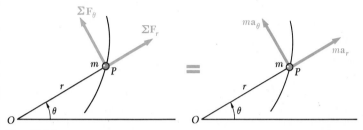

Fig. 12.14

$$\Sigma F_r = ma_r \qquad \Sigma F_\theta = ma_\theta \qquad (12.15)$$

Substituting for a_r and a_θ from (11.33), we have

$$\Sigma F_r = m(\ddot{r} - r\dot{\theta}^2) \qquad (12.16)$$
$$\Sigma F_\theta = m(r\ddot{\theta} + 2\dot{r}\dot{\theta}) \qquad (12.17)$$

The equations obtained may be solved for two unknowns.

Dynamic Equilibrium.   As an alternate method of solution, we may add the inertia vector $(ma)_{\text{rev}}$ to the forces acting on the particle and express that the system obtained is balanced. In practice, it is found convenient to resolve the inertia vector into two components. Resolving, for example, the inertia vector into its tangential and normal components, we obtain the system of vectors shown in Fig. 12.15. The tangential component of the inertia vector provides a measure of the resistance the particle offers to a change in speed, while its normal component (also called *centrifugal force*) represents the tendency of the particle to leave its curved path. We should note that either of these two components may be zero under special conditions: (1) if the particle starts from rest, its initial velocity is zero and the normal component of the inertia vector is zero at $t = 0$; (2) if the particle moves at constant speed along its path, the tangential component of the inertia vector is zero and only its normal component needs to be considered.

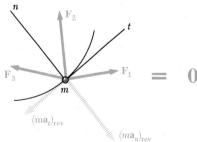

Fig. 12.15

12.8. Motion under a Central Force.   When the only force acting on a particle *P* is a force $\mathbf{F}$ directed toward or away from a fixed point *O*, the particle is said to be moving *under*

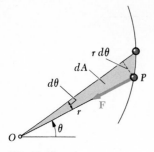

Fig. 12.16

a central force, and the point O is referred to as the *center of force* (Fig. 12.16). Selecting O as the origin of coordinates and using radial and transverse components, we note that the transverse component $\mathbf{F}_\theta$ of the force $\mathbf{F}$ is zero. Thus, Eq. (12.17) yields

$$r\ddot{\theta} + 2\dot{r}\dot{\theta} = 0 \qquad \text{or} \qquad \frac{1}{r}\frac{d}{dt}(r^2\dot{\theta}) = 0$$

It follows that, for any finite value of r, we must have

$$r^2\dot{\theta} = h \tag{12.18}$$

where h is a constant. Equation (12.18) may be given an interesting geometric interpretation. Observing from Fig. 12.16 that the radius vector OP sweeps an infinitesimal area $dA = \frac{1}{2}r^2\,d\theta$ as it rotates through an angle $d\theta$, and defining the *areal velocity* of the particle as the quotient dA/dt, we note that the left-hand member of Eq. (12.18) represents twice the areal velocity of the particle. We thus conclude that, *when a particle moves under a central force, its areal velocity is constant.*†

12.9. Newton's Law of Gravitation. The gravitational force exerted by the sun on a planet, or by the earth on an orbiting satellite, is an important example of a central force. In this section we shall learn how to determine the magnitude of a gravitational force.

In his *law of universal gravitation*, Newton states that two particles at a distance r from each other and, respectively, of mass M and m attract each other with equal and opposite forces $\mathbf{F}$ and $\mathbf{F}'$ directed along the line joining the particles (Fig. 12.17). The common magnitude F of the two forces is

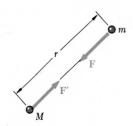

Fig. 12.17

$$F = G\frac{Mm}{r^2} \tag{12.19}$$

where G is a universal constant, called the *constant of gravitation*. Experiments show that the value of G is $(6.673 \pm 0.003) \times 10^{-11}$ m^3/kg · s^2 in SI units, or approximately 3.44×10^{-8} ft^4/lb · s^4 in U.S. customary units. While gravitational forces exist between any pair of bodies, their effect is appreciable only when one of the bodies has a very large mass. The effect of gravitational forces is apparent in the case of the motion of a planet about the sun, of satellites orbiting about the earth, or of bodies falling on the surface of the earth.

† It will be seen in Sec. 14.10 that the left-hand member of Eq. (12.18) also represents the *angular momentum per unit mass* of the particle P. Thus, Eq. (12.18) may be considered as a statement of the *conservation of angular momentum* of a particle moving under a central force.

Since the force exerted by the earth on a body of mass m located on or near its surface is defined as the weight **W** of the body, we may substitute the magnitude $W = mg$ of the weight for F, and the radius R of the earth for r, in Eq. (12.19). We obtain

$$W = mg = \frac{GM}{R^2}m \qquad \text{or} \qquad g = \frac{GM}{R^2} \qquad (12.20)$$

where M is the mass of the earth. Since the earth is not truly spherical, the distance R from the center of the earth depends upon the point selected on its surface, and the values of W and g will thus vary with the altitude and latitude of the point considered. Another reason for the variation of W and g with the latitude is that a system of axes attached to the earth does not constitute a newtonian frame of reference (see Sec. 12.1). A more accurate definition of the weight of a body should therefore include a component representing the centrifugal force due to the rotation of the earth. Values of g at sea level vary from 9.781 m/s^2 or 32.09 ft/s^2 at the equator to 9.833 m/s^2 or 32.26 ft/s^2 at the poles.†

The force exerted by the earth on a body of mass m located in space at a distance r from its center may be found from Eq. (12.19). The computations will be somewhat simplified if we note that, according to Eq. (12.20), the product of the constant of gravitation G and of the mass M of the earth may be expressed as

$$GM = gR^2 \qquad (12.21)$$

where g and the radius R of the earth will be given their average values $g = 9.81 \text{ m/s}^2$ and $R = 6.37 \times 10^6 \text{ m}$ in SI units,‡ or $g = 32.2 \text{ ft/s}^2$ and $R = (3960 \text{ mi})(5280 \text{ ft/mi})$ in U.S. customary units.

The discovery of the law of universal gravitation has often been attributed to the fact that Newton, after observing an apple falling from a tree, had reflected that the earth must attract an apple and the moon in much the same way. While it is doubtful that this incident actually took place, it may be said that Newton would not have formulated his law if he had not first perceived that the acceleration of a falling body must have the same cause as the acceleration which keeps the moon in its orbit. This basic concept of continuity of the gravitational attraction is more easily understood now, when the gap between the apple and the moon is being filled with long-range ballistic missiles and artificial earth satellites.

---

† A formula expressing g in terms of the latitude ϕ was given in Prob. 12.2.

‡ The value of R is easily found if one recalls that the circumference of the earth is $2\pi R = 40 \times 10^6$ m.

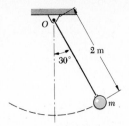

SAMPLE PROBLEM 12.4

The bob of a 2-m pendulum describes an arc of circle in a vertical plane. If the tension in the cord is 2.5 times the weight of the bob for the position shown, find the velocity and acceleration of the bob in that position.

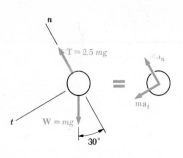

Solution. The weight of the bob is $W = mg$; the tension in the cord is thus $2.5\,mg$. Recalling that $\mathbf{a}_n$ is directed toward O and assuming $\mathbf{a}_t$ as shown, we apply Newton's second law and obtain

$+\swarrow \Sigma F_t = ma_t{:}$ $mg \sin 30° = ma_t$

$\qquad\qquad a_t = g \sin 30° = +4.90 \text{ m/s}^2$ $\mathbf{a}_t = 4.90 \text{ m/s}^2 \swarrow$ ◄

$+\nwarrow \Sigma F_n = ma_n{:}$ $2.5\,mg - mg \cos 30° = ma_n$

$\qquad\qquad a_n = 1.634\,g = +16.03 \text{ m/s}^2$ $\mathbf{a}_n = 16.03 \text{ m/s}^2 \nwarrow$ ◄

Since $a_n = v^2/\rho$, we have $v^2 = \rho a_n = (2 \text{ m})(16.03 \text{ m/s}^2)$

$\qquad\qquad v = \pm 5.66 \text{ m/s}$ $\mathbf{v} = 5.66 \text{ m/s} \;\text{(up or down)}$ ◄

SAMPLE PROBLEM 12.5

Determine the rated speed of a highway curve of radius $\rho = 400$ ft banked through an angle $\theta = 18°$. The rated speed of a banked curved road is the speed at which a car should travel if no lateral friction force is to be exerted on its wheels.

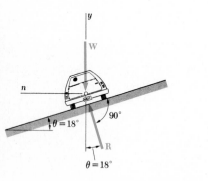

Solution. The car travels in a *horizontal* circular path of radius ρ. The normal component $\mathbf{a}_n$ of the acceleration is directed toward the center of the path; its magnitude is $a_n = v^2/\rho$, where v is the speed of the car in ft/s. The mass m of the car is W/g, where W is the weight of the car. Since no lateral friction force is to be exerted on the car, the reaction $\mathbf{R}$ of the road is shown perpendicular to the roadway. Applying Newton's second law, we write

$+\uparrow \Sigma F_y = 0{:}$ $R \cos \theta - W = 0$ $R = \dfrac{W}{\cos \theta}$ (1)

$\xleftarrow{+} \Sigma F_n = ma_n{:}$ $R \sin \theta = \dfrac{W}{g} a_n$ (2)

Substituting for R from (1) into (2), and recalling that $a_n = v^2/\rho$:

$$\frac{W}{\cos \theta} \sin \theta = \frac{W}{g} \frac{v^2}{\rho} \qquad v^2 = g\rho \tan \theta$$

Substituting the given data, $\rho = 400$ ft and $\theta = 18°$, into this equation, we obtain

$$v^2 = (32.2 \text{ ft/s}^2)(400 \text{ ft}) \tan 18°$$
$$v = 64.7 \text{ ft/s} \qquad v = 44.1 \text{ mi/h}$$ ◄

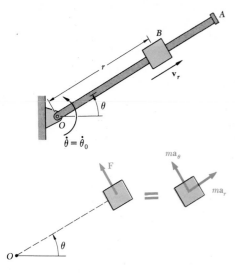

SAMPLE PROBLEM 12.6

A block B of mass m may slide freely on a frictionless arm OA which rotates in a horizontal plane at a constant rate $\dot{\theta}_0$. Knowing that B is released at a distance r_0 from O, express as a function of r, (a) the component v_r of the velocity of B along OA, (b) the magnitude of the horizontal force $\mathbf{F}$ exerted on B by the arm OA.

Solution. Since all other forces are perpendicular to the plane of the figure, the only force shown acting on B is the force $\mathbf{F}$ perpendicular to OA.

Equations of Motion. Using radial and transverse components:

$$+\nearrow \Sigma F_r = ma_r: \qquad\qquad 0 = m(\ddot{r} - r\dot{\theta}^2) \qquad\qquad (1)$$

$$+\nwarrow \Sigma F_\theta = ma_\theta: \qquad\qquad F = m(r\ddot{\theta} + 2\dot{r}\dot{\theta}) \qquad\qquad (2)$$

a. Component v_r of Velocity. Since $v_r = \dot{r}$, we have

$$\ddot{r} = \dot{v}_r = \frac{dv_r}{dt} = \frac{dv_r}{dr}\frac{dr}{dt} = v_r\frac{dv_r}{dr}$$

Substituting for $\ddot{r}$ into (1), recalling that $\dot{\theta} = \dot{\theta}_0$, and separating the variables:

$$v_r\, dv_r = \dot{\theta}_0^2 r\, dr$$

Multiplying by 2, and integrating from 0 to v_r and from r_0 to r:

$$v_r^2 = \dot{\theta}_0^2(r^2 - r_0^2) \qquad v_r = \dot{\theta}_0(r^2 - r_0^2)^{1/2} \quad \blacktriangleleft$$

b. Horizontal Force F. Making $\dot{\theta} = \dot{\theta}_0$, $\ddot{\theta} = 0$, $\dot{r} = v_r$ in Eq. (2), and substituting for v_r the expression obtained in part *a*:

$$F = 2m\dot{\theta}_0(r^2 - r_0^2)^{1/2}\dot{\theta}_0 \qquad F = 2m\dot{\theta}_0^2(r^2 - r_0^2)^{1/2} \quad \blacktriangleleft$$

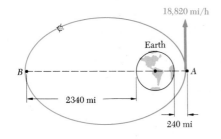

18,820 mi/h

SAMPLE PROBLEM 12.7

A satellite is launched in a direction parallel to the surface of the earth with a velocity of 18,820 mi/h from an altitude of 240 mi. Determine the velocity of the satellite as it reaches its maximum altitude of 2340 mi. It is recalled that the radius of the earth is 3960 mi.

Solution. Since the satellite is moving under a central force directed toward the center O of the earth, we may use Eq. (12.18) and write

$$r_A^2\dot{\theta}_A = r_B^2\dot{\theta}_B = h \qquad\qquad (1)$$

But at A and B the velocity of the satellite reduces to its transverse component; we have $v_A = r_A\dot{\theta}_A$ and $v_B = r_B\dot{\theta}_B$, and Eq. (1) yields

$$r_A v_A = r_B v_B$$

$$v_B = v_A\frac{r_A}{r_B} = (18,820 \text{ mi/h})\frac{3960 \text{ mi} + 240 \text{ mi}}{3960 \text{ mi} + 2340 \text{ mi}}$$

$$v_B = 12,550 \text{ mi/h} \quad \blacktriangleleft$$

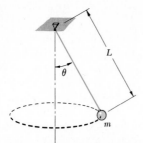

Fig. P12.40 and P12.41

PROBLEMS

12.40 A small ball of mass $m = 5$ kg is attached to a cord of length $L = 2$ m and is made to revolve in a horizontal circle at a constant speed v_0. Knowing that the cord forms an angle $\theta = 40°$ with the vertical, determine (a) the tension in the cord, (b) the speed v_0 of the ball.

12.41 A small ball of mass $m = 5$ kg is made to revolve in a horizontal circle as shown. Knowing that the maximum allowable tension in the cord is 100 N, determine (a) the maximum allowable velocity if $L = 2$ m, (b) the corresponding value of the angle θ.

12.42 Two wires AC and BC are each tied to a sphere at C. The sphere is made to revolve in a horizontal circle at a constant speed v. Determine the range of values of the speed v for which *both* wires are taut.

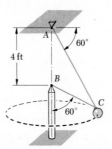

Fig. P12.42 and P12.43

12.43 Two wires AC and BC are each tied to a 10-lb sphere. The sphere is made to revolve in a horizontal circle at a constant speed of 12 ft/s. Determine the tension in each wire.

12.44 A 3-kg ball is swung in a vertical circle at the end of a cord of length $l = 0.8$ m. Knowing that when $\theta = 60°$ the tension in the cord is 25 N, determine the instantaneous velocity and acceleration of the ball.

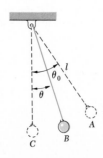

Fig. P12.44 and P12.45

12.45 A ball of weight W is released with no velocity from position A and oscillates in a vertical plane at the end of a cord of length l. Determine (a) the tangential component of the acceleration in position B in terms of the angle θ, (b) the velocity in position B in terms of θ, θ_0, and l, (c) the tension in the cord in terms of W and θ_0 when the ball passes through its lowest position C, (d) the value of θ_0 if the tension in the cord is $T = 2W$ when the ball passes through position C.

12.46 A small sphere of weight W is held as shown by two wires AB and CD. Wire AB is then cut. Determine (a) the tension in wire CD before AB was cut, (b) the tension in wire CD and the acceleration of the sphere just after AB has been cut.

12.47 A man swings a bucket full of water in a vertical plane in a circle of radius 0.75 m. What is the smallest velocity that the bucket should have at the top of the circle if no water is to be spilled?

12.48 A 175-lb pilot flies a small plane in a vertical loop of 300-ft radius. Determine the speed of the plane at points A and B, knowing that at point A the pilot experiences weightlessness and that at point B the pilot's apparent weight is 600 lb.

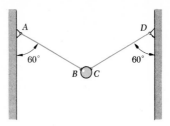

Fig. P12.46

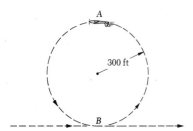

Fig. P12.48

12.49 Three automobiles are proceeding at a speed of 50 mi/h along the road shown. Knowing that the coefficient of friction between the tires and the road is 0.60, determine the tangential deceleration of each automobile if its brakes are suddenly applied and the wheels skid.

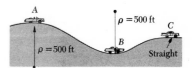

Fig. P12.49

12.50 A series of small packages, being moved by a conveyor belt at a constant speed v, passes over an idler roller as shown. Knowing that the coefficient of friction between the packages and the belt is 0.75, determine the maximum value of v for which the packages do not slip with respect to the belt.

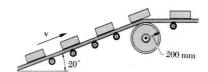

Fig. P12.50

12.51 Express the minimum and maximum safe speeds, with respect to skidding, of a car traveling on a banked road, in terms of the radius r of the curve, the banking angle θ, and the friction angle ϕ between the tires and the pavement.

12.52 A man on a motorcycle takes a turn on a flat unbanked road at 72 km/h. If the radius of the turn is 50 m, determine the minimum value of the coefficient of friction between the tires and the road which will ensure no skidding.

12.53 What angle of banking should be given to the road in Prob. 12.52 if the man on the motorcycle is to be able to take the turn at 72 km/h with a coefficient of friction $\mu = 0.30$?

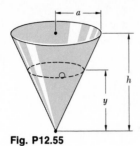

Fig. P12.55

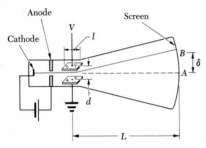

Fig. P12.56 and P12.58

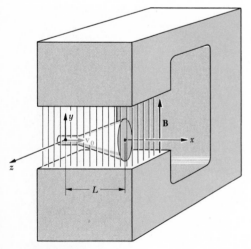

Fig. P12.59

12.54 A stunt driver proposes to drive a small automobile on the vertical wall of a circular pit of radius 50 ft. Knowing that the coefficient of friction between the tires and the wall is 0.75, determine the minimum speed at which the stunt can be performed.

12.55 A small ball rolls at a speed v_0 along a horizontal circle inside the circular cone shown. Express the speed v_0 in terms of the height y of the path above the apex of the cone.

12.56 In the cathode-ray tube shown, electrons emitted by the cathode and attracted by the anode pass through a small hole in the anode and keep traveling in a straight line with a speed v_0 until they strike the screen at A. However, if a difference of potential V is established between the two parallel plates, each electron will be subjected to a force $\mathbf{F}$ perpendicular to the plates while it travels between the plates and will strike the screen at point B at a distance δ from A. The magnitude of the force $\mathbf{F}$ is $F = eV/d$, where $-e$ is the charge of the electron and d is the distance between the plates. Derive an expression for the deflection δ in terms of V, v_0, the charge $-e$ of the electron, its mass m, and the dimensions d, l, and L.

12.57 In Prob. 12.56, determine the smallest allowable value of the ratio d/l in terms of e, m, v_0, and V if the electrons are not to strike the positive plate.

12.58 A manufacturer wishes to design a new cathode-ray tube which will be only half as long as his current model. If the size of the screen is to remain the same, how should the length l of the plates be modified if all the other characteristics of the circuit are to remain unchanged?

12.59 A cathode-ray tube emitting electrons with a velocity $\mathbf{v}_0$ is placed as shown between the poles of a large electromagnet which creates a uniform magnetic field of strength $\mathbf{B}$. Determine the coordinates of the point where the electron beam strikes the tube screen when no difference of potential exists between the plates. It is known that an electron (mass m and charge $-e$) traveling with a velocity $\mathbf{v}$ at a right angle to the lines of force of a magnetic field of strength $\mathbf{B}$ is subjected to a force $\mathbf{F}$ perpendicular to $\mathbf{B}$ and $\mathbf{v}$, and of magnitude $F = eBv$; the sense of the force is such that $\mathbf{B}$, $\mathbf{v}$, and $\mathbf{F}$ form a right-handed triad.

12.60 The two-dimensional motion of particle B is defined by the relations $r = t^2 - \frac{1}{3}t^3$ and $\theta = 2t^2$, where r is expressed in meters, t in seconds, and θ in radians. If the particle has a mass of 2 kg and moves in a horizontal plane, determine the radial and transverse components of the force acting on the particle when (a) $t = 0$, (b) $t = 1$ s.

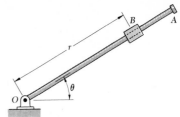

12.61 For the motion defined in Prob. 12.60, determine the radial and transverse components of the force acting on the 2-kg particle as it returns to the origin at $t = 3$ s.

Fig. P12.60

12.62 The two-dimensional motion of a particle is defined by the relations $r = 2 \sin \pi t$ and $\theta = \frac{1}{2}\pi t$, where r is expressed in feet, t in seconds, and θ in radians. If the particle weighs 3 lb and moves in a horizontal plane, determine the radial and transverse components of the force acting on the particle when (a) $t = \frac{1}{2}$ s, (b) $t = 1$ s.

12.63 A block B of mass m may slide on the frictionless arm OA which rotates in a horizontal plane at a constant rate $\dot{\theta}_0$. As the arm rotates, the cord wraps around a *fixed* drum of radius b and pulls the block toward O with a speed $b\dot{\theta}_0$. Express as a function of m, r, b, and $\dot{\theta}_0$, (a) the tension T in the cord, (b) the magnitude of the horizontal force $\mathbf{Q}$ exerted on B by the arm OA.

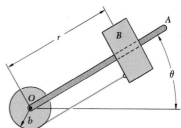

Fig. P12.63

12.64 Solve Prob. 12.63, knowing that the weight of the block is 3 lb and that $r = 30$ in., $b = 2$ in., and $\dot{\theta}_0 = 6$ rad/s.

12.65 In Prob. 12.63, determine the ratio r/b for which the tension T is equal to the magnitude of the horizontal force $\mathbf{Q}$.

12.66 While aiming at a moving target, a man rotates his rifle clockwise in a horizontal plane at the rate of 15° per second. Assuming that he can maintain the motion as the rifle is fired, determine the horizontal force exerted by the barrel on a 45-g bullet just before it leaves the barrel with a muzzle velocity of 550 m/s.

12.67 A particle moves under a central force in a path defined by the equation $r = r_0/\cos n\theta$, where n is a positive constant. Using Eq. (12.18), show that the radial and transverse components of the velocity are $v_r = nv_0 \sin n\theta$ and $v_\theta = v_0 \cos n\theta$, where v_0 is the velocity of the particle for $\theta = 0$. What is the motion of the particle when $n = 0$ and when $n = 1$?

12.68 For the particle and motion of Prob. 12.67, show that the radial and transverse components of the acceleration are $a_r = (n^2 - 1)(v_0^2/r_0) \cos^3 n\theta$ and $a_\theta = 0$.

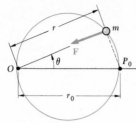

Fig. P12.69

12.69 A particle moves under a central force in a circular path of diameter r_0 which passes through the center of force O. Show that its speed is $v = v_0/\cos^2 \theta$, where v_0 is the speed of the particle at point P_0 directly across the circle from O. [*Hint.* Use Eq. (12.18) with $r = r_0 \cos \theta$].

12.70 If a particle of mass m is attached to the end of a very light circular rod as shown in (*1*), the rod exerts on the mass a force **F** of magnitude $F = kr$ directed toward the origin O, as shown in (*2*). The path of the particle is observed to be an ellipse with axes $a = 6$ in. and $b = 2$ in. (*a*) Knowing that the speed of the particle at A is 8 in./s, determine the speed at B. (*b*) Further knowing that the constant k/m is equal to $16 \, \text{s}^{-2}$, determine the radius of curvature of the path at A and at B.

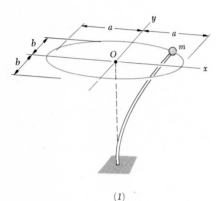

(1)

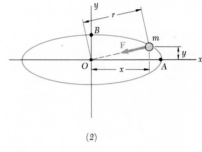

(2)

Fig. P12.70

12.71 Determine the mass of the earth from Newton's law of gravitation, knowing that it takes 94.14 min for a satellite to describe a circular orbit 300 mi above the surface of the earth.

12.72 Show that the radius r of the moon's orbit may be determined from the radius R of the earth, the acceleration of gravity g at the surface of the earth, and the time τ required by the moon to revolve once around the earth. Compute r knowing that $\tau = 27.3$ days.

12.73 Determine the time required for a spacecraft to describe a circular orbit 30 km above the moon's surface. It is known that the radius of the moon is 1740 km and that its mass is 81.3 times smaller than the mass of the earth.

12.74 Denoting by ρ the mean mass density of a planet, show that the periodic time of a satellite in a very low circular orbit is $(3\pi/G\rho)^{1/2}$, where G is the constant of gravitation.

12.75   A spacecraft describes a circular orbit of 6000-mi radius around the earth. In order to transfer it to a larger circular orbit of 24,000-mi radius, the spacecraft is first placed on an elliptic path AB by firing its engine as it passes through A, thus increasing its velocity by 3810 mi/h. By how much should the spacecraft's velocity be increased as it reaches B to insert it in the larger circular orbit?

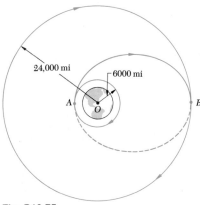

Fig. P12.75

12.76   An Apollo spacecraft describes a circular orbit of 2400-km radius around the moon with a velocity of 5140 km/h. In order to transfer it to a smaller circular orbit of 2000-km radius, the spacecraft is first placed on an elliptic path AB by reducing its velocity to 4900 km/h as it passes through A. Determine (a) the velocity of the spacecraft as it approaches B on the elliptic path, (b) the value to which its velocity must be reduced at B to insert it in the smaller circular orbit.

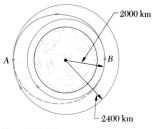

Fig. P12.76

12.77   Solve Prob. 12.76, assuming that the Apollo spacecraft is to be transferred from the orbit of 2400-km radius to a circular orbit of 1800-km radius and that its velocity is reduced to 4760 km/h as it passes through A.

12.78   Solve Prob. 12.75, assuming that the spacecraft is to be transferred from the orbit of 6000-mi radius to a circular orbit of 13,500-mi radius and that its velocity is increased by 2540 mi/h as it passes through A.

*12.10. Trajectory of a Particle under a Central Force.

Consider a particle P moving under a central force **F**. We propose to obtain the differential equation which defines its trajectory.

Assuming that the force **F** is directed toward the center of force O, we note that ΣF_r and ΣF_θ reduce respectively to $-F$ and zero in Eqs. (12.16) and (12.17). We therefore write

$$m(\ddot{r} - r\dot{\theta}^2) = -F \tag{12.22}$$
$$m(r\ddot{\theta} + 2\dot{r}\dot{\theta}) = 0 \tag{12.23}$$

But, as we saw in Sec. 12.8, the last equation yields

$$r^2\dot{\theta} = h \quad \text{or} \quad r^2 \frac{d\theta}{dt} = h \tag{12.24}$$

Equation (12.24) may be used to eliminate the independent variable t from Eq. (12.22). Solving Eq. (12.24) for $\dot{\theta}$ or $d\theta/dt$, we have

$$\dot{\theta} = \frac{d\theta}{dt} = \frac{h}{r^2} \tag{12.25}$$

from which it follows that

$$\dot{r} = \frac{dr}{dt} = \frac{dr}{d\theta}\frac{d\theta}{dt} = \frac{h}{r^2}\frac{dr}{d\theta} = -h\frac{d}{d\theta}\left(\frac{1}{r}\right) \tag{12.26}$$

$$\ddot{r} = \frac{d\dot{r}}{dt} = \frac{d\dot{r}}{d\theta}\frac{d\theta}{dt} = \frac{h}{r^2}\frac{d\dot{r}}{d\theta}$$

or, substituting for $\dot{r}$ from (12.26),

$$\ddot{r} = \frac{h}{r^2}\frac{d}{d\theta}\left[-h\frac{d}{d\theta}\left(\frac{1}{r}\right)\right]$$

$$\ddot{r} = -\frac{h^2}{r^2}\frac{d^2}{d\theta^2}\left(\frac{1}{r}\right) \tag{12.27}$$

Substituting for $\dot{\theta}$ and $\ddot{r}$ from (12.25) and (12.27), respectively, into Eq. (12.22), and introducing the function $u = 1/r$, we obtain after reductions

$$\frac{d^2u}{d\theta^2} + u = \frac{F}{mh^2u^2} \tag{12.28}$$

In deriving Eq. (12.28), the force **F** was assumed directed toward O. The magnitude F should therefore be positive if **F** is actually directed toward O (attractive force) and negative if **F** is directed away from O (repulsive force). If F is a known function of r and thus of u, Eq. (12.28) is a differential equation in u and θ. This differential equation defines the trajectory followed by the particle under the central force **F**. The equation of the trajectory will be obtained by solving the differential equation (12.28) for u as a function of θ and determining the constants of integration from the initial conditions.

*12.11. Application to Space Mechanics.   After the last stage of their launching rockets has burned out, earth satellites and other space vehicles are subjected only to the gravitational pull of the earth. Their motion may therefore be determined from Eqs. (12.24) and (12.28), which govern the motion of a particle under a central force, after F has been replaced by the expression obtained for the force of gravitational attraction.† Setting in Eq. (12.28)

$$F = \frac{GMm}{r^2} = GMmu^2$$

where M = mass of earth
   m = mass of space vehicle
   r = distance from center of earth to vehicle
   $u = 1/r$

we obtain the differential equation

$$\frac{d^2u}{d\theta^2} + u = \frac{GM}{h^2} \qquad (12.29)$$

where the right-hand member is observed to be a constant.

The solution of the differential equation (12.29) is obtained by adding the particular solution $u = GM/h^2$ to the general solution $u = C \cos(\theta - \theta_0)$ of the corresponding homogeneous equation (i.e., the equation obtained by setting the right-hand member equal to zero). Choosing the polar axis so that $\theta_0 = 0$, we write

$$\frac{1}{r} = u = \frac{GM}{h^2} + C \cos \theta \qquad (12.30)$$

Equation (12.30) is the equation of a *conic section* (ellipse, parabola, or hyperbola) in the polar coordinates r and θ. The origin O of the coordinates, which is located at the center of the earth, is a *focus* of this conic section, and the polar axis is one of its axes of symmetry (Fig. 12.18).

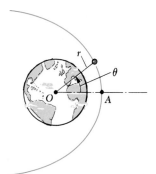

Fig. 12.18

---

† It is assumed that the space vehicles considered here are attracted only by the earth and that their mass is negligible compared to the mass of the earth. If a vehicle moves very far from the earth, its path may be affected by the attraction of the sun, the moon, or another planet.

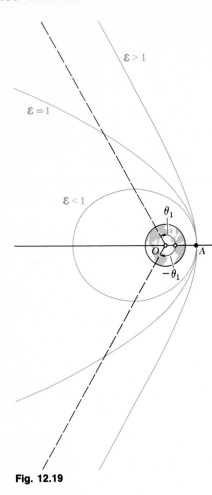

$\varepsilon > 1$

$\varepsilon = 1$

$\varepsilon < 1$

θ_1

O A

$-\theta_1$

Fig. 12.19

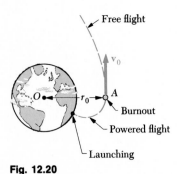

Free flight

v_0

O r_0 A

Burnout

Powered flight

Launching

Fig. 12.20

The ratio of the constants C and GM/h^2 defines the *eccentricity* ε of the conic section; we write

$$\varepsilon = \frac{C}{GM/h^2} = \frac{Ch^2}{GM} \qquad (12.31)$$

Three cases may be distinguished:

1. $\varepsilon > 1$, or $C > GM/h^2$: There are two values θ_1 and $-\theta_1$ of the polar angle, defined by $\cos \theta_1 = -GM/Ch^2$, for which the right-hand member of Eq. (12.30) becomes zero. For both of these values, the radius vector r becomes infinite; the conic section is a *hyperbola* (Fig. 12.19).
2. $\varepsilon = 1$, or $C = GM/h^2$: The radius vector becomes infinite for $\theta = 180°$; the conic section is a *parabola*.
3. $\varepsilon < 1$, or $C < GM/h^2$: The radius vector remains finite for every value of θ; the conic section is an *ellipse*. In the particular case when $\varepsilon = C = 0$, the length of the radius vector is constant; the conic section is a *circle*.

We shall see now how the constants C and GM/h^2 which characterize the trajectory of a space vehicle may be determined from the position and the velocity of the space vehicle at the beginning of its free flight. We shall assume, as it is generally the case, that the powered phase of its flight has been programmed in such a way that, as the last stage of the launching rocket burns out, the vehicle has a velocity parallel to the surface of the earth (Fig. 12.20). In other words, we shall assume that the space vehicle begins its free flight at the vertex A of its trajectory.†

Denoting respectively by r_0 and v_0 the radius vector and speed of the vehicle at the beginning of its free flight, we observe, since the velocity reduces to its transverse component, that $v_0 = r_0\dot{\theta}_0$. Recalling Eq. (12.24), we express the constant h as

$$h = r_0^2\dot{\theta}_0 = r_0 v_0 \qquad (12.32)$$

The value obtained for h may be used to determine the constant GM/h^2. We also note that the computation of this constant will be simplified if we use the relation indicated in Sec. 12.9,

$$GM = gR^2 \qquad (12.21)$$

where R is the radius of the earth ($R = 6.37 \times 10^6$ m or 3960 mi) and g the acceleration of gravity at the surface of the earth.

†Problems involving oblique launchings will be considered in Sec. 14.11.

The constant C will be determined by setting $\theta = 0$, $r = r_0$ in Eq. (12.30); we obtain

$$C = \frac{1}{r_0} - \frac{GM}{h^2} \qquad (12.33)$$

Substituting for h from (12.32), we may then easily express C in terms of r_0 and v_0.

Let us now determine the initial conditions corresponding to each of the three fundamental trajectories indicated above. Considering first the parabolic trajectory, we set C equal to GM/h^2 in Eq. (12.33) and eliminate h between Eqs. (12.32) and (12.33). Solving for v_0, we obtain

$$v_0 = \sqrt{\frac{2GM}{r_0}}$$

We may easily check that a larger value of the initial velocity corresponds to a hyperbolic trajectory, and a smaller value to an elliptic orbit. Since the value of v_0 obtained for the parabolic trajectory is the smallest value for which the space vehicle does not return to its starting point, it is called the *escape velocity*. We write therefore

$$v_{\text{esc}} = \sqrt{\frac{2GM}{r_0}} \qquad \text{or} \qquad v_{\text{esc}} = \sqrt{\frac{2gR^2}{r_0}} \qquad (12.34)$$

if we make use of Eq. (12.21). We note that the trajectory will be (1) hyperbolic if $v_0 > v_{\text{esc}}$; (2) parabolic if $v_0 = v_{\text{esc}}$; (3) elliptic if $v_0 < v_{\text{esc}}$.

Among the various possible elliptic orbits, one is of special interest, the *circular orbit*, which is obtained when $C = 0$. The value of the initial velocity corresponding to a circular orbit is easily found to be

$$v_{\text{circ}} = \sqrt{\frac{GM}{r_0}} \qquad \text{or} \qquad v_{\text{circ}} = \sqrt{\frac{gR^2}{r_0}} \qquad (12.35)$$

if Eq. (12.21) is taken into account. We may note from Fig. 12.21 that, for values of v_0 comprised between v_{circ} and v_{esc}, point A where free flight begins is the point of the orbit closest to the earth; this point is called the *perigee*, while point A', which is farthest away from the earth, is known as the *apogee*. For values of v_0 smaller than v_{circ}, point A becomes the apogee, while point A'', on the other side of the orbit, becomes the perigee. For values of v_0 much smaller than v_{circ}, the trajectory of the space vehicle intersects the surface of the earth; in such a case, the vehicle does not go into orbit.

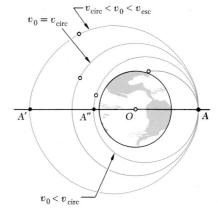

Fig. 12.21

Ballistic missiles, which are designed to hit the surface of the earth, also travel along elliptic trajectories. In fact, we should now realize that any object projected in vacuum with an initial velocity v_0 smaller than v_{esc} will move along an elliptic path. It is only when the distances involved are small that the gravitational field of the earth may be assumed uniform, and that the elliptic path may be approximated by a parabolic path, as was done earlier (Sec. 11.10) in the case of conventional projectiles.

Periodic Time. An important characteristic of the motion of an earth satellite is the time required by the satellite to describe its orbit. This time is known as the *periodic time* of the satellite and is denoted by τ. We first observe, in view of the definition of the areal velocity (Sec. 12.8), that τ may be obtained by dividing the area inside the orbit by the areal velocity. Since the area of an ellipse is equal to πab, where a and b denote, respectively, the semimajor and semiminor axes, and since the areal velocity is equal to $h/2$, we write

$$\tau = \frac{2\pi ab}{h} \qquad (12.36)$$

While h may be readily determined from r_0 and v_0 in the case of a satellite launched in a direction parallel to the surface of the earth, the semiaxes a and b are not directly related to the initial conditions. Since, on the other hand, the values r_0 and r_1 of r corresponding to the perigee and apogee of the orbit may easily be determined from Eq. (12.30), we shall express the semiaxes a and b in terms of r_0 and r_1.

Consider the elliptic orbit shown in Fig. 12.22. The earth's center is located at O and coincides with one of the two foci of the ellipse, while the points A and A' represent, respectively, the perigee and apogee of the orbit. We easily check that

$$r_0 + r_1 = 2a$$

and thus $$a = \tfrac{1}{2}(r_0 + r_1) \qquad (12.37)$$

Recalling that the sum of the distances from each of the foci to any point of the ellipse is constant, we write

$$O'B + BO = O'A + OA = 2a \qquad \text{or} \qquad BO = a$$

On the other hand, we have $CO = a - r_0$. We may therefore write

$$b^2 = (BC)^2 = (BO)^2 - (CO)^2 = a^2 - (a - r_0)^2$$
$$b^2 = r_0(2a - r_0) = r_0 r_1$$

and thus $$b = \sqrt{r_0 r_1} \qquad (12.38)$$

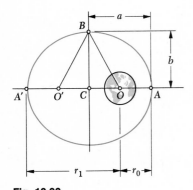

Fig. 12.22

Formulas (12.37) and (12.38) indicate that the semimajor and semiminor axes of the orbit are respectively equal to the arithmetic and geometric means of the maximum and minimum values of the radius vector. Once r_0 and r_1 have been determined, the lengths of the semiaxes may thus be easily computed and substituted for a and b in formula (12.36).

*12.12. Kepler's Laws of Planetary Motion.

The equations governing the motion of an earth satellite may be used to describe the motion of the moon around the earth. In that case, however, the mass of the moon is not negligible compared to the mass of the earth, and the results obtained are not entirely accurate.

The theory developed in the preceding sections may also be applied to the study of the motion of the planets around the sun. While another error is introduced by neglecting the forces exerted by the planets on each other, the approximation obtained is excellent. Indeed, the properties expressed by Eq. (12.30), where M now represents the mass of the sun, and by Eq. (12.24) had been discovered by the German astronomer Johann Kepler (1571–1630) from astronomical observations of the motion of the planets, even before Newton had formulated his fundamental theory.

Kepler's three *laws of planetary motion* may be stated as follows:

1. Each planet describes an ellipse, with the sun located at one of its foci.
2. The radius vector drawn from the sun to a planet sweeps equal areas in equal times.
3. The squares of the periodic times of the planets are proportional to the cubes of the semimajor axes of their orbits.

The first law states a particular case of the result established in Sec. 12.11, while the second law expresses that the areal velocity of each planet is constant (see Sec. 12.8). Kepler's third law may also be derived from the results obtained in Sec. 12.11.†

† See Prob. 12.95.

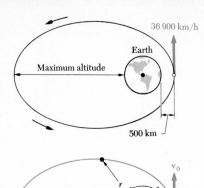

36 900 km/h

Earth

Maximum altitude

500 km

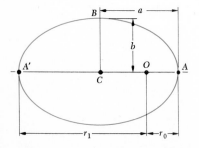

v_0

r

θ

r_1

A'

R

A

r_0

SAMPLE PROBLEM 12.8

A satellite is launched in a direction parallel to the surface of the earth with a velocity of 36 900 km/h from an altitude of 500 km. Determine (a) the maximum altitude reached by the satellite, (b) the periodic time of the satellite.

a. **Maximum Altitude.** After launching, the satellite is subjected only to the gravitational attraction of the earth; its motion is thus governed by Eq. (12.30).

$$\frac{1}{r} = \frac{GM}{h^2} + C \cos \theta \qquad (1)$$

Since the radial component of the velocity is zero at the point of launching A, we have $h = r_0 v_0$. Recalling that the radius of the earth is $R = 6370$ km, we compute

$$r_0 = 6370 \text{ km} + 500 \text{ km} = 6870 \text{ km} = 6.87 \times 10^6 \text{ m}$$

$$v_0 = 36\,900 \text{ km/h} = \frac{3.69 \times 10^7 \text{ m}}{3.6 \times 10^3 \text{ s}} = 1.025 \times 10^4 \text{ m/s}$$

$$h = r_0 v_0 = (6.87 \times 10^6 \text{ m})(1.025 \times 10^4 \text{ m/s}) = 7.04 \times 10^{10} \text{ m}^2/\text{s}$$
$$h^2 = 4.96 \times 10^{21} \text{ m}^4/\text{s}^2$$

Since $GM = gR^2$, where R is the radius of the earth, we have

$$GM = gR^2 = (9.81 \text{ m/s}^2)(6.37 \times 10^6 \text{ m})^2 = 3.98 \times 10^{14} \text{ m}^3/\text{s}^2$$

$$\frac{GM}{h^2} = \frac{3.98 \times 10^{14} \text{ m}^3/\text{s}^2}{4.96 \times 10^{21} \text{ m}^4/\text{s}^2} = 8.03 \times 10^{-8} \text{ m}^{-1}$$

Substituting this value into (1), we obtain

$$\frac{1}{r} = 8.03 \times 10^{-8} + C \cos \theta \qquad (2)$$

Noting that at point A we have $\theta = 0$ and $r = r_0 = 6.87 \times 10^6$ m, we compute the constant C.

$$\frac{1}{6.87 \times 10^6 \text{ m}} = 8.03 \times 10^{-8} + C \cos 0° \qquad C = 6.53 \times 10^{-8} \text{ m}^{-1}$$

At A', the point on the orbit farthest from the earth, we have $\theta = 180°$. Using (2), we compute the corresponding distance r_1.

$$\frac{1}{r_1} = 8.03 \times 10^{-8} + (6.53 \times 10^{-8}) \cos 180°$$

$$r_1 = 0.667 \times 10^8 \text{ m} = 66\,700 \text{ km}$$

$$\textit{Maximum altitude} = 66\,700 \text{ km} - 6370 \text{ km} = 60\,300 \text{ km} \blacktriangleleft$$

b. **Periodic Time.** Since A and A' are the perigee and apogee, respectively, of the elliptic orbit, we use Eqs. (12.37) and (12.38) and compute the semimajor and semiminor axes of the orbit.

$$a = \tfrac{1}{2}(r_0 + r_1) = \tfrac{1}{2}(6.87 + 66.7)(10^6) \text{ m} = 36.8 \times 10^6 \text{ m}$$
$$b = \sqrt{r_0 r_1} = \sqrt{(6.87)(66.7)} \times 10^6 \text{ m} = 21.4 \times 10^6 \text{ m}$$

$$\tau = \frac{2\pi ab}{h} = \frac{2\pi(36.8 \times 10^6 \text{ m})(21.4 \times 10^6 \text{ m})}{7.04 \times 10^{10} \text{ m}^2/\text{s}}$$

$$\tau = 7.03 \times 10^4 \text{ s} = 1171 \text{ min} = 19 \text{ h } 31 \text{ min} \blacktriangleleft$$

B

a

b

A'

O

A

C

r_1

r_0

PROBLEMS

12.79 (a) Knowing that the radius of the moon is 1080 mi and that its mass is 0.01230 times the mass of the earth, determine the speed of an Apollo spacecraft describing a circular orbit 420 mi above the moon's surface. (b) To what value should the speed be increased if the spacecraft is to leave the moon's gravitation field? (Neglect the effect of the earth's gravity.)

12.80 A spacecraft is describing a circular orbit of 5000-mi radius around the earth when its engine is suddenly fired, increasing the speed of the spacecraft by 3000 mi/h. Determine the greatest distance from the center of the earth reached by the spacecraft.

12.81 A space vehicle approaches the planet Venus along a hyperbolic trajectory of eccentricity $\varepsilon = 2$. As the vehicle reaches point A, which is the point of the trajectory closest to the planet, retrorockets are fired to slow the vehicle and place it in a circular orbit. Knowing that the distance from the center O of the planet to A is 6000 mi, and that the mass of Venus is 0.82 times the mass of the earth, determine the velocity of the vehicle (a) as it approaches A, (b) after the retrorockets have been fired.

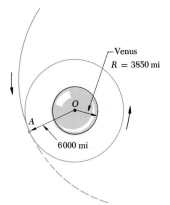

Fig. P12.81

12.82 The maximum altitude a_1 of an unmanned earth satellite is observed to be 8300 km, while its minimum altitude a_2 is known to be 225 km. Determine the maximum and minimum values of its velocity.

12.83 The speed of an earth satellite as it passes through its perigee A is 36×10^3 km/h. Knowing that the altitude of the perigee is $a_2 = 230$ km, determine (a) the maximum altitude reached by the satellite, (b) the corresponding minimum speed of the satellite.

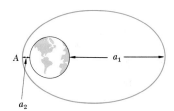

Fig. P12.82 and P12.83

12.84 After completing their moon-exploration mission, the two astronauts forming the crew of an Apollo lunar excursion module (LEM) prepare to rejoin the command module which is orbiting the moon at an altitude of 85 mi. They fire the LEM's engine, bring it along a curved path to a point A, 5 mi above the moon's surface, and shut off the engine. Knowing that the LEM is moving at that time in a direction parallel to the moon's surface and that it will coast along an elliptic path to a rendezvous at B with the command module, determine (a) the speed of the LEM at engine shutoff, (b) the relative velocity with which the command module will approach the LEM at B. (The radius of the moon is 1080 miles and its mass is 0.01230 times the mass of the earth.)

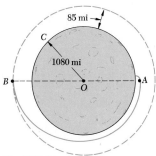

Fig. P12.84

12.85 Solve Prob. 12.84, assuming that the Apollo command module is orbiting the moon at an altitude of 55 mi.

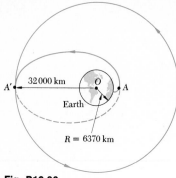

Fig. P12.86

12.86 In order to place a satellite in a circular orbit of radius 32×10^3 km around the earth, the satellite is first projected horizontally from A at an altitude of 500 km into an elliptic path whose apogee A' is at a distance of 32×10^3 km from the center of the earth. Auxiliary rockets are fired as the satellite reaches A' in order to place it in its final orbit. Determine (a) the initial velocity of the satellite at A, (b) the increase in velocity resulting from the firing of the rockets at A'.

12.87 Determine the time required for the LEM of Prob. 12.84 to travel from A to B.

12.88 Determine the periodic time of the spacecraft of Prob. 12.80 on its elliptic orbit.

12.89 Determine the periodic time of the satellite of Prob. 12.82.

12.90 Referring to Prob. 12.86, determine the time elapsed between the end of the launching of the satellite at A and the firing of the auxiliary rockets at A' which will place the satellite in its final orbit.

12.91 A spacecraft describes a circular orbit at an altitude of 3200 km above the earth's surface. Preparatory to reentry it reduces its speed to a value $v_0 = 5400$ m/s, thus placing itself on an elliptic trajectory. Determine the value of θ defining the point B where splashdown will occur. (*Hint.* Point A is the apogee of the elliptic trajectory.)

12.92 A spacecraft describes a circular orbit at an altitude of 3200 km above the earth's surface. Preparatory to reentry it places itself on an elliptic trajectory by reducing its speed to a value v_0. Determine v_0 so that splashdown will occur at a point B corresponding to $\theta = 120°$. (See hint of Prob. 12.91.)

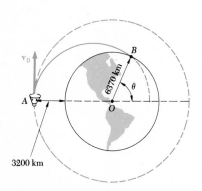

Fig. P12.91 and P12.92

12.93 Upon the LEM's return to the command module, the Apollo spacecraft of Prob. 12.84 is turned around so that the LEM faces to the rear. After completing a full orbit, i.e., as the craft passes again through B, the LEM is cast adrift and crashes on the moon's surface at point C. Determine the velocity of the LEM relative to the command module as it is cast adrift, knowing that the angle BOC is 90°. (*Hint.* Point B is the apogee of the elliptic crash trajectory.)

12.94 Upon the LEM's return to the command module, the Apollo spacecraft of Prob. 12.84 is turned around so that the LEM faces to the rear. After completing a full orbit, i.e., as the craft passes again through B, the LEM is cast adrift with a velocity of 600 ft/s relative to the command module. Determine the point C where the LEM will crash on the moon's surface. (See hint of Prob. 12.93).

12.95 Derive Kepler's third law of planetary motion from Eqs. (12.30) and (12.36).

12.96 Two space stations S_1 and S_2 are describing coplanar circular counterclockwise orbits of radius r_0 and $8r_0$, respectively, around the earth. It is desired to send a vehicle from S_1 to S_2. The vehicle is to be launched in a direction tangent to the orbit of S_1 and is to reach S_2 with a velocity tangent to the orbit of S_2. After a short powered phase, the vehicle will travel in free flight from S_1 to S_2. (*a*) Determine the launching velocity (velocity of the vehicle relative to S_1) in terms of the velocity v_0 of S_1. (*b*) Briefly describe the docking operation with S_2. (*c*) Determine the angle θ defining the required position of S_2 relative to S_1 at the time of launching.

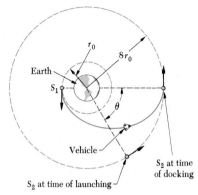

Fig. P12.96

*12.97 A space vehicle is launched in a direction parallel to the earth's surface from point A at an altitude of 340 mi. Knowing that the trajectory of the space vehicle is parabolic, determine (*a*) the launching velocity v_0 of the vehicle, (*b*) the angle θ described by the vehicle as it crosses the orbit of the moon at point B, 239,000 mi away from the center of the earth, (*c*) the velocity of the vehicle at B (magnitude and angle ϕ), (*d*) the time elapsed as the vehicle travels from A to B.

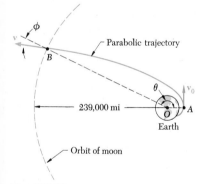

Fig. P12.97

REVIEW PROBLEMS

12.98 In a manufacturing process, disks are moved from one elevation to another by the lifting arm shown. Knowing that the coefficient of friction between the disks and the arm is 0.20, determine the magnitude of the downward acceleration **a** for which the disks slide on the arm.

12.99 Solve Prob. 12.98, assuming that the acceleration **a** is directed upward.

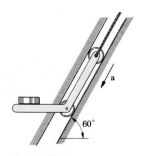

Fig. P12.98

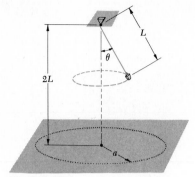

Fig. P12.100

12.100 A bucket is attached to a rope of length $L = 1.2$ m and is made to revolve in a horizontal circle. Drops of water leaking from the bucket fall and strike the floor along the perimeter of a circle of radius a. Determine the radius a when $\theta = 30°$.

12.101 Determine the radius a in Prob. 12.100, assuming that the speed of the bucket is 5 m/s. (The angle θ is not 30° in this case.)

12.102 Determine the required tension T if the acceleration of the 500-lb cylinder is to be (*a*) 6 ft/s² upward, (*b*) 6 ft/s² downward.

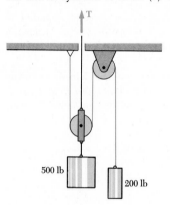

Fig. P12.102 and P12.103

12.103 Determine the acceleration of the 200-lb cylinder if (*a*) $T = 300$ lb, (*b*) $T = 800$ lb.

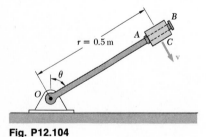

Fig. P12.104

12.104 The rod *OAB* rotates in a vertical plane at the constant rate $\dot\theta = 3$ rad/s, and collar *C* is free to slide on the rod between two stops *A* and *B*. Knowing that the distance between the stops is only slightly larger than the collar and neglecting the effect of friction, determine the range of values of θ for which the collar is in contact with stop *A*.

12.105 (*a*) Express the rated speed of a banked road in terms of the radius *r* of the curve and the banking angle θ. (*b*) What is the apparent weight of an automobile traveling at the rated speed? (See Sample Prob. 12.5 for the definition of rated speed.)

12.106 The assembly shown rotates about a vertical axis at a constant rate. Knowing that the coefficient of friction between the small block A and the cylindrical wall is 0.20, determine the lowest speed v for which the block will remain in contact with the wall.

12.107 Denoting by v_t the terminal speed of an object dropped from a great height, determine the distance the object will fall before its speed reaches the value $\frac{1}{2}v_t$. Assume that the frictional resistance of the air is proportional to the square of the speed of the object.

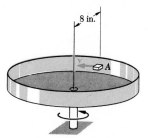

Fig. P12.106

12.108 A spacecraft is describing a circular orbit of radius r_0 with a speed v_0 around an unspecified celestial body of center O, when its engine is suddenly fired, increasing the speed of the spacecraft from v_0 to αv_0, where $1 < \alpha^2 < 2$. Show that the maximum distance r_{max} from O reached by the spacecraft depends only upon r_0 and α, and express the ratio r_{max}/r_0 as a function of α.

12.109 A space vehicle approaches the planet Mars along a parabolic trajectory. As the vehicle reaches point A, which is the point of the trajectory closest to the planet, retrorockets are fired to slow the vehicle and place it in an elliptic path which will bring it to a tangential landing at B. Knowing that the distance from the center O of the planet to A is 12.5×10^3 km, that the radius of the planet is 3400 km, and that its mass is 0.108 times the mass of the earth, determine the velocity of the vehicle (a) as it approaches A, (b) after the retrorockets have been fired, (c) as it lands at B.

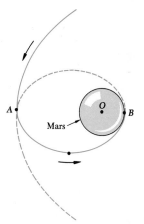

Fig. P12.109

Chapter
13

Kinetics of Particles: Work and Energy

13.1. Introduction. In the preceding chapter, problems dealing with the motion of particles were solved through the use of the fundamental equation of motion $\mathbf{F} = m\mathbf{a}$. Given a particle acted upon by a force $\mathbf{F}$, we could solve this equation for the acceleration $\mathbf{a}$; then, by applying the principles of kinematics, we could determine from $\mathbf{a}$ the velocity and position of the particle at any time.

If the equation $\mathbf{F} = m\mathbf{a}$ and the principles of kinematics are combined, two additional methods of analysis may be obtained, the *method of work and energy* and the *method of impulse and momentum*. The advantage of these methods lies in the fact that they make the determination of the acceleration unnecessary. Indeed, the method of work and energy relates directly force, mass, velocity, and displacement, while the method of impulse and momentum relates force, mass, velocity, and time.

The method of work and energy is treated in the present chapter. It is based on two important concepts, the concept of the *work of a force* and the concept of the *kinetic energy of a particle*. These concepts are defined in the following sections.

13.2. Work of a Force. We shall first define the terms *displacement* and *work* as they are used in mechanics.† Consider

---

† The definition of work was given in Sec. 10.1, and the basic properties of the work of a force were outlined in Secs. 10.1 and 10.5. For convenience, we repeat here the portions of this material which relate to the kinetics of particles.

a particle which moves from a point A to a neighboring point A' (Fig. 13.1a). The small vector which joins A to A' is called the *displacement* of the particle and is denoted by the differential $d\mathbf{r}$, where $\mathbf{r}$ represents the position vector of A. Now, let us assume that a force $\mathbf{F}$, of magnitude F and forming an angle α with $d\mathbf{r}$, is acting on the particle. The *work of the force* $\mathbf{F}$ *during the displacement* $d\mathbf{r}$, of magnitude ds, is defined as the scalar quantity

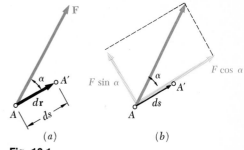

$$dU = F\,ds\,\cos\alpha \qquad (13.1)$$

Fig. 13.1

Being a *scalar quantity*, work has a magnitude and a sign, but no direction. We also note that work should be expressed in units obtained by multiplying units of length by units of force. Thus, if U.S. customary units are used, work should be expressed in ft · lb or in · lb. If SI units are used, work should be expressed in N · m. The unit of work N · m is called a *joule* (J).† Recalling the conversion factors indicated in Sec. 12.2, we write

$$1\text{ ft}\cdot\text{lb} = (1\text{ ft})(1\text{ lb}) = (0.3048\text{ m})(4.448\text{ N}) = 1.356\text{ J}$$

It appears from (13.1) that the work dU may be considered as the product of the magnitude ds of the displacement and of the component $F\cos\alpha$ of the force $\mathbf{F}$ in the direction of $d\mathbf{r}$ (Fig. 13.1b). If this component and the displacement $d\mathbf{r}$ have the same sense, the work is positive; if they have opposite senses, the work is negative. Three particular cases are of special interest. If the force $\mathbf{F}$ has the same direction as $d\mathbf{r}$, the work dU reduces to $F\,ds$. If $\mathbf{F}$ has a direction opposite to that of $d\mathbf{r}$, the work is $dU = -F\,ds$. Finally, if $\mathbf{F}$ is perpendicular to $d\mathbf{r}$, the work dU is zero.

The work of $\mathbf{F}$ during a *finite* displacement of the particle from A_1 to A_2 (Fig. 13.2a) is obtained by integrating Eq. (13.1) from s_1 to s_2. This work, denoted by $U_{1\to2}$, is

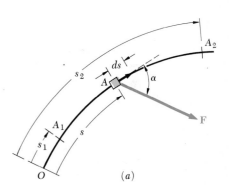

(a)

$$U_{1\to2} = \int_{s_1}^{s_2} (F\cos\alpha)\,ds \qquad (13.2)$$

where the variable of integration s measures the distance traveled by the particle along the path. The work $U_{1\to2}$ is thus represented by the area under the curve obtained by plotting $F\cos\alpha$ against s (Fig. 13.2b).

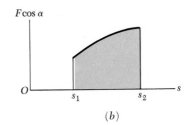

(b)

Fig. 13.2

† The joule (J) is the SI unit of *energy*, whether in mechanical form (work, potential energy, kinetic energy) or in chemical, electrical, or thermal form. We should note that, even though N · m = J, the moment of a force must be expressed in N · m, and not in joules, since the moment of a force is not a form of energy.

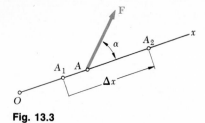

Fig. 13.3

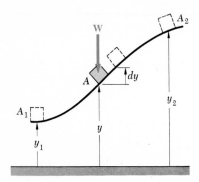

Fig. 13.4

Work of a Constant Force in Rectilinear Motion. When a particle moving in a straight line is acted upon by a force **F** of constant magnitude and of constant direction (Fig. 13.3), formula (13.2) yields

$$U_{1\to 2} = (F \cos \alpha)\, \Delta x \qquad (13.3)$$

where α = angle the force forms with direction of motion
Δx = displacement from A_1 to A_2

Work of a Weight. The work dU of a force **F** during a displacement $d\mathbf{r}$ may also be considered as the product of the magnitude F of the force **F** and of the component $ds \cos \alpha$ of the displacement $d\mathbf{r}$ along **F**. This view is particularly useful in the computation of the work of the weight **W** of a body. The work dU of **W** is equal to the product of W and of the vertical displacement of the center of gravity G of the body. With the y axis chosen upward, the work of **W** during a finite displacement (Fig. 13.4) is obtained by writing

$$dU = -W\, dy$$

$$U_{1\to 2} = -\int_{y_1}^{y_2} W\, dy = Wy_1 - Wy_2 \qquad (13.4)$$

or
$$U_{1\to 2} = -W(y_2 - y_1) = -W\, \Delta y \qquad (13.4')$$

where Δy is the vertical displacement from A_1 to A_2. The work of the weight **W** is thus equal to *the product of W and of the vertical displacement of the center of gravity of the body.* The work is *positive* when $\Delta y < 0$, that is, *when the body moves down.*

Work of the Force Exerted by a Spring. Consider a body A attached to a fixed point B by a spring; it is assumed that the spring is undeformed when the body is at A_0 (Fig. 13.5a). Experimental evidence shows that the magnitude of the force **F** exerted by the spring on body A is proportional to the deflection x of the spring measured from the position A_0. We have

$$F = kx \qquad (13.5)$$

where k is the *spring constant,* expressed in N/m or kN/m if SI units are used and in lb/ft or lb/in. if U.S. customary units are used.†

---

† The relation $F = kx$ is correct under static conditions only. Under dynamic conditions, formula (13.5) should be modified to take the inertia of the spring into account. However, the error introduced by using the relation $F = kx$ in the solution of kinetics problems is small if the mass of the spring is small compared with the other masses in motion.

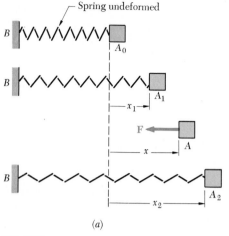

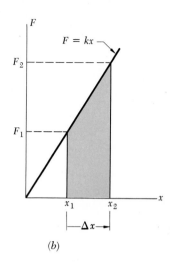

(a)

(b)

Fig. 13.5

The work of the force **F** exerted by the spring during a finite displacement of the body from $A_1(x = x_1)$ to $A_2(x = x_2)$ is obtained by writing

$$dU = -F\,dx = -kx\,dx$$

$$U_{1\to2} = -\int_{x_1}^{x_2} kx\,dx = \tfrac{1}{2}kx_1^2 - \tfrac{1}{2}kx_2^2 \qquad (13.6)$$

Care should be taken to express k and x in consistent units. For example, if U.S. customary units are used, k should be expressed in lb/ft and x in feet, or k in lb/in. and x in inches; in the first case, the work is obtained in ft · lb, in the second case, in in · lb. We note that the work of the force **F** exerted by the spring on the body is *positive* when $x_2 < x_1$, i.e., *when the spring is returning to its undeformed position.*

Since Eq. (13.5) is the equation of a straight line of slope k passing through the origin, the work $U_{1\to2}$ of **F** during the displacement from A_1 to A_2 may be obtained by evaluating the area of the trapezoid shown in Fig. 13.5b. This is done by computing F_1 and F_2 and multiplying the base Δx of the trapezoid by its mean height $\tfrac{1}{2}(F_1 + F_2)$. Since the work of the force **F** exerted by the spring is positive for a negative value of Δx, we write

$$U_{1\to2} = -\tfrac{1}{2}(F_1 + F_2)\,\Delta x \qquad (13.6')$$

Formula (13.6′) is usually more convenient to use than (13.6) and affords fewer chances of confusing the units involved.

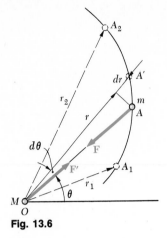

Fig. 13.6

Work of a Gravitational Force. We saw in Sec. 12.9 that two particles at distance r from each other and, respectively, of mass M and m, attract each other with equal and opposite forces **F** and **F′** directed along the line joining the particles, and of magnitude

$$F = G\frac{Mm}{r^2}$$

Let us assume that the particle M occupies a fixed position O while the particle m moves along the path shown in Fig. 13.6. The work of the force **F** exerted on the particle m during an infinitesimal displacement of the particle from A to $A′$ may be obtained by multiplying the magnitude F of the force by the radial component dr of the displacement. Since **F** is directed toward O, the work is negative and we write

$$dU = -F\,dr = -G\frac{Mm}{r^2}\,dr$$

The work of the gravitational force **F** during a finite displacement from $A_1(r = r_1)$ to $A_2(r = r_2)$ is therefore

$$U_{1 \to 2} = -\int_{r_1}^{r_2} \frac{GMm}{r^2}\,dr = \frac{GMm}{r_2} - \frac{GMm}{r_1} \qquad (13.7)$$

The formula obtained may be used to determine the work of the force exerted by the earth on a body of mass m at a distance r from the center of the earth, when r is larger than the radius R of the earth. The letter M represents then the mass of the earth; recalling the first of the relations (12.20), we may thus replace the product GMm in Eq. (13.7) by WR^2, where R is the radius of the earth ($R = 6.37 \times 10^6$ m or 3960 mi) and W the value of the weight of the body at the surface of the earth.

A number of forces frequently encountered in problems of kinetics *do no work.* They are forces applied to fixed points ($ds = 0$) or acting in a direction perpendicular to the displacement ($\cos \alpha = 0$). Among the forces which do no work are the following: the reaction at a frictionless pin when the body supported rotates about the pin, the reaction at a frictionless surface when the body in contact moves along the surface, the reaction at a roller moving along its track, and the weight of a body when its center of gravity moves horizontally.

13.3. Kinetic Energy of a Particle. Principle of Work and Energy. Consider a particle of mass m acted upon by a force **F** and moving along a path which is either rectilinear or curved (Fig. 13.7). Expressing Newton's second law in terms

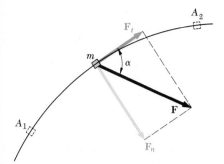

Fig. 13.7

of the tangential components of the force and of the acceleration (see Sec. 12.7), we write

$$F_t = ma_t \quad \text{or} \quad F\cos\alpha = m\frac{dv}{dt} \qquad (13.8)$$

where α = angle the force forms with direction of motion
$\qquad v$ = speed of particle

The speed was shown in Sec. 11.9 to be equal to the rate of change of the distance s traveled by the particle along its path; we write $v = ds/dt$. Solving for dt and substituting into (13.8), we obtain

$$F\cos\alpha = m\frac{dv}{ds}v \qquad (13.9)$$

$$(F\cos\alpha)\,ds = mv\,dv$$

Integrating from A_1, where $s = s_1$ and $v = v_1$, to A_2, where $s = s_2$ and $v = v_2$, we write

$$\int_{s_1}^{s_2}(F\cos\alpha)\,ds = m\int_{v_1}^{v_2}v\,dv = \tfrac{1}{2}mv_2^2 - \tfrac{1}{2}mv_1^2 \quad (13.10)$$

The left-hand member of Eq. (13.10) represents the work $U_{1\to2}$ of the force $\mathbf{F}$ exerted on the particle during the displacement from A_1 to A_2; as indicated in Sec. 13.2, the work $U_{1\to2}$ is a scalar quantity. The expression $\tfrac{1}{2}mv^2$ is also a scalar quantity; it is defined as the kinetic energy of the particle and is denoted by T. We write

$$T = \tfrac{1}{2}mv^2 \qquad (13.11)$$

Substituting into (13.10), we have

$$U_{1\to2} = T_2 - T_1 \qquad (13.12)$$

which expresses that, when a particle moves from A_1 to A_2 under the action of a force **F**, *the work of the force* **F** *is equal to the change in kinetic energy of the particle.* This is known as the *principle of work and energy.* Rearranging the terms in (13.12), we write

$$T_1 + U_{1 \to 2} = T_2 \tag{13.13}$$

Thus, *the kinetic energy of the particle at* A_2 *may be obtained by adding to its kinetic energy at* A_1 *the work done during the displacement from* A_1 *to* A_2 *by the force* **F** *exerted on the particle.* As Newton's second law from which it is derived, the principle of work and energy applies only with respect to a newtonian frame of reference (Sec. 12.1). The speed v used to determine the kinetic energy T should therefore be measured with respect to a newtonian frame of reference.

Since both work and kinetic energy are scalar quantities, their sum may be computed as an ordinary algebraic sum, the work $U_{1 \to 2}$ being considered as positive or negative according to the direction of **F**. When several forces act on the particle, the expression $U_{1 \to 2}$ represents the total work of the forces acting on the particle; it is obtained by adding algebraically the work of the various forces.

As noted above, the kinetic energy of a particle is a scalar quantity. It further appears from the definition $T = \frac{1}{2}mv^2$ that the kinetic energy is always positive, regardless of the direction of motion of the particle. Considering the particular case when $v_1 = 0$, $v_2 = v$, and substituting $T_1 = 0$, $T_2 = T$ into (13.12), we observe that the work done by the forces acting on the particle is equal to T. Thus, the kinetic energy of a particle moving with a speed v represents the work which must be done to bring the particle from rest to the speed v. Substituting $T_1 = T$ and $T_2 = 0$ into (13.12), we also note that, when a particle moving with a speed v is brought to rest, the work done by the forces acting on the particle is $-T$. Assuming that no energy is dissipated into heat, we conclude that the work done by the forces exerted *by the particle* on the bodies which cause it to come to rest is equal to T. Thus, the kinetic energy of a particle also represents *the capacity to do work associated with the speed of the particle.*

The kinetic energy is measured in the same units as work, i.e., in joules if SI units are used, and in ft · lb if U.S. customary units are used. We check that, in SI units,

$$T = \tfrac{1}{2}mv^2 = \text{kg}(\text{m/s})^2 = (\text{kg} \cdot \text{m/s}^2)\text{m} = \text{N} \cdot \text{m} = \text{J}$$

while, in customary units,

$$T = \tfrac{1}{2}mv^2 = (\text{lb} \cdot \text{s}^2/\text{ft})(\text{ft/s})^2 = \text{lb} \cdot \text{ft}$$

13.4. Applications of the Principle of Work and Energy.

The application of the principle of work and energy greatly simplifies the solution of many problems involving forces, displacements, and velocities. Consider, for example, the pendulum OA consisting of a bob A of weight W attached to a cord of length l (Fig. 13.8a). The pendulum is released with no initial velocity from a horizontal position OA_1 and allowed to swing in a vertical plane. We wish to determine the speed of the bob as it passes through A_2, directly under O.

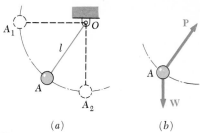

Fig. 13.8

We first determine the work done during the displacement from A_1 to A_2 by the forces acting on the bob. We draw a free-body diagram of the bob, showing all the *actual* forces acting on it, i.e., the weight W and the force P exerted by the cord (Fig. 13.8b). (An inertia vector is not an actual force and *should not* be included in the free-body diagram.) We note that the force P does no work, since it is normal to the path; the only force which does work is thus the weight W. The work of W is obtained by multiplying its magnitude W by the vertical displacement l (Sec. 13.2); since the displacement is downward, the work is positive. We therefore write $U_{1 \to 2} = Wl$.

Considering, now, the kinetic energy of the bob, we find $T_1 = 0$ at A_1 and $T_2 = \frac{1}{2}(W/g)v_2^2$ at A_2. We may now apply the principle of work and energy; recalling formula (13.13), we write

$$T_1 + U_{1 \to 2} = T_2 \qquad 0 + Wl = \frac{1}{2}\frac{W}{g}v_2^2$$

Solving for v_2, we find $v_2 = \sqrt{2gl}$. We note that the speed obtained is that of a body falling freely from a height l.

The example we have considered illustrates the following advantages of the method of work and energy:

1. In order to find the speed at A_2, there is no need to determine the acceleration in an intermediate position A and to integrate the expression obtained from A_1 to A_2.
2. All quantities involved are scalars and may be added directly, without using x and y components.
3. Forces which do no work are eliminated from the solution of the problem.

What is an advantage in one problem, however, may become a disadvantage in another. It is evident, for instance, that the

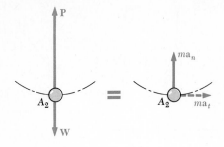

Fig. 13.9

method of work and energy cannot be used to directly determine an acceleration. We also note that it should be supplemented by the direct application of Newton's second law in order to determine a force which is normal to the path of the particle, since such a force does no work. Suppose, for example, that we wish to determine the tension in the cord of the pendulum of Fig. 13.8a as the bob passes through A_2. We draw a free-body diagram of the bob in that position (Fig. 13.9) and express Newton's second law in terms of tangential and normal components. The equations $\Sigma F_t = ma_t$ and $\Sigma F_n = ma_n$ yield, respectively, $a_t = 0$ and

$$P - W = ma_n = \frac{W}{g}\frac{v^2}{l}$$

But the speed at A_2 was determined earlier by the method of work and energy. Substituting $v^2 = 2gl$ and solving for P, we write

$$P = W + \frac{W}{g}\frac{2gl}{l} = 3W$$

13.5. Systems of Particles. When a problem involves several particles, each particle may be considered separately and the principle of work and energy may be applied to each particle. Adding the kinetic energies of all the particles, and considering the work of all the forces involved, we may also write the equation of work and energy for the entire system. We have

$$T_1 + U_{1\rightarrow2} = T_2 \qquad (13.13)$$

where T represents the arithmetic sum of the kinetic energies of the particles forming the system (all terms are positive) and $U_{1\rightarrow2}$ the work of all the forces acting on the various particles, whether these forces are *internal* or *external* from the point of view of the system as a whole.

The method of work and energy is particularly useful in solving problems involving a system of bodies connected by *inextensible cords or links*. In this case, the internal forces occur by pairs of equal and opposite forces, and the points of application of the forces in each pair *move through equal distances*. As a result, the work of the internal forces is zero and $U_{1\rightarrow2}$ reduces to the work of the *external forces only* (see Sample Prob. 13.2).

An automobile weighing 4000 lb is driven down a 5° incline at a speed of 60 mi/h when the brakes are applied, causing a constant total braking force (applied by the road on the tires) of 1500 lb. Determine the distance traveled by the automobile as it comes to a stop.

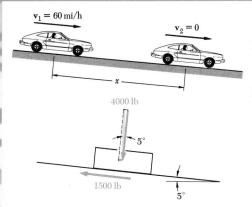

Solution. *Kinetic Energy*

Position *1*: $v_1 = \left(60\frac{\text{mi}}{\text{h}}\right)\left(\frac{5280\text{ ft}}{1\text{ mi}}\right)\left(\frac{1\text{ h}}{3600\text{ s}}\right) = 88\text{ ft/s}$

$$T_1 = \tfrac{1}{2}mv_1^2 = \tfrac{1}{2}(4000/32.2)(88)^2 = 481,000\text{ ft} \cdot \text{lb}$$

Position *2*: $v_2 = 0$ $T_2 = 0$

Work $U_{1\to2} = -1500x + (4000\sin 5°)x = -1151x$

Principle of Work and Energy

$$T_1 + U_{1\to2} = T_2$$
$$481,000 - 1151x = 0 \qquad x = 418\text{ ft} \blacktriangleleft$$

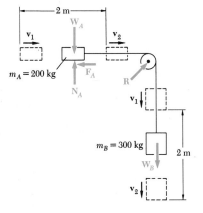

Two blocks are joined by an inextensible cable as shown. If the system is released from rest, determine the velocity of block A after it has moved 2 m. Assume that μ equals 0.25 between block A and the plane and that the pulley is weightless and frictionless.

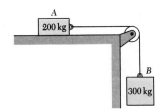

Solution. Since the cable is inextensible, the work done by the forces exerted by the cable will cancel if we consider the two-block system.

Kinetic Energy. Position *1*: $v_1 = 0$ $T_1 = 0$

Position *2*: $v_2 = v$ $T_2 = \tfrac{1}{2}m_A v^2 + \tfrac{1}{2}m_B v^2$

$$T_2 = \tfrac{1}{2}(200\text{ kg})v^2 + \tfrac{1}{2}(300\text{ kg})v^2 = \tfrac{1}{2}(500\text{ kg})v^2$$

Work. During the motion, only $\mathbf{W}_B$ and $\mathbf{F}_A$ do work. Since

$$W_B = (300\text{ kg})(9.81\text{ m/s}^2) = 2940\text{ N}$$

$$F_A = \mu N_A = \mu W_A = 0.25(200\text{ kg})(9.81\text{ m/s}^2) = 490\text{ N}$$

we have

$$U_{1\to2} = W_B(2\text{ m}) - F_A(2\text{ m})$$
$$= (2940\text{ N})(2\text{ m}) - (490\text{ N})(2\text{ m}) = 4900\text{ J}$$

Principle of Work and Energy

$$T_1 + U_{1\to2} = T_2: \qquad 0 + 4900\text{ J} = \tfrac{1}{2}(500\text{ kg})v^2 \qquad v = 4.43\text{ m/s}$$

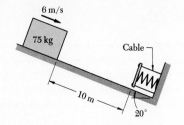

6 m/s

75 kg

Cable

10 m

20°

A spring is used to stop a 75-kg package which is moving down a 20° incline. The spring has a constant $k = 25\,\text{kN/m}$, and is held by cables so that it is initially compressed 100 mm. If the velocity of the package is 6 m/s when it is 10 m from the spring, determine the maximum additional deformation of the spring in bringing the package to rest. Assume $\mu = 0.20$.

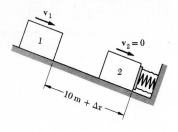

$\mathbf{v}_1$

1

$\mathbf{v}_2 = 0$

2

10 m + Δx

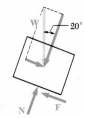

W

20°

N   F

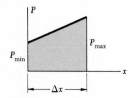

P

P_{max}

P_{min}

Δx

x

Kinetic Energy

Position 1: $v_1 = 6\ \text{m/s}$

$$T_1 = \tfrac{1}{2}mv_1^2 = \tfrac{1}{2}(75\ \text{kg})(6\ \text{m/s})^2 = 1350\ \text{N} \cdot \text{m} = 1350\ \text{J}$$

Position 2 (maximum spring deformation):

$$v_2 = 0 \qquad T_2 = 0$$

Work. We assume that when the package is brought to rest the additional deflection of the spring is Δx. The component of the weight parallel to the plane and the friction force act through the entire displacement, i.e., through $10\ \text{m} + \Delta x$. The total work done by these forces is

$$
\begin{aligned}
U_{1\to2} &= W_t(10\ \text{m} + \Delta x) - F(10\ \text{m} + \Delta x)\\
&= (W\sin 20°)(10\ \text{m} + \Delta x) - 0.20(W\cos 20°)(10\ \text{m} + \Delta x)\\
&= 0.1541W(10\ \text{m} + \Delta x)
\end{aligned}
$$

or, since $W = mg = (75\ \text{kg})(9.81\ \text{m/s}^2) = 736\ \text{N}$,

$$U_{1\to2} = 0.1541(736\ \text{N})(10\ \text{m} + \Delta x) = 1134\ \text{J} + (113.4\ \text{N})\,\Delta x$$

In addition, during the compression of the spring, the variable force **P** exerted by the spring does an amount of negative work equal to the area under the force-deflection curve of the spring force.

$$P_{\min} = kx = (25\ \text{kN/m})(100\ \text{mm}) = (25\,000\ \text{N/m})(0.100\ \text{m}) = 2500\ \text{N}$$

$$
\begin{aligned}
P_{\max} &= P_{\min} + k\,\Delta x\\
&= 2500\ \text{N} + (25\,000\ \text{N/m})\,\Delta x
\end{aligned}
$$

$$U_{1\to2} = -\tfrac{1}{2}(P_{\min} + P_{\max})\,\Delta x = -(2500\ \text{N})\,\Delta x - (12\,500\ \text{N/m})(\Delta x)^2$$

The total work is thus

$$U_{1\to2} = 1134\ \text{J} + (113.4\ \text{N})\,\Delta x - (2500\ \text{N})\,\Delta x - (12\,500\ \text{N/m})(\Delta x)^2$$

Principle of Work and Energy

$$T_1 + U_{1\to2} = T_2$$

$$1350 + 1134 + 113.4\,\Delta x - 2500\,\Delta x - 12\,500(\Delta x)^2 = 0$$

$$(\Delta x)^2 + 0.1909\,\Delta x - 0.1987 = 0$$

$$\Delta x = 0.360\ \text{m} \qquad \Delta x = 360\ \text{mm} \blacktriangleleft$$

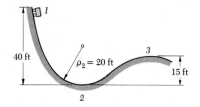

A 2000-lb car starts from rest at point *1* and moves without friction down the track shown. (*a*) Determine the force exerted by the track on the car at point *2*, where the radius of curvature of the track is 20 ft. (*b*) Determine the minimum safe value of the radius of curvature at point *3*.

a. Force Exerted by the Track at Point 2. The principle of work and energy is used to determine the velocity of the car as it passes through point *2*.

Kinetic Energy: $\qquad T_1 = 0 \qquad T_2 = \frac{1}{2}mv_2^2 = \frac{1}{2}\frac{W}{g}v_2^2$

Work. The only force which does work is the weight **W**. Since the vertical displacement from point *1* to point *2* is 40 ft downward, the work of the weight is

$$U_{1\to2} = +W(40\,\text{ft})$$

Principle of Work and Energy

$$T_1 + U_{1\to2} = T_2 \qquad 0 + W(40\,\text{ft}) = \frac{1}{2}\frac{W}{g}v_2^2$$

$$v_2^2 = 80g = 80(32.2) \qquad v_2 = 50.8\,\text{ft/s}$$

Newton's Second Law at Point 2. The acceleration $\mathbf{a}_n$ of the car at point *2* has a magnitude $a_n = v_2^2/\rho$ and is directed upward. Since the external forces acting on the car are **W** and **N**, we write

$$+\uparrow\Sigma F_n = ma_n: \qquad -W + N = ma_n$$
$$= \frac{W}{g}\frac{v_2^2}{\rho}$$
$$= \frac{W}{g}\frac{80g}{20}$$
$$N = 5W \qquad N = 10{,}000\,\text{lb}\uparrow \quad \blacktriangleleft$$

b. Minimum Value of ρ at Point 3. *Principle of Work and Energy.* Applying the principle of work and energy between point *1* and point *3*, we obtain

$$T_1 + U_{1\to3} = T_3 \qquad 0 + W(25\,\text{ft}) = \frac{1}{2}\frac{W}{g}v_3^2$$

$$v_3^2 = 50g = 50(32.2) \qquad v_3 = 40.1\,\text{ft/s}$$

Newton's Second Law at Point 3. The minimum safe value of ρ occurs when $N = 0$. In this case, the acceleration $\mathbf{a}_n$, of magnitude $a_n = v_3^2/\rho$, is directed downward, and we write

$$+\downarrow\Sigma F_n = ma_n: \qquad W = \frac{W}{g}\frac{v_3^2}{\rho}$$
$$= \frac{W}{g}\frac{50g}{\rho} \qquad\qquad \rho = 50\,\text{ft} \quad \blacktriangleleft$$

PROBLEMS

13.1 A 100-kg satellite was placed in a circular orbit 2000 km above the surface of the earth. At this elevation the acceleration of gravity is 5.68 m/s². Determine the kinetic energy of the satellite, knowing that its orbital speed is 24.8×10^3 km/h.

13.2 A stone which weighs 8 lb is dropped from a height h and strikes the ground with a velocity of 75 ft/s. (a) Find the kinetic energy of the stone as it strikes the ground and the height h from which it was dropped. (b) Solve part a, assuming that the same stone is dropped on the moon. (Acceleration of gravity on the moon = 5.31 ft/s².)

13.3 Using the method of work and energy, solve Prob. 12.7.

13.4 A 20-kg package is projected up a 20° incline with an initial velocity of 12 m/s. The coefficient of friction between the incline and the package is 0.15. Determine (a) the maximum distance that the package will move up the incline, (b) the velocity of the package when it returns to its original position.

13.5 A 15-lb package is placed on the chute shown at point A. The coefficient of friction between the package and the chute is 0.15. If the package is released from rest at A, determine the velocity of the package when it reaches the bottom of the chute B.

13.6 A 15-lb package slides down the chute and out along the floor until it comes to rest at point C. If $\mu = 0.15$ and if the initial velocity of the package at A is 3 ft/s directed down the chute, determine the distance d from the bottom of the chute to point C.

13.7 The 2-kg collar was moving down the rod with a velocity of 3 m/s when a force **P** was applied to the horizontal cable. Assuming negligible friction between the collar and the rod, determine the magnitude of the force **P** if the collar stopped after moving 1.2 m more down the rod.

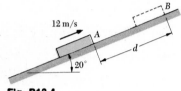

Fig. P13.4

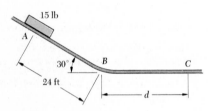

Fig. P13.5 and P13.6

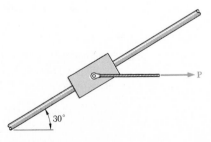

Fig. P13.7

13.8 Solve Prob. 13.7, assuming a coefficient of friction of 0.20 between the collar and the rod.

13.9 The total mass of loading car A and its load is 3500 kg. The car is connected to a 1000-kg counterweight and is at rest when a constant 22-kN force is applied as shown. (*a*) If the force acts through the entire motion, what is the speed of the car after it has traveled 30 m. (*b*) If after the car has moved a distance x the 22-kN force is removed, the car will coast to rest. After what distance x should the force be removed if the car is to come to rest after a total movement of 30 m?

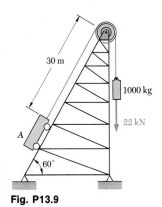

30 m

1000 kg

22 kN

A

60°

Fig. P13.9

13.10 The system shown is at rest when the 50-lb force is applied to block A. Neglecting the effect of friction, determine the velocity of block A after it has moved 9 ft.

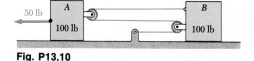

50 lb A B

100 lb   100 lb

Fig. P13.10

13.11 Solve Prob. 13.10, assuming that the coefficient of friction between the blocks and the horizontal plane is 0.20.

13.12 Three 20-kg packages rest on a belt which passes over a pulley and is attached to a 40-kg block. Knowing that the coefficient of friction between the belt and the horizontal surface and also between the belt and the packages is 0.50, determine the speed of package B as it falls off the belt at E.

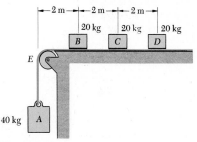

2 m   2 m   2 m

20 kg   20 kg   20 kg

B C D

E

40 kg A

Fig. P13.12

13.13 In Prob. 13.12, determine the speed of package C as it falls off the belt at E.

13.14 Two blocks A and B, of mass 5 kg and 6 kg, respectively, are connected by a cord which passes over pulleys as shown. A collar C is placed on block A and the system is released from rest. After the blocks have moved 0.9 m, the collar C is removed and the blocks continue to move. Knowing that collar C has a mass of 4 kg, determine the speed of block A just before it strikes the ground.

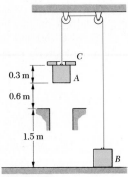

Fig. P13.14

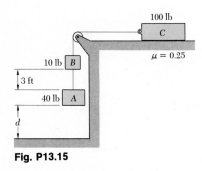

Fig. P13.15

13.15 The system shown is released from rest. After moving a distance d, block A strikes the ground and the cable between A and B becomes slack. Knowing that $d = 5$ ft, determine the velocity of block B as is strikes block A.

13.16 In Prob. 13.15, determine the smallest distance d for which the system will come to rest with block B touching block A.

13.17 Using the method of work and energy, solve Prob. 12.15a.

13.18 Using the method of work and energy, solve Prob. 12.14.

13.19 Using the method of work and energy, solve Prob. 12.17c.

13.20 Using the method of work and energy, solve Prob. 12.16c.

13.21 Two types of energy-absorbing fenders designed to be used on a pier are statically loaded. The force-deflection curve for each type of fender is given in the graph. Determine the maximum deflection of each fender when a 100-ton ship moving at 1 mi/h strikes the fender and is brought to rest.

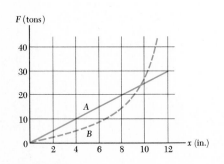

Fig. P13.21

13.22 A 5000-kg airplane lands on an aircraft carrier and is caught by an arresting cable which is characterized by the force-deflection diagram shown. Knowing that the landing speed of the plane is 144 km/h, determine (a) the distance required for the plane to come to rest, (b) the maximum rate of deceleration of the plane.

13.23 A 2-kg block is at rest on a spring of constant 400 N/m. A 4-kg block is held above the 2-kg block so that it just touches it, and released. Determine (a) the maximum velocity attained by the blocks, (b) the maximum force exerted on the blocks.

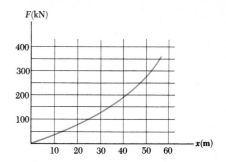

Fig. P13.22

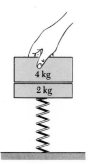

Fig. P13.23

13.24 As the bracket ABC is slowly rotated, the 6-kg block starts to slide toward the spring when $\theta = 15°$. The maximum deflection of the spring is observed to be 50 mm. Determine the values of the coefficients of static and kinetic friction.

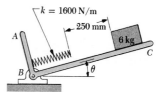

Fig. P13.24

13.25 A 30-ton railroad car was stopped by the bumper shown. Knowing that the maximum deflection of the bumper was 6 in., determine the speed of the car just before it struck the bumper.

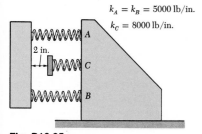

Fig. P13.25

13.26 A railroad car weighing 50,000 lb starts from rest and coasts down a 1-percent incline for a distance of 50 ft. It is stopped by a bumper having a spring constant of 8000 lb/in. (a) What is the speed of the car at the bottom of the incline? (b) How many inches will the spring be compressed?

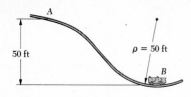

Fig. P13.27 and P13.28

13.27 A roller coaster starts from rest at A and rolls down the track shown. Assuming no energy loss and knowing that the radius of curvature of the track at B is 50 ft, determine the apparent weight of a 150-lb passenger at B.

13.28 A roller coaster is released with no velocity at A and rolls down the track shown. The brakes are suddenly applied as the car passes through point B, causing the wheels of the car to slide on the track ($\mu = 0.25$). Assuming no energy loss between A and B and knowing that the radius of curvature of the track at B is 50 ft, determine the normal and tangential components of the acceleration of the car just after the brakes have been applied.

13.29 A small package of mass m is projected into a vertical return loop at A with a velocity $\mathbf{v_0}$. The package travels without friction along a circle of radius r and is deposited on a horizontal surface at C. For each of the two loops shown, determine (a) the smallest velocity $\mathbf{v_0}$ for which the package will reach the horizontal surface at C, (b) the corresponding force exerted by the loop on the package as it passes point B.

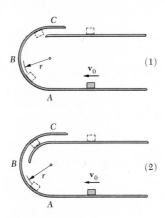

Fig. P13.29

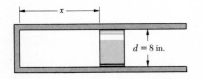

Fig. P13.31

13.30 In Prob. 13.29, it is desired to have the package deposited on the horizontal surface at C with a speed of 2 m/s. Knowing that $r = 0.6$ m, (a) show that this requirement cannot be fulfilled by the first loop, (b) determine the required initial velocity $\mathbf{v_0}$ when the second loop is used.

13.31 An 8-in.-diameter piston weighing 10 lb slides without friction in a cylinder. When the piston is at a distance $x = 14$ in. from the end of the cylinder, the pressure in the cylinder is atmospheric ($p_a = 14.7$ lb/in²). If the pressure varies inversely as the volume, find the work done in moving the piston until $x = 6$ in.

13.32   The piston of Prob. 13.31 is moved to the left and released with no velocity when $x = 6$ in. Neglecting friction, determine (a) the maximum velocity attained by the piston, (b) the maximum value of the coordinate x.

13.33   An object is released with no velocity at an altitude equal to the radius of the earth. Neglecting air resistance, determine the velocity of the object as it strikes the earth. Give the answer in both SI and U.S. customary units.

13.34   A rocket is fired vertically from the ground. Knowing that at burnout the rocket is 80 km above the ground and has a velocity of 5000 m/s, determine the highest altitude it will reach.

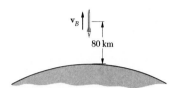

Fig. P13.34 and P13.35

13.35   A rocket is fired vertically from the ground. What should be its velocity $\mathbf{v}_B$ at burnout, 80 km above the ground, if it is to reach an altitude of 1000 km?

13.36   An object is released with no initial velocity at an altitude of 1000 mi. Neglecting air resistance, determine the velocity of the object (a) after it has fallen 500 mi, (b) as it strikes the ground.

13.6. Potential Energy.†   Let us consider again a body of weight $\mathbf{W}$ which moves along a curved path from a point A_1 of elevation y_1 to a point A_2 of elevation y_2 (Fig. 13.4). We recall from Sec. 13.2 that the work of the weight $\mathbf{W}$ during this displacement is

$$U_{1\to2} = Wy_1 - Wy_2 \qquad (13.4)$$

The work of $\mathbf{W}$ may thus be obtained by subtracting the value of the function Wy corresponding to the second position of the body from its value corresponding to the first position. The work of $\mathbf{W}$ is independent of the actual path followed; it depends only upon the initial and final values of the function Wy. This function is called the *potential energy* of the body with respect to the *force of gravity* $\mathbf{W}$ and is denoted by V_g. We write

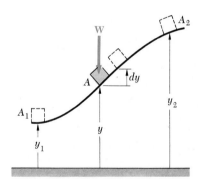

Fig. 13.4 (repeated)

$$U_{1\to2} = (V_g)_1 - (V_g)_2 \quad \text{with } V_g = Wy \qquad (13.14)$$

We note that if $(V_g)_2 > (V_g)_1$, i.e., *if the potential energy increases* during the displacement (as in the case considered here), *the work $U_{1\to2}$ is negative.* If, on the other hand, the work of $\mathbf{W}$ is positive, the potential energy decreases. Therefore, the potential energy V_g of the body provides a measure of the work which may be done by its weight $\mathbf{W}$. Since only the *change* in potential energy,

† Most of the material in this section has already been considered in Sec. 10.6.

and not the actual value of V_g, is involved in formula (13.14), an arbitrary constant may be added to the expression obtained for V_g. In other words, the level, or datum, from which the elevation y is measured may be chosen arbitrarily. Note that potential energy is expressed in the same units as work, i.e., in joules if SI units are used, and in ft · lb or in · lb if U.S. customary units are used.

It should be noted that the expression just obtained for the potential energy of a body with respect to gravity is valid only as long as the weight **W** of the body may be assumed to remain constant, i.e., as long as the displacements of the body are small compared to the radius of the earth. In the case of a space vehicle, however, we should take into consideration the variation of the force of gravity with the distance r from the center of the earth. Using the expression obtained in Sec. 13.2 for the work of a gravitational force, we write (Fig. 13.6)

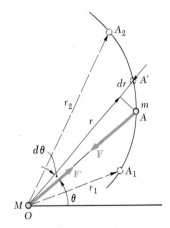

Fig. 13.6 (*repeated*)

$$U_{1 \to 2} = \frac{GMm}{r_2} - \frac{GMm}{r_1} \tag{13.7}$$

The work of the force of gravity may therefore be obtained by subtracting the value of the function $-GMm/r$ corresponding to the second position of the body from its value corresponding to the first position. Thus, the expression which should be used for the potential energy V_g when the variation in the force of gravity cannot be neglected is

$$V_g = -\frac{GMm}{r} \tag{13.15}$$

Taking the first of the relations (12.20) into account, we write V_g in the alternate form

$$V_g = -\frac{WR^2}{r} \tag{13.15'}$$

where R is the radius of the earth and W the value of the weight of the body at the surface of the earth. When either of the relations (13.15) and (13.15') is used to express V_g, the distance r should, of course, be measured from the center of the earth.† Note that V_g is always negative and that it approaches zero for very large values of r.

Consider, now, a body attached to a spring and moving from a position A_1, corresponding to a deflection x_1 of the spring, to a position A_2, corresponding to a deflection x_2 (Fig. 13.5). We

†The expressions given for V_g in (13.15) and (13.15') are valid only when $r \geqq R$, that is, when the body considered is above the surface of the earth.

recall from Sec. 13.2 that the work of the force **F** exerted by the spring on the body is

$$U_{1\to2} = \tfrac{1}{2}kx_1^2 - \tfrac{1}{2}kx_2^2 \qquad (13.6)$$

The work of the elastic force is thus obtained by subtracting the value of the function $\tfrac{1}{2}kx^2$ corresponding to the second position of the body from its value corresponding to the first position. This function is denoted by V_e and is called the *potential energy* of the body with respect to the *elastic force* **F**. We write

$$U_{1\to2} = (V_e)_1 - (V_e)_2 \qquad \text{with } V_e = \tfrac{1}{2}kx^2 \qquad (13.16)$$

and observe that, during the displacement considered, the work of the force **F** exerted by the spring on the body is negative and the potential energy V_e increases. We should note that the expression obtained for V_e is valid only if the deflection of the spring is measured from its undeformed position. On the other hand, formula (13.16) may be used even when the spring is rotated about its fixed end (Fig. 13.10*a*). The work of the elastic

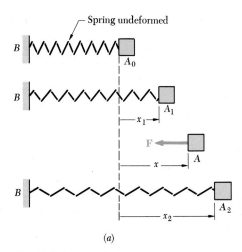

Spring undeformed

(*a*)

Fig. 13.5 (repeated)

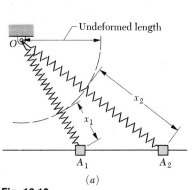

(*a*)

Fig. 13.10

force depends only upon the initial and final deflections of the spring (Fig. 13.10*b*).

The concept of potential energy may be used when forces other than gravity forces and elastic forces are involved. It remains valid as long as the elementary work dU of the force considered is an *exact differential*. It is then possible to find a function V, called potential energy, such that

$$dU = -dV$$

Integrating over a finite displacement, we obtain the general relationship

$$U_{1 \rightarrow 2} = V_1 - V_2 \qquad (13.17)$$

which expresses that *the work of the force is independent of the path followed and is equal to minus the change in potential energy.* A force which satisfies Eq. (13.17) is said to be a *conservative force.*

13.7. Conservation of Energy. We saw in the preceding section that the work of a conservative force, such as the weight of a particle or the force exerted by a spring, may be expressed as a change in potential energy. When a particle, or a system of particles, moves under the action of conservative forces, the principle of work and energy stated in Sec. 13.3 may be expressed in a modified form. Substituting for $U_{1 \rightarrow 2}$ from (13.17) into (13.12), we write

$$V_1 - V_2 = T_2 - T_1$$
$$T_1 + V_1 = T_2 + V_2 \qquad (13.18)$$

Formula (13.18) indicates that, when a system of particles moves under the action of conservative forces, *the sum of the kinetic energy and of the potential energy of the system remains constant.*† The sum $T + V$ is called the *total mechanical energy* of the system and is denoted by E.

Consider, for example, the pendulum analyzed in Sec. 13.4, which is released with no velocity from A_1 and allowed to swing in a vertical plane (Fig. 13.11). Measuring the potential energy from the level of A_2, we have, at A_1,

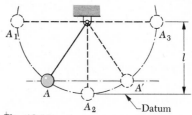

Fig. 13.11

---

† When the particles of a system move with respect to each other under the action of internal forces, the potential energy of the system must include the potential energy corresponding to the internal forces.

$$T_1 = 0 \qquad V_1 = Wl \qquad T_1 + V_1 = Wl$$

Recalling that, at A_2, the speed of the pendulum is $v_2 = \sqrt{2gl}$, we have

$$T_2 = \tfrac{1}{2} mv_2^2 = \frac{1}{2}\frac{W}{g}(2\,gl) = Wl \qquad V_2 = 0$$

$$T_2 + V_2 = Wl$$

We thus check that the total mechanical energy $E = T + V$ of the pendulum is the same at A_1 and A_2. While the energy is entirely potential at A_1, it becomes entirely kinetic at A_2 and, as the pendulum keeps swinging to the right, the kinetic energy is transformed back into potential energy. At A_3, we shall have $T_3 = 0$ and $V_3 = Wl$.

Since the total mechanical energy of the pendulum remains constant and since its potential energy depends only upon its elevation, the kinetic energy of the pendulum will have the same value at any two points located on the same level. Thus, the speed of the pendulum is the same at A and at A' (Fig. 13.11). This result may be extended to the case of a particle moving along any given path, regardless of the shape of the path, as long as the only forces acting on the particle are its weight and the normal reaction of the path. The particle of Fig. 13.12, for example, which slides in a vertical plane along a frictionless track, will have the same speed at A, A', and A''.

While the weight of a particle and the force exerted by a spring are conservative forces, *friction forces are nonconservative forces*. In other words, *the work of a friction force cannot be expressed as a change in potential energy*. The work of a friction force depends upon the path followed by its point of application; and while the work $U_{1 \to 2}$ defined by (13.17) is positive or negative according to the sense of motion, *the work of a friction force is always negative*. It follows that, when a mechanical system involves friction, its total mechanical energy does not remain constant but decreases. The mechanical energy of the system, however, is not lost; it is transformed into heat, and the sum of the *mechanical energy* and of the *thermal energy* of the system remains constant.

Other forms of energy may also be involved in a system. For instance, a generator converts mechanical energy into *electric energy*; a gasoline engine converts *chemical energy* into mechanical energy; a nuclear reactor converts *mass* into thermal energy. If all forms of energy are considered, the energy of any system may be considered as constant and the principle of conservation of energy remains valid under all conditions.

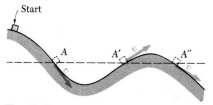

Fig. 13.12

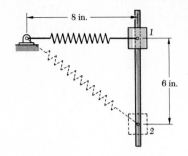

A 20-lb collar slides without friction along a vertical rod as shown. The spring attached to the collar has an undeformed length of 4 in. and a constant of 3 lb/in. If the collar is released from rest in position 1, determine its velocity after it has moved 6 in. to position 2.

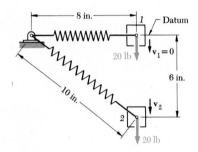

*Position 1. **Potential Energy.*** The elongation of the spring is $x_1 = 8$ in. $- 4$ in. $= 4$ in., and we have

$$V_e = \tfrac{1}{2}kx_1^2 = \tfrac{1}{2}(3 \text{ lb/in.})(4 \text{ in.})^2 = 24 \text{ in} \cdot \text{lb}$$

Choosing the datum as shown, we have $V_g = 0$. Therefore,

$$V_1 = V_e + V_g = 24 \text{ in} \cdot \text{lb} = 2 \text{ ft} \cdot \text{lb}$$

Kinetic Energy. Since the velocity in position 1 is zero, $T_1 = 0$.

*Position 2. **Potential Energy.*** The elongation of the spring is $x_2 = 10$ in. $- 4$ in. $= 6$ in., and we have

$$V_e = \tfrac{1}{2}kx_2^2 = \tfrac{1}{2}(3 \text{ lb/in.})(6 \text{ in.})^2 = 54 \text{ in} \cdot \text{lb}$$
$$V_g = Wy = (20 \text{ lb})(-6 \text{ in.}) = -120 \text{ in} \cdot \text{lb}$$

Therefore,

$$V_2 = V_e + V_g = 54 - 120 = -66 \text{ in} \cdot \text{lb}$$
$$= -5.5 \text{ ft} \cdot \text{lb}$$

Kinetic Energy

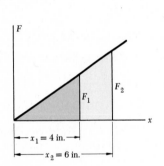

$$T_2 = \tfrac{1}{2}mv_2^2 = \frac{1}{2}\frac{20}{32.2}v_2^2 = 0.311v_2^2$$

Conservation of Energy. Applying the principle of conservation of energy between positions 1 and 2, we write

$$T_1 + V_1 = T_2 + V_2$$
$$0 + 2 \text{ ft} \cdot \text{lb} = 0.311v_2^2 - 5.5 \text{ ft} \cdot \text{lb}$$
$$v_2 = \pm 4.91 \text{ ft/s}$$

$$v_2 = 4.91 \text{ ft/s} \downarrow \quad \blacktriangleleft$$

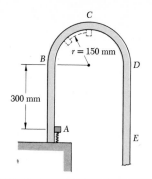

The 200-g pellet is released from rest at A when the spring is compressed 75 mm and travels around the loop $ABCDE$. Determine the smallest value of the spring constant for which the pellet will travel around the loop and will at all times remain in contact with the loop.

Required Speed at Point C. As the pellet passes through the highest point C, its potential energy with respect to gravity is maximum; thus, at the same point its kinetic energy and its speed are minimum. Since the pellet must remain in contact with the loop, the force **N** exerted on the pellet by the loop must be equal to, or greater than, zero. Setting **N** = 0, we compute the smallest possible speed v_C.

$$+\downarrow \Sigma F_n = ma_n: \qquad W = ma_n \qquad mg = ma_n \qquad a_n = g$$

$$a_n = \frac{v_C^2}{r}: \qquad v_C^2 = ra_n = rg = (0.150 \text{ m})(9.81 \text{ m/s}^2) = 1.472 \text{ m}^2/\text{s}^2$$

Position _1_. Potential Energy. Since the spring is compressed 0.075 m from its undeformed position, we have

$$V_e = \tfrac{1}{2}kx^2 = \tfrac{1}{2}k(0.075 \text{ m})^2 = (0.00281 \text{ m}^2)k$$

Choosing the datum at A, we have $V_g = 0$; therefore

$$V_1 = V_e + V_g = (0.00281 \text{ m}^2)k$$

Kinetic Energy. Since the pellet is released from rest, $v_A = 0$ and we have $T_1 = 0$.

Position _2_. Potential Energy. The spring is now undeformed; thus $V_e = 0$. Since the pellet is 0.450 m above the datum, and since $W = (0.200 \text{ kg})(9.81 \text{ m/s}^2) = 1.962 \text{ N}$, we have

$$V_g = Wy = (1.962 \text{ N})(0.450 \text{ m}) = 0.883 \text{ N} \cdot \text{m} = 0.883 \text{ J}$$
$$V_2 = V_e + V_g = 0.883 \text{ J}$$

Kinetic Energy. Using the value of v_C^2 obtained above, we write

$$T_2 = \tfrac{1}{2}mv_C^2 = \tfrac{1}{2}(0.200 \text{ kg})(1.472 \text{ m}^2/\text{s}^2) = 0.1472 \text{ N} \cdot \text{m} = 0.1472 \text{ J}$$

Conservation of Energy. Applying the principle of conservation of energy between positions _1_ and _2_, we write

$$T_1 + V_1 = T_2 + V_2$$
$$0 + (0.00281 \text{ m}^2)k = 0.1472 \text{ J} + 0.883 \text{ J}$$
$$k = 367 \text{ J/m}^2 = 367 \text{ N/m}$$

The required minimum value of k is therefore

$$k = 367 \text{ N/m} \quad \blacktriangleleft$$

PROBLEMS

13.37 The spring AB is of constant k and is unstretched when $\theta = 0$. Determine the potential energy of the system with respect (a) to the spring, (b) to gravity. (Place datum at C.)

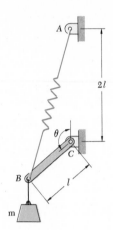

Fig. P13.37

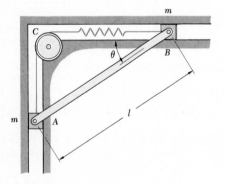

Fig. P13.38

13.38 A slender rod AB of negligible mass is attached to blocks A and B, each of mass m. The constant of the spring is k and the spring is undeformed when AB is horizontal. Determine the potential energy of the system with respect (a) to the spring, (b) to gravity. (Place datum at B.)

13.39 The spring AB is of constant 3 lb/in. and is attached to the 2-lb collar B which moves freely along the horizontal rod. The unstretched length of the spring is 5 in. If the collar is released from rest in the position shown, determine the maximum velocity attained by the collar.

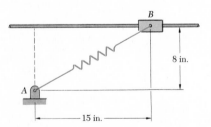

Fig. P13.39

13.40 In Prob. 13.39, determine the weight of the collar B for which the maximum velocity is 15 ft/s.

13.41 A collar of mass 1.5 kg is attached to a spring and slides without friction along a circular rod which lies in a *horizontal* plane. The spring is undeformed when the collar is at C and the constant of the spring is 400 N/m. If the collar is released from rest at B, determine the velocity of the collar as it passes through point C.

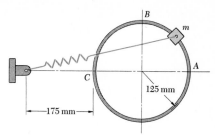

Fig. P13.41

13.42 A toy spring gun is used to shoot bullets of mass 30 g vertically upward. The undeformed length of the spring is 125 mm; it is compressed to a length of 25 mm when the gun is ready to be shot and expands to a length of 75 mm as the bullet leaves the gun. A force of 35 N is required to maintain the spring in firing position when the length of the spring is 25 mm. Determine (a) the velocity of the bullet as it leaves the gun, (b) the maximum height reached by the bullet.

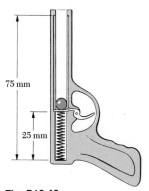

Fig. P13.42

13.43 The collar of Prob. 13.41 has a continuous, although non-uniform, motion along the rod. If the speed of the collar at A is to be half of its speed at C, determine (a) the required speed at C, (b) the corresponding speed at B.

13.44 An elastic cord is stretched between two points A and B, located 16 in. apart in the same horizontal plane. When stretched directly between A and B, the tension in the cord is 10 lb. The cord is then stretched as shown until its midpoint C has moved through 6 in. to C'; a force of 60 lb is required to hold the cord at C'. A 0.20-lb pellet is placed at C', and the cord is released. Determine the speed of the pellet as it passes through C.

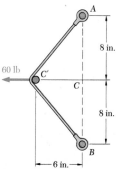

Fig. P13.44

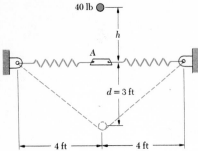

Fig. P13.45

13.45 Two springs are attached to a piece of cloth A of negligible weight, as shown. The initial tension in each spring is 120 lb, and the spring constant of each spring is $k = 10$ lb/in. A 40-lb ball is released from a height h above A; the ball hits the cloth, causing it to move through a maximum distance $d = 3$ ft, as shown by the dashed lines. Determine the height h.

13.46 The tension in the spring is zero when the arm ABC is horizontal ($\phi = 0$). If the 50-kg block is released when $\phi = 0$, determine the speed of the block (a) when $\phi = 90°$, (b) when $\phi = 180°$.

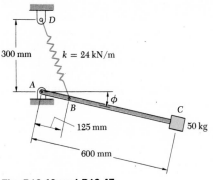

Fig. P13.46 and P13.47

13.47 The 50-kg block is released from rest when $\phi = 0$. If the speed of the block when $\phi = 90°$ is to be 2.5 m/s, determine the required value of the initial tension in the spring.

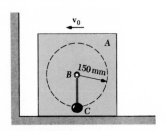

Fig. P13.48

13.48 The sphere C and the block A are both moving to the left with a velocity v_0 when the block is suddenly stopped by the wall. Determine the smallest velocity v_0 for which the sphere C will swing in a full circle about the pivot B (a) if BC is a slender rod of negligible weight, (b) if BC is a cord.

13.49 The collar of Prob. 13.41 is released from rest at point A. Determine the horizontal component of the force exerted by the rod on the collar as the collar passes through point B. Show that the force is independent of the mass of the collar.

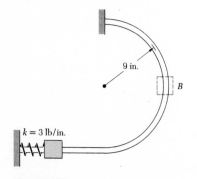

Fig. P13.50

13.50 A $\frac{1}{4}$-lb collar may slide without friction on a rod in a vertical plane. The collar is released from rest when the spring is compressed 1.5 in. As the collar passes through point B, determine (a) the speed of the collar, (b) the force exerted by the rod on the collar.

13.51 The pendulum shown is released from rest at A and swings through 90° before the cord touches the fixed peg B. Determine the smallest value of a for which the pendulum bob will describe a circle about the peg.

13.52 A small block is released at A with zero velocity and moves along the frictionless guide to point B where it leaves the guide with a horizontal velocity. Knowing that $h = 8$ ft and $b = 3$ ft, determine (a) the speed of the block as it strikes the ground at C, (b) the corresponding distance c.

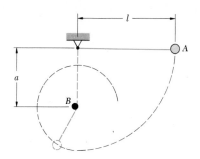

Fig. P13.51

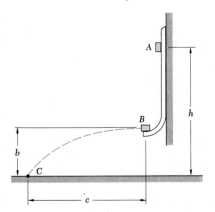

Fig. P13.52

13.53 Assuming a given height h in Prob. 13.52, (a) show that the speed at C is independent of the height b, (b) determine the height b for which the distance c is maximum and the corresponding value of c.

13.54 A bag is gently pushed off the top of a wall at A and swings in a vertical plane at the end of a 4-m rope which can withstand a maximum tension equal to twice the weight of the bag. (a) Determine the difference in elevation h between point A and point B where the rope will break. (b) How far from the vertical wall will the bag strike the floor?

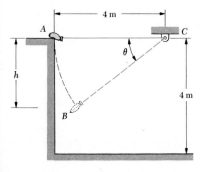

Fig. P13.54

13.55 A delicate instrument weighing 8 lb is placed on a spring of length l so that its base is just touching the undeformed spring. The instrument is then inadvertently released from that position. Determine the maximum deflection x of the spring and the maximum force exerted by the spring if the constant of the spring is $k = 10$ lb/in.

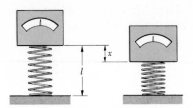

Fig. P13.55

13.56 Nonlinear springs are classified as hard or soft, depending upon the curvature of their force-deflection curves (see figure). Solve Prob. 13.55, assuming (a) that a hard spring is used, for which $F = 10x(1 + 0.1x^2)$, (b) that a soft spring is used, for which $F = 10x(1 - 0.1x^2)$.

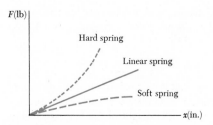

Fig. P13.56

13.57 Determine the escape velocity of a missile, i.e., the velocity with which it should be fired from the surface of the earth if it is to reach an infinite distance from the earth. Give the answer in both SI and U.S. customary units. Show that the result obtained is independent of the firing angle.

13.58 How much energy per kilogram should be imparted to a satellite in order to place it in a circular orbit at an altitude of (a) 500 km, (b) 5000 km?

13.59 A lunar excursion module (LEM) has been used in the Apollo moon-landing missions to save fuel by making it unnecessary to launch the entire Apollo spacecraft from the moon's surface on its return trip to the earth. Check the effectiveness of this approach by computing the energy per pound required for a spacecraft to escape the gravitational field of the moon if the spacecraft starts (a) from the moon's surface, (b) from a circular orbit 60 mi above the moon's surface. Neglect the effect of the earth's gravitational field. (The radius of the moon is 1080 mi and its mass is 0.01230 times the mass of the earth.)

13.60 A satellite of mass m describes a circular orbit around the earth. Express (a) its potential energy, (b) its kinetic energy, and (c) its total energy as a function of the radius r of the orbit. Denote by R the radius of the earth, by g the acceleration of gravity at the surface of the earth, and assume the potential energy of the satellite to be zero on the launching pad.

13.8. Power and Efficiency. *Power* is defined as the time rate at which work is done. In the selection of a motor or engine, power is a much more important criterion than the actual amount of work to be performed. A small motor or a large power plant may both be used to do a given amount of work; but the small motor may require a month to do the work done by the power plant in a matter of minutes. If ΔU is the work done during the time interval Δt, then the average power during this time interval is

$$\text{Average power} = \frac{\Delta U}{\Delta t}$$

Letting Δt approach zero, we obtain at the limit

$$\text{Power} = \frac{dU}{dt} \tag{13.19}$$

Since $dU = (F \cos \alpha)\, ds$ and $v = ds/dt$, the power may also be expressed as follows:

$$\text{Power} = \frac{dU}{dt} = \frac{(F \cos \alpha)\, ds}{dt}$$

$$\text{Power} = (F \cos \alpha)v \tag{13.20}$$

where v = magnitude of velocity of point of application of force **F**

α = angle between **F** and velocity **v**

Since power was defined as the time rate at which work is done, it should be expressed in units obtained by dividing units of work by the unit of time. Thus, if SI units are used, power should be expressed in J/s; this unit is called a *watt* (W). We have

$$1\,W = 1\,J/s = 1\,N \cdot m/s$$

If U.S. customary units are used, power should be expressed in ft · lb/s or in *horsepower* (hp), with the latter defined as

$$1\,hp = 550\,ft \cdot lb/s$$

Recalling from Sec. 13.2 that 1 ft · lb = 1.356 J, we verify that

$$1\,ft \cdot lb/s = 1.356\,J/s = 1.356\,W$$
$$1\,hp = 550(1.356\,W) = 746\,W = 0.746\,kW$$

The *mechanical efficiency* of a machine was defined in Sec. 10.4 as the ratio of the output work to the input work:

$$\eta = \frac{\text{output work}}{\text{input work}} \tag{13.21}$$

This definition is based on the assumption that work is done at a constant rate. The ratio of the output to the input work is therefore equal to the ratio of the rates at which output and input work are done, and we have

$$\eta = \frac{\text{power output}}{\text{power input}} \tag{13.22}$$

Because of energy losses due to friction, the output work is always smaller than the input work, and, consequently, the power output is always smaller than the power input. The mechanical efficiency of a machine, therefore, is always less than 1.

When a machine is used to transform mechanical energy into electric energy, or thermal energy into mechanical energy, its *overall efficiency* may be obtained from formula (13.22). The overall efficiency of a machine is always less than 1; it provides a measure of all the various energy losses involved (losses of electric or thermal energy as well as frictional losses). We should note that it is necessary, before using formula (13.22), to express the power output and the power input in the same units.

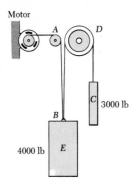

Motor

A D

C 3000 lb

B

4000 lb E

The elevator shown weighs 4000 lb when fully loaded. It is connected to a 3000-lb counterweight C and is powered by an electric motor. Determine the power required when the elevator (*a*) is moving upward at a constant speed of 20 ft/s, (*b*) has an instantaneous velocity of 20 ft/s upward and an upward acceleration of 3 ft/s².

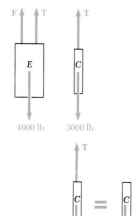

F T T

E C

4000 lb   3000 lb

T

C = C

          $m_C a$

3000 lb

F T

E = E $m_E a$

4000 lb

Solution.   Since **F** and **v** have the same direction, the power is equal to Fv. We must first determine the force **F** exerted by cable AB on the elevator in each of the two given situations.

Force F.   The forces acting on the elevator and on the counterweight are shown in the adjoining sketches.

a. Uniform Motion.   We have $a = 0$; both bodies are in equilibrium.

Free Body C: $+\uparrow\Sigma F_y = 0$: $T - 3000\ \text{lb} = 0$
Free Body E: $+\uparrow\Sigma F_y = 0$: $F + T - 4000\ \text{lb} = 0$

Eliminating T, we find: $F = 1000\ \text{lb}$

b. Accelerated Motion.   We have $a = 3\ \text{ft/s}^2$. The equations of motion are

Free Body C: $+\downarrow\Sigma F_y = m_C a$: $3000 - T = \dfrac{3000}{32.2}\,3$

Free Body E: $+\uparrow\Sigma F_y = m_E a$: $F + T - 4000 = \dfrac{4000}{32.2}\,3$

Eliminating T: $F = 1000 + \dfrac{7000}{32.2}\,3 = 1652\ \text{lb}$

Power.   Substituting the given values of v and the values found for F into the expression for the power, we have

a. $Fv = (1000\ \text{lb})(20\ \text{ft/s}) = 20{,}000\ \text{ft} \cdot \text{lb/s}$

$$\text{Power} = (20{,}000\ \text{ft} \cdot \text{lb/s})\frac{1\ \text{hp}}{550\ \text{ft} \cdot \text{lb/s}} = 36.4\ \text{hp} \ \blacktriangleleft$$

b. $Fv = (1652\ \text{lb})(20\ \text{ft/s}) = 33{,}040\ \text{ft} \cdot \text{lb/s}$

$$\text{Power} = \frac{33{,}040}{550} = 60.1\ \text{hp} \ \blacktriangleleft$$

PROBLEMS

13.61 A 70-kg man and an 80-kg man run up a flight of stairs in 5 s. If the flight of stairs is 4 m high, determine the average power required by each man.

13.62 A utility hoist can lift its maximum allowable load of 7000 lb at the rate of 65 ft/min. Knowing that the hoist is run by a 20-hp engine, determine the overall efficiency of the hoist.

13.63 A 3500-lb automobile starts from rest and accelerates uniformly to a speed of 60 mi/h in 30 s. Knowing that the rolling resistance of the automobile is 100 lb/ton and that the road is horizontal, determine the power required as a function of time.

13.64 A 1500-kg automobile travels 200 m while being accelerated at a uniform rate from 50 to 75 km/h. During the entire motion, the automobile is traveling on a horizontal road, and the rolling resistance is equal to 2 percent of the weight of the automobile. Determine (a) the maximum power required, (b) the power required to maintain a constant speed of 75 km/h.

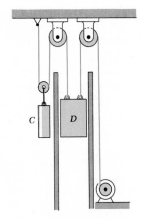

C   D

Fig. P13.65 and P13.66

13.65 The dumbwaiter D and its counterweight C weigh 600 lb each. Determine the power required when the dumbwaiter (a) is moving upward at a constant speed of 15 ft/s, (b) has an instantaneous velocity of 15 ft/s upward and an upward acceleration of 2 ft/s^2.

13.66 The dumbwaiter D and its counterweight C weigh 600 lb each. Knowing that the motor is delivering to the system 10 hp at the instant the speed of the dumbwaiter is 15 ft/s upward, determine the acceleration of the dumbwaiter.

13.67 A chair-lift is designed to transport 900 skiers per hour from the base A to the summit B. The average mass of a skier is 75 kg, and the average speed of the lift is 80 m/min. Determine (a) the average power required, (b) the required capacity of the motor if the mechanical efficiency is 85 percent and if a 300-percent overload is to be allowed.

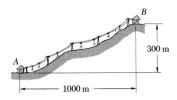

B

300 m

A

1000 m

Fig. P13.67

13.68 Crushed stone is moved from a quarry at A to a construction site at B at the rate of 2000 Mg per 8-h period. An electric generator is attached to the system in order to maintain a constant belt speed. Knowing that the efficiency of the belt-generator system is 0.65, determine the average power developed by the generator (a) if the belt speed is 0.75 m/s, (b) if the belt speed is 2 m/s.

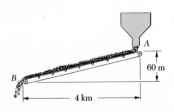

A

60 m

B

4 km

Fig. P13.68

13.69 The fluid transmission of a truck of mass m permits the engine to deliver an essentially constant power P to the driving wheels. Determine the time elapsed and the distance traveled as the speed is increased from v_0 to v_1.

13.70 The fluid transmission of a 20-ton truck permits the engine to deliver an essentially constant power of 100 hp to the driving wheels. Determine the time required and the distance traveled as the speed of the truck is increased (*a*) from 30 to 45 mi/h, (*b*) from 45 to 60 mi/h.

13.71 The frictional resistance of a ship is known to vary directly as the 1.75 power of the speed *v* of the ship. A single tugboat at full power can tow the ship at a constant speed of 5 km/h by exerting a constant force of 200 kN. Determine (*a*) the power developed by the tugboat, (*b*) the maximum speed at which two tugboats, capable of delivering the same power, can tow the ship.

13.72 Determine the speed at which the single tugboat of Prob. 13.71 will tow the ship if the tugboat is developing half of its maximum power.

REVIEW PROBLEMS

13.73 Two cylinders are suspended from an inextensible cable as shown. If the system is released from rest, determine the maximum velocity attained by the 10-lb cylinder.

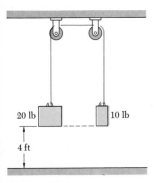

20 lb   10 lb

4 ft

Fig. P13.73

13.74 In Prob. 13.73 determine the maximum height above the floor to which the 10-lb cylinder will rise.

13.75 Starting from rest, a 1500-kg automobile is accelerated at a constant rate of 1.25 m/s² until it reaches a speed of 60 km/h and then travels at a constant speed. During the entire motion, the automobile is traveling on a horizontal road, and the rolling resistance is equal to 1.8 percent of the weight of the automobile. Determine the power required as a function of time.

13.76 Two blocks *A* and *B* connected by a cord are released from rest in the position shown. Neglecting the effect of friction, determine the velocity of block *B* (*a*) as it leaves the table, (*b*) as it strikes the floor.

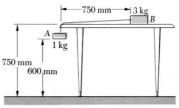

750 mm ──── 3 kg
                        B
A
    1 kg
750 mm
  600 mm

Fig. P13.76

13.77 An elevator travels upward at a constant speed of 2 m/s. A boy riding the elevator throws a 0.8-kg stone upward with a speed of 4 m/s *relative* to the elevator. Determine (*a*) the work done by the boy in throwing the stone, (*b*) the difference in the values of the kinetic energy of the stone before and after it was thrown. (*c*) Why are the values obtained in parts *a* and *b* not the same?

13.78 A ball of mass m attached to an inextensible cord rotates in a vertical circle of radius r. Show that the difference between the maximum value T_{max} of the tension in the cord and its minimum value T_{min} is independent of the speed at which the ball rotates, and determine $T_{max} - T_{min}$.

13.79 The escalator shown is designed to transport 9000 persons per hour at a constant speed of 90 ft/min. Assuming an average weight of 150 lb per person, determine (*a*) the average power required, (*b*) the required capacity of the motor if the mechanical efficiency is 75 percent and if a 250-percent overload is to be allowed.

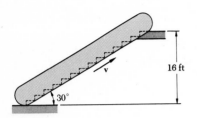

16 ft

30°

Fig. P13.79

13.80 Two blocks are joined by an inextensible cable as shown. If the system is released from rest, determine the velocity of block A after it has moved 2 m. Assume that μ equals 0.25 between block A and the plane and neglect the mass and friction of the pulleys.

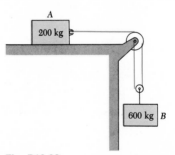

A

200 kg

600 kg B

Fig. P13.80

13.81 The 0.5-lb pellet is released when the spring is compressed 6 in. and travels without friction around the vertical loop *ABCD*. Determine the force exerted by the loop on the pellet (*a*) at point A, (*b*) at point B, (*c*) at point C.

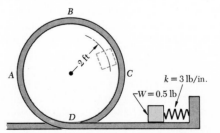

B

2 ft

A

C

$k = 3$ lb/in.

W = 0.5 lb

D

Fig. P13.81

13.82 In Prob. 13.81, determine the smallest allowable deflection of the spring if the pellet is to travel around the entire loop without leaving the track.

13.83 A 5-kg collar slides without friction along a rod which forms an angle of 30° with the vertical. The spring is unstretched when the collar is at A. If the collar is released from rest at A, determine the value of the spring constant k for which the collar has zero velocity at B.

13.84 In Prob. 13.83, determine the value of the spring constant k for which the velocity of the collar at B is 1.5 m/s.

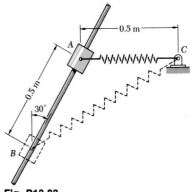

Fig. P13.83

Chapter

14

Kinetics of Particles: Impulse and Momentum

14.1. Principle of Impulse and Momentum. A third basic method for the solution of problems dealing with the motion of particles will be considered in this chapter. This method is based on the principle of impulse and momentum and is of particular interest in problems involving force, mass, velocity, and time.

Consider a particle of mass m acted upon by a force $\mathbf{F}$. Expressing Newton's second law $\mathbf{F} = m\mathbf{a}$ in terms of the x and y components of the force and of the acceleration, we write

$$F_x = ma_x \qquad F_y = ma_y$$

$$F_x = m\frac{dv_x}{dt} \qquad F_y = m\frac{dv_y}{dt}$$

Since the mass m of the particle is constant, we have

$$F_x = \frac{d}{dt}(mv_x) \qquad F_y = \frac{d}{dt}(mv_y) \tag{14.1}$$

or, in vector form,

$$\mathbf{F} = \frac{d}{dt}(m\mathbf{v}) \tag{14.2}$$

The vector $m\mathbf{v}$ is called the *linear momentum*, or simply the *momentum*, of the particle. It has the same direction as the velocity of the particle and its magnitude is expressed in $N \cdot s$ or in $lb \cdot s$. We check that, with SI units,

$$mv = (kg)(m/s) = (kg \cdot m/s^2)s = N \cdot s$$

while, with U.S. customary units,

$$mv = (lb \cdot s^2/ft)(ft/s) = lb \cdot s$$

Equation (14.2) expresses that *the force $\mathbf{F}$ acting on the particle is equal to the rate of change of the momentum of the particle.* It is in this form that the second law of motion was originally stated by Newton.

Multiplying both sides of Eqs. (14.1) by dt and integrating from a time t_1 to a time t_2, we write

$$F_x \, dt = d(mv_x) \qquad F_y \, dt = d(mv_y)$$

$$\int_{t_1}^{t_2} F_x \, dt = (mv_x)_2 - (mv_x)_1 \qquad \int_{t_1}^{t_2} F_y \, dt = (mv_y)_2 - (mv_y)_1$$

$$(mv_x)_1 + \int_{t_1}^{t_2} F_x \, dt = (mv_x)_2 \qquad (mv_y)_1 + \int_{t_1}^{t_2} F_y \, dt = (mv_y)_2$$

$$(14.3)$$

or, in vector form,

$$m\mathbf{v}_1 + \int_{t_1}^{t_2} \mathbf{F} \, dt = m\mathbf{v}_2 \qquad (14.4)$$

The integral in Eq. (14.4) is a vector known as the *linear impulse,* or simply the *impulse,* of the force $\mathbf{F}$ during the interval of time considered. From Eqs. (14.3), we note that the components of the impulse of the force $\mathbf{F}$ are equal, respectively, to the areas under the curves obtained by plotting the components F_x and F_y against t (Fig. 14.1). In the case of a force $\mathbf{F}$ of constant magnitude and direction, the impulse is represented by the vector $\mathbf{F}(t_2 - t_1)$, which has the same direction as $\mathbf{F}$. We easily check that the magnitude of the impulse of a force is expressed in $N \cdot s$ or in $lb \cdot s$.

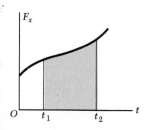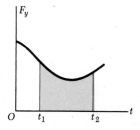

Fig. 14.1

Equation (14.4) expresses that, when a particle is acted upon by a force **F** during a given time interval, *the final momentum* $m\mathbf{v}_2$ *of the particle may be obtained by adding vectorially its initial momentum* $m\mathbf{v}_1$ *and the impulse of the force* **F** *during the time interval considered* (Fig. 14.2). We write

$$m\mathbf{v}_1 + \mathbf{Imp}_{1\rightarrow 2} = m\mathbf{v}_2 \tag{14.5}$$

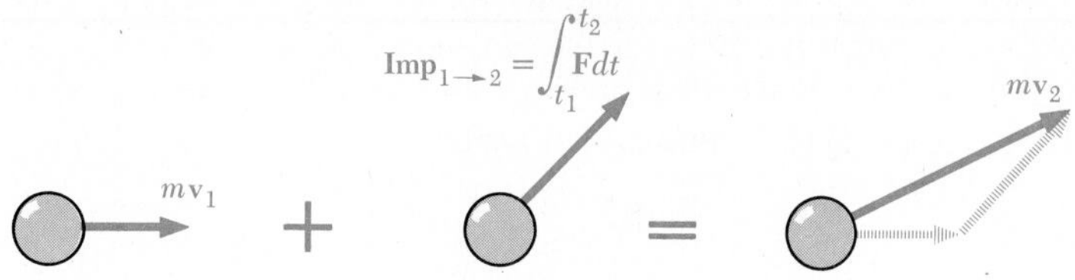

Fig. 14.2

We note that, while kinetic energy and work are scalar quantities, momentum and impulse are vector quantities. To obtain an analytic solution, it is thus necessary to replace Eq. (14.5) by the two corresponding component equations (14.3).

When several forces act on a particle, the impulse of each of the forces must be considered. We have

$$m\mathbf{v}_1 + \Sigma\, \mathbf{Imp}_{1\rightarrow 2} = m\mathbf{v}_2 \tag{14.6}$$

Again, the equation obtained represents a relation between vector quantities; in the actual solution of a problem, it should be replaced by the two corresponding component equations.

14.2. Impulsive Motion. In some problems, a very large force may act during a very short time interval on a particle and produce a definite change in momentum. Such a force is called an *impulsive force* and the resulting motion an *impulsive motion*. For example, when a baseball is struck, the contact between bat and ball takes place during a very short time interval Δt. But the average value of the force **F** exerted by the bat on the ball is very large, and the resulting impulse $\mathbf{F}\,\Delta t$ is large enough to change the sense of motion of the ball (Fig. 14.3).

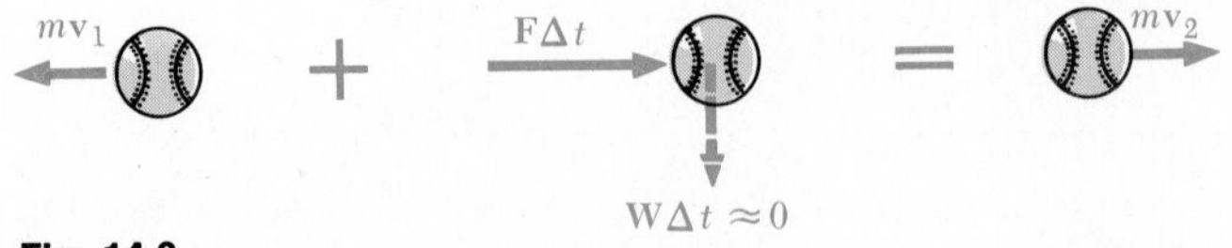

Fig. 14.3

When impulsive forces act on a particle, Eq. (14.6) becomes

$$m\mathbf{v}_1 + \Sigma \mathbf{F}\,\Delta t = m\mathbf{v}_2 \tag{14.7}$$

Any force which is not an impulsive force may be neglected, since the corresponding impulse $\mathbf{F}\,\Delta t$ is very small. *Nonimpulsive forces* include the weight of the body, the force exerted by a spring, or any other force which is *known* to be small compared with an impulsive force. Unknown reactions may or may not be impulsive; their impulse should therefore be included in Eq. (14.7) as long as it has not been proved negligible. The impulse of the weight of the baseball considered above, for example, may be neglected. If the motion of the bat is analyzed, the impulse of the weight of the bat may also be neglected. The impulses of the reactions of the player's hands on the bat, however, should be included; these impulses will not be negligible if the ball is incorrectly hit.

14.3. Systems of Particles. When a problem involves the motion of several particles, each particle may be considered separately and Eq. (14.6) may be written for each particle. We may also add vectorially the momenta of all the particles and the impulses of all the forces involved. We write then

$$\Sigma m\mathbf{v}_1 + \Sigma\,\mathbf{Imp}_{1\to2} = \Sigma m\mathbf{v}_2$$

But since the internal forces occur by pairs of equal and opposite forces having the same line of action, their sum is zero and we need consider only the impulses of the external forces. We have

$$\Sigma m\mathbf{v}_1 + \Sigma\,\mathbf{Ext\ Imp}_{1\to2} = \Sigma m\mathbf{v}_2 \tag{14.8}$$

Formula (14.8) may be written in a modified form if the mass center of the system of particles is considered. From Sec. 12.5, we recall that the mass center of the system is the point G of coordinates $\bar{x}$ and $\bar{y}$ defined by the equations

$$(\Sigma m)\bar{x} = \Sigma mx \qquad (\Sigma m)\bar{y} = \Sigma my \tag{14.9}$$

Differentiating (14.9) with respect to t, we have

$$(\Sigma m)\bar{v}_x = \Sigma mv_x \qquad (\Sigma m)\bar{v}_y = \Sigma mv_y \tag{14.10}$$

Writing (14.8) in terms of x and y components, and substituting for Σmv_x and Σmv_y from (14.10), we obtain

$$(\Sigma m)(\bar{v}_x)_1 + \Sigma \int_{t_1}^{t_2} (F_x)_{ext}\,dt = (\Sigma m)(\bar{v}_x)_2$$

$$(\Sigma m)(\bar{v}_y)_1 + \Sigma \int_{t_1}^{t_2} (F_y)_{ext}\,dt = (\Sigma m)(\bar{v}_y)_2 \tag{14.11}$$

We note that Eqs. (14.11) are identical with the equations we would obtain for a particle of mass Σm acted upon by all the external forces. We thus check that the mass center of a system of particles moves as if the entire mass of the system and all the external forces were concentrated at that point.

The equations derived in this section relate only the components of the impulses and momenta of a system of particles. By considering the *moments* as well as the *components* of the vectors involved, we shall obtain in Sec. 14.12 a more meaningful formulation of the principle of impulse and momentum for a system of particles.

14.4. Conservation of Momentum. We shall consider now the motion of a system of particles when the sum of the impulses of the external forces is zero. Equation (14.8) reduces then to

$$\Sigma m v_1 = \Sigma m v_2 \qquad (14.12)$$

Thus, *when the sum of the impulses of the external forces acting on a system of particles is zero, the total momentum of the system remains constant.*

Introducing again the mass center G of the system, and using formulas (14.10), Eq. (14.12) reduces to

$$\bar{v}_1 = \bar{v}_2 \qquad (14.13)$$

Thus, when the sum of the impulses of the external forces acting on a system of particles is zero, *the mass center of the system moves with a constant velocity.*

Two distinct cases of conservation of momentum are frequently encountered:

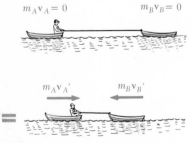

$m_A v_A = 0$ $m_B v_B = 0$

$m_A v_A'$ $m_B v_B'$

Fig. 14.4

1. *The external forces acting on the system during the interval of time considered are balanced.* No matter how long the time interval is, we have Σ **Ext Imp**$_{1\to2} = 0$ and formula (14.12) applies. Consider, for example, two boats, of mass m_A and m_B, initially at rest, which are being pulled together (Fig. 14.4). If the resistance of the water is neglected, the only external forces acting on the boats are their weights and the buoyant forces exerted on them. Since these forces are balanced, we write

$$\Sigma m v_1 = \Sigma m v_2$$
$$0 = m_A v_A' + m_B v_B'$$

where v_A' and v_B' represent the velocities of the boats after a finite interval of time. The equation obtained indicates that the boats move in opposite directions (toward each other) with velocities inversely proportional to their masses. We also note that the mass center of the two boats, which was initially at rest, remains in the same position.

2. *The interval of time considered is very short, and all the external forces are nonimpulsive.* Again we have Σ **Ext Imp**$_{1\to2}$ = 0, and formula (14.12) applies. Consider, for example, a bullet of mass m_A fired with a velocity $\mathbf{v}_A$ into a wooden sphere of mass m_B suspended from an inextensible wire and initially at rest (Fig. 14.5). The bullet penetrates the sphere and imparts to it a velocity $\mathbf{v}'$, which we propose to determine. After the sphere has been hit, the tension in the wire becomes zero and the combined weight of the sphere and bullet is unbalanced. But the impulse of the weight may be neglected since the time interval is very short. We write, therefore,

Fig. 14.5

$$\Sigma m\mathbf{v}_1 = \Sigma m\mathbf{v}_2$$
$$m_A\mathbf{v}_A + 0 = (m_A + m_B)\mathbf{v}'$$

The equation obtained may be solved for $\mathbf{v}'$; we note that $\mathbf{v}'$ will have the same direction as $\mathbf{v}_A$.

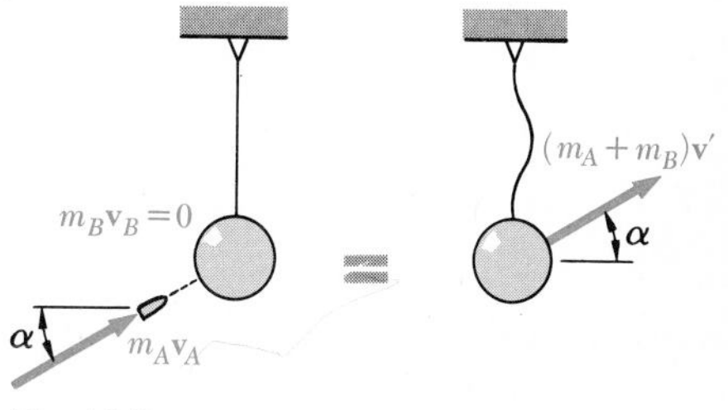

Let us now consider the case when the bullet is fired downward into the sphere (Fig. 14.6). Since the wire is inextensible, it will prevent any downward motion of the sphere and exert on it a reaction **P**. This reaction is unknown and should therefore be assumed impulsive until proved to the contrary. Writing the general formula (14.8) in terms of horizontal and vertical components, and observing again that the impulse of the weight is negligible, we have

$$\Sigma m\mathbf{v}_1 + \Sigma \textbf{ Ext Imp}_{1\to2} = \Sigma m\mathbf{v}_2$$

$\xrightarrow{+}$ x components: $\quad m_A v_A \cos\alpha + 0 = (m_A + m_B)v'$

$+\uparrow y$ components: $\quad -m_A v_A \sin\alpha + P\,\Delta t = 0$

The first equation expresses that *the x component of the momentum is conserved;* it may be used to determine v'. The second equation indicates that the *y component of the linear momentum is not conserved;* this equation may be used to determine the magnitude $P\,\Delta t$ of the impulse of the force exerted by the wire. Thus, the momentum of the system considered is not conserved, and Eq. (14.12) does not hold, except in the x direction.

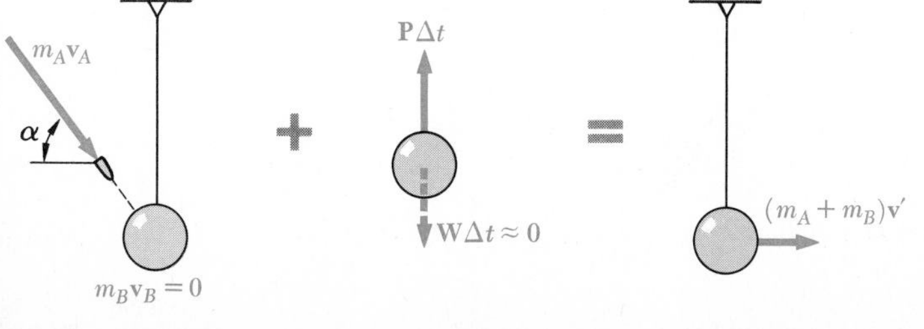

Fig. 14.6

SAMPLE PROBLEM 14.1

An automobile weighing 4000 lb is driven down a 5° incline at a speed of 60 mi/h when the brakes are applied, causing a constant total braking force (applied by the road on the tires) of 1500 lb. Determine the time required for the automobile to come to a stop.

Solution. We apply the principle of impulse and momentum. Since each force is constant in magnitude and direction, each corresponding impulse is equal to the product of the force and of the time interval t.

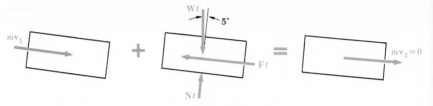

$$m\mathbf{v}_1 + \Sigma \, \mathbf{Imp}_{1\rightarrow2} = m\mathbf{v}_2$$

$+ \searrow x$ components: $mv_1 + (W \sin 5°)t - Ft = 0$

$$(4000/32.2)(88 \text{ ft/s}) + (4000 \sin 5°)t - 1500t = 0$$

$$t = 9.49 \text{ s} \quad \blacktriangleleft$$

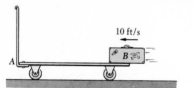

SAMPLE PROBLEM 14.2

An airline employee tosses a 30-lb suitcase with a horizontal velocity of 10 ft/s onto a 70-lb baggage carrier. Knowing that the carrier can roll freely and is initially at rest, determine the velocity of the carrier after the suitcase has slid to a relative stop on the carrier.

Solution. We apply the principle of impulse and momentum to the carrier-suitcase system. Since the impulses of the internal forces cancel out, and since there are no horizontal external forces, the total momentum of the carrier and suitcase is conserved.

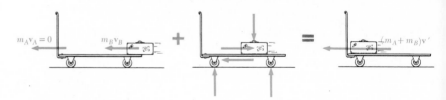

$$m_A\mathbf{v}_A + m_B\mathbf{v}_B = (m_A + m_B)\mathbf{v}'$$

$\xleftrightarrow{} x$ components: $0 + \dfrac{30}{g}(10 \text{ ft/s}) = \dfrac{70 + 30}{g}v'$

$$v' = 3 \text{ ft/s} \quad \blacktriangleleft$$

SAMPLE PROBLEM 14.3

An old 2000-kg gun fires a 10-kg shell with an initial velocity of 600 m/s at an angle of 30°. The gun rests on a horizontal surface and is free to move horizontally. Assuming that the barrel of the gun is rigidly attached to the frame (no recoil mechanism) and that the shell leaves the barrel 6 ms after firing, determine the recoil velocity of the gun and the resultant **R** of the vertical impulsive forces exerted by the ground on the gun.

Solution. We apply the principle of impulse and momentum to the system consisting of the gun and the shell. Since the time interval $\Delta t = 6$ ms $= 0.006$ s is very short, we neglect all nonimpulsive forces and consider only the impulse **R** Δt.

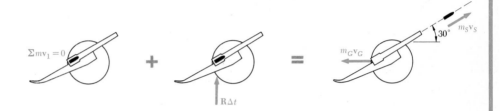

$$\Sigma m\mathbf{v}_1 + \Sigma \textbf{ Ext Imp}_{1\to2} = \Sigma m\mathbf{v}_2$$

$\xrightarrow{+} x$ components:
$$0 + 0 = -m_G v_G + m_S v_S \cos 30°$$
$$0 = -(2000 \text{ kg})v_G + (10 \text{ kg})(600 \text{ m/s}) \cos 30°$$
$$v_G = +2.60 \text{ m/s}$$
$$v_G = 2.60 \text{ m/s} \leftarrow \quad \blacktriangleleft$$

$+\uparrow y$ components:
$$0 + R \, \Delta t = m_S v_S \sin 30°$$
$$R(0.006 \text{ s}) = (10 \text{ kg})(600 \text{ m/s}) \sin 30°$$
$$R = +500\,000 \text{ N}$$
$$R = 500 \text{ kN} \uparrow \quad \blacktriangleleft$$

The total force exerted by the ground on the gun during the time interval Δt should include the static reaction due to the weight of the gun as well as the impulsive reaction **R**. Since the weight of the gun is $(2000 \text{ kg})(9.81 \text{ m/s}^2) = 19.62$ kN, the magnitude of this total force is 500 kN + 19.62 kN = 520 kN.

The high value obtained for the magnitude of **R** clearly indicates the need in modern guns for a recoil mechanism which will allow the barrel to move and will bring it to rest over a period of time substantially longer than Δt. Although the total vertical impulse must remain constant, the longer time interval will result in a smaller value for the magnitude of the reaction **R**.

PROBLEMS

14.1 A 3000-lb automobile is moving at a speed of 45 mi/h when the brakes are fully applied, causing all four wheels to skid. Determine the time required to stop the automobile (*a*) on concrete ($\mu = 0.75$), (*b*) on ice ($\mu = 0.08$).

14.2 A 60,000-ton ocean liner has an initial velocity of 2 mi/h. Neglecting the frictional resistance of the water, determine the time required to bring the liner to rest by using a single tugboat which exerts a constant force of 50,000 lb.

14.3 and 14.4 The initial velocity of the 50-kg car is 5 m/s to the left. Determine the time *t* at which the car has (*a*) no velocity, (*b*) a velocity of 5 m/s to the right.

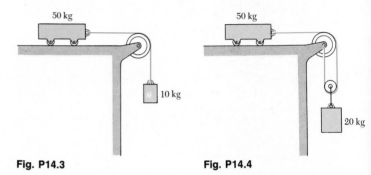

Fig. P14.3   Fig. P14.4

14.5 Using the principle of impulse and momentum, solve Prob. 12.17*b*.

14.6 Using the principle of impulse and momentum, solve Prob. 12.16*b*.

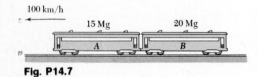

Fig. P14.7

14.7 A light train made of two cars travels at 100 km/h. The mass of car *A* is 15 Mg, and the mass of car *B* is 20 Mg. When the brakes are applied, a constant braking force of 25 kN is applied to each car. Determine (*a*) the time required for the train to stop after the brakes are applied, (*b*) the force in the coupling between the cars while the train is slowing down.

14.8 Solve Prob. 14.7, assuming that a constant braking force of 25 kN is applied to car *B* but that the brakes on car *A* are not applied.

14.9 A 150-lb block initially at rest is acted on by a force **P** which varies as shown. Knowing that $\mu = 0.20$, determine the velocity (*a*) at $t = 4$ s, (*b*) at $t = 7$ s.

14.10 In Prob. 14.9, determine (*a*) the maximum velocity reached by the block and the corresponding time, (*b*) the time at which the block comes to rest.

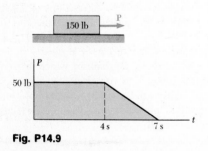

Fig. P14.9

14.11   A 20-kg block is initially at rest and is subjected to a force **P** which varies as shown. Neglecting the effect of friction, determine (*a*) the maximum speed attained by the block, (*b*) the speed of the block at $t = 1.5$ s.

14.12   Solve Prob. 14.11, assuming that $\mu = 0.25$ between the block and the surface.

14.13   A gun of mass 50 Mg is designed to fire a 250-kg shell with an initial velocity of 600 m/s. Determine the average force required to hold the gun motionless if the shell leaves the gun 0.02 s after being fired.

14.14   A 6000-kg plane lands on the deck of an aircraft carrier at a speed of 200 km/h relative to the carrier and is brought to a stop in 3.0 s. Determine the average horizontal force exerted by the carrier on the plane (*a*) if the carrier is at rest, (*b*) if the carrier is moving at a speed of 15 knots in the same direction as the airplane. (1 knot = 0.514 m/s.)

14.15   A 4-oz baseball is pitched with a velocity of 50 ft/s toward a batter. After the ball is hit by the bat *B*, it has a velocity of 150 ft/s in the direction shown. If the bat and ball are in contact 0.02 s, determine the average impulsive force exerted on the ball during the impact.

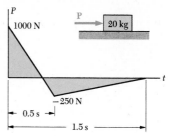

Fig. P14.11

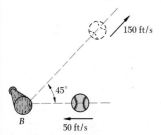

Fig. P14.15

14.16   A 175-lb man dives off the end of a pier with an initial velocity of 8 ft/s in the direction shown. Determine the horizontal and vertical components of the average force exerted on the pier during the 0.75 s that the man takes to leave the pier.

14.17   Determine the initial recoil velocity of an 8-lb rifle which fires a $\frac{3}{4}$-oz bullet with a velocity of 1800 ft/s.

14.18   A 2-oz rifle bullet is fired horizontally with a velocity of 1200 ft/s into an 8-lb block of wood which can move freely in the horizontal direction. Determine the final velocity of the block.

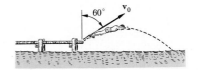

Fig. P14.16

14.19 Two men dive horizontally and to the right off the end of a 140-kg boat. The boat is initially at rest, and each man weighs 70 kg. If each man dives so that his relative horizontal velocity with respect to the boat is 4 m/s, determine (a) the velocity of the boat after the men dive simultaneously, (b) the velocity of the boat after one man dives and the velocity of the boat after the second man dives.

14.20 A 65-Mg engine coasting at 6 km/h strikes, and is automatically coupled with, a 10-Mg flat car which carries a 25-Mg load. The load is *not* securely fastened to the car but may slide along the floor ($\mu = 0.20$). Knowing that the car was at rest with its brakes released and that the coupling takes place instantaneously, determine the velocity of the engine (a) immediately after the coupling, (b) after the load has slid to a stop relative to the car.

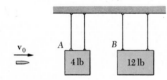

Fig. P14.20

14.21 A $\frac{3}{4}$-oz bullet is fired in a horizontal direction through block A and becomes embedded in block B. The bullet causes A and B to start moving with velocities of 7 and 5 ft/s, respectively. Determine (a) the initial velocity $\mathbf{v}_0$ of the bullet, (b) the velocity of the bullet as it travels from block A to block B.

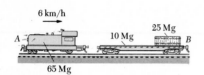

Fig. P14.21

14.22 A 30-ton railroad car is to be coupled to a second car which weighs 50 tons. If initially the speed of the 30-ton car is 1 mi/h and the 50-ton car is at rest, determine (a) the final speed of the coupled cars, (b) the average impulsive force acting on each car if the coupling is completed in 0.40 s.

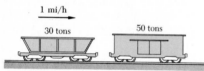

Fig. P14.22

14.23 Solve Prob. 14.22, assuming that, initially, the 30-ton car is at rest and the 50-ton car has a speed of 1 mi/h.

14.24 A 10-kg package is discharged from a conveyor belt with a velocity of 3 m/s and lands in a 25-kg cart. Knowing that the cart is initially at rest and may roll freely, determine the final velocity of the cart.

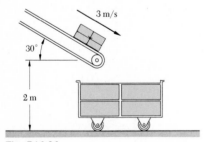

Fig. P14.24

14.25 Solve Prob. 14.24, assuming that the single 10-kg package is replaced by two 5-kg packages. The first 5-kg package comes to relative rest in the cart before the second package strikes the cart.

∗14.26 A machine part is forged in a small drop forge. The hammer weighs 400 lb and is dropped from a height of 5 ft. Determine the initial impulse exerted on the machine part, assuming that the 1000-lb anvil (a) is resting directly on hard ground, (b) is supported by springs.

14.5. Impact. A collision between two bodies which occurs in a very small interval of time, and during which the two bodies exert on each other relatively large forces, is called an *impact*. The common normal to the surfaces in contact during the impact is called the *line of impact*. If the mass centers of the two colliding bodies are located on this line, the impact is a *central impact*. Otherwise, the impact is said to be *eccentric*. We shall limit our present study to that of the central impact of two particles and postpone until later the analysis of the eccentric impact of two rigid bodies (Sec. 18.6).

If the velocities of the two particles are directed along the line of impact, the impact is said to be a *direct impact* (Fig. 14.7a). If, on the other hand, either or both particles move along a line other than the line of impact, the impact is said to be an *oblique impact* (Fig. 14.7b).

14.6. Direct Central Impact. Consider two particles A and B, of mass m_A and m_B, which are moving in the same straight line and to the right with known velocities $\mathbf{v}_A$ and $\mathbf{v}_B$ (Fig. 14.8a). If $\mathbf{v}_A$ is larger than $\mathbf{v}_B$, particle A will eventually strike particle B. Under the impact, the two particles will *deform* and, at the end of the period of deformation, they will have the same velocity $\mathbf{u}$ (Fig. 14.8b). A period of *restitution* will then take place, at the end of which, depending upon the magnitude of the impact forces and upon the materials involved, the two particles either will have regained their original shape or will stay permanently deformed. Our purpose here is to determine the velocities $\mathbf{v}_A'$ and $\mathbf{v}_B'$ of the particles at the end of the period of restitution (Fig. 14.8c).

Considering first the system of the two particles as a whole, we note that the only impulsive forces acting during the impact are internal forces. Thus, *the total momentum of the system is conserved*, and we write

$$m_A\mathbf{v}_A + m_B\mathbf{v}_B = m_A\mathbf{v}_A' + m_B\mathbf{v}_B' \qquad (14.14)$$

Since all the velocities considered are directed along the same axis, we may replace the equation obtained by the following relation involving only scalar components:

$$m_A v_A + m_B v_B = m_A v_A' + m_B v_B' \qquad (14.15)$$

A positive value for any of the scalar quantities v_A, v_B, v_A', or v_B' means that the corresponding vector is directed to the right; a negative value indicates that the corresponding vector is directed to the left.

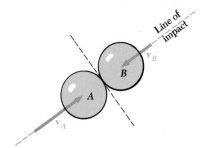

(a) Direct central impact

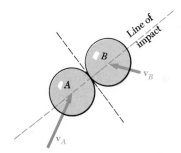

(b) Oblique central impact

Fig. 14.7

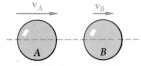

(a) Before impact

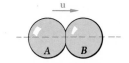

(b) At maximum deformation

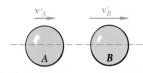

(c) After impact

Fig. 14.8

To obtain the velocities $\mathbf{v}'_A$ and $\mathbf{v}'_B$, it is necessary to establish a second relation between the scalars v'_A and v'_B. For this purpose, we shall consider now the motion of particle A during the period of deformation and apply the principle of impulse and momentum. Since the only impulsive force acting on A during this period is the force $\mathbf{P}$ exerted by B (Fig. 14.9a), we write, using again scalar components,

$$m_A v_A - \int P\,dt = m_A u \qquad (14.16)$$

where the integral extends over the period of deformation.

(a) Period of deformation

(b) Period of restitution

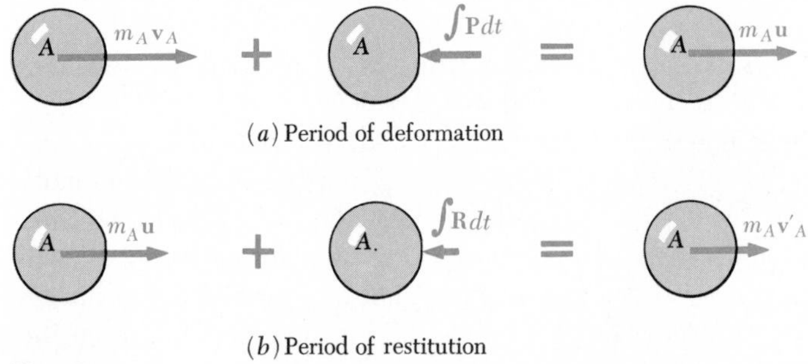

Fig. 14.9

Considering now the motion of A during the period of restitution, and denoting by $\mathbf{R}$ the force exerted by B on A during this period (Fig. 14.9b), we write

$$m_A u - \int R\,dt = m_A v'_A \qquad (14.17)$$

where the integral extends over the period of restitution.

In general, the force $\mathbf{R}$ exerted on A during the period of restitution differs from the force $\mathbf{P}$ exerted during the period of deformation, and the magnitude $\int R\,dt$ of its impulse is smaller than the magnitude $\int P\,dt$ of the impulse of $\mathbf{P}$. The ratio of the magnitudes of the impulses corresponding respectively to the period of restitution and to the period of deformation is called the *coefficient of restitution* and is denoted by e. We write

$$e = \frac{\int R\,dt}{\int P\,dt} \qquad (14.18)$$

The value of the coefficient e is always between 0 and 1 and depends to a large extent on the two materials involved. However, it also varies considerably with the impact velocity and the shape and size of the two colliding bodies.

Solving Eqs. (14.16) and (14.17) for the two impulses and substituting into (14.18), we write

$$e = \frac{u - v'_A}{v_A - u} \tag{14.19}$$

A similar analysis of particle B leads to the relation

$$e = \frac{v'_B - u}{u - v_B} \tag{14.20}$$

Since the quotients in (14.19) and (14.20) are equal, they are also equal to the quotient obtained by adding, respectively, their numerators and their denominators. We have, therefore,

$$e = \frac{(u - v'_A) + (v'_B - u)}{(v_A - u) + (u - v_B)} = \frac{v'_B - v'_A}{v_A - v_B}$$

and

$$v'_B - v'_A = e(v_A - v_B) \tag{14.21}$$

Since $v'_B - v'_A$ represents the relative velocity of the two particles after impact and $v_A - v_B$ their relative velocity before impact, formula (14.21) expresses that *the relative velocity of the two particles after impact may be obtained by multiplying their relative velocity before impact by the coefficient of restitution.* This property is used to determine experimentally the value of the coefficient of restitution of two given materials.

The velocities of the two particles after impact may now be obtained by solving Eqs. (14.15) and (14.21) simultaneously for v'_A and v'_B, i.e., by using the principle of conservation of momentum and the concept of coefficient of restitution. It is recalled that the derivation of Eqs. (14.15) and (14.21) was based on the assumption that particle B is located to the right of A, and that both particles are initially moving to the right. If particle B is initially moving to the left, the scalar v_B should be considered negative. The same sign convention holds for the velocities after impact: a positive sign for v'_A will indicate that particle A moves to the right after impact, and a negative sign that it moves to the left.

Two particular cases of impact are of special interest:

1. $e = 0$, *Perfectly Plastic Impact.* When $e = 0$, Eq. (14.21) yields $v'_B = v'_A$. There is no period of restitution, and both particles stay together after impact. Substituting $v'_B = v'_A = v'$ into Eq. (14.15), which expresses that the momentum of the system is conserved, we write

$$m_A v_A + m_B v_B = (m_A + m_B)v' \tag{14.22}$$

This equation may be solved for the common velocity v' of the two particles after impact.

2. $e = 1$, *Perfectly Elastic Impact.* When $e = 1$, Eq. (14.21) reduces to

$$v'_B - v'_A = v_A - v_B \qquad (14.23)$$

which expresses that the relative velocities before and after impact are equal. The impulses received by each particle during the period of deformation and during the period of restitution are equal. The particles move away from each other after impact with the same velocity with which they approached each other before impact. The velocities v'_A and v'_B may be obtained by solving Eqs. (14.15) and (14.23) simultaneously.

It is worth noting that, *in the case of a perfectly elastic impact, the energy of the system,* as well as its momentum, *is conserved.* Equations (14.15) and (14.23) may be written as follows:

$$m_A(v_A - v'_A) = m_B(v'_B - v_B) \qquad (14.15')$$
$$v_A + v'_A = v_B + v'_B \qquad (14.23')$$

Multiplying (14.15′) and (14.23′) member by member, we have

$$m_A(v_A - v'_A)(v_A + v'_A) = m_B(v'_B - v_B)(v'_B + v_B)$$
$$m_A v_A^2 - m_A(v'_A)^2 = m_B(v'_B)^2 - m_B v_B^2$$

Rearranging the terms in the equation obtained, and multiplying by $\frac{1}{2}$, we write

$$\tfrac{1}{2} m_A v_A^2 + \tfrac{1}{2} m_B v_B^2 = \tfrac{1}{2} m_A(v'_A)^2 + \tfrac{1}{2} m_B(v'_B)^2$$

which expresses that the kinetic energy of the system is conserved. It should be noted, however, that *in the general case of impact,* i.e., when e is not equal to 1, *the energy of the system is not conserved.* This may be shown in any given case by comparing the kinetic energies before and after impact. The lost kinetic energy is in part transformed into heat and in part spent in generating elastic waves within the two colliding bodies.

14.7. Oblique Central Impact. Let us now consider the case when the velocities of the two colliding particles are *not* directed along the line of impact (Fig. 14.10). As indicated in Sec. 14.5, the impact is said to be *oblique.* Since the velocities $\mathbf{v}'_A$ and $\mathbf{v}'_B$ of the particles after impact are unknown in direction as well as in magnitude, their determination will require the use of four independent equations.

We choose x and y axes, respectively, along the line of impact and along the common tangent to the surfaces in contact. Assuming that the particles are perfectly *smooth and frictionless,*

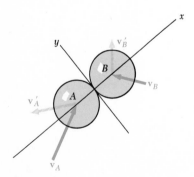

Fig. 14.10

we observe that the only impulsive forces acting on the particles during the impact are internal forces directed along the x axis. We may therefore express that:

1. The y component of the momentum of particle A is conserved.
2. The y component of the momentum of particle B is conserved.
3. The x component of the total momentum of the particles is conserved.
4. The x component of the relative velocity of the two particles after impact is obtained by multiplying the x component of their relative velocity before impact by the coefficient of restitution.

We thus obtain four independent equations which may be solved for the components of the velocities of A and B after impact. This method of solution is illustrated in Sample Prob. 14.6.

14.8. Problems Involving Energy and Momentum.

We have now at our disposal three different methods for the solution of kinetics problems: the direct application of Newton's second law, $\Sigma \mathbf{F} = m\mathbf{a}$, the method of work and energy, and the method of impulse and momentum. To derive maximum benefit from these three methods, we should be able to choose the method best suited for the solution of a given problem. We should also be prepared to use different methods for solving the various parts of a problem when such a procedure seems advisable.

We have already seen that the method of work and energy is in many cases more expeditious than the direct application of Newton's second law. As indicated in Sec. 13.4, however, the method of work and energy has limitations, and it must sometimes be supplemented by the use of $\Sigma \mathbf{F} = m\mathbf{a}$. This is the case, for example, when we wish to determine an acceleration or a normal force.

There is generally no great advantage in using the method of impulse and momentum for the solution of problems involving no impulsive forces. It will usually be found that the equation $\Sigma \mathbf{F} = m\mathbf{a}$ yields a solution just as fast and that the method of work and energy, if it applies, is more rapid and more convenient. However, the method of impulse and momentum is the only practicable method in problems of impact. A solution based on the direct application of $\Sigma \mathbf{F} = m\mathbf{a}$ would be unwieldy, and the method of work and energy cannot be used since impact (unless perfectly elastic) involves a loss of mechanical energy.

Many problems involve only conservative forces, except for a short impact phase during which impulsive forces act. The solution of such problems may be divided into several parts. While the part corresponding to the impact phase calls for the use of the method of impulse and momentum and of the relation between relative velocities, the other parts may usually be solved by the method of work and energy. The use of the equation $\Sigma \mathbf{F} = m\mathbf{a}$ will be necessary, however, if the problem involves the determination of a normal force.

Consider, for example, a pendulum A, of weight $\mathbf{W}_A$ and length l, which is released with no velocity from a position A_1 (Fig. 14.11a). The pendulum swings freely in a vertical plane and hits a second pendulum B, of weight $\mathbf{W}_B$ and same length l, which is initially at rest. After the impact (with coefficient of restitution e), pendulum B swings through an angle θ that we wish to determine.

The solution of the problem may be divided into three parts:

1. *Pendulum A Swings from A_1 to A_2.* The principle of conservation of energy may be used to determine the velocity $(\mathbf{v}_A)_2$ of the pendulum at A_2 (Fig. 14.11b).
2. *Pendulum A Hits Pendulum B.* Using the principle of conservation of momentum and the relation between the relative velocities of the two pendulums, we determine the velocities $(\mathbf{v}_A)_3$ and $(\mathbf{v}_B)_3$ of the pendulums after impact (Fig. 14.11c).
3. *Pendulum B Swings from B_3 to B_4.* Applying the principle of conservation of energy, we determine the maximum elevation y_4 reached by pendulum B (Fig. 14.11d). The angle θ may then be determined by trigonometry.

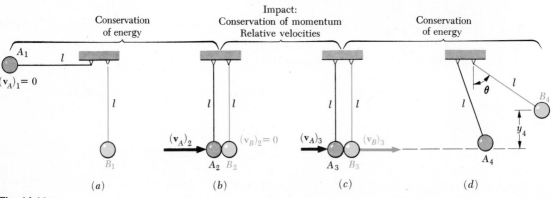

Fig. 14.11

We note that the method of solution just described should be supplemented by the use of $\Sigma \mathbf{F} = m\mathbf{a}$ if the tensions in the cords holding the pendulums are to be determined.

A 20-Mg railroad car moving at a speed of 0.5 m/s to the right collides with a 35-Mg car which is at rest. If after the collision the 35-Mg car is observed to move to the right at a speed of 0.3 m/s, determine the coefficient of restitution between the two cars.

Solution. We consider the system consisting of the two cars and express that the total momentum is conserved.

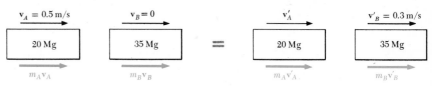

$$m_A \mathbf{v}_A + m_B \mathbf{v}_B = m_A \mathbf{v}'_A + m_B \mathbf{v}'_B$$
$$(20 \text{ Mg})(+0.5 \text{ m/s}) + (35 \text{ Mg})(0) = (20 \text{ Mg})v'_A + (35 \text{ Mg})(+0.3 \text{ m/s})$$
$$v^t_A = -0.025 \text{ m/s} \qquad \mathbf{v}'_A = 0.025 \text{ m/s} \leftarrow$$

The coefficient of restitution is obtained by writing

$$e = \frac{v'_B - v'_A}{v_A - v_B} = \frac{+0.3 - (-0.025)}{+0.5 - 0} = \frac{0.325}{0.5} \qquad e = 0.65 \quad \blacktriangleleft$$

A ball is thrown against a frictionless vertical wall. Immediately before the ball strikes the wall, its velocity has a magnitude v and forms an angle of 30° with the horizontal. Knowing that $e = 0.90$, determine the magnitude and direction of the velocity of the ball as it rebounds from the wall.

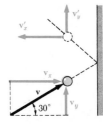

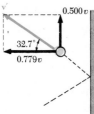

Solution. We resolve the initial velocity of the ball into components

$$v_x = v \cos 30° = 0.866v \qquad v_y = v \sin 30° = 0.500v$$

Vertical Motion. Since the wall is frictionless, no vertical impulsive force acts on the ball during the time it is in contact with the wall. The vertical component of the momentum, and hence the vertical component of the velocity, of the ball is thus unchanged:

$$\mathbf{v}'_y = \mathbf{v}_y = 0.500v \uparrow$$

Horizontal Motion. Since the mass of the wall (and earth) is essentially infinite, there is no point in expressing that the total momentum of the ball-wall system is conserved. Using the relation between relative velocities, we write

$$0 - v'_x = e(v_x - 0)$$
$$v'_x = -0.90(0.866v) = -0.779v \qquad v'_x = 0.779v \leftarrow$$

Resultant Motion. Adding vectorially the components $\mathbf{v}'_x$ and $\mathbf{v}'_y$,

$$v' = 0.926v \; \diagdown \; 32.7° \quad \blacktriangleleft$$

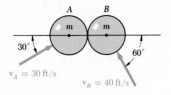

The magnitude and direction of the velocities of two friction-less balls before they strike each other are as shown. Assuming $e = 0.90$, determine the magnitude and direction of the velocity of each ball after the impact.

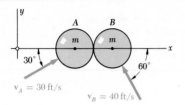

Solution. The impulsive forces acting between the balls during the impact are directed along a line joining the centers of the balls called the *line of impact*. Choosing x and y axes, respectively, parallel and perpendicular to the line of impact and directed as shown, we write

$$(v_A)_x = v_A \cos 30° = +26.0 \text{ ft/s}$$
$$(v_A)_y = v_A \sin 30° = +15.0 \text{ ft/s}$$
$$(v_B)_x = -v_B \cos 60° = -20.0 \text{ ft/s}$$
$$(v_B)_y = v_B \sin 60° = +34.6 \text{ ft/s}$$

Principle of Impulse and Momentum. In the adjoining sketches we show in turn the initial momenta, the impulsive reactions, and the final momenta.

Motion Perpendicular to the Line of Impact. Considering only the y components, we apply the principle of impulse and momentum to each ball *separately*. Since no vertical impulsive force acts during the impact, the vertical component of the momentum, and hence the vertical component of the velocity, of each ball is unchanged.

$$(v'_A)_y = 15.0 \text{ ft/s} \uparrow \qquad (v'_B)_y = 34.6 \text{ ft/s} \uparrow$$

Motion Parallel to the Line of Impact. In the x direction, we consider the two balls as a *single* system and note that, by Newton's third law, the internal impulses $\mathbf{F}\,\Delta t$ and $\mathbf{F}'\,\Delta t$ cancel. Applying the principle of conservation of momentum, we write

$$m_A(v_A)_x + m_B(v_B)_x = m_A(v'_A)_x + m_B(v'_B)_x$$
$$m(26.0) + m(-20.0) = m(v'_A)_x + m(v'_B)_x$$
$$(v'_A)_x + (v'_B)_x = 6.0 \quad (1)$$

Using the relation between relative velocities, we write

$$(v'_B)_x - (v'_A)_x = e[(v_A)_x - (v_B)_x]$$
$$(v'_B)_x - (v'_A)_x = (0.90)[26.0 - (-20.0)]$$
$$(v'_B)_x - (v'_A)_x = 41.4 \quad (2)$$

Solving Eqs. (1) and (2) simultaneously, we obtain

$$(v'_A)_x = -17.7 \qquad\qquad (v'_B)_x = +23.7$$
$$(v'_A)_x = 17.7 \text{ ft/s} \leftarrow \qquad (v'_B)_x = 23.7 \text{ ft/s} \rightarrow$$

Resultant Motion. Adding vectorially the velocity components of each ball, we obtain

$$v'_A = 23.2 \text{ ft/s} \searrow 40.3° \qquad v'_B = 41.9 \text{ ft/s} \nearrow 55.6° \quad \blacktriangleleft$$

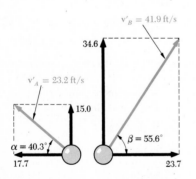

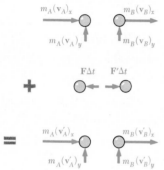

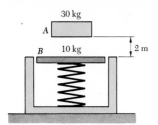

30 kg

A

B   10 kg   ↕ 2 m

A 30-kg block is dropped from a height of 2 m onto the 10-kg **pan** of a spring scale. Assuming the impact to be perfectly plastic, determine the maximum deflection of the pan. The constant of the **spring** is $k = 20 \text{ kN/m}$.

Solution. The impact between the block and the pan *must* be treated separately; therefore we divide the solution into three parts.

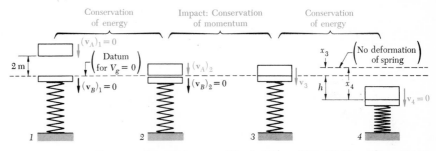

Conservation of Energy. Block: $W_A = (30 \text{ kg})(9.81 \text{ m/s}^2) = 294 \text{ N}$

$$T_1 = \tfrac{1}{2}m_A(v_A)_1^2 = 0 \qquad V_1 = W_A y = (294 \text{ N})(2 \text{ m}) = 588 \text{ J}$$
$$T_2 = \tfrac{1}{2}m_A(v_A)_2^2 = \tfrac{1}{2}(30 \text{ kg})(v_A)_2^2 \qquad V_2 = 0$$
$$T_1 + V_1 = T_2 + V_2: \qquad 0 + 588 \text{ J} = \tfrac{1}{2}(30 \text{ kg})(v_A)_2^2 + 0$$
$$(v_A)_2 = +6.26 \text{ m/s} \qquad (\mathbf{v}_A)_2 = 6.26 \text{ m/s} \downarrow$$

Impact: Conservation of Momentum. Since the impact is perfectly plastic, $e = 0$; the block and pan move together after the impact.

$$m_A(v_A)_2 + m_B(v_B)_2 = (m_A + m_B)v_3$$
$$(30 \text{ kg})(6.26 \text{ m/s}) + 0 = (30 \text{ kg} + 10 \text{ kg})v_3$$
$$v_3 = +4.70 \text{ m/s} \qquad \mathbf{v}_3 = 4.70 \text{ m/s} \downarrow$$

Conservation of Energy. Initially the spring supports the weight W_B of the pan; thus the initial deflection of the spring is

$$x_3 = \frac{W_B}{k} = \frac{(10 \text{ kg})(9.81 \text{ m/s}^2)}{20 \times 10^3 \text{ N/m}} = \frac{98.1 \text{ N}}{20 \times 10^3 \text{ N/m}} = 4.91 \times 10^{-3} \text{ m}$$

Denoting by x_4 the total maximum deflection of the spring, we **write**

$T = KE.$

$V = P.E.$

$$T_3 = \tfrac{1}{2}(m_A + m_B)v_3^2 = \tfrac{1}{2}(30 \text{ kg} + 10 \text{ kg})(4.70 \text{ m/s})^2 = 442 \text{ J}$$
$$V_3 = V_g + V_e = 0 + \tfrac{1}{2}kx_3^2 = \tfrac{1}{2}(20 \times 10^3)(4.91 \times 10^{-3})^2 = 0.241 \text{ J}$$
$$T_4 = 0$$
$$V_4 = V_g + V_e = (W_A + W_B)(-h) + \tfrac{1}{2}kx_4^2 = -(392)h + \tfrac{1}{2}(20 \times 10^3)x_4^2$$

Noting that the displacement of the pan is $h = x_4 - x_3$, we write

$$T_3 + V_3 = T_4 + V_4:$$
$$442 + 0.241 = 0 - 392(x_4 - 4.91 \times 10^{-3}) + \tfrac{1}{2}(20 \times 10^3)x_4^2$$
$$x_4 = 0.230 \text{ m} \qquad h = x_4 - x_3 = 0.230 \text{ m} - 4.91 \times 10^{-3} \text{ m}$$
$$h = 0.225 \text{ m} \qquad h = 225 \text{ mm} \blacktriangleleft$$

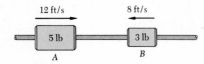

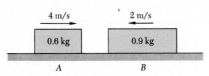

Fig. P14.27 and P14.28

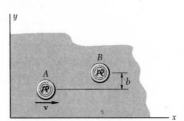

Fig. P14.29 and P14.30

PROBLEMS

14.27 The coefficient of restitution between the two collars is known to be 0.70; determine (*a*) their velocities after impact, (*b*) the energy loss during the impact.

14.28 Solve Prob. 14.27, assuming that the velocity of collar *B* is 8 ft/s to the right.

14.29 Two steel blocks slide without friction on a horizontal surface; immediately before impact their velocities are as shown. Knowing that $e = 0.75$, determine (*a*) their velocities after impact, (*b*) the energy loss during impact.

14.30 The velocities of two steel blocks before impact are as shown. If after the impact the velocity of block *B* is observed to be 2.5 m/s to the right, determine the coefficient of restitution between the two blocks.

14.31 Two identical pucks *A* and *B*, of 80-mm diameter, may move freely on a hockey rink. Puck *B* is at rest and puck *A* has an initial velocity **v** as shown. (*a*) Knowing that $b = 40$ mm and $e = 0.80$, determine the velocity of each puck after the impact. (*b*) Show that if $e = 1$, the final velocities of the pucks form a right angle for all values of *b*.

Fig. P14.31

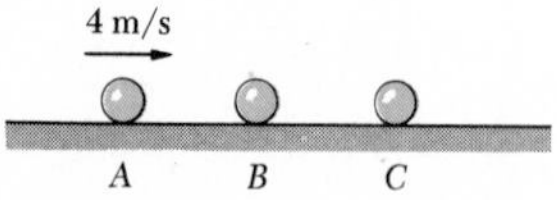

14.32 Two identical balls *B* and *C* are at rest when ball *B* is struck by a ball *A* of the same mass, moving with a velocity of 4 m/s. This causes a series of collisions between the various balls. Knowing that $e = 0.40$, determine the velocity of each ball after *all* collisions have taken place.

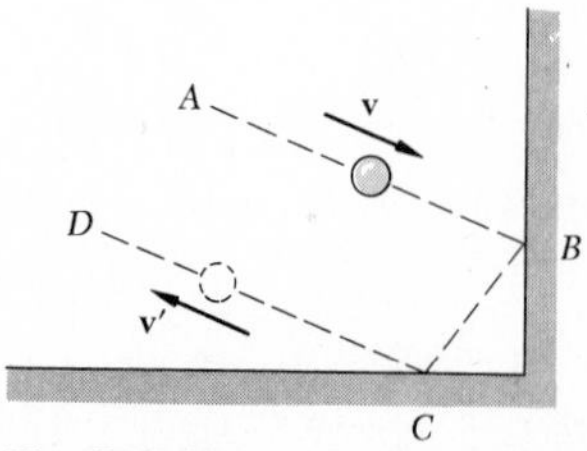

Fig. P14.32

14.33 A ball is thrown into a 90° corner with an initial velocity **v**. Denoting the coefficient of restitution by *e*, show that the final velocity **v′** is of magnitude *ev* and that the initial and final paths *AB* and *CD* are parallel.

Fig. P14.33

14.34   A steel ball falling vertically strikes a rigid plate A and rebounds horizontally as shown. Denoting by e the coefficient of restitution, determine (a) the required angle θ, (b) the magnitude of the velocity $\mathbf{v}_1$.

14.35   The ball of Prob. 14.34 strikes a second rigid plate B and rebounds vertically as shown. Determine (a) the required angle β, (b) the magnitude of the velocity $\mathbf{v}_2$ with which the ball leaves plate B.

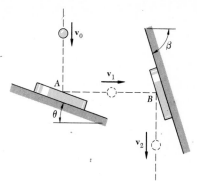

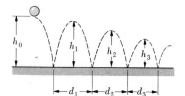

Fig. P14.34

14.36   A ball is dropped from a height $h_0 = 40$ in. onto a smooth floor. The coefficient of restitution is known to be 0.85 and the length of the first bounce is observed to be $d_1 = 12$ in. Determine (a) the height h_2 of the second bounce, (b) the length d_2 of the second bounce.

14.37   A ball is dropped onto a smooth floor and allowed to bounce several times as shown. Derive an expression for the coefficient of restitution in terms of (a) the heights of two successive bounces h_n and h_{n+1}, (b) the lengths of two successive bounces d_n and d_{n+1}, (c) the durations of two successive bounces t_n and t_{n+1}.

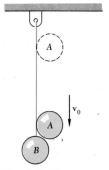

Fig. P14.36, P14.37, and P14.38

14.38   A ball is dropped from a height $h_0 = 48$ in. onto a smooth floor. Knowing that for the first bounce $h_1 = 40$ in. and $d_1 = 16$ in., determine (a) the coefficient of restitution, (b) the height and length of the second bounce.

*14.39   A 2-kg sphere moving to the left with a velocity of 10 m/s strikes the inclined surface of a 5-kg block which is at rest. The block rests on rollers and may move freely in the horizontal direction. Knowing that $e = 0.75$, determine the velocities of the block and of the sphere immediately after impact.

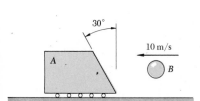

Fig. P14.39

Fig. P14.40

*14.40   Ball B is suspended by an inextensible cord. An identical ball A is released from rest when it is just touching the cord and acquires a velocity $\mathbf{v}_0$ before striking ball B. Assuming $e = 1$, determine the velocity of each ball immediately after impact.

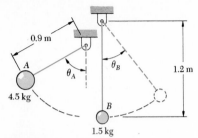

Fig. P14.41 and P14.42

14.41 The 4.5-kg sphere A strikes the 1.5-kg sphere B. Knowing that $e = 0.90$, determine the angle θ_A at which A must be released if the maximum angle θ_B reached by B is to be $90°$.

14.42 The 4.5-kg sphere A is released from rest when $\theta_A = 60°$ and strikes the 1.5-kg sphere B. Knowing that $e = 0.90$, determine (a) the highest position to which sphere B will rise, (b) the maximum tension which will occur in the cord holding B.

14.43 The 4-lb sphere is released from rest when $\theta_A = 90°$. The coefficient of restitution between the sphere and the block is 0.70. Knowing that the coefficient of friction between the block and the horizontal surface is 0.30, determine how far the block will move after the impact.

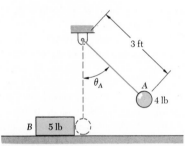

Fig. P14.43 and P14.44

14.44 The 4-lb sphere is released from rest when $\theta_A = 60°$. It is observed that the velocity of the sphere is zero after the impact and that the block moves 3 ft before coming to rest. Determine (a) the coefficient of restitution between the sphere and block, (b) the coefficient of friction between the block and the horizontal surface.

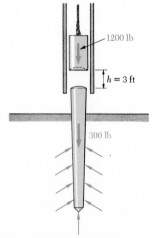

Fig. P14.45 and P14.46

14.45 It is desired to drive the 300-lb pile into the ground until the resistance to its penetration is 20,000 lb. Each blow of the 1200-lb hammer is the result of a 3-ft free fall onto the top of the pile. Determine how far the pile will be driven into the ground by a single blow when the 20,000-lb resistance is achieved. Assume that the impact is perfectly plastic.

14.46 The 1200-lb hammer of a drop-hammer pile driver falls from a height of 3 ft onto the top of a 300-lb pile. The pile is driven 4 in. into the ground. Assuming perfectly plastic impact, determine the average resistance of the ground to penetration.

14.47 Cylinder A is dropped 2 m onto cylinder B, which is resting on a spring of constant $k = 3\,\text{kN/m}$. Assuming a perfectly plastic impact, determine (a) the maximum deflection of cylinder B, (b) the energy loss during the impact.

14.48 The efficiency η of a drop-hammer pile driver may be defined as the ratio of the kinetic energy available after impact to the kinetic energy immediately before impact. Denoting by r the ratio of the pile mass m_p to the hammer mass m_h, and assuming perfectly plastic impact, show that $\eta = 1/(1 + r)$.

14.49 A bumper is designed to protect a 1600-kg automobile from damage when it hits a rigid wall at speeds up to 12 km/h. Assuming perfectly plastic impact, determine (a) the energy absorbed by the bumper during the impact, (b) the speed at which the automobile can hit another 1600-kg automobile without incurring any damage, if the other automobile is at rest and is similarly protected.

14.50 Solve Prob. 14.49, assuming a coefficient of restitution $e = 0.50$. Show that the answer to part b is independent of e.

14.51 A small rivet connecting two pieces of sheet metal is being clinched by hammering. Determine the energy absorbed by the rivet under each blow, knowing that the head of the hammer weighs 1.5 lb and that it strikes the rivet with a velocity of 20 ft/s. Assume that the anvil is supported by springs and (a) is infinite in weight (rigid support), (b) weighs 10 lb.

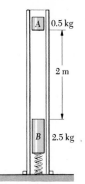

Fig. P14.47

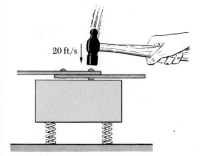

Fig. P14.51

14.52 In Prob. 14.51, determine the weight of the hammer required to deliver to the rivet an amount of energy equal to $8\,\text{ft}\cdot\text{lb}$ per blow if the hammer strikes the rivet with a velocity of 20 ft/s. Consider both types of support.

*14.53 A mass m_A moving to the right with a velocity v_A strikes a second mass m_B which is at rest. Derive an expression for the kinetic-energy loss during the impact of the two masses. Assume that the balls strike each other squarely, and denote the coefficient of restitution by e.

Fig. P14.53

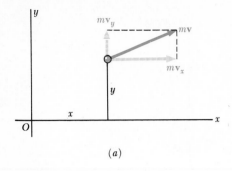

(a)

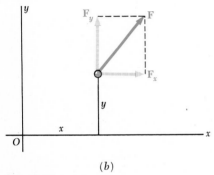

(b)

Fig. 14.12

14.9. Angular Momentum. Consider a particle of mass m moving in the xy plane. As we saw in Sec. 14.1, the momentum of the particle at a given instant is defined as the vector $m\mathbf{v}$ obtained by multiplying the velocity $\mathbf{v}$ of the particle by its mass m. The moment about O of the vector $m\mathbf{v}$ is called the *moment of momentum*, or the *angular momentum*, of the particle about O at that instant and is denoted by H_O. Resolving the vector $m\mathbf{v}$ into its x and y components (Fig. 14.12a), and recalling that counterclockwise is positive, we write

$$H_O = x(mv_y) - y(mv_x)$$
$$H_O = m(xv_y - yv_x) \tag{14.24}$$

We shall now compute the derivative of the angular momentum H_O with respect to t; we have

$$\dot{H}_O = m\frac{d}{dt}(xv_y - yv_x)$$
$$= m(\dot{x}v_y + x\dot{v}_y - \dot{y}v_x - y\dot{v}_x)$$

or, since $\dot{x} = v_x$ and $\dot{y} = v_y$,

$$\dot{H}_O = m(x\dot{v}_y - y\dot{v}_x) \tag{14.25}$$

But, according to Newton's second law, we have

$$m\dot{v}_x = ma_x = F_x \qquad m\dot{v}_y = ma_y = F_y$$

where F_x and F_y are the rectangular components of the force $\mathbf{F}$ acting on the particle. Equation (14.25) may therefore be written in the form

$$\dot{H}_O = xF_y - yF_x$$

or, since the right-hand member of the equation obtained represents the moment M_O of $\mathbf{F}$ about O (Fig. 14.12b),

$$M_O = \dot{H}_O \tag{14.26}$$

14.10. Conservation of Angular Momentum. If the moment M_O about O of the force $\mathbf{F}$ acting on the particle of the preceding section is zero for every value of t, Eq. (14.26) yields

$$\dot{H}_O = 0$$

for any t; integrating with respect to t, we obtain

$$H_O = \text{constant} \tag{14.27}$$

Thus, if the moment about O of the force $\mathbf{F}$ acting on a particle is zero for every value of t, *the angular momentum of the particle about O is conserved.*

Clearly, the angular momentum of a particle about a fixed point O will be conserved if the resultant of the forces acting on the particle is zero. But the angular momentum of the particle about O will also be conserved when the particle is acted upon by an unbalanced force $\mathbf{F}$ *if the force* $\mathbf{F}$ *is a central force whose line of action passes through* O (Fig. 14.13).

The angular momentum H_O of a particle was expressed in Eq. (14.24) in terms of the rectangular coordinates of the particle and of the rectangular components of its momentum. In the case of a particle moving under a central force $\mathbf{F}$, it is found more convenient to use polar coordinates. Recalling that H_O is, by definition, the moment about O of the momentum $m\mathbf{v}$ of the particle, and denoting by ϕ the angle formed by the radius vector and the tangent to the path of the particle, we write Eq. (14.27) in the form

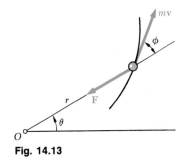

Fig. 14.13

$$H_O = r(mv \sin \phi) = \text{constant} \tag{14.28}$$

Observing that $v \sin \phi$ represents the transverse component v_θ of the velocity, which is equal to $r\dot{\theta}$ (Sec. 11.13), we may also write

$$H_O = mr^2\dot{\theta} = \text{constant} \tag{14.29}$$

Comparing Eq. (14.29) and Eq. (12.18), we note that, when a particle moves under a central force, its angular momentum H_O about the center of force O is equal to the product of its mass and of the constant h introduced in Sec. 12.8. It follows that the angular momentum of the particle is equal to twice the product of its mass and of its areal velocity; thus, stating that the areal velocity of the particle is constant is equivalent to stating that its angular momentum is conserved.

14.11. Application to Space Mechanics. The principles of conservation of angular momentum and of conservation of energy may be used to solve many problems concerning the motion of earth satellites and other space vehicles.

Consider, for example, a space vehicle moving under the earth's gravitational force. We shall assume that it begins its free flight at point P_0 at a distance r_0 from the center of the earth, with a velocity $\mathbf{v}_0$ forming an angle ϕ_0 with the radius vector OP_0 (Fig. 14.14). Let P be a point of the trajectory described by the vehicle; we denote by r the distance from O to P, by $\mathbf{v}$ the velocity of the vehicle at P, and by ϕ the angle formed by $\mathbf{v}$ and the radius vector OP. Applying the principle of conservation of angular momentum about O between P_0 and P, we write

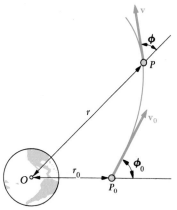

Fig. 14.14

$$r_0 m v_0 \sin \phi_0 = rmv \sin \phi \tag{14.30}$$

Recalling expression (13.15) obtained for the potential energy due to a gravitational force, we apply the principle of conservation of energy between P_0 and P and write

$$T_0 + V_0 = T + V$$

$$\tfrac{1}{2}mv_0^2 - \frac{GMm}{r_0} = \tfrac{1}{2}mv^2 - \frac{GMm}{r} \qquad (14.31)$$

where M is the mass of the earth.

Equation (14.31) may be solved for the magnitude v of the velocity of the vehicle at P when the distance r from O to P is known; Eq. (14.30) may then be used to determine the angle ϕ that the velocity forms with the radius vector OP.

Equations (14.30) and (14.31) may also be used to determine the maximum and minimum values of r in the case of a satellite launched from P_0 in a direction forming an angle ϕ_0 with the vertical OP_0 (Fig. 14.15). The desired values of r are obtained by making $\phi = 90°$ in (14.30) and eliminating v between Eqs. (14.30) and (14.31).

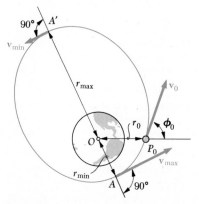

Fig. 14.15

It should be noted that the application of the principles of conservation of energy and of conservation of angular momentum leads to a more fundamental formulation of the problems of space mechanics than the method indicated in Sec. 12.11. In all cases involving oblique launchings, it will also result in much simpler computations. And while the method of Sec. 12.11 must be used when the actual trajectory or the periodic time of a space vehicle is to be determined, the calculations will be simplified if the conservation principles are first used to compute the maximum and minimum values of the radius vector r.

14.12. Principle of Impulse and Momentum for a System of Particles. The moment of momentum, or angular momentum, about O of a system of particles is defined as the sum of the moments about O of the momenta of the various particles of the system; we write

$$H_O = \Sigma m(xv_y - yv_x) \qquad (14.32)$$

Since Eq. (14.26) is satisfied for each particle, we may write that the sum of the moments of all the forces acting on the various particles of the system is equal to the rate of change of the angular momentum of the system. However, since the internal forces occur by pairs of equal and opposite forces having the same line of action, the sum of their moments is zero and we need consider only the sum of the moments of the external forces. We thus write

$$\Sigma(M_O)_{\text{ext}} = \dot{H}_O \qquad (14.33)$$

where H_O is the angular momentum of the system defined by relation (14.32).

We saw in Sec. 14.1 that the linear impulse of a force $\mathbf{F}$, of components F_x and F_y, during a given time interval may be defined by the integrals of F_x and F_y over that time interval. We shall now, in a similar way, define the *angular impulse* of a couple $\mathbf{M}$ over a time interval from t_1 to t_2 as the integral

$$\int_{t_1}^{t_2} M \, dt$$

of the moment M of the couple over the given time interval. This new concept, as we shall see presently, will enable us to express in a more general form the principle of impulse and momentum for a system of particles.

Consider a system of particles and the *external forces* acting on the various particles. Each external force $\mathbf{F}$ may be replaced by an equal force $\mathbf{F}$ attached at the origin O, plus a couple $\mathbf{M_O}$ of moment equal to the moment about O of the original force (Chap. 3). Now we shall draw three separate sketches (Fig. 14.16). The first and third sketches will show the systems of vectors representing the momenta of the various particles at a

(a) $\qquad\qquad\qquad\qquad$ (b) $\qquad\qquad\qquad\qquad$ (c) $\qquad\qquad$ **Fig. 14.16** time t_1 and a time t_2, respectively. The second sketch will show the resultant of the vectors representing the linear impulses of the external forces $\mathbf{F}$ during the given time interval, as well as the sum of the couples representing the angular impulses of the couples $\mathbf{M}_O$ during the same time interval. We recall from Sec. 14.3 that the sums of the x and y components of the vectors shown in parts a and b of Fig. 14.16 must be respectively equal to the sums of the x and y components of the vectors in part c of the same figure. Indeed, rewriting Eq. (14.8) in terms of x and y components, we have

$$\sum (mv_x)_1 + \sum \int_{t_1}^{t_2} (F_x)_\text{ext}\, dt = \sum (mv_x)_2$$

$$(14.34)$$

$$\sum (mv_y)_1 + \sum \int_{t_1}^{t_2} (F_y)_\text{ext}\, dt = \sum (mv_y)_2$$

Returning now to Eq. (14.33) and integrating both members over the time interval from t_1 to t_2, we write

$$(H_O)_1 + \sum \int_{t_1}^{t_2} (M_O)_\text{ext}\, dt = (H_O)_2 \qquad (14.35)$$

The equation obtained expresses that the sum of the moments about O of the vectors shown in parts a and b of Fig. 14.16 must be equal to the sum of the moments about O of the vectors shown in part c of the same figure. Together, Eqs. (14.34) and (14.35) thus express that *the momenta of the particles at time t_1 and the impulses of the external forces from t_1 to t_2 form a system of vectors equipollent to the system of the momenta of the particles at time t_2*.[†] We write

$$\textbf{Syst Momenta}_1 + \textbf{Syst Ext Imp}_{1 \to 2} = \textbf{Syst Momenta}_2 \quad (14.36)$$

This more general statement of the principle of impulse and momentum for a system of particles will prove particularly useful for the study of the motion of rigid bodies (Chap. 18).

When no external force acts on the particles of a system, the integrals in Eqs. (14.34) and (14.35) are zero, and part b of Fig. 14.16 may be omitted. The system of the momenta of the particles at time t_1 is equipollent to the system of the momenta at time t_2:

$$\textbf{Syst Momenta}_1 = \textbf{Syst Momenta}_2 \qquad (14.37)$$

[†] Since we are dealing here with a system of particles, and not with a single rigid body, we cannot conclude that the two systems of vectors are actually equivalent. Hence we use gray plus and equals signs in Fig. 14.16 (cf. Sec. 12.4).

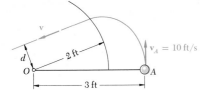

SAMPLE PROBLEM 14.8

A ball weighing 0.5 lb is attached to a fixed point O by means of an elastic cord of constant $k = 10$ lb/ft and of undeformed length equal to 2 ft. The ball slides on a horizontal frictionless surface. If the ball is placed at point A, 3 ft from O, and is given an initial velocity of 10 ft/s in a direction perpendicular to OA, determine (a) the speed of the ball after the cord has become slack, (b) the closest distance d that the ball will come to O.

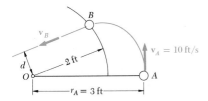

Solution.   The force exerted by the cord on the ball passes through the fixed point O, and its work may be expressed as a change in potential energy. It is therefore a conservative central force, and both the total energy of the ball and its angular momentum about O are conserved between points A and B. After the cord has become slack at B, the resultant force acting on the ball is zero. The ball, therefore, will move in a straight line at a constant speed v. The straight line is the line of action of $\mathbf{v}_B$ and the speed v is equal to v_B.

   a. Conservation of Energy.

At point A: $T_A = \frac{1}{2}mv_A^2 = \frac{1}{2}\dfrac{0.5\ \text{lb}}{32.2\ \text{ft/s}^2}(10\ \text{ft/s})^2 = 0.776\ \text{ft}\cdot\text{lb}$

$V_A = \frac{1}{2}kx_A^2 = \frac{1}{2}(10\ \text{lb/ft})(3\ \text{ft} - 2\ \text{ft})^2 = 5\ \text{ft}\cdot\text{lb}$

At point B: $T_B = \frac{1}{2}mv_B^2 = \frac{1}{2}\dfrac{0.5}{32.2}v_B^2 = 0.00776 v_B^2$

$V_B = 0$

Applying the principle of conservation of energy between points A and B, we write

$$T_A + V_A = T_B + V_B$$
$$0.776 + 5 = 0.00776 v_B^2$$
$$v_B^2 = 744 \qquad v = v_B = 27.3\ \text{ft/s} \quad \blacktriangleleft$$

   b. Conservation of Angular Momentum About O.   Since r_A and d represent the perpendicular distances to $\mathbf{v}_A$ and $\mathbf{v}_B$, respectively, we write

$$r_A(mv_A) = d(mv_B)$$
$$(3\ \text{ft})\left(\frac{0.5}{g}\right)(10\ \text{ft/s}) = d\left(\frac{0.5}{g}\right)(27.3\ \text{ft/s})$$
$$d = 1.099\ \text{ft} \quad \blacktriangleleft$$

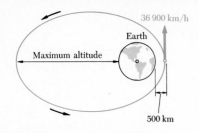

36 900 km/h

Earth

Maximum altitude

500 km

A satellite is launched in a direction parallel to the surface of the earth with a velocity of 36 900 km/h from an altitude of 500 km. Determine (*a*) the maximum altitude reached by the satellite, (*b*) the maximum allowable error in the direction of launching if the satellite is to go into orbit and to come not closer than 200 km to the surface of the earth.

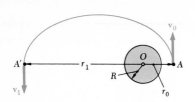

a. **Maximum Altitude.** We denote by A' the point of the orbit farthest from the earth and by r_1 the corresponding distance from the center of the earth. Since the satellite is in free flight between A and A', we apply the principle of conservation of energy.

$$T_A + V_A = T_{A'} + V_{A'}$$

$$\tfrac{1}{2}mv_0^2 - \frac{GMm}{r_0} = \tfrac{1}{2}mv_1^2 - \frac{GMm}{r_1} \tag{1}$$

Since the only force acting on the satellite is the force of gravity, which is a central force, the angular momentum of the satellite about O is conserved. Considering points A and A', we write

$$r_0 m v_0 = r_1 m v_1 \qquad v_1 = v_0 \frac{r_0}{r_1} \tag{2}$$

Substituting this expression for v_1 into Eq. (1) and dividing each term by the mass m, we obtain after rearranging the terms,

$$\tfrac{1}{2}v_0^2 \left(1 - \frac{r_0^2}{r_1^2}\right) = \frac{GM}{r_0}\left(1 - \frac{r_0}{r_1}\right) \qquad 1 + \frac{r_0}{r_1} = \frac{2GM}{r_0 v_0^2} \tag{3}$$

Recalling that the radius of the earth is $R = 6370$ km, we compute

$$r_0 = 6370 \text{ km} + 500 \text{ km} = 6870 \text{ km} = 6.87 \times 10^6 \text{ m}$$
$$v_0 = 36\,900 \text{ km/h} = (3.69 \times 10^7 \text{ m})/(3.6 \times 10^3 \text{ s}) = 1.025 \times 10^4 \text{ m/s}$$
$$GM = gR^2 = (9.81 \text{ m/s}^2)(6.37 \times 10^6 \text{ m})^2 = 3.98 \times 10^{14} \text{ m}^3/\text{s}^2$$

Substituting these values into (3), we obtain $r_1 = 66.8 \times 10^6$ m

Maximum altitude $= 66.8 \times 10^6 \text{ m} - 6.37 \times 10^6 \text{ m} = 60.4 \times 10^6 \text{ m}$
$$= 60\,400 \text{ km} \quad \blacktriangleleft$$

b. **Allowable Error in Direction of Launching.** The satellite is launched from P_0 in a direction forming an angle ϕ_0 with the vertical OP_0. The value of ϕ_0 corresponding to $r_{\min} = 6370$ km $+ 200$ km $= 6570$ km is obtained by applying the principles of conservation of energy and of conservation of angular momentum between P_0 and A.

$$\tfrac{1}{2}mv_0^2 - \frac{GMm}{r_0} = \tfrac{1}{2}mv_{\max}^2 - \frac{GMm}{r_{\min}} \tag{4}$$

$$r_0 m v_0 \sin \phi_0 = r_{\min} m v_{\max} \tag{5}$$

Solving (5) for $v_{\max}$ and then substituting for $v_{\max}$ into (4), we may solve (4) for $\sin \phi_0$. Using the values of v_0 and GM computed in part *a* and noting that $r_0/r_{\min} = 6870/6570 = 1.0457$, we find

$$\sin \phi_0 = 0.9801 \qquad \phi_0 = 90° \pm 11.5° \qquad \text{Allowable error} = \pm 11.5° \quad \blacktriangleleft$$

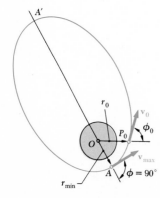

PROBLEMS

14.54 A 2-kg sphere is attached to an elastic cord of constant 150 N/m which is undeformed when the sphere is located at the origin O. Knowing that in the position shown v_A is perpendicular to OP and has a magnitude of 10 m/s, determine (a) the maximum distance from the origin attained by the sphere, (b) the corresponding speed of the sphere.

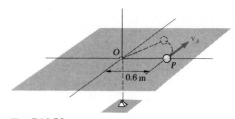

Fig. P14.54

14.55 In Prob. 14.54, determine the required initial speed v_A if the maximum distance from the origin attained by the sphere is to be 1.5 m.

14.56 In Sample Prob. 14.8, determine the required magnitude of v_A if the ball is to pass at a distance $d = 4$ in. from point O. Assume that the direction of v_A is not changed.

14.57 In Sample Prob. 14.8, determine the smallest magnitude of v_A for which the elastic cord will remain taut at all times.

14.58 Collar B weighs 10 lb and is attached to a spring of constant 50 lb/ft and of undeformed length equal to 18 in. The system is set in motion with $r = 12$ in., $v_\theta = 16$ ft/s, and $v_r = 0$. Neglecting the mass of the rod and the effect of friction, determine the radial and transverse components of the velocity of the collar when $r = 21$ in.

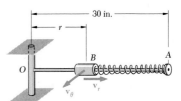

Fig. P14.58

14.59 For the motion described in Prob. 14.58, determine (a) the maximum distance between the origin and the collar, (b) the corresponding velocity.

14.60 through 14.63 Using the principles of conservation of energy and of conservation of angular momentum, solve the following problems:

    14.60 Prob. 12.82.
    14.61 Prob. 12.84.
    14.62 Prob. 12.86.
    14.63 Prob. 12.85.

14.64 Upon the LEM's return to the command module, the Apollo spacecraft of Prob. 12.84 is turned around so that the LEM faces to the rear. The LEM is then cast adrift with a velocity of 600 ft/s relative to the command module. Determine the magnitude and direction (angle ϕ formed with the vertical OC) of the velocity v_C of the LEM just before it crashes at C on the moon's surface.

14.65 Determine the magnitude and direction (angle ϕ formed with the vertical OB) of the velocity v_B of the spacecraft of Prob. 12.91 just before splashdown at B. Neglect the effect of the atmosphere.

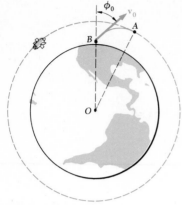

Fig. P14.67

14.66 To what value v_0 should the speed of the spacecraft of Prob. 12.92 be reduced preparatory to reentry if its velocity v_B just before splashdown at B is to form an angle $\phi = 30°$ with the vertical OB? Neglect the effect of the atmosphere.

14.67 A spacecraft is launched from the ground to rendezvous with a skylab orbiting the earth at a constant altitude of 1040 mi. The spacecraft has reached at burnout an altitude of 40 miles and a velocity v_0 of magnitude 20,000 ft/s. What is the angle ϕ_0 that v_0 should form with the vertical OB if the trajectory of the spacecraft is to be tangent at A to the orbit of the skylab?

14.68 In Prob. 14.67, determine the maximum and minimum values of v_0, and the corresponding values of ϕ_0, for which the trajectory of the spacecraft will be tangent to the orbit of the skylab.

14.69 At engine burnout a satellite has reached an altitude of 2400 km and has a velocity v_0 of magnitude 8100 m/s forming an angle $\phi_0 = 76°$ with the vertical. Determine the maximum and minimum heights reached by the satellite.

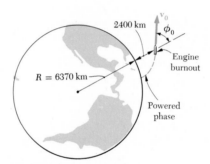

Fig. P14.69 and P14.70

14.70 At engine burnout a satellite has reached an altitude of 2400 km and has a velocity v_0 of magnitude 8100 m/s. For what range of values of the angle ϕ_0, formed by v_0 and the vertical, will the satellite go into a permanent orbit? (Assume that if the satellite gets closer than 300 km from the earth's surface, it will soon burn up.)

14.71 A satellite is projected into space with a velocity v_0 at a distance r_0 from the center of the earth by the last stage of its launching rocket. The velocity v_0 was designed to send the satellite into a circular orbit of radius r_0. However, owing to a malfunction of control, the satellite is not projected horizontally but at an angle α with the horizontal and, as a result, is propelled into an elliptic orbit. Determine the maximum and minimum values of the distance from the center of the earth to the satellite.

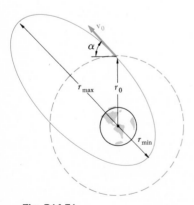

Fig. P14.71

*14.72 Using the answers obtained in Prob. 14.71, show that the intended circular orbit and the resulting elliptic orbit intersect at the ends of the minor axis of the elliptic orbit.

14.73 A spacecraft of mass m describes a circular orbit of radius r_1 around the earth. (a) Show that the additional energy ΔE which must be imparted to the spacecraft to transfer it to a circular orbit of larger radius r_2 is

$$\Delta E = \frac{GMm(r_2 - r_1)}{2r_1 r_2}$$

where M is the mass of the earth. (b) Further show that, if the transfer from one circular orbit to the other is executed by placing the spacecraft on a transitional semielliptic path AB, the amounts of energy ΔE_A and ΔE_B which must be imparted at A and B are respectively proportional to r_2 and r_1:

$$\Delta E_A = \frac{r_2}{r_1 + r_2} \Delta E \qquad \Delta E_B = \frac{r_1}{r_1 + r_2} \Delta E$$

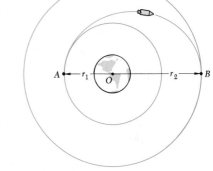

Fig. P14.73

14.74 Show that the total energy E of a satellite of mass m describing an elliptic orbit is $E = -GMm/(r_1 + r_2)$, where M is the mass of the earth, and r_1 and r_2 represent, respectively, the maximum and minimum distance of the orbit to the center of the earth. (It is recalled that the gravitational potential energy of a satellite was defined as being zero at an infinite distance from the earth.)

*14.75 A missile is fired from the ground with a velocity $\mathbf{v}_0$ of magnitude $v_0 = \sqrt{gR}$, forming an angle ϕ_0 with the vertical. (a) Express the maximum height d reached by the missile in terms of ϕ_0. (b) Show that the angle 2α subtending the trajectory BAC of the missile is equal to $2\phi_0$ and explain what happens when ϕ_0 approaches 90°. [*Hint.* Use Eq. (12.30) of Sec. 12.11 to solve part b, noting that $\theta = 180°$ for point A and $\theta = 180° - \alpha$ for point B.]

14.76 A missile is fired from the ground with an initial velocity $\mathbf{v}_0$ forming an angle ϕ_0 with the vertical. If the missile is to reach a maximum altitude equal to the radius of the earth, (a) show that the required angle ϕ_0 is defined by the relation

$$\sin \phi_0 = 2\sqrt{1 - \frac{1}{2}\left(\frac{v_{esc}}{v_0}\right)^2}$$

where v_{esc} is the escape velocity, (b) determine the maximum and minimum allowable values of v_0.

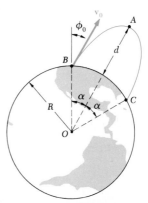

Fig. P14.75 and P14.76

*14.13. Variable Systems of Particles.   All the systems of particles considered so far consisted of well-defined particles. These systems did not gain or lose any particles during their motion. In a large number of engineering applications, however, it is necessary to consider *variable systems of particles,* i.e., systems which are continuously gaining or losing particles, or doing both at the same time. Consider, for example, a hydraulic turbine. Its analysis involves the determination of the forces exerted by a stream of water on rotating blades, and we note that the particles of water in contact with the blades form an everchanging system which continuously acquires and loses particles. Rockets furnish another example of variable systems, since their propulsion depends upon the continuous ejection of fuel particles.

We recall that all the kinetics principles established so far were derived for constant systems of particles, which neither gain nor lose particles. We must therefore find a way to reduce the analysis of a variable system of particles to that of an auxiliary constant system. The procedure to follow is indicated in Secs. 14.14 and 14.15 for two broad categories of applications.

*14.14. Steady Stream of Particles.   Consider a steady stream of particles, such as a stream of water diverted by a fixed vane or a flow of air through a duct or through a blower. In order to determine the resultant of the forces exerted on the particles in contact with the vane, duct, or blower, we isolate these particles and denote by S the system thus defined (Fig. 14.17). We observe that S is a variable system of particles, since it continuously gains particles flowing in and loses an equal number of particles flowing out. Therefore, the kinetics principles that have been established so far cannot be directly applied to S.

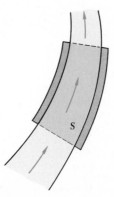

Fig. 14.17

However, we may easily define an auxiliary system of particles which does remain constant for a short interval of time Δt. Consider at time t the system S *plus* the particles which will enter S during the interval of time Δt (Fig. 14.18a). Next, consider at time $t + \Delta t$ the system S *plus* the particles which have left S during the interval Δt (Fig. 14.18c). Clearly, *the same particles are involved in both cases,* and we may apply to these particles the principle of impulse and momentum. Since the total mass m of the system S remains constant, the particles entering the system and those leaving the system in the time Δt must have the same mass Δm. Denoting by $\mathbf{v}_A$ and $\mathbf{v}_B$, respectively, the velocities of the particles entering S at A and leaving S at B, we represent the momentum of the particles entering S by $(\Delta m)\mathbf{v}_A$ (Fig. 14.18a) and the momentum of the particles leaving S by $(\Delta m)\mathbf{v}_B$ (Fig. 14.18c). We also represent the momenta of the particles forming S and the impulses of the forces exerted on S by the appropriate vectors, and indicate by gray plus and equals signs that the system of the momenta and impulses in parts a and b of Fig. 14.18 is equipollent to the system of the momenta in part c of the same figure.

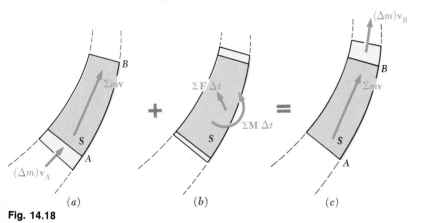

Fig. 14.18

Since the resultant $\Sigma m\mathbf{v}$ of the momenta of the particles of S is found on both sides of the equals sign, it may be omitted. We conclude that *the system formed by the momentum $(\Delta m)\mathbf{v}_A$ of the particles entering S in the time Δt and the impulses of the forces exerted on S during that time is equipollent to the momentum $(\Delta m)\mathbf{v}_B$ of the particles leaving S in the same time Δt.* We may therefore write

$$(\Delta m)\mathbf{v}_A + \Sigma \mathbf{F}\,\Delta t = (\Delta m)\mathbf{v}_B \qquad (14.38)$$

A similar equation may be obtained by taking the moments of the vectors involved (see Sample Prob. 14.10). Dividing all terms of Eq. (14.38) by Δt and letting Δt approach zero, we obtain at the limit

$$\Sigma \mathbf{F} = \frac{dm}{dt}(\mathbf{v}_B - \mathbf{v}_A) \tag{14.39}$$

where $\mathbf{v}_B - \mathbf{v}_A$ represents the difference between the *vectors* $\mathbf{v}_B$ and $\mathbf{v}_A$.

If SI units are used, dm/dt is expressed in kg/s and the velocities in m/s; we check that both members of Eq. (14.39) are expressed in the same units (newtons). If U.S. customary units are used, dm/dt must be expressed in slugs/s and the velocities in ft/s; we check again that both members of the equation are expressed in the same units (pounds).†

The principle we have established may be used to analyze a large number of engineering applications. Some of the most common are indicated below.

Fluid Stream Diverted by a Vane. If the vane is fixed, the method of analysis given above may be applied directly to find the force $\mathbf{F}$ exerted by the vane on the stream. We note that $\mathbf{F}$ is the only force which needs to be considered since the pressure in the stream is constant (atmospheric pressure). The force exerted by the stream on the vane will be equal and opposite to $\mathbf{F}$. If the vane moves with a constant velocity, the stream is not steady. However, it will appear steady to an observer moving with the vane. We should therefore choose a system of axes moving with the vane. Since this system of axes is not accelerated, Eq. (14.38) may still be used, but $\mathbf{v}_A$ and $\mathbf{v}_B$ must be replaced by the *relative velocities* of the stream with respect to the vane (see Sample Prob. 14.11).

Fluid Flowing through a Pipe. The force exerted by the fluid on a pipe transition such as a bend or a contraction may be determined by considering the system of particles S in contact with the transition. Since, in general, the pressure in the flow will vary, we should also consider the forces exerted on S by the adjoining portions of the fluid.

---

† It is often convenient to express the mass rate of flow dm/dt as the product ρQ, where ρ is the density of the stream (mass per unit volume) and Q its volume rate of flow (volume per unit time). If SI units are used, ρ is expressed in kg/m³ (for instance, $\rho = 1000 \text{ kg/m}^3$ for water) and Q in m³/s. However, if U.S. customary units are used, ρ will generally have to be computed from the corresponding specific weight γ (weight per unit volume), $\rho = \gamma/g$. Since γ is expressed in lb/ft³ (for instance, $\gamma = 62.4 \text{ lb/ft}^3$ for water), ρ is obtained in slugs/ft³. The volume rate of flow Q is expressed in ft³/s.

Jet Engine. In a jet engine, air enters with no velocity through the front of the engine and leaves through the rear with a high velocity. The energy required to accelerate the air particles is obtained by burning fuel. While the exhaust gases contain burned fuel, the mass of the fuel is small compared with the mass of the air flowing through the engine and usually may be neglected. Thus, the analysis of a jet engine reduces to that of an air stream. This stream may be considered as a steady stream if all velocities are measured with respect to the airplane. The air stream shall be assumed, therefore, to enter the engine with a velocity **v** of magnitude equal to the speed of the airplane and to leave with a velocity **u** equal to the relative velocity of the exhaust gases (Fig. 14.19). Since the intake and exhaust

Fig. 14.19

pressures are nearly atmospheric, the only external force which needs to be considered is the force exerted by the engine on the air stream. This force is equal and opposite to the thrust.†

Fan. We consider the system of particles S shown in Fig. 14.20. The velocity v_A of the particles entering the system is assumed equal to zero, and the velocity v_B of the particles leaving the system is the velocity of the *slipstream*. The rate of flow may be obtained by multiplying v_B by the cross-sectional

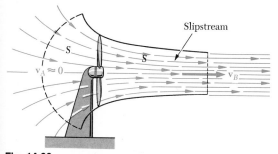

Slipstream

Fig. 14.20

† Note that, if the airplane is accelerated, it cannot be used as a newtonian frame of reference. The same result will be obtained for the thrust, however, by using a reference frame at rest with respect to the atmosphere, since the air particles will then be observed to enter the engine with no velocity and to leave it with a velocity of magnitude $u - v$.

area of the slipstream. Since the pressure all around S is atmospheric, the only external force acting on S is the thrust of the fan.

Airplane Propeller. In order to obtain a steady stream of air, velocities should be measured with respect to the airplane. Thus, the air particles will be assumed to enter the system with a velocity **v** of magnitude equal to the speed of the airplane and to leave with a velocity **u** equal to the relative velocity of the slipstream.

*14.15. Systems Gaining or Losing Mass.** We shall now analyze a different type of variable system of particles, namely, a system which gains mass by continuously absorbing particles or loses mass by continuously expelling particles. Consider the system S shown in Fig. 14.21. Its mass, equal to m at the instant t, increases by Δm in the interval of time Δt. In order to apply the principle of impulse and momentum to the analysis of this system, we must consider at time t the system

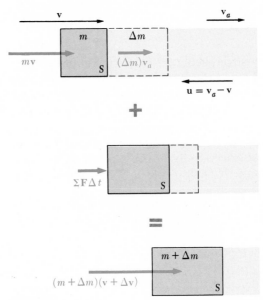

Fig. 14.21

S *plus* the particles of mass Δm which S absorbs during the time interval Δt. The velocity of S at time t is denoted by **v**, and its velocity at time $t + \Delta t$ is denoted by $\mathbf{v} + \Delta\mathbf{v}$, while the absolute velocity of the particles which are absorbed is denoted by $\mathbf{v}_a$. Applying the principle of impulse and momentum, we write

$$m\mathbf{v} + (\Delta m)\mathbf{v}_a + \Sigma\mathbf{F}\,\Delta t = (m + \Delta m)(\mathbf{v} + \Delta\mathbf{v})$$

Solving for the sum $\Sigma \mathbf{F} \, \Delta t$ of the impulses of the external forces acting on S (excluding the forces exerted by the particles being absorbed), we have

$$\Sigma \mathbf{F} \, \Delta t = m \, \Delta \mathbf{v} + \Delta m(\mathbf{v} - \mathbf{v}_a) + (\Delta m)(\Delta \mathbf{v}) \qquad (14.40)$$

Introducing the *relative velocity* $\mathbf{u}$ with respect to S of the particles which are absorbed, we write $\mathbf{u} = \mathbf{v}_a - \mathbf{v}$ and note, since $v_a < v$, that the relative velocity $\mathbf{u}$ is directed to the left, as shown in Fig. 14.21. Neglecting the last term in Eq. (14.40), which is of the second order, we write

$$\Sigma \mathbf{F} \, \Delta t = m \, \Delta \mathbf{v} - (\Delta m)\mathbf{u}$$

Dividing through by Δt and letting Δt approach zero, we have at the limit†

$$\Sigma \mathbf{F} = m \, \frac{d\mathbf{v}}{dt} - \frac{dm}{dt} \, \mathbf{u} \qquad (14.41)$$

Rearranging the terms, we obtain the equation

$$\Sigma \mathbf{F} + \frac{dm}{dt} \, \mathbf{u} = m \, \frac{d\mathbf{v}}{dt} \qquad (14.42)$$

which shows that the action on S of the particles being absorbed is equivalent to a thrust of magnitude $(dm/dt)u$ which tends to slow down the motion of S, since the relative velocity $\mathbf{u}$ of the particles is directed to the left. If SI units are used, dm/dt is expressed in kg/s, the relative velocity u in m/s, and the corresponding thrust in newtons. If U.S. customary units are used, dm/dt must be expressed in slugs/s and u in ft/s; the corresponding thrust will then be expressed in pounds. ‡

The equations obtained may also be used to determine the motion of a system S losing mass. In this case, the rate of change of mass is negative, and the action on S of the particles being expelled is equivalent to a thrust in the direction of $-\mathbf{u}$, that is, in the direction opposite to that in which the particles are being expelled. A *rocket* represents a typical case of a system continuously losing mass (see Sample Prob. 14.12).

†When the absolute velocity $\mathbf{v}_a$ of the particles absorbed is zero, we have $\mathbf{u} = -\mathbf{v}$, and formula (14.41) becomes

$$\Sigma \mathbf{F} = \frac{d}{dt}(m\mathbf{v})$$

Comparing the formula obtained to Eq. (14.2) of Sec. 14.1, we observe that Newton's second law may be applied to a system gaining mass, *provided that the particles absorbed are initially at rest.* It may also be applied to a system losing mass, *provided that the velocity of the particles expelled is zero* with respect to the frame of reference selected.

‡See footnote on page 568.

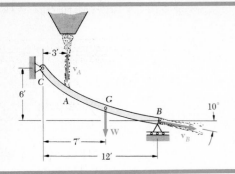

Grain falls from a hopper onto a chute CB at the rate of 240 lb/s. It hits the chute at A with a velocity of 20 ft/s and leaves at B with a velocity of 15 ft/s, forming an angle of 10° with the horizontal. Knowing that the combined weight of the chute and of the grain it supports is a force **W** of magnitude 600 lb applied at G, determine the reaction at the roller-support B and the components of the reaction at the hinge C.

Solution. We apply the principle of impulse and momentum for the time interval Δt to the system consisting of the chute, the grain it supports, and the amount of grain which hits the chute in the interval Δt. Since the chute does not move, it has no momentum. We also note that the sum $\Sigma m\mathbf{v}$ of the momenta of the particles supported by the chute is the same at t and $t + \Delta t$ and thus may be omitted.

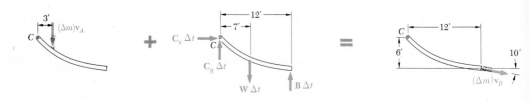

Since the system formed by the momentum $(\Delta m)\mathbf{v}_A$ and the impulses is equipollent to the momentum $(\Delta m)\mathbf{v}_B$, we write

$\xrightarrow{+} x$ components: $C_x \Delta t = (\Delta m)v_B \cos 10°$ (1)

$+\uparrow y$ components: $-(\Delta m)v_A + C_y \Delta t - W \Delta t + B \Delta t$

$\qquad\qquad\qquad\qquad = -(\Delta m)v_B \sin 10°$ (2)

$+\curvearrowleft$ moments about C: $-3(\Delta m)v_A - 7(W \Delta t) + 12(B \Delta t)$

$\qquad\qquad\qquad = 6(\Delta m)v_B \cos 10° - 12(\Delta m)v_B \sin 10°$ (3)

Using the given data, $W = 600$ lb, $v_A = 20$ ft/s, $v_B = 15$ ft/s, $\Delta m/\Delta t = 240/32.2 = 7.45$ slugs/s, and solving Eq. (3) for B and Eq. (1) for C_x:

$\qquad 12B = 7(600) + 3(7.45)(20) + 6(7.45)(15)(\cos 10° - 2 \sin 10°)$

$\qquad 12B = 5075 \qquad B = 423$ lb $B = 423$ lb ↑ ◀

$\qquad\qquad C_x = (7.45)(15) \cos 10° = 110.1$ lb $C_x = 110.1$ lb → ◀

Substituting for B and solving Eq. (2) for C_y:

$\qquad\qquad C_y = 600 - 423 + (7.45)(20 - 15 \sin 10°) = 307$ lb

$\qquad\qquad\qquad\qquad\qquad C_y = 307$ lb ↑ ◀

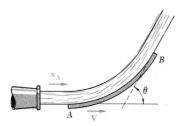

A nozzle discharges a stream of water of cross-sectional area A with a velocity v_A. The stream is deflected by a *single* blade which moves to the right with a constant velocity $\mathbf{V}$. Assuming that the water moves along the blade at constant speed, determine (a) the components of the force $\mathbf{F}$ exerted by the blade on the stream, (b) the velocity $\mathbf{V}$ for which maximum power is developed.

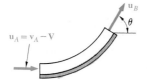

a. **Components of Force Exerted on Stream.** We choose a coordinate system which moves with the blade at a constant velocity $\mathbf{V}$. The particles of water strike the blade with a relative velocity $\mathbf{u}_A = \mathbf{v}_A - \mathbf{V}$ and leave the blade with a relative velocity $\mathbf{u}_B$. Since the particles move along the blade at a constant speed, the relative velocities $\mathbf{u}_A$ and $\mathbf{u}_B$ have the same magnitude u. Denoting the density of water by ρ, the mass of the particles striking the blade during the time interval Δt is $\Delta m = A\rho(v_A - V)\,\Delta t$; an equal mass of particles leaves the blade during Δt. We apply the principle of impulse and momentum to the system formed by the particles in contact with the blade and by those striking the blade in the time Δt.

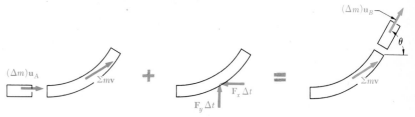

Recalling that $\mathbf{u}_A$ and $\mathbf{u}_B$ have the same magnitude u, and omitting the momentum $\Sigma m\mathbf{v}$ which appears on both sides, we write

$\xrightarrow{+}\ x$ components: $\qquad (\Delta m)u - F_x\,\Delta t = (\Delta m)u\cos\theta$

$+\!\uparrow y$ components: $\qquad\qquad\quad +F_y\,\Delta t = (\Delta m)u\sin\theta$

Substituting $\Delta m = A\rho(v_A - V)\,\Delta t$ and $u = v_A - V$, we obtain

$$F_x = A\rho(v_A - V)^2(1 - \cos\theta) \leftarrow \qquad F_y = A\rho(v_A - V)^2\sin\theta\uparrow \quad \blacktriangleleft$$

b. **Velocity of Blade for Maximum Power.** The power is obtained by multiplying the velocity V of the blade by the component F_x of the force exerted by the stream on the blade.

$$\text{Power} = F_x V = A\rho(v_A - V)^2(1 - \cos\theta)V$$

Differentiating the power with respect to V and setting the derivative equal to zero, we obtain

$$\frac{d(\text{power})}{dV} = A\rho(v_A^2 - 4v_A V + 3V^2)(1 - \cos\theta) = 0$$

$$V = v_A \qquad V = \tfrac{1}{3}v_A \qquad \text{For maximum power } V = \tfrac{1}{3}v_A \to \quad \blacktriangleleft$$

Note. These results are valid only when a *single* blade deflects the stream. Different results are obtained when a series of blades deflects the stream, as in a Pelton-wheel turbine. (See Prob. 14.97.)

A rocket of initial mass m_0 (including shell and fuel) is fired vertically at time $t = 0$. The fuel is consumed at a constant rate $q = dm/dt$ and is expelled at a constant speed u relative to the rocket. Derive an expression for the velocity of the rocket at time t, neglecting the resistance of the air.

Solution. At time t, the mass of the rocket shell and remaining fuel is $m = m_0 - qt$, and the velocity is $\mathbf{v}$. During the time interval Δt, a mass of fuel $\Delta m = q\,\Delta t$ is expelled with a speed u relative to the rocket. Denoting by $\mathbf{v}_e$ the absolute velocity of the expelled fuel, we apply the principle of impulse and momentum between time t and time $t + \Delta t$.

 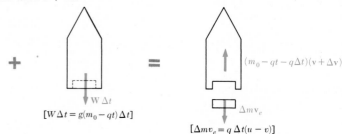

We write

$$(m_0 - qt)v - g(m_0 - qt)\,\Delta t$$
$$= (m_0 - qt - q\,\Delta t)(v + \Delta v) - q\,\Delta t(u - v)$$

Dividing through by Δt, and letting Δt approach zero, we obtain

$$-g(m_0 - qt) = (m_0 - qt)\frac{dv}{dt} - qu$$

Separating variables and integrating from $t = 0$, $v = 0$ to $t = t$, $v = v$,

$$dv = \left(\frac{qu}{m_0 - qt} - g\right)dt \qquad \int_0^v dv = \int_0^t \left(\frac{qu}{m_0 - qt} - g\right)dt$$

$$v = [-u\ln(m_0 - qt) - gt]_0^t \qquad v = u\ln\frac{m_0}{m_0 - qt} - gt \quad \blacktriangleleft$$

Remark. The mass remaining at time t_f, after all the fuel has been expended, is equal to the mass of the rocket shell $m_s = m_0 - qt_f$, and the maximum velocity attained by the rocket is $v_m = u\ln(m_0/m_s) - gt_f$. Assuming that the fuel is expelled in a relatively short period of time, the term gt_f is small and we have $v_m \approx u\ln(m_0/m_s)$. In order to escape the gravitational field of the earth, a rocket must reach a velocity of 11 180 m/s. Assuming $u = 2200$ m/s and $v_m = 11\,180$ m/s, we obtain $m_0/m_s = 161$. Thus, to project each kilogram of the rocket shell into space, it is necessary to consume more than 161 kg of fuel if a propellant yielding $u = 2200$ m/s is used.

PROBLEMS

Note. In the following problems use $\rho = 1000\,\text{kg/m}^3$ for the density of water in SI units, and $\gamma = 62.4\,\text{lb/ft}^3$ for its specific weight in U.S. customary units. (See footnote on page 568.)

14.77 A stream of water of cross-sectional area A and velocity v_1 strikes a plate which is held motionless by a force **P**. Determine the magnitude of **P** when $A = 500\,\text{mm}^2$, $v_1 = 40\,\text{m/s}$, and $V = 0$.

14.78 A stream of water of cross-sectional area A and velocity v_1 strikes a plate which moves to the right with a velocity **V**. Determine the magnitude of the force **P** required to hold the plate when $A = 2\,\text{in.}^2$, $v_1 = 120\,\text{ft/s}$, and $V = 30\,\text{ft/s}$.

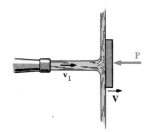

Fig. P14.77 and P14.78

14.79 Water flows in a continuous sheet from between two plates A and B with a velocity **v**. The stream is split into two equal streams 1 and 2 by a vane attached to plate C. Denoting the total rate of flow by Q, determine the force exerted by the stream on plate C.

14.80 Water flows in a continuous sheet from between two plates A and B with a velocity **v**. The stream is split into two parts by a smooth horizontal plate C. Denoting the total rate of flow by Q, determine the rate of flow of each of the resulting streams. (*Hint.* The plate C can exert only a vertical force on the water.)

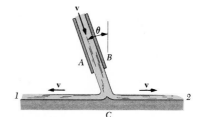

Fig. P14.79

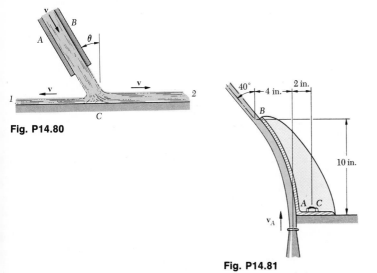

Fig. P14.80

Fig. P14.81

14.81 The nozzle shown discharges 200 gal/min of water with a velocity v_A of 100 ft/s. The stream is deflected by the fixed vane AB. Determine the force-couple system which must be applied at C in order to hold the vane in place ($1\,\text{ft}^3 = 7.48\,\text{gal}$).

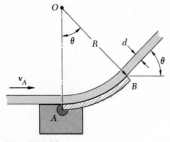

Fig. P14.82

14.82 A stream of water of cross-sectional area A and velocity $\mathbf{v}_A$ is deflected by a vane AB in the shape of an arc of circle of radius R. Knowing that the vane is welded to a fixed support at A, determine the components of the force-couple system exerted by the support on the vane.

14.83 Knowing that the blade AB of Sample Prob. 14.11 is in the shape of an arc of circle, show that the resultant force $\mathbf{F}$ exerted by the blade on the stream is applied at the midpoint C of the arc AB. (*Hint.* First show that the line of action of $\mathbf{F}$ must pass through the center O of the circle.)

14.84 The nozzle shown discharges water at the rate of $1.2 \text{ m}^3/\text{min}$. Knowing that at both A and B the stream of water moves with a velocity of magnitude 25 m/s and neglecting the weight of the vane, determine the components of the reactions at C and D.

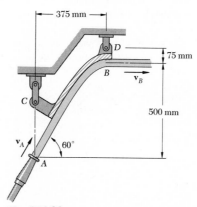

Fig. P14.84

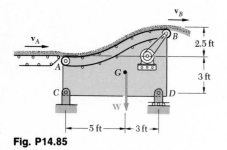

Fig. P14.85

14.85 The final component of a conveyor system receives sand at the rate of 180 lb/s at A and discharges it at B. The sand is moving horizontally at A and B with a velocity of magnitude $v_A = v_B = 12 \text{ ft/s}$. Knowing that the combined weight of the component and of the sand it supports is $W = 800 \text{ lb}$, determine the reactions at C and D.

14.86 Solve Prob. 14.85, assuming the velocity of the belt of the final component of the conveyor system is increased in such a way that, while the sand is still received with a velocity $\mathbf{v}_A$ of 12 ft/s, it is discharged with a velocity $\mathbf{v}_B$ of 24 ft/s.

14.87 A jet airplane scoops in air at the rate of 200 lb/s and discharges it with a velocity of 2000 ft/s relative to the airplane. If the speed of the airplane is 600 mi/h, determine (*a*) the propulsive force developed, (*b*) the horsepower actually used to propel the airplane, (*c*) the horsepower developed by the engine.

14.88 The total drag due to air friction of a jet airplane traveling at 1000 km/h is 16 kN. Knowing that the exhaust velocity is 600 m/s relative to the airplane, determine the mass of air which must pass through the engine per second to maintain the speed of 1000 km/h in level flight.

14.89 While cruising in horizontal flight at a speed of 800 km/h, a 9000-kg jet airplane scoops in air at the rate of 70 kg/s and discharges it with a velocity of 600 m/s relative to the airplane. (*a*) Determine the total drag due to air friction. (*b*) Assuming that the drag is proportional to the square of the speed, determine the horizontal cruising speed if the flow of air through the jet is increased by 10 percent, i.e., to 77 kg/s.

14.90 The maximum speed of a jet airliner is known to be 600 mi/h. Each of the four engines discharges air with a velocity of 2000 ft/s relative to the plane. Assuming that the drag due to air resistance is proportional to the square of the speed, determine the maximum cruising speed of the airliner when only two of the engines are in operation.

Fig. P14.90

14.91 A 9000-kg jet airplane maintains a constant speed of 900 km/h while climbing at an angle $\alpha = 5°$. The airplane scoops in air at the rate of 80 kg/s and discharges it with a velocity of 700 m/s relative to the airplane. If the pilot changes to a horizontal flight and the same engine conditions are maintained, determine (*a*) the initial acceleration of the plane, (*b*) the maximum horizontal speed attained. Assume that the drag due to air friction is proportional to the square of the speed.

Fig. P14.91

14.92 In order to shorten the distance required for landing, a jet airplane is equipped with movable vanes which partially reverse the direction of the air discharged by each of its engines. Each engine scoops in air at the rate of 200 lb/s and discharges it with a velocity of 2000 ft/s relative to the engine. At an instant when the speed of the airplane is 120 mi/h, determine the reversed thrust provided by each of the engines.

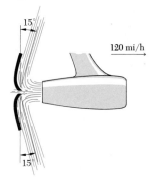

Fig. P14.92

14.93 An unloaded helicopter of weight 1800 lb produces a slip-stream of 28-ft diameter. Assuming that air weighs 0.076 lb/ft³, determine the vertical component of the velocity of the air in the slipstream when the helicopter is hovering in midair.

14.94 The helicopter shown weighs 1800 lb and can produce a maximum downward air speed of 45 ft/s in the 28-ft diameter slipstream. Determine the maximum load which the helicopter can carry while hovering in midair.

Fig. P14.93 and P14.94

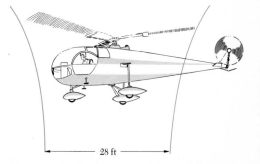

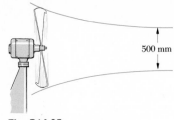

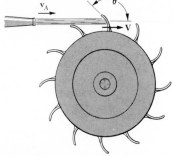

Fig. P14.95

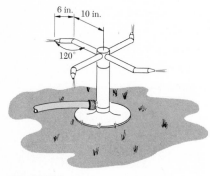

Fig. P14.97

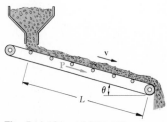

Fig. P14.98

14.95 The slipstream of a fan has a diameter of 500 mm and a velocity of 10 m/s relative to the fan. Assuming $\rho = 1.21$ kg/m³ for air and neglecting the velocity of approach of the air, determine the force required to hold the fan motionless.

14.96 The propeller of an airplane produces a thrust of 4000 N when the airplane is at rest on the ground and has a slipstream of 2-m diameter. Assuming $\rho = 1.21$ kg/m³ for air, determine (a) the speed of the air in the slipstream, (b) the volume of air passing through the propeller per second, (c) the kinetic energy imparted per second to the air of the slipstream.

14.97 In a Pelton-wheel turbine, a stream of water is deflected by a series of blades so that the rate at which water is deflected by the blades is equal to the rate at which water issues from the nozzle ($\Delta m/\Delta t = A\rho v_A$). Using the same notation as in Sample Prob. 14.11, (a) determine the velocity $\mathbf{V}$ of the blades for which maximum power is developed, (b) derive an expression for the maximum power, (c) derive an expression for the mechanical efficiency.

14.98 A garden sprinkler has four rotating arms, each of which consists of two horizontal straight sections of pipe forming an angle of 120°. Each arm discharges water at the rate of 2.50 gal/min with a velocity of 40 ft/s relative to the arm. Knowing that the friction between the moving and stationary parts of the sprinkler is equivalent to a couple of moment $M = 0.200$ lb · ft, determine the constant rate at which the sprinkler rotates (1 ft³ = 7.48 gal).

14.99 Solve Prob. 14.98, assuming that the flow is increased and that each arm now discharges water at the rate of 3.00 gal/min with a velocity of 48 ft/s relative to the arm.

14.100 The kinetic energy available per second from the stream of Sample Prob. 14.11 is $\frac{1}{2}Q\rho v_A^2$, where Q is the flow of water per second, $Q = Av_A$. (a) Derive an expression for the mechanical efficiency of the single blade. (b) Determine the maximum possible value of the efficiency.

14.101 Gravel falls with practically zero velocity onto a conveyor belt at the constant rate $q = dm/dt$. A force $\mathbf{P}$ is applied to the belt to maintain a constant speed v. Derive an expression for the angle θ for which the force $\mathbf{P}$ is zero.

14.102 Gravel falls with practically zero velocity onto a conveyor belt at the constant rate $q = dm/dt$. (a) Determine the magnitude of the force $\mathbf{P}$ required to maintain a constant belt speed v, when $\theta = 0$. (b) Show that the kinetic energy acquired by the gravel in a given time interval is equal to half the work done in that interval by the force $\mathbf{P}$. Explain what happens to the other half of the work done by $\mathbf{P}$.

Fig. P14.101 and P14.102

14.103 The ends of a chain of mass m per unit length lie in piles at A and at C; when released, the chain moves over the pulley at B. Determine the required initial speed v for which the chain will move at a constant speed. Neglect axle friction.

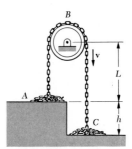

14.104 A chain of mass m per unit length and total length l lies in a pile on the floor. At time $t = 0$ a force **P** is applied and the chain is raised with a constant velocity **v**. Determine (a) the required magnitude of the force **P**, (b) the magnitude of the upward force exerted by the floor as the chain is raised.

Fig. P14.103

Fig. P14.104

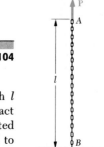

14.105 A chain of mass m per unit length and total length l is held at end A by a force **P**. If end B of the chain is just in contact with the floor at time $t = 0$ and the velocity of the chain is **v** directed downward, determine (a) the magnitude of the force **P** required to lower the chain with the constant velocity **v**, (b) the magnitude of the upward force exerted by the floor as the chain is lowered.

Fig. P14.105

14.106 A moving railroad car, of mass m_0 when empty, is loaded by dropping sand vertically into it from a stationary chute at the rate $q = dm/dt$. At the same time, however, sand is leaking out through the floor of the car at the lesser rate q'. Determine the magnitude of the horizontal force **P** required to keep the car moving at a constant speed v while being loaded.

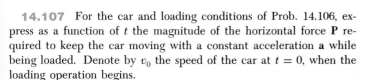

Fig. P14.106

14.107 For the car and loading conditions of Prob. 14.106, express as a function of t the magnitude of the horizontal force **P** required to keep the car moving with a constant acceleration **a** while being loaded. Denote by v_0 the speed of the car at $t = 0$, when the loading operation begins.

14.108 A railroad car, of mass m_0 when empty and moving freely on a horizontal track, is loaded by dropping sand vertically into it from a stationary chute at the rate $q = dm/dt$. Determine the velocity and acceleration of the car as functions of t. Denote by v_0 the speed of the car at $t = 0$, when the loading operation begins.

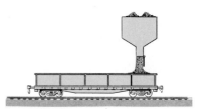

Fig. P14.108

14.109 If the car of Prob. 14.106 moves freely (**P** = 0), determine its velocity and acceleration as functions of t. Denote by v_0 the speed of the car at $t = 0$, when the loading operation begins.

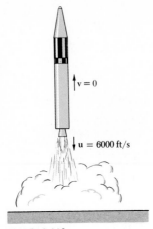

Fig. P14.110

14.110 A test rocket is designed to hover motionless above the ground. The shell of the rocket weighs 2500 lb, and the initial fuel load is 7500 lb. The fuel is burned and ejected with a velocity of 6000 ft/s. Determine the required rate of fuel consumption (*a*) when the rocket is fired, (*b*) as the last particle of fuel is being consumed.

14.111 An experimental space vehicle has a gross weight of 4000 lb, including 3000 lb of fuel. The fuel is consumed at the rate of 150 lb/s and is ejected with a relative velocity of 7000 ft/s. If the vehicle is fired vertically, determine its acceleration (*a*) as the engine is fired, (*b*) as the last particle of fuel is being consumed.

14.112 A rocket of gross mass 1000 kg, including 900 kg of fuel, is fired vertically when $t = 0$. Knowing that fuel is consumed at the rate of 10 kg/s and ejected with a relative velocity of 2100 m/s, determine the acceleration and velocity of the rocket when (*a*) $t = 0$, (*b*) $t = 45$ s, (*c*) $t = 90$ s.

14.113 An experimental rocket has a gross mass of 2500 kg, including 2200 kg of fuel. Knowing that the consumed fuel is ejected with a relative velocity of 2000 m/s, determine the maximum speed attained by the rocket when fired vertically if the fuel is consumed at the rate of (*a*) 50 kg/s, (*b*) 100 kg/s.

14.114 The rocket of Prob. 14.112 is redesigned as a two-stage rocket consisting of rockets *A* and *B*, each of gross mass 500 kg, including 450 kg of fuel. The fuel is again consumed at the rate of 10 kg/s and is ejected with a relative velocity of 2100 m/s. Knowing that, when rocket *A* expells its last particle of fuel, its shell is released and rocket *B* is fired, determine (*a*) the speed when rocket *A* is released, (*b*) the maximum speed attained by rocket *B*.

Fig. P14.114

14.115 A spacecraft is launched vertically by a two-stage rocket. When the speed is 9000 mi/h the first-stage-rocket casing is released and the second-stage rocket is fired. Fuel is consumed at the rate of 250 lb/s and ejected with a relative velocity of 6500 ft/s. Knowing that the combined weight of the second-stage rocket and spacecraft is 30,000 lb, including 27,000 lb of fuel, determine the maximum speed which can be attained by the spacecraft.

14.116 Determine the distance between the spacecraft and the first-stage-rocket casing of Prob. 14.115 as the last particle of fuel is being expelled by the second-stage rocket.

14.117 For the rocket of Sample Prob. 14.12, derive an expression for the height of the rocket as a function of the time *t*.

REVIEW PROBLEMS

14.118 Collar B has an initial velocity of 2 m/s. It strikes collar A causing a series of impacts involving the collars and the fixed support at C. Assuming $e = 1$ for all impacts and neglecting friction, determine (a) the number of impacts which will occur, (b) the final velocity of B, (c) the final position of A.

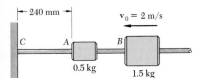

Fig. P14.118

14.119 The jet engine shown scoops in air at A at the rate of 145 lb/s and discharges it at B with a velocity of 2000 ft/s relative to the airplane. Determine the magnitude and line of action of the propulsive thrust developed by the engine when the speed of the airplane is (a) 300 mi/h, (b) 600 mi/h.

14.120 Solve Prob. 14.119, including the effect of the fuel which is consumed by the engine at the rate of 3 lb/s.

14.121 A 5-kg sphere is moving with a velocity of 60 m/s when it explodes into two fragments. Immediately after the explosion the fragments are observed to travel in the directions shown and the speed of fragment A is observed to be 90 m/s. Determine (a) the mass of fragment A, (b) the speed of fragment B.

14.122 Show that the values v_1 and v_2 of the speed of a satellite at the perigee A and the apogee A' of an elliptic orbit are defined by the relations

$$v_1^2 = \frac{2GM}{r_1 + r_2}\frac{r_2}{r_1} \qquad v_2^2 = \frac{2GM}{r_1 + r_2}\frac{r_1}{r_2}$$

14.123 A space vehicle equipped with a retrorocket, which may expel fuel with a relative velocity $\mathbf{u}$, is moving with a velocity $\mathbf{v}_0$. Denoting by m_s the net mass of the vehicle and by m_f the mass of the unexpended fuel, determine the minimum ratio m_f/m_s for which the velocity of the vehicle can be reduced to zero.

14.124 A chain of length l and mass m per unit length falls through a small hole in a plate. Initially, when y is very small, the chain is at rest. In each case shown, determine (a) the acceleration of the first link A as a function of y, (b) the velocity of the chain as the last link passes through the hole. In case 1 assume that the individual links are at rest until they fall through the hole; in case 2 assume that at any instant all links have the same speed. Ignore the effect of friction.

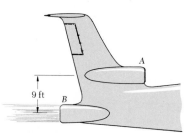

Fig. P14.119

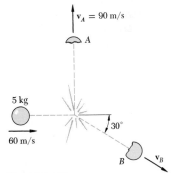

Fig. P14.121

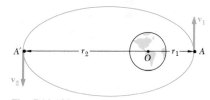

Fig. P14.122

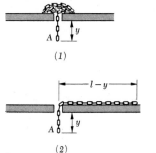

Fig. P14.124

Fig. P14.125

14.125   Three identical balls A, B, and C may roll freely on a horizontal surface. Balls B and C are at rest and in contact when struck by ball A, which was moving to the right with a velocity $\mathbf{v}_0$. Assuming $e = 1$, determine the final velocity of ball A assuming (a) that the path of A is perfectly centered and that A strikes B and C simultaneously, (b) that the path of A is not perfectly centered and that A strikes B slightly before it strikes C.

14.126   A 1-oz bullet is fired with a velocity of 1600 ft/s into block A, which weighs 10 lb. The coefficient of friction between block A and the cart BC is 0.50. Knowing that the cart weighs 8 lb and can roll freely, determine (a) the final velocity of the cart and block, (b) the final position of the block on the cart.

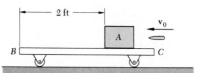

Fig. P14.126

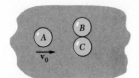

Fig. P14.127

14.127   A booster rocket is attached to the fuselage of an airplane weighing 40,000 lb. The rocket fuel is consumed at the rate of 10 lb/s and is ejected with a relative velocity of 5000 ft/s. Determine the additional propulsive thrust available while the rocket is being fired (a) if the speed of the airplane is 400 mi/h, (b) if the airplane is at rest on the ground.

14.128   In order to test the resistance of a chain to impact, the chain is suspended from a 100-kg block supported by two columns. A rod attached to the last link of the chain is then hit by a 25-kg cylinder dropped from a 1.5-m height. Determine the initial impulse exerted on the chain, assuming that the impact is perfectly plastic and that the columns supporting the dead weight (a) are perfectly rigid, (b) are equivalent to two perfectly elastic springs. (c) Determine the energy absorbed by the chain in parts a and b.

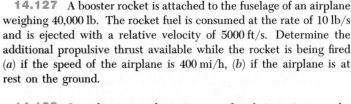

1.5 m

Fig. P14.128

14.129   The 20-Mg truck and the 40-Mg railroad flatcar are both at rest with their brakes released. An engine bumps the flatcar and causes the flatcar alone to start moving with a velocity of 1 m/s to the right. Assuming $e = 1$ between the truck and the ends of the flatcar, determine the velocities of the truck and of the flatcar after end A strikes the truck. Describe the subsequent motion of the system. Neglect the effect of friction.

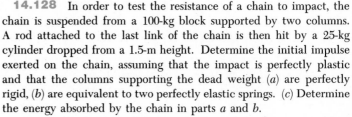

Fig. P14.129

Chapter
15

Kinematics of Rigid Bodies

15.1. Various Types of Plane Motion. In this chapter, we shall study the kinematics of *rigid bodies*. We shall investigate the relations existing between the time, the positions, the velocities, and the accelerations of the various particles forming a rigid body. Our study will be limited to that of the *plane motion* of a rigid body, which is characterized by the fact that each particle of the body remains at a constant distance from a fixed reference plane. Thus all particles of the body move in parallel planes, and the motion of the body may be represented by the motion of a representative slab in the reference plane.

The various types of plane motion may be grouped as follows:

1. *Translation.* A motion is said to be a translation if any straight line drawn on the body keeps the same direction during the motion. It may also be observed that in a translation all the particles forming the body move along parallel paths. If these paths are straight lines, the motion is said to be a *rectilinear translation* (Fig. 15.1a); if the paths are curved lines, the motion is a *curvilinear translation* (Fig. 15.1b).

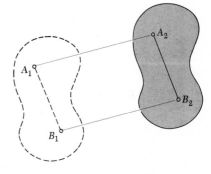

(*a*) Rectilinear translation

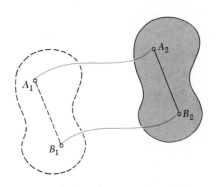

(*b*) Curvilinear translation

Fig. 15.1

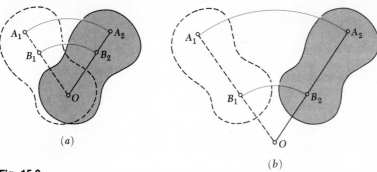

(a)

Fig. 15.2

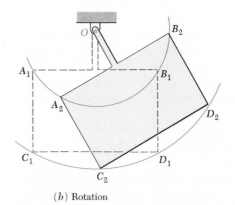

(b)

2. *Rotation.* A motion is said to be a rotation when the particles of the representative slab move along concentric circles. If the common center O of the circles is located on the slab, the corresponding point remains fixed (Fig. 15.2a). In many cases of rotation, however, the common center O is located outside the slab, and none of the points of the slab remains fixed (Fig. 15.2b).

Rotation should not be confused with certain types of curvilinear translation. For example, the plate shown in Fig. 15.3a is in curvilinear translation, with all its particles moving along *parallel* circles, while the plate shown in Fig. 15.3b is in rotation, with all its particles moving along *concentric* circles.

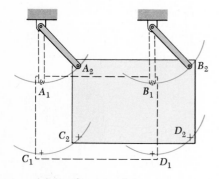

(a) Curvilinear translation

Fig. 15.3

(b) Rotation

In the first case, any given straight line drawn on the plate will maintain the same direction, while, in the second case, point O remains fixed.

3. *General Plane Motion.* Any plane motion which is neither a translation nor a rotation is referred to as a general plane motion. Two examples of general plane motion are given in Fig. 15.4.

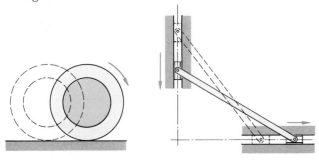

(*a*) Rolling wheel   (*b*) Sliding rod

Fig. 15.4 Examples of general plane motion.

15.2. Translation. Consider a rigid slab in translation (either rectilinear or curvilinear translation), and denote by A and B the positions of two of its particles at time t. At time $t + \Delta t$, the two particles will occupy new positions A' and B' (Fig. 15.5*a*). Now, from the very definition of a translation, the line $A'B'$ must be parallel to AB. Besides, since the slab is rigid, we must have $A'B' = AB$. Thus, $ABB'A'$ is a parallelogram, and the two vectors $\Delta \mathbf{r}_A$ and $\Delta \mathbf{r}_B$ representing the displacements of the two particles during the time interval Δt must have the same magnitude and direction. We have, therefore, $\Delta \mathbf{r}_A / \Delta t = \Delta \mathbf{r}_B / \Delta t$ and, at the limit, as Δt approaches zero,

$$\mathbf{v}_A = \mathbf{v}_B \qquad (15.1)$$

Since relation (15.1) holds at any time, the vectors $\Delta \mathbf{v}_A$ and $\Delta \mathbf{v}_B$ representing the change in velocity of the two particles during the time interval Δt must be equal. We have, therefore, $\Delta \mathbf{v}_A / \Delta t = \Delta \mathbf{v}_B / \Delta t$ and, at the limit, as Δt approaches zero,

$$\mathbf{a}_A = \mathbf{a}_B \qquad (15.2)$$

Thus, *when a rigid slab is in translation, all the points of the slab have the same velocity and the same acceleration at any given instant* (Fig. 15.5*b* and *c*). In the case of curvilinear translation, the velocity and acceleration change in direction as well as in magnitude at every instant. In the case of rectilinear translation, all particles of the slab move along parallel straight lines, and their velocity and acceleration keep the same direction during the entire motion.

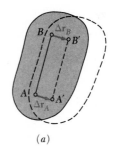

(*a*)

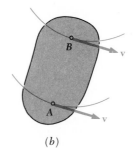

(*b*)

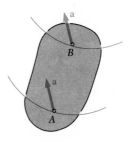

(*c*)

Fig. 15.5

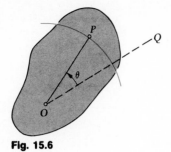

Fig. 15.6

15.3. Rotation. Consider a rigid slab which rotates about a fixed axis perpendicular to the plane of the slab and intersecting it at point O. Let P be a point of the slab (Fig. 15.6). The position of the slab will be entirely defined if the angle θ the line OP forms with a fixed direction OQ is given; the angle θ is known as the *angular coordinate* of the slab. The angular coordinate is defined as positive when counterclockwise and will be expressed in radians (rad) or, occasionally, in degrees (°) or revolutions (rev). We recall that

$$1 \text{ rev} = 2\pi \text{ rad} = 360°$$

In Chap. 11, the first and second derivatives, respectively, of the coordinate x of a particle were used to define the velocity and acceleration of a particle moving in a straight line. Similarly, the first derivative of the angular coordinate θ of a rotating slab may be used to define the *angular velocity* ω of the slab,

$$\omega = \frac{d\theta}{dt} \tag{15.3}$$

and its second derivative may be used to define the *angular acceleration* α of the slab,

$$\alpha = \frac{d\omega}{dt} = \frac{d^2\theta}{dt^2} \tag{15.4}$$

An alternate form for the angular acceleration is obtained by solving (15.3) for dt and substituting into (15.4). We write $dt = d\theta/\omega$ and

$$\alpha = \omega \frac{d\omega}{d\theta} \tag{15.5}$$

The angular velocity ω is expressed in rad/s or in rpm (revolutions per minute). The angular acceleration α is generally expressed in rad/s².

Since the angular velocity and the angular acceleration of a slab have a *sense*, as well as a *magnitude*, they should be considered as *vector quantities*.† We may, for instance, represent the angular velocity of the slab of Fig. 15.6 by a vector $\boldsymbol{\omega}$ of magnitude ω directed along the axis of rotation of the slab. This vector will be made to point out of the plane of the figure if the angular velocity is counterclockwise, and into the plane of the figure if the angular velocity is clockwise. The angular

† It may be shown, in the more general case of a rigid body rotating simultaneously about axes having different directions, that angular velocities and angular accelerations obey the parallelogram law of addition.

acceleration may similarly be represented by a vector $\boldsymbol{\alpha}$ of magnitude α directed along the axis of rotation and pointing out of, or into, the plane of the figure, depending upon the sense of the angular acceleration. However, since in the cases considered in this chapter the axis of rotation will always be perpendicular to the plane of the figure, we need to specify only the sense and the magnitude of the angular velocity and of the angular acceleration. This may be done by using the scalar quantities ω and α, together with a plus or minus sign. As in the case of the angular coordinate θ, the plus sign corresponds to counterclockwise, and the minus sign to clockwise.

It should be observed that the values obtained for the angular velocity and the angular acceleration of a slab are independent of the choice of the reference point P. If a point P' were chosen instead of P (Fig. 15.7), the angular coordinate would be θ', which differs from θ only by the constant angle POP'. The successive derivatives of θ and θ' are thus equal, and the same values are obtained for ω and α.

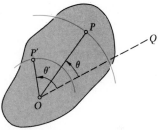

Fig. 15.7

The motion of a slab rotating about a fixed point O is said to be *known* when its coordinate θ may be expressed as a known function of t. In practice, however, the rotation of a slab is seldom defined by a relation between θ and t. More often, the conditions of motion will be specified by the type of angular acceleration that the slab possesses. For example, α may be given as a function of t, or as a function of θ, or as a function of ω. Equations (15.3) to (15.5) may then be used to determine the angular velocity ω and the angular coordinate θ, the procedure to follow being the same as that described in Sec. 11.3 for the rectilinear motion of a particle.

Two particular cases of rotation are frequently encountered:

1. *Uniform Rotation.* This case is characterized by the fact that the angular acceleration is zero. The angular velocity ω is thus constant, and the angular coordinate is given by the formula

$$\theta = \theta_0 + \omega t \qquad (15.6)$$

2. *Uniformly Accelerated Rotation.* In this case, the angular acceleration α is constant. The following formulas relating angular velocity, angular coordinate, and time may then be derived in a manner similar to that described in Sec. 11.5. The similitude between the formulas derived here and those obtained for the rectilinear uniformly accelerated motion of a particle is easily noted.

$$\omega = \omega_0 + \alpha t \qquad (15.7)$$
$$\theta = \theta_0 + \omega_0 t + \tfrac{1}{2}\alpha t^2 \qquad (15.8)$$
$$\omega^2 = \omega_0^2 + 2\alpha(\theta - \theta_0) \qquad (15.9)$$

It should be emphasized that formula (15.6) may be used only when $\alpha = 0$, and formulas (15.7) to (15.9) only when $\alpha =$ constant. In any other case, the general formulas (15.3) to (15.5) should be used.

15.4. Linear and Angular Velocity, Linear and Angular Acceleration in Rotation. Consider again a point P on a rotating slab (Fig. 15.8a). Point P describes a circle

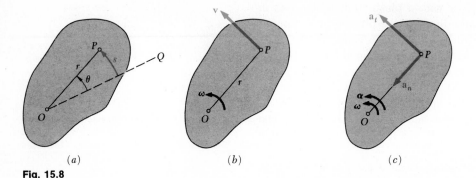

Fig. 15.8

of center O when the slab describes a full rotation. The arc s that P describes while the slab rotates through an angle θ depends not only upon θ but also upon the distance r from O to P. Expressing θ in radians and r and s in the same units of length, we write

$$s = r\theta \tag{15.10}$$

Differentiating, we obtain the following relation between the magnitude v of the *linear velocity* of P (Fig. 15.8b) and the *angular velocity* ω of the slab,

$$\frac{ds}{dt} = r\frac{d\theta}{dt} \qquad v = r\omega \tag{15.11}$$

While the value of the angular velocity ω is independent of the choice of P, the magnitude v of the linear velocity clearly depends upon the distance r. Recalling the definition of the tangential and normal components of the *linear acceleration* of a point P (Sec. 11.12), we write (Fig. 15.8c)

$$a_t = \frac{dv}{dt} = r\frac{d\omega}{dt} \qquad a_t = r\alpha \tag{15.12}$$

$$a_n = \frac{v^2}{r} = \frac{(r\omega)^2}{r} \qquad a_n = r\omega^2 \tag{15.13}$$

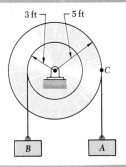

3 ft ─┐ ┌─ 5 ft

C

B   A

A pulley and two loads are connected by inextensible cords as shown. Load A has a constant acceleration of 10 ft/s² and an initial velocity of 15 ft/s, both directed upward. Determine (a) the number of revolutions executed by the pulley in 3 s, (b) the velocity and position of load B after 3 s, (c) the acceleration of point C on the rim of the pulley at $t = 0$.

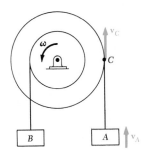

v_C

ω

C

B   A v_A

a. Motion of Pulley. Since the cord connecting the pulley to load A is inextensible, the velocity of C is equal to the velocity of A and the tangential component of the acceleration of C is equal to the acceleration of A.

$$(\mathbf{v}_C)_0 = (\mathbf{v}_A)_0 = 15 \text{ ft/s} \uparrow \qquad (\mathbf{a}_C)_t = \mathbf{a}_A = 10 \text{ ft/s}^2 \uparrow$$

Noting that the distance from C to the center of the pulley is 5 ft, we write

$$(v_C)_0 = r\omega_0 \qquad 15 \text{ ft/s} = (5 \text{ ft})\omega_0 \qquad \omega_0 = 3 \text{ rad/s } \text{↻}$$
$$(a_C)_t = r\alpha \qquad 10 \text{ ft/s}^2 = (5 \text{ ft})\alpha \qquad \alpha = 2 \text{ rad/s}^2 \text{↻}$$

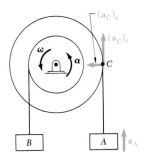

$(a_C)_n$

$(a_C)_t$

ω α C

B   A a_A

From the equations for uniformly accelerated motion, we obtain, for $t = 3$ s,

$$\omega = \omega_0 + \alpha t = 3 \text{ rad/s} + (2 \text{ rad/s}^2)(3 \text{ s}) = 9 \text{ rad/s}$$
$$\omega = 9 \text{ rad/s } \text{↻}$$
$$\theta = \omega_0 t + \tfrac{1}{2}\alpha t^2 = (3 \text{ rad/s})(3 \text{ s}) + \tfrac{1}{2}(2 \text{ rad/s}^2)(3 \text{ s})^2$$
$$\theta = 18 \text{ rad}$$

$$\text{Number of revolutions} = (18 \text{ rad})\left(\frac{1 \text{ rev}}{2\pi \text{ rad}}\right) = 2.86 \text{ rev} \quad \blacktriangleleft$$

b. Motion of Load B. Using the following relations between the linear and angular motion, with $r = 3$ ft, we write

$$v_B = r\omega = (3 \text{ ft})(9 \text{ rad/s}) \qquad\qquad v_B = 27 \text{ ft/s} \downarrow \quad \blacktriangleleft$$
$$s_B = r\theta = (3 \text{ ft})(18 \text{ rad}) \qquad\qquad s_B = 54 \text{ ft} \downarrow \quad \blacktriangleleft$$

c. Acceleration of Point C at $t = 0$. The tangential component of the acceleration is

$$(\mathbf{a}_C)_t = \mathbf{a}_A = 10 \text{ ft/s}^2 \uparrow$$

Since, at $t = 0$, $\omega_0 = 3$ rad/s, the normal component of the acceleration is

$$(a_C)_n = r_C\omega_0^2 = (5 \text{ ft})(3 \text{ rad/s})^2 \qquad (\mathbf{a}_C)_n = 45 \text{ ft/s}^2 \leftarrow$$

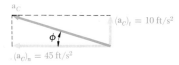

a_C

ϕ

$(a_C)_t = 10 \text{ ft/s}^2$

$(a_C)_n = 45 \text{ ft/s}^2$

The magnitude and direction of the total acceleration are obtained by writing

$$\tan \phi = (10 \text{ ft/s}^2)/(45 \text{ ft/s}^2) \qquad \phi = 12.5°$$
$$a_C \sin 12.5° = 10 \text{ ft/s}^2 \qquad a_C = 46.1 \text{ ft/s}^2$$
$$\mathbf{a}_C = 46.1 \text{ ft/s}^2 \text{ ⦨ } 12.5° \quad \blacktriangleleft$$

PROBLEMS

15.1   The motion of a cam is defined by the relation $\theta = t^3 - 2t^2 - 4t + 10$, where θ is expressed in radians and t in seconds. Determine the angular coordinate, the angular velocity, and the angular acceleration of the cam when (a) $t = 0$, (b) $t = 3$ s.

15.2   For the cam of Prob. 15.1 determine (a) the time at which the angular velocity is zero, (b) the corresponding angular coordinate and angular acceleration.

15.3   The motion of a disk rotating in an oil bath is defined by the relation $\theta = \theta_0(1 - e^{-t/3})$, where θ is expressed in radians and t in seconds. Knowing that $\theta_0 = 1.20$ rad, determine the angular coordinate, velocity, and acceleration of the disk when (a) $t = 0$, (b) $t = 3$ s, (c) $t = \infty$.

15.4   The motion of an oscillating crank is defined by the relation $\theta = \theta_0 \cos{(2\pi t/T)}$, where θ is expressed in radians and t in seconds. Knowing that $\theta_0 = 0.75$ rad and $T = 1.20$ s, determine the maximum angular velocity and the maximum angular acceleration of the crank.

15.5   The rotor of a steam turbine is rotating at a speed of 720 rpm when the steam supply is suddenly cut off. It is observed that 5 min are required for the rotor to come to rest. Assuming uniformly accelerated motion, determine (a) the angular acceleration, (b) the total number of revolutions that the rotor executes before coming to rest.

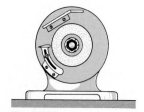

Fig. P15.6

15.6   A small grinding wheel is attached to the shaft of an electric motor which has a rated speed of 1800 rpm. When the power is turned on, the unit reaches its rated speed in 4 s, and when the power is turned off, the unit coasts to rest in 50 s. Assuming uniformly accelerated motion, determine the number of revolutions that the motor executes (a) in reaching its rated speed, (b) in coasting to rest.

15.7   The rotor of an electric motor has a speed of 1200 rpm when the power is cut off. The rotor is then observed to come to rest after executing 520 revolutions. Assuming uniformly accelerated motion, determine (a) the angular acceleration, (b) the time required for the rotor to come to rest.

15.8   The angular acceleration of the rotor of an electric motor is directly proportional to the time t. The rotor starts at $t = 0$ with no initial velocity; after 3 s, the rotor has completed 6 revolutions. Write the equations of motion for the rotor, and determine the angular velocity at $t = 2$ s.

15.9 The earth makes one complete revolution about the sun in 365.24 days. Assuming that the orbit of the earth is circular and has a radius of 93,000,000 mi, determine the velocity and acceleration of the earth.

15.10 The earth makes one complete revolution on its axis in 23.93 h. Knowing that the mean radius of the earth is 3960 mi, determine the linear velocity and acceleration of a point on the surface of the earth (*a*) at the equator, (*b*) at Philadelphia, latitude 40° north, (*c*) at the North Pole.

15.11 A small block *B* rests on a horizontal plate which rotates about a fixed vertical axis. If the plate is initially at rest at $t = 0$ and is accelerated at the constant rate α, derive an expression (*a*) for the total acceleration of the block at time *t*, (*b*) for the angle between the total acceleration and the radius *AB* at time *t*.

15.12 It is known that the static-friction force between block *B* and the plate will be exceeded and that the block will start sliding on the plate when the total acceleration of the block reaches 5 m/s². If the plate starts from rest at $t = 0$ and is accelerated at the constant rate of 6 rad/s², determine the time *t* and the angular velocity of the plate when the block starts sliding, assuming $r = 100$ mm.

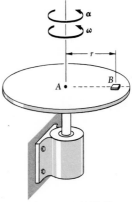

Fig. P15.11 and P15.12

15.13 At a given instant, the acceleration of the rack is 15 in./s² directed downward, and the velocity of the rack is 6 in./s directed upward. Determine (*a*) the angular acceleration and the angular velocity of the gear, (*b*) the total acceleration of the gear tooth in contact with the rack.

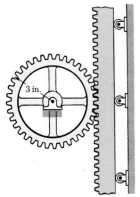

Fig. P15.13 and P15.14

15.14 The gear-and-rack system shown is initially at rest. If the acceleration of the rack is constant and equal to 0.5 ft/s² directed upward, determine (*a*) the angular velocity of the gear after the rack has moved 3 ft, (*b*) the time required for the gear to reach an angular velocity of 40 rpm.

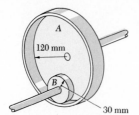

Fig. P15.15 and P15.16

15.15 The friction wheel B executes 100 revolutions about its fixed shaft during the time interval t, while its angular velocity is being increased uniformly from 200 to 600 rpm. Knowing that wheel B rolls without slipping on the inside rim of wheel A, determine (a) the angular acceleration of wheel A, (b) the time interval t.

15.16 The friction wheel B rotates at the constant angular velocity of 200 rpm about its fixed shaft. Knowing that wheel B rolls without slipping on the inside rim of wheel A, determine (a) the angular velocity of wheel A, (b) the linear accelerations of the points of A and B which are in contact.

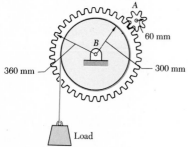

Fig. P15.17 and P15.18

15.17 A load is to be raised 6 m by the hoisting system shown. Assuming gear A is initially at rest, accelerates uniformly to a speed of 120 rpm in 5 s, and then maintains a constant speed of 120 rpm, determine (a) the number of revolutions executed by gear A in raising the load, (b) the time required to raise the load.

15.18 The system shown starts from rest at $t = 0$ and accelerates uniformly. Knowing that at $t = 4$ s the velocity of the load is 4.8 m/s downward, determine (a) the angular acceleration of gear A, (b) the number of revolutions executed by gear A during the 4-s interval.

15.19 The two pulleys shown may be operated with the V belt in any of three positions. If the angular acceleration of shaft A is 8 rad/s² and if the system is initially at rest, determine the time required for shaft B to reach a speed of 300 rpm with the belt in each of the three positions.

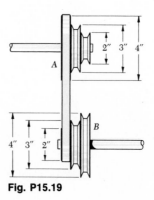

Fig. P15.19

15.20 A simple friction drive consists of two disks A and B. Initially, disk B has a clockwise angular velocity of 500 rpm, and disk A is at rest. It is known that disk B will coast to rest in 60 s. However, rather than waiting until both disks are at rest to bring them together, disk A is given a constant angular acceleration of 3 rad/s² counter-clockwise. Determine (a) at what time the disks may be brought together if they are not to slip, (b) the angular velocity of each disk as contact is made.

15.21 Disk B is at rest when it is brought into contact with disk A which is rotating freely at 500 rpm clockwise. After 5 s of slippage, during which each disk has a constant angular acceleration, disk A reaches a final angular velocity of 300 rpm clockwise. Determine the angular acceleration of each disk during the period of slippage.

*15.22 In a continuous printing process, paper is drawn into the presses at a constant speed v. Denoting by r the radius of paper on the roll at any given time and by b the thickness of the paper, derive an expression for the angular acceleration of the paper roll.

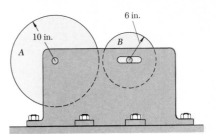

Fig. P15.20 and P15.21

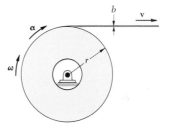

Fig. P15.22

15.5. General Plane Motion. As indicated in Sec. 15.1, we understand by general plane motion a plane motion which is neither a translation nor a rotation. As we shall presently see, however, *a general plane motion may always be considered as the sum of a translation and a rotation.*

Consider, for example, a wheel rolling on a straight track (Fig. 15.9). Over a certain interval of time, two given points A and B will have moved, respectively, from A_1 to A_2 and from B_1 to B_2. The same result could be obtained through a translation which would bring A and B into A_2 and B_1' (the line AB remaining vertical), followed by a rotation about A bringing B into B_2. Although the original rolling motion differs from the combination of translation and rotation when these motions are taken in succession, the original motion may be completely duplicated by a combination of simultaneous translation and rotation.

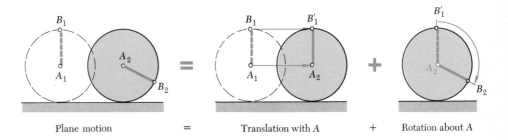

Plane motion = Translation with A + Rotation about A

Fig. 15.9

Another example of plane motion is given in Fig. 15.10, which represents a rod whose extremities slide, respectively, along a horizontal and a vertical track. This motion may be replaced by a translation in a horizontal direction and a rotation about A (Fig. 15.10a) or by a translation in a vertical direction and a rotation about B (Fig. 15.10b).

In general, any plane motion of a rigid slab may be replaced by a translation in which all the particles of the slab move along paths parallel to the path actually followed by some arbitrary reference point A, and by a rotation about the reference point A. We saw in Sec. 11.11 that, when two particles A and B move in a plane, the absolute motion of B may be obtained by combining the motion of A and the relative motion of B with respect to A. We recall that by "absolute motion of B" we mean the motion of B with respect to fixed axes, while by "relative motion

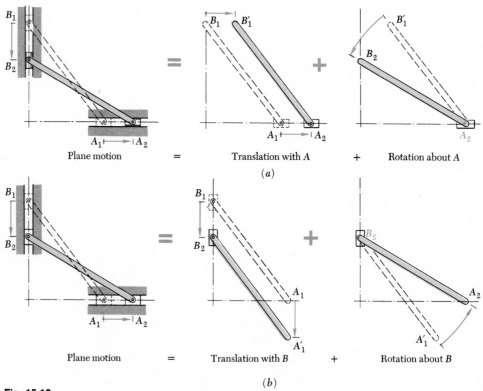

Plane motion   =   Translation with A +   Rotation about A

(a)

Plane motion   =   Translation with B +   Rotation about B

(b)

Fig. 15.10

of B with respect to A" we mean the motion of B with respect to a system of axes whose origin moves with A but which does not rotate. Now, in the present case, the two particles belong to the same rigid body. Particle B must thus remain at a constant distance from A and, *to an observer moving with A, but not rotating, particle B will appear to describe an arc of circle centered at A.* Considering successively each particle of the slab, and resolving its motion into the motion of A and its relative motion about A, we thus conclude that any plane motion of a rigid slab is the sum of a translation with A and a rotation about A.

15.6. Absolute and Relative Velocity in Plane Motion.

We saw in the preceding section that any plane motion of a slab may be replaced by a translation defined by the motion of an arbitrary reference point A, and by a rotation about A. The absolute velocity $\mathbf{v}_B$ of a particle B of the slab is obtained from the relative-velocity formula derived in Sec. 11.11,

$$\mathbf{v}_B = \mathbf{v}_A + \mathbf{v}_{B/A} \qquad (15.14)$$

where the right-hand member represents a vector sum. The velocity $\mathbf{v}_A$ corresponds to the translation of the slab with A, while the relative velocity $\mathbf{v}_{B/A}$ is associated with the rotation of the slab about A and is measured with respect to axes centered at A and of fixed orientation (Fig. 15.11). Denoting by r the

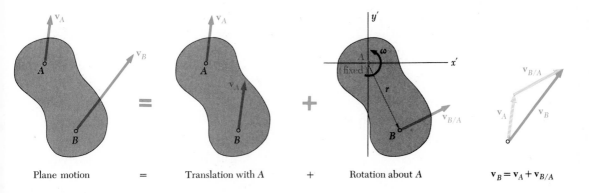

| Plane motion | = | Translation with A | + | Rotation about A | $\mathbf{v}_B = \mathbf{v}_A + \mathbf{v}_{B/A}$ |

Fig. 15.11

distance from A to B, and by ω the magnitude of the angular velocity ω of the slab with respect to axes of fixed orientation, we find that the magnitude of the relative velocity $\mathbf{v}_{B/A}$ is

$$v_{B/A} = r\omega \tag{15.15}$$

As an example, we shall consider again the rod AB of Fig. 15.10. Assuming that the velocity $\mathbf{v}_A$ of end A is known, we propose to find the velocity $\mathbf{v}_B$ of end B and the angular velocity ω of the rod, in terms of the velocity $\mathbf{v}_A$, the length l, and the angle θ. Choosing A as reference point, we express that the given motion is equivalent to a translation with A and a rotation about A (Fig. 15.12). The absolute velocity of B must therefore be equal to the vector sum

$$\mathbf{v}_B = \mathbf{v}_A + \mathbf{v}_{B/A} \tag{15.14}$$

We note that, while the direction of $\mathbf{v}_{B/A}$ is known, its magnitude $l\omega$ is unknown. However, this is compensated by the fact that the direction of $\mathbf{v}_B$ is known. We may therefore complete the diagram of Fig. 15.12. Solving for the magnitudes v_B and ω, we write

$$v_B = v_A \tan \theta \qquad \omega = \frac{v_{B/A}}{l} = \frac{v_A}{l \cos \theta} \tag{15.16}$$

The same result may be obtained by using B as a point of reference. Resolving the given motion into a translation with

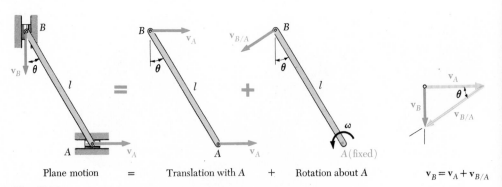

Plane motion $\quad=\quad$ Translation with A $\quad+\quad$ Rotation about A $\qquad \mathbf{v}_B = \mathbf{v}_A + \mathbf{v}_{B/A}$

Fig. 15.12

B and a rotation about B (Fig. 15.13), we write the equation

$$\mathbf{v}_A = \mathbf{v}_B + \mathbf{v}_{A/B} \qquad (15.17)$$

which is represented graphically in Fig. 15.13. We note that $\mathbf{v}_{A/B}$ and $\mathbf{v}_{B/A}$ have the same magnitude $l\omega$ but opposite sense. The sense of the relative velocity depends, therefore, upon the point of reference which has been selected and should be carefully ascertained from the appropriate diagram (Fig. 15.12 or 15.13).

Finally, we observe that the angular velocity ω of the rod in its rotation about B is the same as in its rotation about A. It is measured in both cases by the rate of change of the angle θ. This result is quite general; we should therefore bear in mind that *the angular velocity ω of a rigid body in plane motion is independent of the reference point.*

Most mechanisms consist, not of one, but of *several* moving parts. When the various parts of a mechanism are pin-connected, its analysis may be carried out by considering each part as a rigid body, while keeping in mind that the points where two parts are connected must have the same absolute velocity (see Sample Prob. 15.3). A similar analysis may be used when gears are involved, since the teeth in contact must also have the same absolute velocity. However, when a mechanism contains parts which slide on each other, the relative velocity of the parts in contact must be taken into account (see Secs. 15.10 and 15.11).

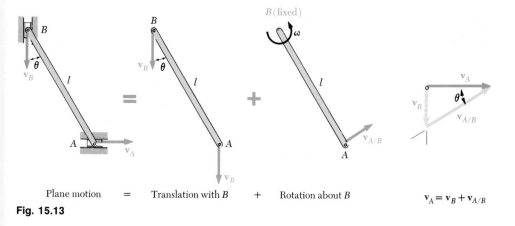

Plane motion = Translation with B + Rotation about B

$$\mathbf{v}_A = \mathbf{v}_B + \mathbf{v}_{A/B}$$

Fig. 15.13

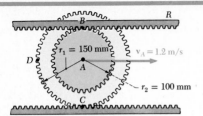

The double gear shown rolls on the stationary lower rack; the velocity of its center A is 1.2 m/s directed to the right. Determine (*a*) the angular velocity of the gear, (*b*) the velocities of the upper rack R and of point D of the gear.

a. **Angular Velocity of the Gear.** Since the gear rolls on the lower rack, its center A moves through a distance equal to the outer circumference $2\pi r_1$ for each full revolution of the gear. Noting that 1 rev $= 2\pi$ rad, we obtain the coordinate x_A in terms of the corresponding angular coordinate θ (in radians) of the gear by a proportion,

$$\frac{x_A}{2\pi r_1} = \frac{\theta}{2\pi} \qquad x_A = r_1\theta$$

Differentiating with respect to the time t and substituting the known values $v_A = 1.2$ m/s and $r_1 = 150$ mm $= 0.150$ m, we obtain

$$v_A = r_1\omega \qquad 1.2 \text{ m/s} = (0.150 \text{ m})\omega \qquad \omega = 8 \text{ rad/s} \circlearrowright \quad \blacktriangleleft$$

b. **Velocities.** The rolling motion is resolved into two component motions: a translation with the center A and a rotation about the center A. In the translation, all points of the gear move with the same velocity $\mathbf{v}_A$. In the rotation, each point P of the gear moves about A with a relative velocity of magnitude $r\omega$, where r is the distance from A.

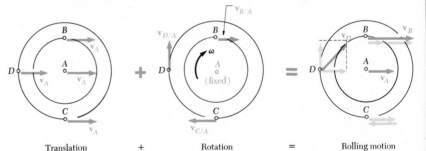

Translation   +   Rotation   =   Rolling motion

Velocity of Upper Rack. The velocity of the upper rack is equal to the velocity of point B; we write

$$\mathbf{v}_R = \mathbf{v}_B = \mathbf{v}_A + \mathbf{v}_{B/A}$$

where
$$\mathbf{v}_A = 1.2 \text{ m/s} \rightarrow$$

$$v_{B/A} = r_2\omega = (0.100 \text{ m})(8 \text{ rad/s}) = 0.8 \text{ m/s} \qquad \mathbf{v}_{B/A} = 0.8 \text{ m/s} \rightarrow$$

Since $\mathbf{v}_A$ and $\mathbf{v}_{B/A}$ are collinear, we add their magnitudes and obtain

$$\mathbf{v}_R = 2 \text{ m/s} \rightarrow \quad \blacktriangleleft$$

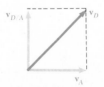

Velocity of Point D: $\qquad \mathbf{v}_D = \mathbf{v}_A + \mathbf{v}_{D/A}$

where
$$\mathbf{v}_A = 1.2 \text{ m/s} \rightarrow$$

$$v_{D/A} = r_1\omega = (0.150 \text{ m})(8 \text{ rad/s}) = 1.2 \text{ m/s} \qquad \mathbf{v}_{D/A} = 1.2 \text{ m/s} \uparrow$$

Adding these velocities vectorially, we obtain

$$\mathbf{v}_D = 1.697 \text{ m/s} \measuredangle 45° \quad \blacktriangleleft$$

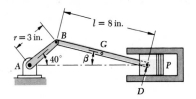

In the engine system shown, the crank AB has a constant clockwise angular velocity of 2000 rpm. For the crank position indicated, determine (a) the angular velocity of the connecting rod BD, (b) the velocity of the piston P.

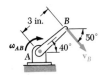

Motion of Crank AB. The crank AB rotates about point A. Expressing ω_{AB} in rad/s and writing $v_B = r\omega_{AB}$, we obtain

$$\omega_{AB} = \left(2000\ \frac{\text{rev}}{\text{min}}\right)\left(\frac{1\ \text{min}}{60\ \text{s}}\right)\left(\frac{2\pi\ \text{rad}}{1\ \text{rev}}\right) = 209\ \text{rad/s}$$

$$v_B = (AB)\omega_{AB} = (3\ \text{in.})(209\ \text{rad/s}) = 627\ \text{in./s}$$

$$v_B = 627\ \text{in./s} \ \searrow\ 50°$$

Motion of Connecting Rod BD. We consider this motion as a general plane motion. Using the law of sines, we compute the angle β between the connecting rod and the horizontal,

$$\frac{\sin 40°}{8\ \text{in.}} = \frac{\sin \beta}{3\ \text{in.}} \qquad \beta = 13.9°$$

The velocity $\mathbf{v}_D$ of the point D where the rod is attached to the piston must be horizontal, while the velocity of point B is equal to the velocity $\mathbf{v}_B$ obtained above. Resolving the motion of BD into a translation with B and a rotation about B, we obtain

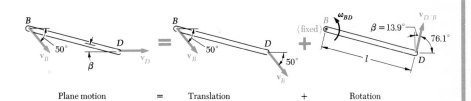

| Plane motion | = | Translation | + | Rotation |

Expressing the relation between the velocities $\mathbf{v}_D$, $\mathbf{v}_B$, and $\mathbf{v}_{D/B}$, we write

$$\mathbf{v}_D = \mathbf{v}_B + \mathbf{v}_{D/B}$$

We draw the vector diagram corresponding to this equation. Recalling that $\beta = 13.9°$, we determine the angles of the triangle and write

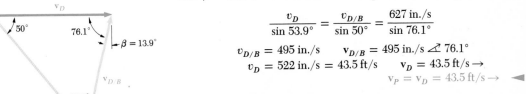

$$\frac{v_D}{\sin 53.9°} = \frac{v_{D/B}}{\sin 50°} = \frac{627\ \text{in./s}}{\sin 76.1°}$$

$$v_{D/B} = 495\ \text{in./s} \qquad \mathbf{v}_{D/B} = 495\ \text{in./s} \ \measuredangle\ 76.1°$$

$$v_D = 522\ \text{in./s} = 43.5\ \text{ft/s} \qquad \mathbf{v}_D = 43.5\ \text{ft/s} \rightarrow$$

$$\mathbf{v}_P = \mathbf{v}_D = 43.5\ \text{ft/s} \rightarrow \quad \blacktriangleleft$$

Since $v_{D/B} = l\omega_{BD}$, we have

$$495\ \text{in./s} = (8\ \text{in.})\omega_{BD} \qquad \omega_{BD} = 61.9\ \text{rad/s} \ \curvearrowleft \quad \blacktriangleleft$$

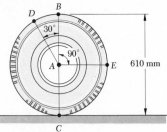

Fig. P15.23

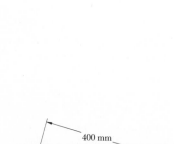

Fig. P15.26

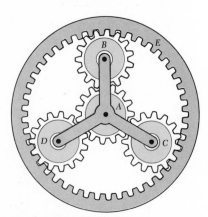

Fig. P15.29 and P15.30

PROBLEMS

15.23 An automobile travels to the right at a constant speed of 50 km/h. If the diameter of a wheel is 610 mm, determine the velocities of points B, C, D, and E on the rim of the wheel.

15.24 Rod AB is 6 ft long and slides with its ends in contact with the floor and the inclined plane. End A moves with a constant velocity of 20 ft/s to the right. At the instant when $\theta = 25°$, determine (a) the angular velocity of the rod, (b) the velocity of end B.

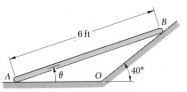

Fig. P15.24

15.25 Solve Prob. 15.24, assuming $\theta = 30°$.

15.26 A 400-mm rod AB is guided by wheels at A and B which roll in the track shown. Knowing that $\theta = 90°$ and that B moves at a constant speed $v_B = 1.2$ m/s, determine (a) the angular velocity of the rod, (b) the velocity of A.

15.27 Solve Prob. 15.26 assuming $\theta = 60°$.

15.28 Solve Prob. 15.26 assuming $\theta = 120°$.

15.29 In the planetary gear system shown, the radius of gears A, B, C, and D is a and the radius of the outer gear E is $3a$. Knowing that the angular velocity of gear A is ω_A clockwise and that the outer gear E is stationary, determine (a) the angular velocity of the spider connecting the planetary gears, (b) the angular velocity of each planetary gear.

15.30 In the planetary gear system shown, the radius of the central gear A is a, the radius of each of the planetary gears is b, and the radius of the outer gear E is $a + 2b$. The angular velocity of gear A is ω_A clockwise and the outer gear is stationary. If the angular velocity of the spider BCD is to be $\omega_A/5$, determine (a) the required value of the ratio b/a, (b) the corresponding angular velocity of each planetary gear.

15.31 Arm AB rotates with an angular velocity of 120 rpm clockwise. If the motion of gear B is to be a curvilinear translation, determine (a) the required angular velocity of gear A, (b) the corresponding velocity of the center of gear B.

15.32 Gear A rotates clockwise with a constant angular velocity of 60 rpm. Knowing that at the same time the arm AB rotates counterclockwise with a constant angular velocity of 30 rpm, determine the angular velocity of gear B.

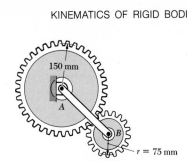

Fig. P15.31 and P15.32

15.33 In the engine system shown, $l = 160$ mm and $b = 60$ mm; the crank AB rotates with a constant angular velocity of 1000 rpm clockwise. Determine the velocity of the piston P and the angular velocity of the connecting rod for the position corresponding to (a) $\theta = 0$, (b) $\theta = 90°$, (c) $\theta = 180°$.

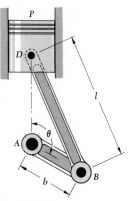

Fig. P15.33

15.34 Crank AB has a constant angular velocity of 12 rad/s clockwise. Determine the angular velocity of rod BD and the velocity of collar D when (a) $\theta = 0$, (b) $\theta = 90°$, (c) $\theta = 180°$.

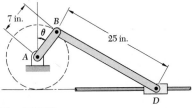

Fig. P15.34

15.35 Solve Prob. 15.34 for the position corresponding to $\theta = 120°$.

15.36 Solve Prob. 15.33 for the position corresponding to $\theta = 60°$.

15.37 through 15.40 In the position shown, bar AB has a constant angular velocity of 3 rad/s counterclockwise. Determine the angular velocity of bars BD and DE.

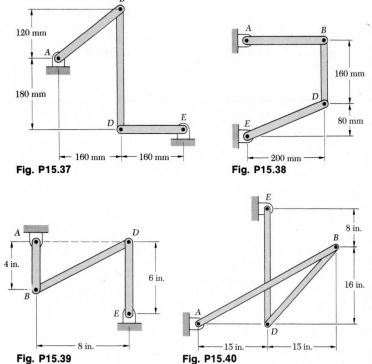

Fig. P15.37   **Fig. P15.38**

Fig. P15.39   **Fig. P15.40**

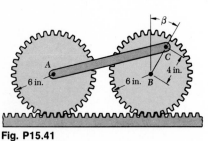

Fig. P15.41

15.41 Two gears, each of 12-in. diameter, are connected by an 18-in. rod AC. Knowing that the center of gear B has a constant velocity of 30 in./s to the right, determine the velocity of the center of gear A and the angular velocity of the connecting rod (*a*) when $\beta = 0$, (*b*) when $\beta = 90°$.

15.42 Solve Prob. 15.41, assuming (*a*) $\beta = 180°$, (*b*) $\beta = 60°$.

15.43 Rod AB of the mechanism shown rotates clockwise with an angular velocity ω_0. Determine in terms of l, ω_0, and θ, (*a*) the velocity of collar D, (*b*) the components of the velocity of point E.

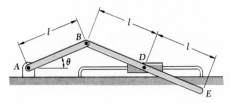

Fig. P15.43 and P15.44

15.44   Collar D of the mechanism shown moves to the right with a velocity of 300 mm/s. Knowing that $l = 200$ mm and $\theta = 30°$, determine (*a*) the angular velocity of rod AB, (*b*) the velocity of point E.

*__15.45__   Assuming that the crank AB of Prob. 15.33 rotates with a constant clockwise angular velocity ω and that $\theta = 0$ at $t = 0$, derive an expression for the velocity of the piston P in terms of the time t.

15.7. Instantaneous Center of Rotation in Plane Motion.

Consider the general plane motion of a slab. We shall show that at any given instant the velocities of the various particles of the slab are the same as if the slab were rotating about a certain axis perpendicular to the plane of the slab, called the *instantaneous axis of rotation*. This axis intersects the plane of the slab at a point C, called the *instantaneous center of rotation* of the slab.

To prove our statement, we first recall that the plane motion of a slab may always be replaced by a translation defined by the motion of an arbitrary reference point A, and by a rotation about A. As far as the velocities are concerned, the translation is characterized by the velocity $\mathbf{v}_A$ of the reference point A and the rotation is characterized by the angular velocity ω of the slab (which is independent of the choice of A). Thus, the velocity $\mathbf{v}_A$ of point A and the angular velocity ω of the slab define completely the velocities of all the other particles of the slab (Fig. 15.14*a*). Now let us assume that $\mathbf{v}_A$ and ω are known and that they are both different from zero. (If $\mathbf{v}_A = 0$, point A is itself the instantaneous center of rotation, and if $\omega = 0$, the slab

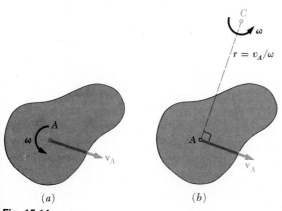

(*a*)   (*b*)

Fig. 15.14

is in translation.) These velocities could be obtained by letting the slab rotate with the angular velocity ω about a point C located on the perpendicular to $\mathbf{v}_A$ at a distance $r = v_A/\omega$ from A as shown in Fig. 15.14b. We check that the velocity of A would be perpendicular to AC and that its magnitude would be $r\omega = (v_A/\omega)\omega = v_A$. Thus the velocities of all the other particles of the slab would be the same as originally defined. Therefore, *as far as the velocities are concerned, the slab seems to rotate about the instantaneous center C* at the instant considered.

The position of the instantaneous center may be defined in two other ways. If the directions of the velocities of two particles A and B of the slab are known, and if they are different, the instantaneous center C is obtained by drawing the perpendicular to $\mathbf{v}_A$ through A and the perpendicular to $\mathbf{v}_B$ through B and determining the point in which these two lines intersect (Fig. 15.15a). If the velocities $\mathbf{v}_A$ and $\mathbf{v}_B$ of two particles A and B are perpendicular to the line AB, and if their magnitudes are known, the instantaneous center may be found by intersecting the line AB with the line joining the extremities of the vectors $\mathbf{v}_A$ and $\mathbf{v}_B$ (Fig. 15.15b). Note that, if $\mathbf{v}_A$ and $\mathbf{v}_B$ were parallel in Fig. 15.15a, or if $\mathbf{v}_A$ and $\mathbf{v}_B$ had the same magnitude in Fig. 15.15b, the instantaneous center C would be at an infinite distance and ω would be zero; the slab would be in translation.

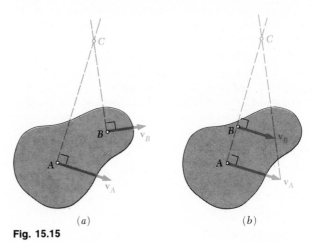

(a) $\qquad\qquad\qquad\qquad$ (b)

Fig. 15.15

To see how the concept of instantaneous center of rotation may be put to use, let us consider again the rod of Sec. 15.6. Drawing the perpendicular to $\mathbf{v}_A$ through A and the perpendicular to $\mathbf{v}_B$ through B (Fig. 15.16), we obtain the instantaneous center C. At the instant considered, the velocities of all the particles of the rod are thus the same as if the rod rotated about

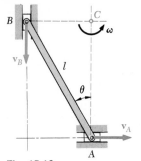

Fig. 15.16

C. Now, if the magnitude v_A of the velocity of A is known, the magnitude ω of the angular velocity of the rod may be obtained by writing

$$\omega = \frac{v_A}{AC} = \frac{v_A}{l \cos \theta}$$

The magnitude of the velocity of B may then be obtained by writing

$$v_B = (BC)\omega = l \sin \theta \, \frac{v_A}{l \cos \theta} = v_A \tan \theta$$

Note that only *absolute* velocities are involved in the computation.

The instantaneous center of a slab in plane motion may be located either on the slab or outside the slab. If it is located on the slab, the particle C coinciding with the instantaneous center at a given instant t must have zero velocity at that instant. The instantaneous-center method for the determination of velocities may thus be considered as a particular case of the method of relative velocities, where C is chosen as point of reference. Since $v_C = 0$, the general formula $v_B = v_C + v_{B/C}$ reduces to $v_B = v_{B/C}$, which means that the velocities of the various particles of the slab may be determined as if the slab were rotating about C.

It should be noted that the instantaneous center of rotation is valid only at a given instant. As the motion of the slab proceeds, other instantaneous centers should be used. The particle C of the slab which coincides with the instantaneous center at time t will generally not coincide with the instantaneous center at time $t + \Delta t$; while its velocity is zero at time t, it will probably be different from zero at time $t + \Delta t$. This means that, in general, the particle C *does not have zero acceleration,* and therefore that the *accelerations* of the various particles of the slab *cannot* be determined as if the slab were rotating about C.

Solve Sample Prob. 15.2, using the method of the instantaneous center of rotation.

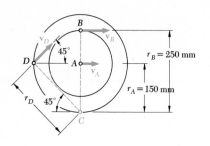

a. Angular Velocity of the Gear. Since the gear rolls on the stationary lower rack, the point of contact C of the gear with the rack has no velocity; point C is therefore the instantaneous center of rotation. We write

$$v_A = r_A\omega \qquad 1.2 \text{ m/s} = (0.150 \text{ m})\omega \qquad \omega = 8 \text{ rad/s} \; \rotatebox{-30}{$\downharpoonleft$} \quad \blacktriangleleft$$

b. Velocities. All points of the gear seem to rotate about the instantaneous center as far as velocities are concerned.

Velocity of Upper Rack. Recalling that $v_R = v_B$, we write

$$v_R = v_B = r_B\omega \qquad v_R = (0.250 \text{ m})(8 \text{ rad/s}) = 2 \text{ m/s}$$
$$v_R = 2 \text{ m/s} \rightarrow \quad \blacktriangleleft$$

Velocity of Point D. Since $r_D = (0.150 \text{ m})\sqrt{2} = 0.212 \text{ m}$, we write

$$v_D = r_D\omega \qquad v_D = (0.212 \text{ m})(8 \text{ rad/s}) = 1.696 \text{ m/s}$$
$$v_D = 1.696 \text{ m/s} \measuredangle 45° \quad \blacktriangleleft$$

Solve Sample Prob. 15.3, using the method of the instantaneous center of rotation.

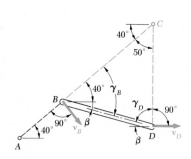

Motion of Crank AB. Referring to Sample Prob. 15.3, we obtain the velocity of point B; $\mathbf{v}_B = 627$ in./s $\searrow 50°$.

Motion of the Connecting Rod BD. We first locate the instantaneous center C by drawing lines perpendicular to the absolute velocities $\mathbf{v}_B$ and $\mathbf{v}_D$. Recalling from Sample Prob. 15.3 that $\beta = 13.9°$ and that $BD = 8$ in., we solve the triangle BCD.

$$\gamma_B = 40° + \beta = 53.9° \qquad \gamma_D = 90° - \beta = 76.1°$$
$$\frac{BC}{\sin 76.1°} = \frac{CD}{\sin 53.9°} = \frac{8 \text{ in.}}{\sin 50°}$$
$$BC = 10.14 \text{ in.} \qquad CD = 8.44 \text{ in.}$$

Since the connecting rod BD seems to rotate about point C, we write

$$v_B = (BC)\omega_{BD}$$
$$627 \text{ in./s} = (10.14 \text{ in.})\omega_{BD}$$
$$\omega_{BD} = 61.9 \text{ rad/s} \; \rotatebox{-30}{$\downharpoonleft$} \quad \blacktriangleleft$$

$$v_D = (CD)\omega_{BD} = (8.44 \text{ in.})(61.9 \text{ rad/s})$$
$$= 522 \text{ in./s} = 43.5 \text{ ft/s}$$
$$v_P = v_D = 43.5 \text{ ft/s} \rightarrow \quad \blacktriangleleft$$

PROBLEMS

15.46 A helicopter moves horizontally in the x direction at a speed of 45 mi/h. Knowing that the main blades rotate clockwise at an angular velocity of 120 rpm, determine the instantaneous axis of rotation of the main blades.

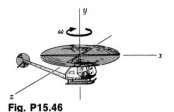

Fig. P15.46

15.47 A double pulley is attached to a slider block by a pin at A. The 50-mm-radius inner pulley is rigidly attached to the 100-mm-radius outer pulley. Knowing that each of the two cords is pulled at a constant speed of 300 mm/s as shown, determine (*a*) the instantaneous center of rotation of the double pulley, (*b*) the velocity of the slider block, (*c*) the length of cord wrapped or unwrapped on each pulley per second.

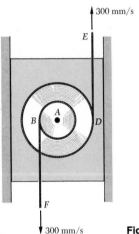

Fig. P15.47

15.48 Solve Sample Prob. 15.2, assuming that the lower rack is not stationary but moves to the left with a velocity of 0.6 m/s.

15.49 A drum, of radius 4.5 in., is mounted on a cylinder, of radius 6 in. A cord is wound around the drum, and its extremity D is pulled to the left at a constant velocity of 3 in./s, causing the cylinder to roll without sliding. Determine (*a*) the angular velocity of the cylinder, (*b*) the velocity of the center of the cylinder, (*c*) the length of cord which is wound or unwound per second.

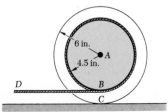

Fig. P15.49

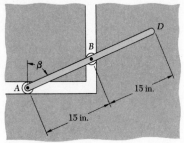

Fig. P15.50

15.50 The rod *ABD* is guided by wheels at *A* and *B* which roll in horizontal and vertical tracks as shown. Knowing that $\beta = 60°$ and that the velocity of *B* is 40 in./s downward, determine (*a*) the angular velocity of the rod, (*b*) the velocity of point *D*.

15.51 Knowing that at the instant shown the velocity of collar *D* is 20 in./s upward, determine (*a*) the angular velocity of rod *AD*, (*b*) the velocity of point *B*, (*c*) the velocity of point *A*.

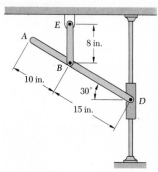

Fig. P15.51

15.52 Solve Prob. 15.50 assuming that $\beta = 30°$.

15.53 Knowing that at the instant shown the angular velocity of crank *AB* is 3 rad/s clockwise, determine (*a*) the angular velocity of link *BD*, (*b*) the velocity of collar *D*, (*c*) the velocity of the midpoint of link *BD*.

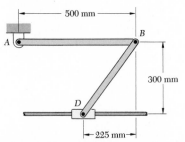

Fig. P15.53 and P15.54

15.54 Knowing that at the instant shown the velocity of collar *D* is 1.5 m/s to the right, determine (*a*) the angular velocities of crank *AB* and link *BD*, (*b*) the velocity of the midpoint of link *BD*.

15.55 Collar A slides downward with a constant velocity $\mathbf{v}_A$. Determine the angle θ corresponding to the position of rod AB for which the velocity of B is horizontal.

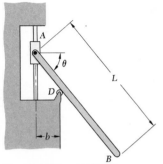

Fig. P15.55 and P15.57

15.56 Two rods AB and BD are connected to three collars as shown. Knowing that collar A moves downward with a constant velocity of 120 mm/s, determine (a) the angular velocity of each rod, (b) the velocity of collar D.

15.57 Collar A slides downward with a constant speed of 16 in./s. Knowing that $b = 2$ in., $L = 10$ in., and $\theta = 60°$, determine (a) the angular velocity of rod AB, (b) the velocity of B.

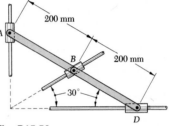

Fig. P15.56

15.58 The rectangular plate is supported by two 6-in. links as shown. Knowing that at the instant shown the angular velocity of link AB is 4 rad/s clockwise, determine (a) the angular velocity of the plate, (b) the velocity of the center of the plate, (c) the velocity of corner F.

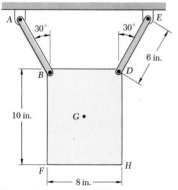

Fig. P15.58 and P15.59

15.59 Knowing that, at the instant shown, the angular velocity of link AB is 4 rad/s clockwise, determine (a) the angular velocity of the plate, (b) the points of the plate for which the magnitude of the velocity is equal to or less than 6 in./s.

15.60 At the instant shown, the velocity of the center of the gear is 200 mm/s to the right. Determine (*a*) the velocity of point *B*, (*b*) the velocity of collar *D*.

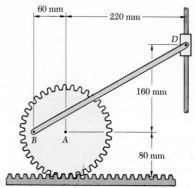

Fig. P15.60

15.61 Using the method of Sec. 15.7, solve Prob. 15.33.

15.62 Using the method of Sec. 15.7, solve Prob. 15.34.

15.63 Using the method of Sec. 15.7, solve Prob. 15.35.

15.64 Using the method of Sec. 15.7, solve Prob. 15.36.

15.65 Using the method of Sec. 15.7, solve Prob. 15.37.

15.66 Using the method of Sec. 15.7, solve Prob. 15.38.

15.67 Using the method of Sec. 15.7, solve Prob. 15.39.

15.68 Using the method of Sec. 15.7, solve Prob. 15.40.

15.69 Using the method of Sec. 15.7, solve Prob. 15.43.

15.70 Using the method of Sec. 15.7, solve Prob. 15.44.

15.71 Using the method of Sec. 15.7, solve Prob. 15.29.

15.72 Using the method of Sec. 15.7, solve Prob. 15.32.

15.8. Absolute and Relative Acceleration in Plane Motion. We saw in Sec. 15.5 that any plane motion may be replaced by a translation defined by the motion of an arbitrary reference point *A*, and by a rotation about *A*. This property was used in Sec. 15.6 to determine the velocity of the various points of a moving slab. We shall now use the same property to determine the acceleration of the points of the slab.

We first recall that the absolute acceleration $\mathbf{a}_B$ of a particle of the slab may be obtained from the relative-acceleration formula derived in Sec. 11.11,

$$\mathbf{a}_B = \mathbf{a}_A + \mathbf{a}_{B/A} \tag{15.18}$$

where the right-hand member represents a vector sum. The acceleration $\mathbf{a}_A$ corresponds to the translation of the slab with A, while the relative acceleration $\mathbf{a}_{B/A}$ is associated with the rotation of the slab about A and is measured with respect to axes of fixed orientation centered at A (Fig. 15.17). Denoting by r the distance from A to B and, respectively, by $\boldsymbol{\omega}$ and $\boldsymbol{\alpha}$ the angular velocity and angular acceleration of the slab with respect to axes of fixed orientation, we find that the relative acceleration $\mathbf{a}_{B/A}$ of B with respect to A consists of two components, a *normal component* $(\mathbf{a}_{B/A})_n$ of magnitude $r\omega^2$ directed toward A, and a *tangential component* $(\mathbf{a}_{B/A})_t$ of magnitude $r\alpha$ perpendicular to the line AB.

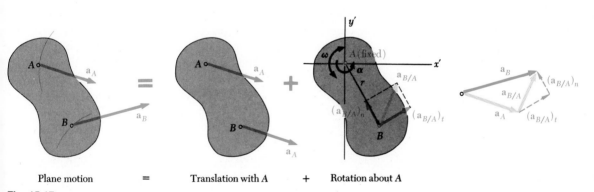

| Plane motion | = | Translation with A | + | Rotation about A |

Fig. 15.17

As an example, we shall consider again the rod AB whose extremities slide, respectively, along a horizontal and a vertical track (Fig. 15.18). Assuming that the velocity $\mathbf{v}_A$ and the acceleration $\mathbf{a}_A$ of A are known, we propose to determine the acceleration $\mathbf{a}_B$ of B and the angular acceleration $\boldsymbol{\alpha}$ of the rod. Choosing A as a reference point, we express that the given motion is equivalent to a translation with A and a rotation about A. The

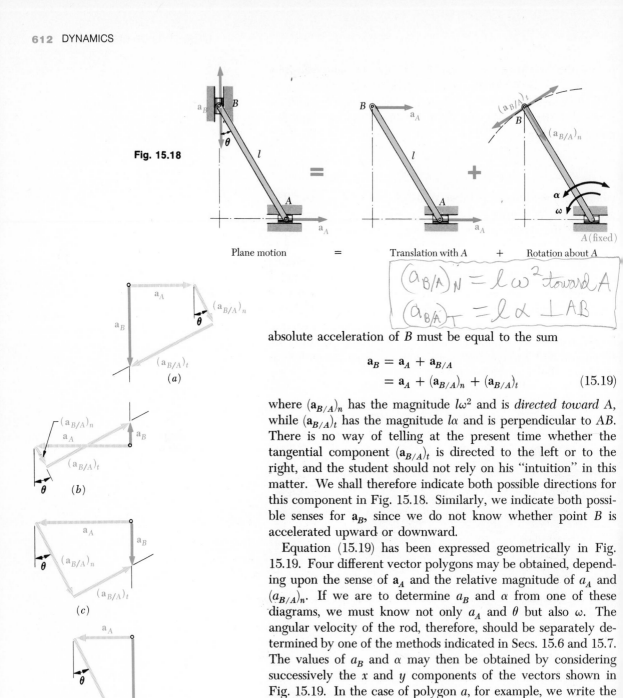

Fig. 15.18

Plane motion = Translation with A + Rotation about A

$$(a_{B/A})_N = l\omega^2 \text{ toward } A$$
$$(a_{B/A})_T = l\alpha \perp AB$$

(a)

(b)

(c)

(d)

Fig. 15.19

absolute acceleration of B must be equal to the sum

$$\mathbf{a}_B = \mathbf{a}_A + \mathbf{a}_{B/A}$$
$$= \mathbf{a}_A + (\mathbf{a}_{B/A})_n + (\mathbf{a}_{B/A})_t \qquad (15.19)$$

where $(\mathbf{a}_{B/A})_n$ has the magnitude $l\omega^2$ and is *directed toward A*, while $(\mathbf{a}_{B/A})_t$ has the magnitude $l\alpha$ and is perpendicular to AB. There is no way of telling at the present time whether the tangential component $(\mathbf{a}_{B/A})_t$ is directed to the left or to the right, and the student should not rely on his "intuition" in this matter. We shall therefore indicate both possible directions for this component in Fig. 15.18. Similarly, we indicate both possible senses for $\mathbf{a}_B$, since we do not know whether point B is accelerated upward or downward.

Equation (15.19) has been expressed geometrically in Fig. 15.19. Four different vector polygons may be obtained, depending upon the sense of $\mathbf{a}_A$ and the relative magnitude of a_A and $(a_{B/A})_n$. If we are to determine a_B and α from one of these diagrams, we must know not only a_A and θ but also ω. The angular velocity of the rod, therefore, should be separately determined by one of the methods indicated in Secs. 15.6 and 15.7. The values of a_B and α may then be obtained by considering successively the x and y components of the vectors shown in Fig. 15.19. In the case of polygon a, for example, we write the equations

$\xrightarrow{+} x$ components: $\qquad 0 = a_A + l\omega^2 \sin\theta - l\alpha \cos\theta$

$+\uparrow y$ components: $\qquad -a_B = -l\omega^2 \cos\theta - l\alpha \sin\theta$

which may be solved for a_B and α. The two unknowns may also be obtained by direct measurement on the vector polygon. In that case, care should be taken to draw first the known vectors $\mathbf{a}_A$ and $(\mathbf{a}_{B/A})_n$.

It is quite evident that the determination of accelerations is considerably more involved than the determination of velocities. Yet, in the example considered here, the extremities A and B of the rod were moving along straight tracks, and the diagrams drawn were relatively simple. If A and B had moved along curved tracks, the accelerations $\mathbf{a}_A$ and $\mathbf{a}_B$ should have been resolved into normal and tangential components and the solution of the problem would have involved six different vectors.

When a mechanism consists of several moving parts which are pin-connected, its analysis may be carried out by considering each part as a rigid body, while keeping in mind that the points where two parts are connected must have the same absolute acceleration (see Sample Prob. 15.7). In the case of meshed gears, the tangential components of the accelerations of the teeth in contact are equal, but their normal components are different.

*15.9. Analysis of Plane Motion in Terms of a Parameter. In the case of certain mechanisms, it is possible to express the coordinates x and y of all the significant points of the mechanism by means of simple analytic expressions containing a single parameter. It may be advantageous in such a case to determine directly the absolute velocity and the absolute acceleration of the various points of the mechanism, since the components of the velocity and of the acceleration of a given point may be obtained by differentiating the coordinates x and y of that point.

Let us consider again the rod AB whose extremities slide, respectively, in a horizontal and a vertical track (Fig. 15.20). The coordinates x_A and y_B of the extremities of the rod may be expressed in terms of the angle θ the rod forms with the vertical,

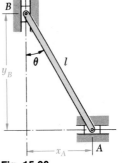

Fig. 15.20

$$x_A = l \sin \theta \qquad y_B = l \cos \theta \qquad (15.20)$$

Differentiating Eqs. (15.20) twice with respect to t, we write

$$v_A = \dot{x}_A = l\dot{\theta} \cos \theta \qquad\qquad v_B = \dot{y}_B = -l\dot{\theta} \sin \theta$$
$$a_A = \ddot{x}_A = -l\dot{\theta}^2 \sin \theta + l\ddot{\theta} \cos \theta$$
$$a_B = \ddot{y}_B = -l\dot{\theta}^2 \cos \theta - l\ddot{\theta} \sin \theta$$

Recalling that $\dot{\theta} = \omega$ and $\ddot{\theta} = \alpha$, we obtain

$$v_A = l\omega \cos \theta \qquad\qquad v_B = -l\omega \sin \theta \qquad (15.21)$$
$$a_A = -l\omega^2 \sin \theta + l\alpha \cos \theta \qquad a_B = -l\omega^2 \cos \theta - l\alpha \sin \theta$$
$$(15.22)$$

We note that a positive sign for v_A or a_A indicates that the velocity $\mathbf{v}_A$ or the acceleration $\mathbf{a}_A$ is directed to the right; a positive sign for v_B or a_B indicates that $\mathbf{v}_B$ or $\mathbf{a}_B$ is directed upward. Equations (15.21) may be used, for example, to determine v_B and ω when v_A and θ are known. Substituting for ω in (15.22), we may then determine a_B and α if a_A is known.

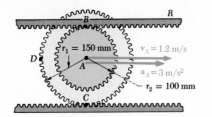

The center of the double gear of Sample Prob. 15.2 has a velocity of 1.2 m/s to the right and an acceleration of 3 m/s² to the right. Determine (a) the angular acceleration of the gear, (b) the acceleration of points B, C, and D of the gear.

a. Angular Acceleration of the Gear. In Sample Prob. 15.2, we found that $x_A = r_1\theta$ and $v_A = r_1\omega$. Differentiating the latter with respect to time, we obtain $a_A = r_1\alpha$.

$$v_A = r_1\omega \qquad 1.2 \text{ m/s} = (0.150 \text{ m})\omega \qquad \omega = 8 \text{ rad/s} \downcurvearrowright$$

$$a_A = r_1\alpha \qquad 3 \text{ m/s}^2 = (0.150 \text{ m})\alpha \qquad \alpha = 20 \text{ rad/s}^2 \downcurvearrowright \blacktriangleleft$$

b. Accelerations. The rolling motion of the gear is resolved into a translation with A and a rotation about A.

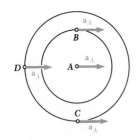

Translation

$+$

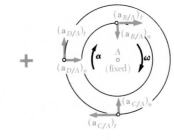

Rotation

$=$

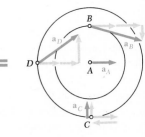

Rolling motion

Acceleration of Point B. Adding vectorially the accelerations corresponding to the translation and to the rotation, we obtain

$$\mathbf{a}_B = \mathbf{a}_A + \mathbf{a}_{B/A} = \mathbf{a}_A + (\mathbf{a}_{B/A})_t + (\mathbf{a}_{B/A})_n$$

where

$$\mathbf{a}_A = 3 \text{ m/s}^2 \rightarrow$$

$$(a_{B/A})_t = r_2\alpha = (0.100 \text{ m})(20 \text{ rad/s}^2) \qquad (\mathbf{a}_{B/A})_t = 2 \text{ m/s}^2 \rightarrow$$

$$(a_{B/A})_n = r_2\omega^2 = (0.100 \text{ m})(8 \text{ rad/s})^2 \qquad (\mathbf{a}_{B/A})_n = 6.4 \text{ m/s}^2 \downarrow$$

We have $\quad \mathbf{a}_B = [3 \text{ m/s}^2 \rightarrow] + [2 \text{ m/s}^2 \rightarrow] + [6.4 \text{ m/s}^2 \downarrow]$

$$\mathbf{a}_B = 8.12 \text{ m/s}^2 \searrow 52.0° \blacktriangleleft$$

Acceleration of Point C

$$\mathbf{a}_C = \mathbf{a}_A + \mathbf{a}_{C/A} = \mathbf{a}_A + (\mathbf{a}_{C/A})_t + (\mathbf{a}_{C/A})_n$$

$$= [3 \text{ m/s}^2 \rightarrow] + [(0.150 \text{ m})(20 \text{ rad/s}^2) \leftarrow] + [(0.150 \text{ m})(8 \text{ rad/s})^2 \uparrow]$$

$$\mathbf{a}_C = 9.60 \text{ m/s}^2 \uparrow \blacktriangleleft$$

Acceleration of Point D

$$\mathbf{a}_D = \mathbf{a}_A + \mathbf{a}_{D/A} = \mathbf{a}_A + (\mathbf{a}_{D/A})_t + (\mathbf{a}_{D/A})_n$$

$$= [3 \text{ m/s}^2 \rightarrow] + [(0.150 \text{ m})(20 \text{ rad/s}^2) \uparrow] + [(0.150 \text{ m})(8 \text{ rad/s})^2 \rightarrow]$$

$$\mathbf{a}_D = 12.95 \text{ m/s}^2 \measuredangle 13.4° \blacktriangleleft$$

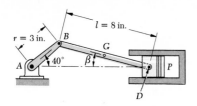

Crank AB of the engine system of Sample Prob. 15.3 has a constant clockwise angular velocity of 2000 rpm. For the crank position shown, determine the angular acceleration of the connecting rod BD and the acceleration of point D.

Motion of Crank AB. Since the crank rotates about A with constant $\omega_{AB} = 2000$ rpm $= 209$ rad/s, we have $\alpha_{AB} = 0$. The acceleration of B is therefore directed toward A and has a magnitude

$$a_B = r\omega_{AB}^2 = (\tfrac{3}{12} \text{ ft})(209 \text{ rad/s})^2 = 10{,}920 \text{ ft/s}^2$$

$$\mathbf{a}_B = 10{,}920 \text{ ft/s}^2 \; \nearrow \; 40°$$

Motion of the Connecting Rod BD. The angular velocity ω_{BD} and the value of β were obtained in Sample Prob. 15.3.

$$\omega_{BD} = 61.9 \text{ rad/s} \; \curvearrowright \qquad \beta = 13.9°$$

The motion of BD is resolved into a translation with B and a rotation about B. The relative acceleration $\mathbf{a}_{D/B}$ is resolved into normal and tangential components.

$$(a_{D/B})_n = (BD)\omega_{BD}^2 = (\tfrac{8}{12} \text{ ft})(61.9 \text{ rad/s})^2 = 2550 \text{ ft/s}^2$$

$$(\mathbf{a}_{D/B})_n = 2550 \text{ ft/s}^2 \; \searrow \; 13.9°$$

$$(a_{D/B})_t = (BD)\alpha_{BD} = (\tfrac{8}{12})\alpha_{BD} = 0.667\alpha_{BD}$$

$$(\mathbf{a}_{D/B})_t = 0.667\alpha_{BD} \; \measuredangle \; 76.1°$$

While $(\mathbf{a}_{B/D})_t$ must be perpendicular to BD, its sense is not known.

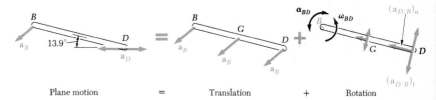

Plane motion	=	Translation	+	Rotation

Noting that the acceleration $\mathbf{a}_D$ must be horizontal, we write

$$\mathbf{a}_D = \mathbf{a}_B + \mathbf{a}_{D/B} = \mathbf{a}_B + (\mathbf{a}_{D/B})_n + (\mathbf{a}_{D/B})_t$$

$$[a_D \leftrightarrow] = [10{,}920 \; \nearrow \; 40°] + [2560 \; \searrow \; 13.9°] + [0.667\alpha_{BD} \; \measuredangle \; 76.1°]$$

Equating x and y components, we obtain the following scalar equations:

$\xrightarrow{+} x$ components:

$$-a_D = -10{,}920 \cos 40° - 2550 \cos 13.9° + 0.667\alpha_{BD} \sin 13.9°$$

$+\uparrow y$ components:

$$0 = -10{,}920 \sin 40° + 2550 \sin 13.9° + 0.667\alpha_{BD} \cos 13.9°$$

Solving the equations simultaneously, we obtain $\alpha_{BD} = +9890$ rad/s^2 and $a_D = +9260$ ft/s^2. The positive signs indicate that the senses shown on the vector polygon are correct; we write

$$\alpha_{BD} = 9890 \text{ rad/s}^2 \; \curvearrowright \qquad \blacktriangleleft$$

$$\mathbf{a}_D = 9260 \text{ ft/s}^2 \leftarrow \qquad \blacktriangleleft$$

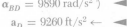

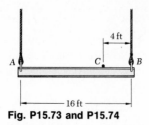

Fig. P15.73 and P15.74

PROBLEMS

15.73 A 16-ft steel beam is lowered by means of two cables unwinding at the same speed from overhead cranes. As the beam approaches the ground, the crane operators apply brakes to slow down the unwinding motion. The deceleration of the cable attached at A is 11 ft/s^2, while that of the cable attached at B is 3 ft/s^2. Determine the acceleration of point C and the angular acceleration of the beam at that instant.

15.74 The acceleration of point C is 2 ft/s^2 upward and the angular acceleration of the beam is 1.5 rad/s^2 clockwise. Knowing that the angular velocity of the beam is zero at the instant considered, determine the acceleration of each cable.

15.75 A 600-mm rod rests on a smooth horizontal table. A force **P** applied as shown produces the following accelerations: $\mathbf{a}_A = 0.8$ m/s^2 to the right, $\boldsymbol{\alpha} = 2$ rad/s^2 clockwise as viewed from above. Determine the acceleration (a) of point B, (b) of point G.

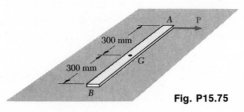

Fig. P15.75

15.76 In Prob. 15.75, determine the point of the rod which (a) has no acceleration, (b) has an acceleration of 0.350 m/s^2 to the right.

15.77 Determine the accelerations of points C and D of the wheel of Prob. 15.23, knowing that the automobile moves at a constant speed of 50 km/h.

15.78 Determine the accelerations of points B and E of the wheel of Prob. 15.23, knowing that the automobile moves at a constant speed of 50 km/h.

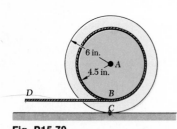

Fig. P15.79

15.79 A drum, of radius 4.5 in., is mounted on a cylinder, of radius 6 in. A cord is wound around the drum and is pulled in such a way that point D has a velocity of 3 in./s and an acceleration of 15 in./s^2, both directed to the left. Assuming that the cylinder rolls without slipping, determine the acceleration (a) of point A, (b) of point B, (c) of point C.

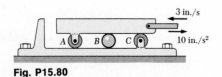

Fig. P15.80

15.80 The moving carriage is supported by two casters A and C, each of $\frac{1}{2}$-in. diameter, and by a $\frac{1}{2}$-in.-diameter ball B. If at a given instant the velocity and acceleration of the carriage are as shown, determine (a) the angular accelerations of the ball and of each caster, (b) the accelerations of the center of the ball and of each caster.

15.81 and 15.82 At the instant shown, the disk rotates with a constant angular velocity ω_0 clockwise. Determine the angular velocities and the angular accelerations of the rods AB and BC.

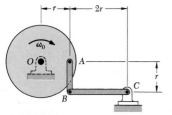

Fig. P15.81

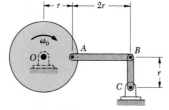

Fig. P15.82

15.83 Crank AB rotates about A with a constant angular velocity of 900 rpm clockwise. Determine the acceleration of the piston P when (a) $\theta = 90°$, (b) $\theta = 180°$.

15.84 Solve Prob. 15.83 when (a) $\theta = 0$, (b) $\theta = 270°$.

15.85 Arm AB rotates with a constant angular velocity of 120 rpm clockwise. Knowing that gear A does not rotate, determine the acceleration of the tooth of gear B which is in contact with gear A.

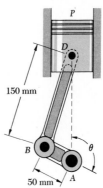

Fig. P15.83

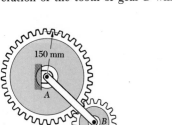

Fig. P15.85

15.86 and 15.87 For the linkage indicated, determine the angular acceleration (a) of bar BD, (b) of bar DE.
     15.86 Linkage of Prob. 15.38.
     15.87 Linkage of Prob. 15.39.

15.88 and 15.89 The end A of the rod AB moves to the right with a constant velocity of 8 ft/s. For the position shown, determine (a) the angular acceleration of the rod, (b) the acceleration of the midpoint G of the rod.

15.90 and 15.91 In the position shown, end A of the rod AB has a velocity of 8 ft/s and an acceleration of 6 ft/s², both directed to the right. Determine (a) the angular acceleration of the rod, (b) the acceleration of the midpoint G of the rod.

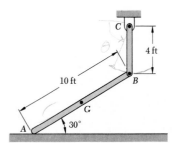

Fig. P15.88 and P15.90

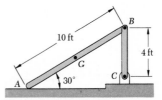

Fig. P15.89 and P15.91

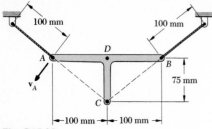

Fig. P15.92

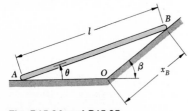

Fig. P15.94 and P15.95

15.92 Point A of the bracket ABCD moves with a velocity of constant magnitude $v_A = 250$ mm/s. For the position shown, determine (a) the angular acceleration of the bracket, (b) the acceleration of point C.

15.93 In Prob. 15.92, determine the acceleration of point D.

*15.94 Rod AB slides with its ends in contact with the floor and the inclined plane. Using the method of Sec. 15.9, derive an expression for the angular velocity of the rod in terms of v_B, θ, l, and β.

*15.95 Derive an expression for the angular acceleration of the rod AB in terms of v_B, θ, l, and β, knowing that the acceleration of point B is zero.

*15.96 The drive disk of the Scotch crosshead mechanism shown has an angular velocity ω and an angular acceleration α, both directed clockwise. Using the method of Sec. 15.9, derive an expression (a) for the velocity of point B, (b) for the acceleration of point B.

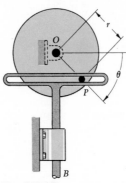

Fig. P15.96

*15.97 A disk of radius r rolls to the right with a constant velocity **v.** Denoting by P the point of the rim in contact with the ground at $t = 0$, derive expressions for the horizontal and vertical components of the velocity of P at any time t. (The curve described by point P is called a *cycloid*.)

*15.98 For the disk of Prob. 15.97, derive expressions for the horizontal and vertical components of the acceleration of P at any time t.

*15.99 Knowing that rod AB rotates with an angular velocity ω and with an angular acceleration α, both counterclockwise, derive expressions for the velocity and acceleration of collar C.

*15.100 Knowing that collar C moves upward with a constant velocity **v_0,** derive expressions for the angular velocity and angular acceleration of rod AB in the position shown.

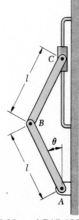

Fig. P15.99 and P15.100

*15.101   Collar *B* slides along rod *OC* and is attached to a sliding block which moves in a vertical slot. Knowing that rod *OC* rotates with an angular velocity ω and with an angular acceleration α, both counterclockwise, derive expressions for the velocity and acceleration of collar *B*.

*15.102   Collar *B* slides along rod *OC* and is attached to a sliding block which moves upward with a constant velocity **v** in a vertical slot. Using the method of Sec. 15.9, derive an expression (*a*) for the angular velocity of rod *OC*, (*b*) for the angular acceleration of rod *OC*.

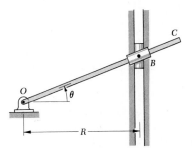

Fig. P15.101 and P15.102

*15.103   The crank *AB* of Prob. 15.33 rotates with a constant clockwise angular velocity ω, and $\theta = 0$ at $t = 0$. Using the method of Sec. 15.9, derive an expression for the velocity of the piston *P* in terms of the time *t*.

*15.104   A slender rod *AB* is attached to a collar at *B* and rests on a smooth circular cylinder of radius *r*. Knowing that the collar *B* moves upward with a constant velocity of magnitude *v*, derive an expression in terms of *r*, θ, and *v* (*a*) for the angular velocity of the rod, (*b*) for the angular acceleration of the rod.

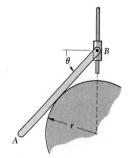

Fig. P15.104

*15.105   Collar *A* slides upward with a constant velocity $\mathbf{v}_A$. Using the method of Sec. 15.9, derive an expression for (*a*) the angular velocity of rod *AB*, (*b*) the components of the velocity of point *B*.

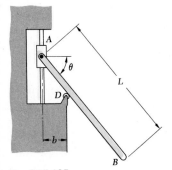

Fig. P15.105

*15.106   In Prob. 15.105, derive an expression for the angular acceleration of rod *AB*.

*15.10. Particle Moving on a Slab in Translation.

Let us consider the motion of a particle *P* which describes a path on a slab *S* which is itself in translation. The motion of *P* may be analyzed, either in terms of its coordinates *x* and *y* with respect to a fixed set of axes, or in terms of its coordinates x_1 and y_1 with respect to a set of axes attached to the slab *S*

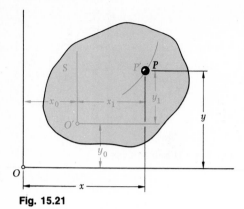

Fig. 15.21

and moving with it (Fig. 15.21). We propose to determine the relation existing between the absolute motion of P with respect to the fixed axes and its relative motion with respect to the axes moving with S.

Denoting by x_0 and y_0 the coordinates of O' with respect to the fixed axes, we write

$$x = x_0 + x_1 \qquad y = y_0 + y_1 \qquad (15.23)$$

Differentiating with respect to t and using dots to indicate time derivatives, we have

$$\dot{x} = \dot{x}_0 + \dot{x}_1 \qquad \dot{y} = \dot{y}_0 + \dot{y}_1 \qquad (15.24)$$

Now, $\dot{x}$ and $\dot{y}$ represent the components of the absolute velocity $\mathbf{v}_P$ of P, while $\dot{x}_1$ and $\dot{y}_1$ represent the components of the velocity $\mathbf{v}_{P/S}$ of P with respect to S. On the other hand, $\dot{x}_0$ and $\dot{y}_0$ represent the components of the velocity of O' or, since the slab S is in translation, the components of the velocity of any other point of S. We may, for example, consider that $\dot{x}_0$ and $\dot{y}_0$ represent the components of the velocity $\mathbf{v}_{P'}$ of the *point P' of the slab S which happens to coincide with the particle P at the instant considered.* We thus write

$$\mathbf{v}_P = \mathbf{v}_{P'} + \mathbf{v}_{P/S} \qquad (15.25)$$

where the right-hand member represents a vector sum. We note that the velocity $\mathbf{v}_P$ reduces to $\mathbf{v}_{P/S}$ if the slab is stopped and if the particle is allowed to keep moving on S; it reduces to the velocity $\mathbf{v}_{P'}$ of the coinciding point P' if the particle P is immobilized on S while S is allowed to keep moving. Thus, formula (15.25) expresses that the absolute velocity of P may be obtained by adding vectorially these two partial velocities.

Differentiating Eqs. (15.24) with respect to t, we obtain

$$\ddot{x} = \ddot{x}_0 + \ddot{x}_1 \qquad \ddot{y} = \ddot{y}_0 + \ddot{y}_1 \qquad (15.26)$$

or, in vector form,

$$\mathbf{a}_P = \mathbf{a}_{P'} + \mathbf{a}_{P/S} \qquad (15.27)$$

where $\mathbf{a}_{P/S}$ = acceleration of P with respect to S

$\mathbf{a}_{P'}$ = acceleration of point P' of S coinciding with P at the instant considered

Formula (15.27) may be interpreted in the same way as formula (15.25).

Formulas (15.25) and (15.27) actually restate the results obtained in Sec. 11.11. Since S is in translation, the x_1 and y_1 axes, respectively, remain parallel to the fixed x and y axes, and the velocity $\mathbf{v}_{P/S}$ is equal to the velocity $\mathbf{v}_{P/P'}$ relative to point P'

as it was defined in Sec. 11.11; similarly, the acceleration $\mathbf{a}_{P/S}$ is equal to the acceleration $\mathbf{a}_{P/P'}$ relative to P'.

∗15.11. Particle Moving on a Rotating Slab. Coriolis Acceleration. We shall now consider the motion of a particle P which describes a path on a slab S which is itself rotating about a fixed point O. The motion of P may be analyzed either in terms of its polar coordinates r and θ with respect to fixed axes or in terms of its coordinates r and θ_1 with respect to axes attached to the slab S and rotating with it (Fig. 15.22). Again we propose to determine the relation existing between the absolute motion of P and its relative motion with respect to S.

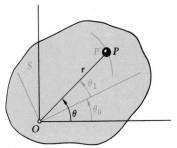

Fig. 15.22

We shall use formulas (11.30) to express the radial and transverse components of the absolute velocity $\mathbf{v}_P$ of P. Observing that $\theta = \theta_0 + \theta_1$, where θ_0 denotes the angular displacement of the slab at the instant considered, we write

$$(v_P)_r = \dot{r} \qquad (v_P)_\theta = r\dot{\theta} = r(\dot{\theta}_0 + \dot{\theta}_1) \qquad (15.28)$$

Considering the particular case when P is immobilized on S and S is allowed to rotate, $\mathbf{v}_P$ reduces to the velocity $\mathbf{v}_{P'}$ of the point P' of the slab S which happens to coincide with P at the instant considered. Making $r = $ constant and $\theta_1 = $ constant in (15.28), we obtain

$$(v_{P'})_r = 0 \qquad (v_{P'})_\theta = r\dot{\theta}_0 \qquad (15.29)$$

Considering now the particular case when the slab is maintained fixed and P is allowed to move, $\mathbf{v}_P$ reduces to the relative velocity $\mathbf{v}_{P/S}$. Making θ_0 constant in (15.28), we obtain, therefore,

$$(v_{P/S})_r = \dot{r} \qquad (v_{P/S})_\theta = r\dot{\theta}_1 \qquad (15.30)$$

Comparing formulas (15.28) to (15.30), we find that

$$(v_P)_r = (v_{P'})_r + (v_{P/S})_r \qquad (v_P)_\theta = (v_{P'})_\theta + (v_{P/S})_\theta$$

or, in vector form,

$$\mathbf{v}_P = \mathbf{v}_{P'} + \mathbf{v}_{P/S} \qquad (15.31)$$

Formula (15.31) expresses that $\mathbf{v}_P$ may be obtained by adding vectorially the velocities corresponding to the two particular cases considered above.

Using formulas (11.33), we now express the radial and transverse components of the absolute acceleration $\mathbf{a}_P$ of P,

$$(a_P)_r = \ddot{r} - r\dot{\theta}^2 = \ddot{r} - r(\dot{\theta}_0 + \dot{\theta}_1)^2$$
$$= \ddot{r} - r(\dot{\theta}_0^2 + 2\dot{\theta}_0\dot{\theta}_1 + \dot{\theta}_1^2) \qquad (15.32)$$

$$(a_P)_\theta = r\ddot{\theta} + 2\dot{r}\dot{\theta}$$
$$= r(\ddot{\theta}_0 + \ddot{\theta}_1) + 2\dot{r}(\dot{\theta}_0 + \dot{\theta}_1) \qquad (15.33)$$

Considering again the particular case when $r = $ constant and $\theta_1 = $ constant (P immobilized on S), Eqs. (15.32) and (15.33) yield

$$(a_{P'})_r = -r\dot{\theta}_0^2 \qquad (a_{P'})_\theta = r\ddot{\theta}_0 \qquad (15.34)$$

Considering now the particular case when $\theta_0 = $ constant (slab fixed), we obtain

$$(a_{P/S})_r = \ddot{r} - r\dot{\theta}_1^2 \qquad (a_{P/S})_\theta = r\ddot{\theta}_1 + 2\dot{r}\dot{\theta}_1 \qquad (15.35)$$

Comparing formulas (15.32) to (15.35), we find that the absolute acceleration $\mathbf{a}_P$ *cannot* be obtained by adding the accelerations $\mathbf{a}_{P'}$ and $\mathbf{a}_{P/S}$ corresponding to the two particular cases considered above. We have instead

$$\mathbf{a}_P = \mathbf{a}_{P'} + \mathbf{a}_{P/S} + \mathbf{a}_c \qquad (15.36)$$

where $\mathbf{a}_c$ is a vector of components

$$(a_c)_r = -2r\dot{\theta}_0\dot{\theta}_1 \qquad (a_c)_\theta = 2\dot{r}\dot{\theta}_0$$

Noting that $\dot{\theta}_0$ represents the angular velocity ω of S, and recalling formulas (15.30), we have

$$(a_c)_r = -2\omega(v_{P/S})_\theta \qquad (a_c)_\theta = 2\omega(v_{P/S})_r \qquad (15.37)$$

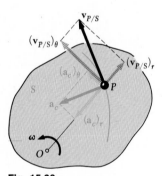

Fig. 15.23

The vector $\mathbf{a}_c$ is called the *complementary acceleration*, or *Coriolis acceleration*, after the French mathematician De Coriolis (1792–1843). The first of formulas (15.37) indicates that the vector $(\mathbf{a}_c)_r$ is obtained by multiplying the vector $(\mathbf{v}_{P/S})_\theta$ by 2ω and rotating it through 90° in the sense of rotation of the slab (Fig. 15.23); the second formula defines $(\mathbf{a}_c)_\theta$ from $(\mathbf{v}_{P/S})_r$ in a similar way. *The Coriolis acceleration $\mathbf{a}_c$ is thus a vector perpendicular to the relative velocity $\mathbf{v}_{P/S}$, and of magnitude equal to $2\omega v_{P/S}$; the sense of $\mathbf{a}_c$ is obtained by rotating the vector $\mathbf{v}_{P/S}$ through 90° in the sense of rotation of S.*

Comparing formulas (15.36) and (11.26), we note that the acceleration $\mathbf{a}_{P/S}$ of P relative to the slab S is *not* equal to the acceleration $\mathbf{a}_{P/P'}$ of P relative to the point P' of S; this is due to the fact that the first acceleration is defined with respect to *rotating axes*, while the second is defined with respect to *axes of fixed orientation*.

The following example will help in understanding the physical meaning of the Coriolis acceleration. Consider a collar P which is made to slide at a constant relative speed u along a rod OB rotating at a constant angular velocity ω about O (Fig. 15.24a). According to formula (15.36), the absolute acceleration of P may be obtained by adding vectorially the acceleration $\mathbf{a}_A$ of the point A of the rod coinciding with P, the relative acceleration $\mathbf{a}_{P/OB}$ of P with respect to the rod, and the Coriolis acceleration $\mathbf{a}_c$. Since the angular velocity ω of the rod is constant, $\mathbf{a}_A$ reduces to its normal component $(\mathbf{a}_A)_n$ of magnitude $r\omega^2$; and since u

is constant, the relative acceleration $\mathbf{a}_{P/OB}$ is zero. According to the definition given above, the Coriolis acceleration is a vector perpendicular to OB, of magnitude $2\omega u$, and directed as shown in the figure. The acceleration of the collar P consists, therefore, of the two vectors shown in Fig. 15.24a. Note that the result obtained may be checked by applying formulas (11.33).

To understand better the significance of the Coriolis acceleration, we shall consider the absolute velocity of P at time t and at time $t + \Delta t$ (Fig. 15.24b). At time t, the velocity may be resolved into its components $\mathbf{u}$ and $\mathbf{v}_A$, and at time $t + \Delta t$ into its components $\mathbf{u}'$ and $\mathbf{v}_{A'}$. Drawing these components from the same origin (Fig. 15.24c), we note that the change in velocity during the time Δt may be represented by the sum of three vectors $\overrightarrow{RR'}$, $\overrightarrow{TT''}$, and $\overrightarrow{T''T'}$. The vector $\overrightarrow{TT''}$ measures the change in direction of the velocity $\mathbf{v}_A$, and the quotient $\overrightarrow{TT''}/\Delta t$ represents the acceleration $\mathbf{a}_A$ when Δt approaches zero. We check that the direction of $\overrightarrow{TT''}$ is that of $\mathbf{a}_A$ when Δt approaches zero and that

$$\lim_{\Delta t \to 0} \frac{TT''}{\Delta t} = \lim_{\Delta t \to 0} v_A \frac{\Delta \theta}{\Delta t} = r\omega\omega = r\omega^2 = a_A$$

The vector $\overrightarrow{RR'}$ measures the change in direction of $\mathbf{u}$ due to the rotation of the rod; the vector $\overrightarrow{T''T'}$ measures the change in magnitude of $\mathbf{v}_A$ due to the motion of P on the rod. The vectors $\overrightarrow{RR'}$ and $\overrightarrow{T''T'}$ result from the *combined effect* of the relative motion of P and of the rotation of the rod; they would vanish if *either* of these two motions stopped. We may easily verify that the sum of these two vectors defines the Coriolis acceleration. Their direction is that of $\mathbf{a}_c$ when Δt approaches zero and, since $RR' = u\,\Delta\theta$ and $T''T' = v_{A'} - v_A = (r + \Delta r)\omega - r\omega = \omega\,\Delta r$, we check that

$$\lim_{\Delta t \to 0}\left(\frac{RR'}{\Delta t} + \frac{T''T'}{\Delta t}\right) = \lim_{\Delta t \to 0}\left(u \frac{\Delta\theta}{\Delta t} + \omega \frac{\Delta r}{\Delta t}\right)$$
$$= u\omega + \omega u = 2\omega u = a_c$$

Formulas (15.31) and (15.36) may be used to analyze the motion of mechanisms which contain parts sliding on each other. They make it possible, for example, to relate the absolute and relative motions of sliding pins and collars (see Sample Probs. 15.8 and 15.9). The concept of Coriolis acceleration is also very useful in the study of long-range projectiles and of other bodies whose motions are appreciably affected by the rotation of the earth. As was pointed out in Sec. 12.1, a system of axes attached to the earth does not truly constitute a newtonian frame of reference; such a system of axes should actually be considered as rotating. The formulas derived in this section will therefore facilitate the study of the motion of bodies with respect to axes attached to the earth.

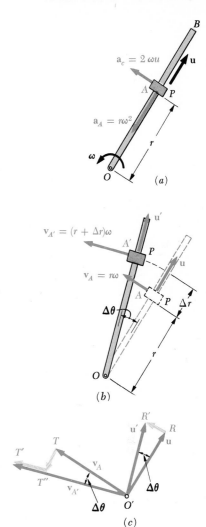

Fig. 15.24

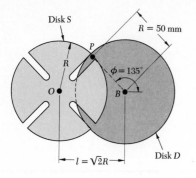

Disk S
$R = 50$ mm
P
R
$\phi = 135°$
O
B
$l = \sqrt{2}R$
Disk D

The Geneva mechanism shown is used in many counting instruments and in other applications where an intermittent rotary motion is required. Disk D rotates with a constant counterclockwise angular velocity ω_D of 10 rad/s. A pin P is attached to disk D and slides along one of several slots cut in disk S. It is desirable that the angular velocity of disk S be zero as the pin enters and leaves each slot; in the case of four slots, this will occur if the distance between the centers of the disks is $l = \sqrt{2}\ R$.

At the instant when $\phi = 150°$, determine (a) the angular velocity of disk S, (b) the velocity of pin P relative to disk S.

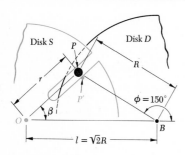

Disk S
P
Disk D
R
r
P'
$\phi = 150°$
β
O
B
$l = \sqrt{2}R$

Solution. We solve triangle OPB, which corresponds to the position $\phi = 150°$. Using the law of cosines, we write

$$r^2 = R^2 + l^2 - 2Rl \cos 30° = 0.551R^2 \qquad r = 0.742R = 37.1 \text{ mm}$$

From the law of sines

$$\frac{\sin \beta}{R} = \frac{\sin 30°}{r} \qquad \sin \beta = \frac{\sin 30°}{0.742} \qquad \beta = 42.4°$$

Since pin P is attached to disk D, and since disk D rotates about point B, the magnitude of the absolute velocity of P is

$$v_P = R\omega_D = (50 \text{ mm})(10 \text{ rad/s}) = 500 \text{ mm/s}$$
$$\mathbf{v}_P = 500 \text{ mm/s} \ \nearrow \ 60°$$

We consider now the motion of pin P along the slot in disk S. Denoting by P' the point of disk S which coincides with P at the instant considered, we write

$$\mathbf{v}_P = \mathbf{v}_{P'} + \mathbf{v}_{P/S}$$

Noting that $\mathbf{v}_{P'}$ is perpendicular to the radius OP and that $\mathbf{v}_{P/S}$ is directed along the slot, we draw the velocity triangle corresponding to the above equation. From the triangle, we compute

$$\gamma = 90° - 42.4° - 30° = 17.6°$$
$$v_{P'} = v_P \sin \gamma = (500 \text{ mm/s}) \sin 17.6°$$
$$\mathbf{v}_{P'} = 151.2 \text{ mm/s} \ \nwarrow \ 42.4°$$
$$v_{P/S} = v_P \cos \gamma = (500 \text{ mm/s}) \cos 17.6°$$
$$\mathbf{v}_{P/S} = 477 \text{ mm/s} \ \nearrow \ 42.4° \ \blacktriangleleft$$

Since $\mathbf{v}_{P'}$ is perpendicular to the radius OP, we write

$$v_{P'} = r\omega_S \qquad 151.2 \text{ mm/s} = (37.1 \text{ mm})\omega_S$$
$$\omega_S = 4.08 \text{ rad/s} \ \downarrow \ \blacktriangleleft$$

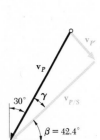

$v_{P'}$
$\mathbf{v}_P$
$30°$
γ
$v_{P/S}$
$\beta = 42.4°$

In the Geneva mechanism of Sample Prob. 15.8, disk D rotates with a constant counterclockwise angular velocity ω_D of 10 rad/s. At the instant when $\phi = 150°$, determine the angular acceleration of disk S.

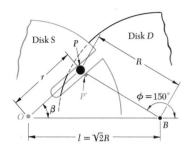

Solution. Referring to Sample Prob. 15.8, we obtain the angular velocity of disk S and the velocity of the pin relative to disk S.

$$\omega_S = 4.08 \text{ rad/s} \downdownarrows$$
$$\beta = 42.4° \qquad \mathbf{v}_{P/S} = 477 \text{ mm/s} \;\nearrow\; 42.4°$$

Since pin P moves with respect to the rotating disk S, we write

$$\mathbf{a}_P = \mathbf{a}_{P'} + \mathbf{a}_{P/S} + \mathbf{a}_c \qquad (1)$$

Each term of this vector equation is investigated separately.

Absolute Acceleration $\mathbf{a}_P$. Since disk D rotates with constant ω, the absolute acceleration $\mathbf{a}_P$ is directed toward B.

$$a_P = R\omega_D^2 = (50 \text{ mm})(10 \text{ rad/s})^2 = 5000 \text{ mm/s}^2$$
$$\mathbf{a}_P = 5000 \text{ mm/s}^2 \;\searrow\; 30°$$

Acceleration $\mathbf{a}_{P'}$ *of the Coinciding Point* P'. The acceleration $\mathbf{a}_{P'}$ of the point P' of disk S which coincides with P at the instant considered is resolved into normal and tangential components. (We recall from Sample Prob. 15.8 that $r = 37.1$ mm.)

$$(a_{P'})_n = r\omega_S^2 = (37.1 \text{ mm})(4.08 \text{ rad/s})^2 = 618 \text{ mm/s}^2$$
$$(\mathbf{a}_{P'})_n = 618 \text{ mm/s}^2 \;\nearrow\; 42.4°$$
$$(a_{P'})_t = r\alpha_S = 37.1\alpha_S \qquad (\mathbf{a}_{P'})_t = 37.1\alpha_S \;\nwarrow\; 42.4°$$

Relative Acceleration $\mathbf{a}_{P/S}$. Since the pin P moves in a straight slot cut in disk S, the relative acceleration $\mathbf{a}_{P/S}$ must be parallel to the slot; i.e., its direction must be $\swarrow$ 42.4°.

Coriolis Acceleration $\mathbf{a}_c$. Rotating the relative velocity $\mathbf{v}_{P/S}$ through 90° in the sense of ω_S, we obtain the direction of the Coriolis component of the acceleration.

$$a_c = 2\omega_S v_{P/S} = 2(4.08 \text{ rad/s})(477 \text{ mm/s}) = 3890 \text{ mm/s}^2$$
$$\mathbf{a}_c = 3890 \text{ mm/s}^2 \;\nwarrow\; 42.4°$$

We rewrite Eq. (1) and substitute the accelerations found above.

$$\mathbf{a}_P = (\mathbf{a}_{P'})_n + (\mathbf{a}_{P'})_t + \mathbf{a}_{P/S} + \mathbf{a}_c$$
$$[5000 \;\searrow\; 30°] = [618 \;\nearrow\; 42.4°] + [37.1\alpha_S \;\nwarrow\; 42.4°]$$
$$+ [a_{P/S} \;\swarrow\; 42.4°] + [3890 \;\nwarrow\; 42.4°]$$

Equating components in a direction perpendicular to the slot:

$$5000 \cos 17.6° = 37.1\alpha_S - 3890$$
$$\alpha_S = 233 \text{ rad/s}^2 \;\downdownarrows \quad \blacktriangleleft$$

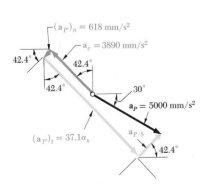

PROBLEMS

15.107 and 15.108   Two rotating rods are connected by a slider block P. The rod attached at B rotates with a constant clockwise angular velocity ω_B. For the given data, determine for the position shown (a) the angular velocity of the rod attached at A, (b) the relative velocity of the slider block P with respect to the rod on which it slides.

15.107 $b = 8$ in., $\omega_B = 6$ rad/s.
15.108 $b = 200$ mm, $\omega_B = 9$ rad/s.

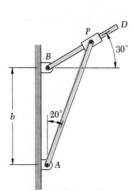
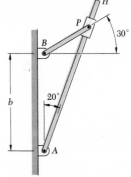

Fig. P15.107 and P15.109   **Fig. P15.108 and P15.110**

15.109 and 15.110   Two rotating rods are connected by a slider block P. The velocity $\mathbf{v}_0$ of the slider block relative to the rod on which it slides is constant and is directed outward. For the given data, determine the angular velocity of each rod for the position shown.

15.109 $b = 200$ mm, $v_0 = 300$ mm/s.
15.110 $b = 8$ in., $v_0 = 12$ in./s.

15.111   Two rods AH and BD pass through smooth holes drilled in a hexagonal block. (The holes are drilled in different planes so that the rods will not hit each other.) Knowing that rod AH rotates counterclockwise at the rate ω, determine the angular velocity of rod BD and the relative velocity of the block with respect to each rod when (a) $\theta = 30°$, (b) $\theta = 15°$.

15.112   Solve Prob. 15.111 when (a) $\theta = 90°$, (b) $\theta = 60°$.

15.113   Four pins slide in four separate slots cut in a circular plate as shown. When the plate is at rest, each pin has a velocity directed as shown and of the same constant magnitude u. If each pin maintains the same velocity in relation to the plate when the plate rotates about O with a constant *clockwise* angular velocity ω, determine the acceleration of each pin.

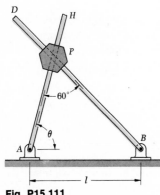

Fig. P15.111

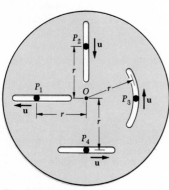

Fig. P15.113

15.114 Solve Prob. 15.113, assuming that the plate rotates about O with a constant *counterclockwise* angular velocity ω.

15.115 Water flows through a straight pipe OB which rotates counterclockwise with an angular velocity of 120 rpm. If the velocity of the water relative to the pipe is 20 ft/s, determine the total acceleration (a) of the particle of water P_1, (b) of the particle of water P_2.

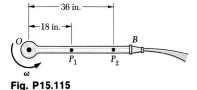

Fig. P15.115

15.116 A block P may slide on the arm OA which rotates at the constant angular velocity ω. As the arm rotates, the cord wraps around a *fixed* drum of radius b and pulls the block toward O with a speed $b\omega$. Determine the magnitude of the acceleration of the block in terms of r, b, and ω.

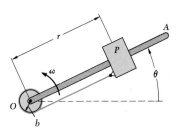

Fig. P15.116

15.117 A pin P slides in a circular slot of radius r_2 which is cut in the plate OE. The velocity of P relative to the plate is of constant magnitude u and is directed as shown. Knowing that the plate rotates counterclockwise with a constant angular velocity ω, derive an expression for (a) the magnitude u for which the acceleration of the pin is zero as it passes through point A, (b) the corresponding magnitude of the acceleration of the pin as it passes through point B.

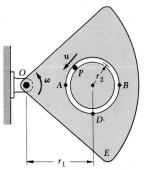

Fig. P15.117 and P15.118

15.118 Pin P slides in the circular slot cut in the plate OE at a constant relative speed $u = 0.5$ m/s as the plate rotates about O at the constant rate $\omega = 6$ rad/s. Knowing that $r_1 = 300$ mm and $r_2 = 100$ mm, determine the acceleration of the pin as it passes through (a) point A, (b) point B, (c) point D.

15.119 A rocket sled is tested on a straight track which is built along a meridian. Knowing that the track is located at a latitude of 40° north, determine the Coriolis acceleration of the sled when its velocity is 1000 km/h due north. (*Hint.* Consider separately the components of the motion parallel and perpendicular to the plane of the equator.)

15.120 A train crosses the parallel 50° north, traveling due north at a constant speed v. Determine the speed of the train if the Coriolis component of its acceleration is 0.01 ft/s². (See hint of Prob. 15.119.)

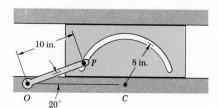

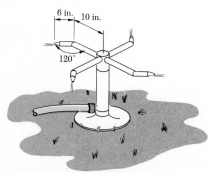

Fig. P15.123

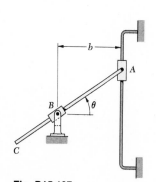

Fig. P15.125

15.121 In Prob. 15.107, determine the angular acceleration of the rod attached at A.

15.122 In Prob. 15.108, determine the angular acceleration of the rod attached at A.

15.123 At the instant shown, the slotted plate slides with a velocity of 10 in./s to the right and has an acceleration of 50 in./s² to the left. Determine the angular acceleration of rod OP.

15.124 At the instant when $\theta = 90°$ in Prob. 15.111, rod AH has a counterclockwise angular velocity ω and no angular acceleration. Determine at that instant (a) the angular acceleration of rod BD, (b) the total acceleration of the block P.

15.125 A garden sprinkler has four rotating arms, each of which consists of two horizontal straight sections of pipe forming an angle of 120°. The sprinkler when operating rotates with a constant angular velocity of 180 rpm. If the velocity of the water relative to the pipe sections is 12 ft/s, determine the magnitude of the total acceleration of a particle of water as it passes the midpoint of (a) the 10-in. section of pipe, (b) the 6-in. section of pipe.

15.126 Water flows through the curved pipe OB, which has a constant radius of 0.375 m and which rotates with a constant counterclockwise angular velocity of 120 rpm. If the velocity of the water relative to the pipe is 12 m/s, determine the total acceleration of the particle of water P.

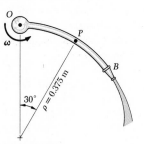

Fig. P15.126

REVIEW PROBLEMS

15.127 Rod AC of length $2b$ is attached to a collar at A and passes through a pivoted collar at B. Knowing that the collar A moves upward with a constant velocity v_A, derive an expression for (a) the angular velocity of rod AC, (b) the velocity of the point of the rod in contact with the pivoted collar B, (c) the angular acceleration of rod AC.

Fig. P15.127

15.128 In Prob. 15.26, determine (*a*) the angular acceleration of the rod, (*b*) the acceleration of A.

15.129 Three gears A, B, and C are pinned at their centers to rod ABC. Knowing that $r_A = 3r_B = 3r_C$ and that gear A does not rotate, determine the angular velocity of gears B and C when the rod ABC rotates clockwise with a constant angular velocity of 10 rpm.

Fig. P15.129

15.130 In Prob. 15.129 it is known that $r_A = 12$ in., $r_B = r_C = 4$ in. Determine the acceleration of the tooth of gear C which is in contact with gear B.

15.131 Three links AB, BC, and BD are connected by a pin B as shown. Knowing that at the instant shown point D has a velocity of 200 mm/s to the right and no acceleration, determine (*a*) the angular acceleration of each link, (*b*) the accelerations of points A and B.

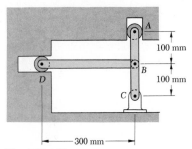

Fig. P15.131

15.132 The eccentric shown consists of a disk of 2-in. radius which revolves about a shaft O located $\frac{1}{2}$ in. from the center of the disk A. Assuming that the disk rotates about O with a constant angular velocity of 1800 rpm clockwise, determine the velocity and acceleration of block B when point A is directly below the shaft O.

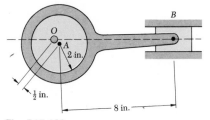

Fig. P15.132

15.133 It takes 0.8 s for the turntable of a 33-rpm record player to reach full speed after being started. Assuming uniformly accelerated motion, determine (*a*) the angular acceleration of the turntable, (*b*) the normal and tangential components of the acceleration of a point on the rim of the 12-in.-diameter turntable just before the speed of 33 rpm is reached, (*c*) the total acceleration of the same point at that time.

15.134 Two collars C and D move along the vertical rod shown. Knowing that the velocity of collar D is 0.210 m/s downward, determine (a) the velocity of collar C, (b) the angular velocity of member AB.

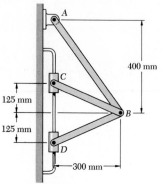

Fig. P15.134

*15.135 Prove for any given position of the mechanism of Prob. 15.134 that the ratio of the magnitudes of the velocities of collars C and D is equal to the ratio of the distances AC and AD.

15.136 The disk shown has a constant angular velocity of 6 rad/s clockwise. For the position shown, determine (a) the angular acceleration of each rod, (b) the acceleration of point C.

15.137 Solve Prob. 15.136 after the disk has rotated 180° and pin B is directly below A.

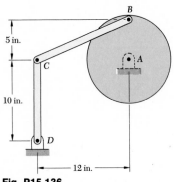

Fig. P15.136

15.138 At the instant shown, the slotted plate slides with a velocity of 0.5 m/s upward and has an acceleration of 2 m/s² downward. Determine the angular velocity and the angular acceleration of rod OP.

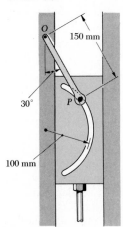

Fig. P15.138

Chapter
16

Kinetics of Rigid Bodies: Forces and Accelerations

16.1. Introduction. In this chapter and in Chaps. 17 and 18, we shall study the *kinetics of rigid bodies*, i.e., the relations existing between the forces acting on a rigid body, the shape and mass of the body, and the motion produced. In Chaps. 12 to 14, we studied similar relations, assuming then that the body could be considered as a particle, i.e., that its mass could be concentrated in one point and that all forces acted at that point. We shall now take the shape of the body into account, as well as the exact location of the points of application of the forces. Besides, we shall be concerned not only with the motion of the body as a whole but also with the rotation of the body about its mass center.

In this chapter, our study of the motion of rigid bodies will be based directly on the equation $\mathbf{F} = m\mathbf{a}$, while in the next two chapters we shall make use of the principles of work and energy and of impulse and momentum. Our study will be limited to that of the *plane motion* of rigid bodies. This motion was defined in Sec. 15.1 as a motion in which each particle of the body remains at a constant distance from a fixed reference plane. In the study of the kinetics of plane motion, the reference plane is chosen so that it contains the mass center of the body.

Throughout most of this chapter, our study will be further limited to that of plane slabs and of bodies which are symmet-

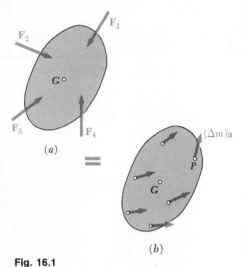

Fig. 16.1

rical with respect to the reference plane.† The study of the motion of three-dimensional bodies which are not symmetrical with respect to the reference plane will be postponed until Sec. 16.6, where problems such as the balancing of rotating shafts will be considered.

16.2. Plane Motion of a Rigid Body. D'Alembert's Principle. Consider a rigid slab of mass m moving under the action of several forces $\mathbf{F}_1$, $\mathbf{F}_2$, $\mathbf{F}_3$, etc., contained in the plane of the slab (Fig. 16.1a). We may regard the slab as being made of a large number of particles and use the results obtained in Sec. 12.4 for a system of particles. It was shown at that time that the system of the external forces acting on the particles is equipollent to the system of the effective forces of the particles. In other words, the sum of the components of the external forces in any given direction is equal to the sum of the components of the effective forces in that direction, and the sum of the moments of the external forces about any given axis is equal to the sum of the moments of the effective forces about that axis.

It follows that the external forces $\mathbf{F}_1$, $\mathbf{F}_2$, $\mathbf{F}_3$, etc., acting on the slab of Fig. 16.1 are equipollent to the system of the effective forces of the particles forming the slab. But since we are now dealing with a rigid body, the two systems of forces are not only equipollent, they are also *equivalent*; i.e., they both have the same effect on the slab (Sec. 3.11). This is shown in Fig. 16.1, where colored equals signs have been used to connect the system of the external forces and the system of the effective forces.

We may thus state that *the external forces acting on a rigid body are equivalent to the effective forces of the various particles forming the body*. This statement is referred to as *D'Alembert's principle*, after the French mathematician Jean le Rond d'Alembert (1717–1783), even though D'Alembert's original statement was written in a somewhat different form.

Again, because we are dealing with a rigid body, we may reduce the system of the effective forces to a force-couple system attached at a point of our choice (Sec. 3.10). By selecting the mass center G of the slab as the reference point, we shall be able to obtain a particularly simple result.

In order to determine the force-couple system at G which is equivalent to the system of the effective forces, we must evaluate the sums of the x and y components of the effective forces and the sum of their moments about G. We first write

$$\Sigma(F_x)_{\text{eff}} = \Sigma(\Delta m)a_x \qquad \Sigma(F_y)_{\text{eff}} = \Sigma(\Delta m)a_y \qquad (16.1)$$

† Or, more generally, bodies which have a principal centroidal axis of inertia perpendicular to the reference plane.

But, according to Eqs. (12.9) of Sec. 12.5, we have

$$\Sigma(\Delta m)a_x = (\Sigma\,\Delta m)\bar{a}_x \qquad \Sigma(\Delta m)a_y = (\Sigma\,\Delta m)\bar{a}_y \qquad (16.2)$$

where $\bar{a}_x$ and $\bar{a}_y$ are the components of the acceleration $\bar{\mathbf{a}}$ of the mass center G. Noting that $\Sigma\,\Delta m$ represents the total mass m of the slab, and substituting from (16.2) into (16.1), we have

$$\Sigma(F_x)_{\text{eff}} = m\bar{a}_x \qquad \Sigma(F_y)_{\text{eff}} = m\bar{a}_y \qquad (16.3)$$

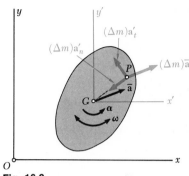

Fig. 16.2

Turning now our attention to the determination of the sum of the moments about G of the effective forces, we recall from Sec. 15.8 that the acceleration $\mathbf{a}$ of any given particle P of the slab (Fig. 16.2) may be expressed as the sum of the acceleration $\bar{\mathbf{a}}$ of the mass center G and of the acceleration $\mathbf{a}'$ of P relative to a frame $Gx'y'$ attached to G and of fixed orientation:

$$\mathbf{a} = \bar{\mathbf{a}} + \mathbf{a}'$$

Resolving $\mathbf{a}'$ into normal and tangential components, we write

$$\mathbf{a} = \bar{\mathbf{a}} + \mathbf{a}'_n + \mathbf{a}'_t \qquad (16.4)$$

Each effective force $(\Delta m)\mathbf{a}$ may thus be resolved into the three component effective forces shown in Fig. 16.2. The effective forces $(\Delta m)\bar{\mathbf{a}}$ obtained in this fashion are associated with a translation of the slab with G, while the effective forces $(\Delta m)\mathbf{a}'_n$ and $(\Delta m)\mathbf{a}'_t$ are associated with a rotation of the slab about G.

Considering first the effective forces $(\Delta m)\bar{\mathbf{a}}$ associated with the translation of the slab, and resolving them into rectangular components (Fig. 16.3), we find that the sum of their moments about G is

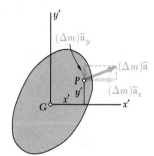

Fig. 16.3

$$+\circlearrowleft\ \Sigma(M_G)_{\text{eff, translation}} = \Sigma x'(\Delta m)\bar{a}_y - \Sigma y'(\Delta m)\bar{a}_x$$
$$= \bar{a}_y\Sigma x'\,\Delta m - \bar{a}_x\Sigma y'\,\Delta m \qquad (16.5)$$

Now, according to the definition of the mass center of a system of particles (Sec. 12.5), we have

$$\Sigma x'\,\Delta m = m\bar{x}' \qquad \Sigma y'\,\Delta m = m\bar{y}' \qquad (16.6)$$

where $\bar{x}'$ and $\bar{y}'$ are the coordinates of the mass center G and m the total mass. But $\bar{x}' = \bar{y}' = 0$, since the origin of the x' and y' coordinate axes was chosen at G, and Eq. (16.5) yields

$$\Sigma(M_G)_{\text{eff, translation}} = 0 \qquad (16.7)$$

Considering now the effective forces $(\Delta m)\mathbf{a}'_n$ and $(\Delta m)\mathbf{a}'_t$ associated with the rotation of the slab about G (Fig. 16.4), we note that the moment of $(\Delta m)\mathbf{a}'_n$ about G is zero and that the moment of $(\Delta m)\mathbf{a}'_t$ about G is $r'(\Delta m)a'_t$. Recalling that $a'_t = r'\alpha$, and considering all the particles of the slab, we have

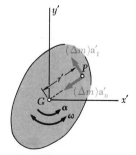

Fig. 16.4

$$+\circlearrowleft\ \Sigma(M_G)_{\text{eff, rotation}} = \Sigma r'(\Delta m)r'\alpha = \alpha\Sigma r'^2\,\Delta m$$

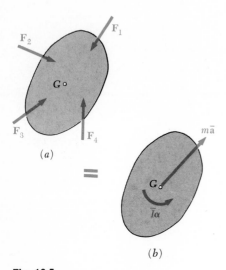

Fig. 16.5

Since the sum $\Sigma r'^2 \Delta m$ represents the moment of inertia $\bar{I}$ of the slab about an axis through G perpendicular to the slab, we write

$$\Sigma(M_G)_{\text{eff, rotation}} = \bar{I}\alpha \qquad (16.8)$$

The sum of the moments about G of the effective forces of the slab is obtained by combining the information provided by Eqs. (16.7) and (16.8). We have

$$\Sigma(M_G)_{\text{eff}} = \bar{I}\alpha \qquad (16.9)$$

Equations (16.3) and (16.9) indicate that the system of the effective forces shown in Fig. 16.1b may be replaced by an equivalent force-couple system consisting of a vector $m\bar{a}$ attached at G and a couple of moment $\bar{I}\alpha$ and of the same sense as the angular acceleration $\boldsymbol{\alpha}$ (Fig. 16.5). Recalling from Sec. 15.3 that the angular acceleration of a rigid body rotating about a fixed axis may be represented by a vector $\boldsymbol{\alpha}$ of magnitude α directed along that axis, we note that the couple obtained may be represented by the vector $\bar{I}\boldsymbol{\alpha}$ of magnitude $\bar{I}\alpha$ directed along the axis of rotation.

In the case of the plane motion of a rigid slab,† D'Alembert's principle may thus be restated as follows: *The external forces acting on the body are equivalent to a force-couple system consisting of a vector $m\bar{a}$ attached at the mass center G of the body and a couple $\bar{I}\boldsymbol{\alpha}$.*

The relation shown in Fig. 16.5 can be expressed algebraically by writing three equations relating respectively the x and y components and the moments about any given point A of the external and effective forces in Fig. 16.5. If the moments are computed about the mass center G of the rigid body, these equations of motion read

$$\Sigma F_x = m\bar{a}_x \qquad \Sigma F_y = m\bar{a}_y \qquad \Sigma M_G = \bar{I}\alpha \qquad (16.10)$$

Translation. When a rigid body is constrained to move in translation, its angular acceleration is identically equal to zero and its effective forces reduce to the vector $m\bar{a}$ attached at G (Fig. 16.6). Thus, the resultant of the external forces acting on a rigid body in translation passes through the mass center of the body and is equal to $m\bar{a}$.

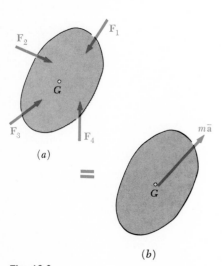

Fig. 16.6

†Or, more generally, in the case of the plane motion of a rigid body which has a principal centroidal axis of inertia perpendicular to the reference plane.

Centroidal Rotation.   When a slab, or, more generally, a body symmetrical with respect to the reference plane, is constrained to rotate about a fixed axis perpendicular to the reference plane and passing through its mass center G, we say that the body is in *centroidal rotation.* Since the acceleration $\bar{\mathbf{a}}$ is identically equal to zero, the effective forces of the body reduce to the couple $\bar{I}\boldsymbol{\alpha}$ (Fig. 16.7). Thus, the external forces acting on a body in centroidal rotation are equivalent to a couple $\bar{I}\boldsymbol{\alpha}$.

General Plane Motion.   Comparing Fig. 16.5 with Figs. 16.6 and 16.7, we observe that, from the point of view of *kinetics,* the most general plane motion of a rigid body symmetrical with respect to the reference plane may be replaced by the sum of a translation and a centroidal rotation. We should note that this statement is more restrictive than the similar statement made earlier from the point of view of *kinematics* (Sec. 15.5), since we now require that the mass center of the body be selected as the reference point.

Referring to Eqs. (16.10), we observe that the first two equations are identical with the equations of motion of a particle of mass m acted upon by the given forces $\mathbf{F}_1$, $\mathbf{F}_2$, $\mathbf{F}_3$, etc. We thus check that *the mass center G of a rigid body in plane motion moves as if the entire mass of the body were concentrated at that point, and as if all the external forces acted on it.* We recall that this result has already been obtained in Sec. 12.5 in the general case of a system of particles, the particles being not necessarily rigidly connected. We should note, however, that the system of the external forces does not, in general, reduce to a single vector $m\bar{\mathbf{a}}$ attached at G. Therefore, in the general case of the plane motion of a rigid body, *the resultant of the external forces acting on the body does not pass through the mass center of the body.*

Finally, we may observe that the last of Eqs. (16.10) would still be valid if the rigid body, while subjected to the same applied forces, were constrained to rotate about a fixed axis through G. Thus, *a rigid body in plane motion rotates about its mass center as if this point were fixed.*

16.3. Solution of Problems Involving the Plane Motion of Rigid Bodies.

We saw in the preceding section that, when a rigid body is in plane motion, there exists a fundamental relation between the forces $\mathbf{F}_1$, $\mathbf{F}_2$, $\mathbf{F}_3$, etc., acting on the body, the acceleration $\bar{\mathbf{a}}$ of its mass center, and the angular acceleration $\boldsymbol{\alpha}$ of the body. This relation, which is represented in Fig. 16.5, may be used to determine the acceleration $\bar{\mathbf{a}}$ and the angular acceleration $\boldsymbol{\alpha}$ produced by a given system of forces acting on a rigid body or, conversely, to determine the forces which produce a given motion of the rigid body.

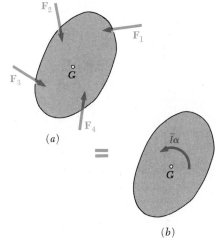

(a)

(b)

Fig. 16.7

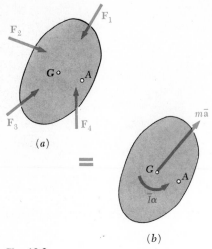

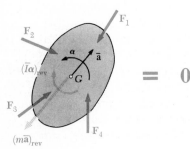

Fig. 16.8

Fig. 16.9

While the three algebraic equations (16.10) may be used to solve problems of plane motion,† our experience in statics suggests that the solution of many problems involving rigid bodies could be simplified by an appropriate choice of the point about which the moments of the forces are computed. It is therefore preferable to remember the relation existing between the forces and the accelerations in the vectorial form shown in Fig. 16.8, and to derive from this fundamental relation the component or moment equations which fit best the solution of the problem under consideration. The equations most frequently used are

$$\Sigma F_x = \Sigma(F_x)_{\text{eff}} \qquad \Sigma F_y = \Sigma(F_y)_{\text{eff}} \qquad \Sigma M_A = \Sigma(M_A)_{\text{eff}} \quad (16.11)$$

We easily verify that the first two equations reduce to the corresponding equations in (16.10). The third equation, however, will generally be different from the corresponding equation in (16.10), since the selection of point A, about which the moments of the external and effective forces are to be computed, will be guided by our desire to simplify the solution of the problem as much as possible. In Sample Prob. 16.1, for example, the three unknown forces $\mathbf{F}_A$, $\mathbf{F}_B$, and $\mathbf{N}_A$ may be eliminated from this equation by equating moments about A. It should be noted, however, that unless A happens to coincide with the mass center G, the determination of $\Sigma(M_A)_{\text{eff}}$ will involve the computation of the moments of the components of the vector $m\bar{\mathbf{a}}$, as well as the computation of the moment of the couple $\bar{I}\alpha$.

The fundamental relation shown in Fig. 16.8 may be presented in an alternate form if we add to the external forces an inertia vector $(m\bar{\mathbf{a}})_{\text{rev}}$ of sense opposite to that of $\bar{\mathbf{a}}$, attached at G, and an inertia couple $(\bar{I}\alpha)_{\text{rev}}$ of moment equal in magnitude to $\bar{I}\alpha$ and of sense opposite to that of α (Fig. 16.9). The system obtained is equivalent to zero, and the rigid body is said to be in dynamic equilibrium.

Whether the principle of equivalence of external and effective forces is directly applied, as in Fig. 16.8, or whether the concept of dynamic equilibrium is introduced, as in Fig. 16.9, the use of free-body diagrams showing vectorially the relationship existing between the forces applied on the rigid body and the resulting linear and angular accelerations presents considerable advantages over the blind application of the formulas (16.10). These advantages may be summarized as follows:

† We recall that the last of Eqs. (16.10) is valid only in the case of the plane motion of a rigid body symmetrical with respect to the reference plane. The plane motion of bodies which are not symmetrical with respect to the reference plane is discussed in Sec. 16.6.

1. First of all, a much clearer understanding of the effect of the forces on the motion of the body will result from the use of a pictorial representation.
2. This approach makes it possible to divide the solution of a dynamics problem into two parts: In the first part, the analysis of the kinematic and kinetic characteristics of the problem leads to the free-body diagrams of Fig. 16.8 or 16.9; in the second part, the diagram obtained is used to analyze by the methods of Chap. 3 the various forces and vectors involved.
3. A unified approach is provided for the analysis of the plane motion of a rigid body, regardless of the particular type of motion involved. While the kinematics of the various motions considered may vary from one case to the other, the approach to the kinetics of the motion is consistently the same. In every case we shall draw a diagram showing the external forces, the vector $m\bar{a}$ associated with the motion of G, and the couple $\bar{I}\alpha$ associated with the rotation of the body about G.
4. The resolution of the plane motion of a rigid body into a translation and a centroidal rotation, which is used here, is a basic concept which may be applied effectively throughout the study of mechanics. We shall use it again in Chap. 17 with the method of work and energy and in Chap. 18 with the method of impulse and momentum.

16.4. Systems of Rigid Bodies.   The method described in the preceding section may also be used in problems involving the plane motion of several connected rigid bodies. A diagram similar to Fig. 16.8 or Fig. 16.9 may be drawn for each part of the system. The equations of motion obtained from these diagrams are solved simultaneously.

In some cases, as in Sample Prob. 16.3, a single diagram may be drawn for the entire system. This diagram should include all the external forces, as well as the vectors $m\bar{a}$ and the couples $\bar{I}\alpha$ associated with the various parts of the system. However, internal forces, such as the forces exerted by connecting cables, may be omitted since they occur in pairs of equal and opposite forces and are thus equipollent to zero. The equations obtained by expressing that the system of the external forces is equipollent to the system of the effective forces may be solved for the remaining unknowns.†

This second approach may not be used in problems involving more than three unknowns, since only three equations of motion are available when a single diagram is used. We shall not elaborate upon this point, since the discussion involved would be completely similar to that given in Sec. 6.12 in the case of the equilibrium of a system of rigid bodies.

† Note that we cannot speak of *equivalent* systems since we are not dealing with a single rigid body.

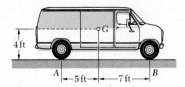

When the forward speed of the truck shown was 30 ft/s, the brakes were suddenly applied, causing all four wheels to stop rotating. It was observed that the truck skidded to rest in 20 ft. Determine the magnitude of the normal reaction and of the friction force at each wheel as the truck skidded to rest.

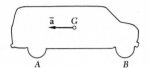

Kinematics of Motion. Choosing the positive sense to the right and using the equations of uniformly accelerated motion, we write

$$\bar{v}_0 = +30 \text{ ft/s} \qquad \bar{v}^2 = \bar{v}_0^2 + 2\bar{a}x \qquad 0 = (30)^2 + 2\bar{a}(20)$$
$$\bar{a} = -22.5 \text{ ft/s}^2 \qquad \bar{a} = 22.5 \text{ ft/s}^2 \leftarrow$$

Equations of Motion. The external forces consist of the weight **W** of the truck and of the normal reactions and friction forces at the wheels. (The vectors N_A and F_A represent the sum of the reactions at the rear wheels, while N_B and F_B represent the sum of the reactions at the front wheels.) Since the truck is in translation, the effective forces reduce to the vector $m\bar{a}$ attached at G. Three equations of motion are obtained by expressing that the system of the external forces is equivalent to the system of the effective forces.

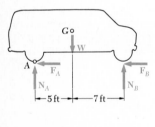

$$+\uparrow \Sigma F_y = \Sigma(F_y)_{\text{eff}}: \qquad N_A + N_B - W = 0$$

Since $F_A = \mu N_A$ and $F_B = \mu N_B$, we find

$$F_A + F_B = \mu(N_A + N_B) = \mu W$$

$$\xrightarrow{+} \Sigma F_x = \Sigma(F_x)_{\text{eff}}: \qquad -(F_A + F_B) = -m\bar{a}$$

$$-\mu W = -\frac{W}{32.2 \text{ ft/s}^2}(22.5 \text{ ft/s}^2)$$

$$\mu = 0.699$$

$$+\circlearrowright \Sigma M_A = \Sigma(M_A)_{\text{eff}}: \qquad -W(5 \text{ ft}) + N_B(12 \text{ ft}) = m\bar{a}(4 \text{ ft})$$

$$-W(5 \text{ ft}) + N_B(12 \text{ ft}) = \frac{W}{32.2 \text{ ft/s}^2}(22.5 \text{ ft/s}^2)(4 \text{ ft})$$

$$N_B = 0.650W$$
$$F_B = \mu N_B = (0.699)(0.650W) \qquad F_B = 0.454W$$

$$+\uparrow \Sigma F_y = \Sigma(F_y)_{\text{eff}}: \qquad N_A + N_B - W = 0$$
$$N_A + 0.650W - W = 0$$
$$N_A = 0.350W$$
$$F_A = \mu N_A = (0.699)(0.350W) \qquad F_A = 0.245W$$

Reactions at Each Wheel. Recalling that the values computed above represent the sum of the reactions at the two front wheels or the two rear wheels, we obtain the magnitude of the reactions at each wheel by writing

$$N_{\text{front}} = \tfrac{1}{2}N_B = 0.325W \qquad N_{\text{rear}} = \tfrac{1}{2}N_A = 0.175W \quad \blacktriangleleft$$
$$F_{\text{front}} = \tfrac{1}{2}F_B = 0.227W \qquad F_{\text{rear}} = \tfrac{1}{2}F_A = 0.122W \quad \blacktriangleleft$$

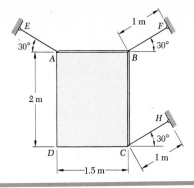

The thin plate $ABCD$ has a mass of 50 kg and is held in position by the three inextensible wires AE, BF, and CH. Wire AE is then cut. Determine both (*a*) the acceleration of the plate, (*b*) the tension in wires BF and CH immediately after wire AE has been cut.

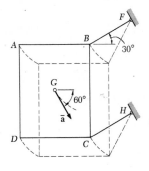

Motion of Plate. After wire AE has been cut, we observe that corners B and C move along parallel circles of radius 1 m centered, respectively, at F and H. The motion of the plate is thus a curvilinear translation; the particles forming the plate move along parallel circles of radius 1 m.

At the instant wire AE is cut, the velocity of the plate is zero; the acceleration $\bar{\mathbf{a}}$ of the mass center G is thus tangent to the circular path which will be described by G.

Equations of Motion. The external forces consist of the weight $\mathbf{W}$ and of the forces $\mathbf{T}_B$ and $\mathbf{T}_C$ exerted by the wires. Since the plate is in translation, the effective forces reduce to the vector $m\bar{\mathbf{a}}$ attached at G and directed along the t axis. Expressing that the system of the external forces is equivalent to the system of the effective forces, we write

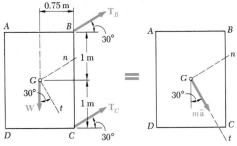

$+\searrow \Sigma F_t = \Sigma(F_t)_{\text{eff}}:$ $\qquad W \cos 30° = m\bar{a}$

$$mg \cos 30° = m\bar{a} \qquad (1)$$
$$\bar{a} = g \cos 30° = (9.81 \text{ m/s}^2) \cos 30°$$
$$\bar{\mathbf{a}} = 8.50 \text{ m/s}^2 \searrow 60° \quad \blacktriangleleft$$

$+\nearrow \Sigma F_n = \Sigma(F_n)_{\text{eff}}:$ $\qquad T_B + T_C - W \sin 30° = 0 \qquad (2)$

$+\circlearrowleft \Sigma M_G = \Sigma(M_G)_{\text{eff}}:$ $\quad (T_B \sin 30°)(0.75 \text{ m}) - (T_B \cos 30°)(1 \text{ m})$
$$+ (T_C \sin 30°)(0.75 \text{ m}) + (T_C \cos 30°)(1 \text{ m}) = 0$$
$$-0.491T_B + 1.241T_C = 0$$
$$T_C = 0.396T_B \qquad (3)$$

Substituting for T_C from (3) into (2), we write

$$T_B + 0.396T_B - W \sin 30° = 0$$
$$T_B = 0.358W$$
$$T_C = 0.396(0.358W) = 0.1418W$$

Noting that $W = mg = (50 \text{ kg})(9.81 \text{ m/s}^2) = 491 \text{ N}$, we have
$$T_B = 175.8 \text{ N} \qquad T_C = 69.6 \text{ N} \quad \blacktriangleleft$$

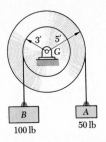

A pulley weighing 120 lb and having a radius of gyration of 4 ft is connected to two blocks as shown. Assuming no axle friction, determine the angular acceleration of the pulley.

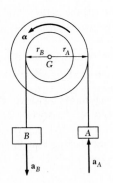

Sense of Motion. Although an arbitrary sense of motion may be assumed (since no friction forces are involved) and later checked by the sign of the answer, we may prefer first to determine the actual sense of rotation of the pulley. We first find the weight of block B required to maintain the equilibrium of the pulley when it is acted upon by the 50-lb block A. We write

$$+\mathord{\circlearrowleft}\ \Sigma M_G = 0: \qquad W_B(3\text{ ft}) - (50\text{ lb})(5\text{ ft}) = 0 \qquad W_B = 83.3\text{ lb}$$

Since block B actually weighs 100 lb, the pulley will rotate counterclockwise.

Kinematics of Motion. Assuming $\boldsymbol{\alpha}$ counterclockwise and noting that $a_A = r_A\alpha$ and $a_B = r_B\alpha$, we obtain

$$\mathbf{a}_A = 5\alpha \uparrow \qquad \mathbf{a}_B = 3\alpha \downarrow$$

Equations of Motion. A single system consisting of the pulley and the two blocks is considered. Forces external to this system consist of the weights of the pulley and the two blocks and of the reaction at G. (The forces exerted by the cables on the pulley and on the blocks are internal to the system considered and cancel out.) Since the motion of the pulley is a centroidal rotation and the motion of each block is a translation, the effective forces reduce to the couple $\bar{I}\boldsymbol{\alpha}$ and the two vectors $m\mathbf{a}_A$ and $m\mathbf{a}_B$. The centroidal moment of inertia of the pulley is

$$\bar{I} = mk^2 = \frac{W}{g}k^2 = \frac{120\text{ lb}}{32.2\text{ ft/s}^2}(4\text{ ft})^2 = 59.6\text{ lb}\cdot\text{ft}\cdot\text{s}^2$$

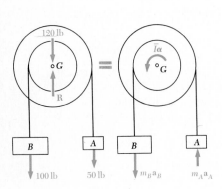

Since the system of the external forces is equipollent to the system of the effective forces, we write

$$+\mathord{\circlearrowleft}\ \Sigma M_G = \Sigma(M_G)_{\text{eff}}:$$
$$(100\text{ lb})(3\text{ ft}) - (50\text{ lb})(5\text{ ft}) = +\bar{I}\alpha + m_B a_B(3\text{ ft}) + m_A a_A(5\text{ ft})$$

$$(100)(3) - (50)(5) = +59.6\alpha + \frac{100}{32.2}(3\alpha)(3) + \frac{50}{32.2}(5\alpha)(5)$$

$$\alpha = +0.396\text{ rad/s}^2 \qquad \alpha = 0.396\text{ rad/s}^2 \mathord{\circlearrowleft} \ \blacktriangleleft$$

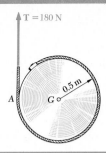

T = 180 N

A cord is wrapped around a homogeneous disk of radius $r = 0.5\,\text{m}$ and mass $m = 15\,\text{kg}$. If the cord is pulled upward with a force $\mathbf{T}$ of magnitude 180 N, determine (a) the acceleration of the center of the disk, (b) the angular acceleration of the disk, (c) the acceleration of the cord.

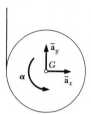

Equations of Motion. We assume that the components $\bar{\mathbf{a}}_x$ and $\bar{\mathbf{a}}_y$ of the acceleration of the center are directed, respectively, to the right and upward and that the angular acceleration of the disk is counter-clockwise. The external forces acting on the disk consist of the weight $\mathbf{W}$ and the force $\mathbf{T}$ exerted by the cord. This system is equivalent to the system of the effective forces, which consists of a vector of components $m\bar{\mathbf{a}}_x$ and $m\bar{\mathbf{a}}_y$ attached at G and a couple $\bar{I}\boldsymbol{\alpha}$. We write

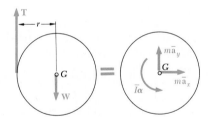

$$\xrightarrow{+}\ \Sigma F_x = \Sigma(F_x)_{\text{eff}}: \qquad 0 = m\bar{a}_x \qquad\qquad \bar{\text{a}}_x = 0 \quad \blacktriangleleft$$

$$+\!\uparrow \Sigma F_y = \Sigma(F_y)_{\text{eff}}: \qquad T - W = m\bar{a}_y$$

$$\bar{a}_y = \frac{T - W}{m}$$

Since $T = 180\,\text{N}$, $m = 15\,\text{kg}$, and $W = (15\,\text{kg})(9.81\,\text{m/s}^2) = 147.1\,\text{N}$, we have

$$\bar{a}_y = \frac{180\,\text{N} - 147.1\,\text{N}}{15\,\text{kg}} = +2.19\,\text{m/s}^2 \qquad \bar{\text{a}}_y = 2.19\,\text{m/s}^2\ \uparrow \quad \blacktriangleleft$$

$$+\!\circlearrowleft \Sigma M_G = \Sigma(M_G)_{\text{eff}}: \qquad -Tr = \bar{I}\alpha$$

$$-Tr = (\tfrac{1}{2}mr^2)\alpha$$

$$\alpha = -\frac{2T}{mr} = -\frac{2(180\,\text{N})}{(15\,\text{kg})(0.5\,\text{m})} = -48.0\,\text{rad/s}^2$$

$$\alpha = 48.0\,\text{rad/s}^2\ \circlearrowright \quad \blacktriangleleft$$

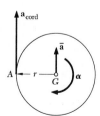

Acceleration of Cord. Since the acceleration of the cord is equal to the tangential component of the acceleration of point A on the disk, we write

$$\mathbf{a}_{\text{cord}} = (\mathbf{a}_A)_t = \bar{\mathbf{a}} + (\mathbf{a}_{A/G})_t$$

$$= [2.19\,\text{m/s}^2\ \uparrow] + [(0.5\,\text{m})(48\,\text{rad/s}^2)\ \uparrow]$$

$$\text{a}_{\text{cord}} = 26.2\,\text{m/s}^2\ \uparrow \quad \blacktriangleleft$$

A hoop of radius r and mass m is placed on a horizontal surface with no linear velocity but with a clockwise angular velocity ω_0. Denoting by μ the coefficient of friction between the hoop and the floor, determine (a) the time t_1 at which the hoop will start rolling without sliding, (b) the linear and angular velocities of the hoop at time t_1.

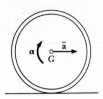

Solution. Since the entire mass is located at a distance r from the center of the hoop, we write $\bar{I} = mr^2$.

Equations of Motion. The positive sense is chosen to the right for $\bar{\mathbf{a}}$ and clockwise for $\boldsymbol{\alpha}$. The external forces acting on the hoop consist of the weight $\mathbf{W}$, the normal reaction $\mathbf{N}$, and the friction force $\mathbf{F}$. While the hoop is sliding, the magnitude of the friction force is $F = \mu N$. The effective forces consist of the vector $m\bar{\mathbf{a}}$ attached at G and the couple $\bar{I}\boldsymbol{\alpha}$. Expressing that the system of the external forces is equivalent to the system of the effective forces, we write

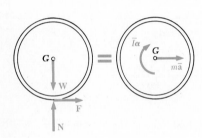

$$+\uparrow \Sigma F_y = \Sigma(F_y)_{\text{eff}}: \quad N - W = 0$$
$$N = W = mg \qquad F = \mu N = \mu mg$$

$$\xrightarrow{+} \Sigma F_x = \Sigma(F_x)_{\text{eff}}: \quad F = m\bar{a} \qquad \mu mg = m\bar{a} \qquad \bar{a} = +\mu g$$

$$+\rotatebox{0}{\raisebox{0pt}{↻}}\, \Sigma M_G = \Sigma(M_G)_{\text{eff}}: \quad -Fr = \bar{I}\alpha$$
$$-(\mu mg)r = (mr^2)\alpha \qquad \alpha = -\frac{\mu g}{r}$$

Kinematics of Motion. As long as the hoop both rolls and slides, its linear and angular motions are uniformly accelerated.

$$t = 0, \bar{v}_0 = 0 \qquad \bar{v} = \bar{v}_0 + \bar{a}t = 0 + \mu gt \tag{1}$$

$$t = 0, \omega = \omega_0 \qquad \omega = \omega_0 + \alpha t = \omega_0 + \left(-\frac{\mu g}{r}\right)t \tag{2}$$

The hoop will start rolling without sliding when the velocity $\mathbf{v}_C$ of the point of contact is zero. At that time, $t = t_1$, point C becomes the instantaneous center of rotation, and we have

$$\bar{v}_1 = r\omega_1$$

$$\mu gt_1 = r\left(\omega_0 - \frac{\mu g}{r}t_1\right) \qquad\qquad t_1 = \frac{r\omega_0}{2\mu g} \quad \blacktriangleleft$$

Substituting for t_1 into (1), we have

$$\bar{v}_1 = \mu gt_1 = \mu g\frac{r\omega_0}{2\mu g} \qquad\qquad \bar{v}_1 = \tfrac{1}{2}r\omega_0 \qquad\qquad \bar{\mathbf{v}}_1 = \tfrac{1}{2}r\omega_0 \rightarrow \quad \blacktriangleleft$$

$$\bar{v}_1 = r\omega_1 \qquad \tfrac{1}{2}r\omega_0 = r\omega_1 \qquad\qquad \omega_1 = \tfrac{1}{2}\omega_0 \qquad\qquad \omega_1 = \tfrac{1}{2}\omega_0 \,\rotatebox{0}{↻} \quad \blacktriangleleft$$

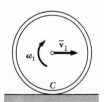

PROBLEMS

16.1 The open tailgate of a truck is held by hinges at A and a chain as shown. Determine the maximum allowable acceleration of the truck if the tailgate is to remain in the position shown.

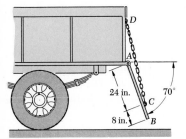

Fig. P16.1

16.2 A uniform rod ABC of mass 8 kg is connected to two collars of negligible mass which slide on smooth horizontal rods located in the same vertical plane. If a force $\mathbf{P}$ of magnitude 40 N is applied at C, determine (a) the acceleration of the rod, (b) the reactions at B and C.

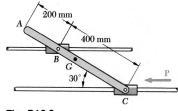

Fig. P16.2

16.3 In Prob. 16.2, determine (a) the required magnitude of $\mathbf{P}$ if the reaction at B is to be 45 N upward, (b) the corresponding acceleration of the rod.

16.4 A sphere of mass m and radius r is placed on a rough horizontal surface and pulled by a horizontal cord attached at A. The sphere is observed to slide to the right without rotating. Knowing that the acceleration of the sphere is $a = g/5$ directed to the right and that $\mu = 0.30$, determine (a) the magnitude of the force $\mathbf{P}$, (b) the distance h from the surface to the cord.

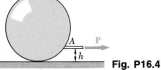

Fig. P16.4

16.5 Cylindrical cans are transported from one elevation to another by the moving horizontal arms shown. Assuming that $\mu = 0.20$ between the cans and the arms, determine (a) the magnitude of the downward acceleration $\mathbf{a}$ for which the cans slide on the horizontal arms, (b) the smallest ratio h/d for which the cans tip before they slide.

16.6 Solve Prob. 16.5, assuming that the acceleration $\mathbf{a}$ of the horizontal arms is directed upward.

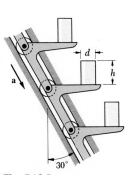

Fig. P16.5

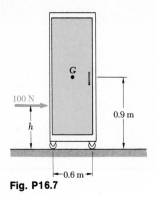

Fig. P16.7

16.7 A 20-kg cabinet is mounted on casters which allow it to move freely ($\mu = 0$) on the floor. If a 100-N force is applied as shown, determine (a) the acceleration of the cabinet, (b) the range of values of h for which the cabinet will not tip.

16.8 Solve Prob. 16.7, assuming that the casters are locked and slide along the rough floor ($\mu = 0.25$).

16.9 A fork-lift truck weighs 2250 lb and is used to lift a crate of weight $W = 2500$ lb. Determine (a) the upward acceleration of the crate for which the reactions at the rear wheels B are zero, (b) the corresponding reaction at each of the front wheels A.

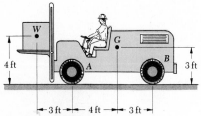

Fig. P16.9 and P16.10

16.10 The fork-lift truck shown weighs 2250 lb and carries a crate of weight $W = 2500$ lb. It is moving to the left with a speed of 10 ft/s when the brakes are applied on all four wheels. Determine the shortest distance in which the truck can be brought to a stop if the crate is not to slide and if the truck is not to tip forward. The coefficient of friction between the crate and the fork lift is 0.30.

16.11 Determine the distance through which the truck of Sample Prob. 16.1 will skid if the rear-wheel brakes fail to operate.

16.12 Determine the distance through which the truck of Sample Prob. 16.1 will skid if the front-wheel brakes fail to operate.

16.13 Knowing that the coefficient of friction between the tires and road is 0.80 for the car shown, determine the maximum possible acceleration on a level road, assuming (a) four-wheel drive, (b) conventional rear-wheel drive, (c) front-wheel drive.

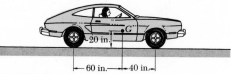

Fig. P16.13

16.14 A man rides a bicycle at a speed of 30 km/h. The distance between axles is 1050 mm, and the mass center of the man and bicycle is located 650 mm behind the front axle and 1000 mm above the ground. If the man applies the brakes on the front wheel only, determine the shortest distance in which he can stop without being thrown over the front wheel.

16.15 The total mass of the loading car and its load is 2500 kg. Neglecting the mass and friction of the wheels, determine (*a*) the minimum tension *T* in the cable for which the upper wheels are lifted from the track, (*b*) the corresponding acceleration of the car.

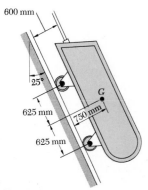

Fig. P16.15

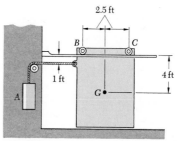

Fig. P16.16

16.16 The 500-lb fire door is supported by wheels *B* and *C* which may roll freely on the horizontal track. The counterweight *A* weighs 100 lb and is connected to the door by the cable shown. If the system is released from rest, determine (*a*) the acceleration of the door, (*b*) the reactions at *B* and *C*.

16.17 The cranks *AB* and *CD* rotate at a constant speed of 240 rpm. For the position $\phi = 30°$, determine the horizontal components of the forces exerted on the 5-kg uniform connecting rod *BC* by the pins *B* and *C*.

16.18 Two uniform rods *AB* and *BC*, each of mass 3 kg, are welded together to form the rigid body *ABC* which is held in the position shown by the wire *AE* and the links *AD* and *BE*. Determine the force in each link immediately after wire *AE* has been cut. Neglect the mass of the links.

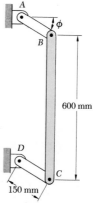

Fig. P16.17

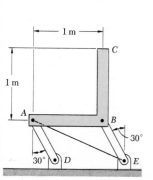

Fig. P16.18

16.19 The retractable shelf shown is supported by two identical linkage-and-spring systems; only one of the systems is shown. A 40-lb machine is placed on the shelf so that half of its weight is supported by the system shown. If the springs are removed and the system is released from rest, determine (a) the acceleration of the machine, (b) the tension in link AB. Neglect the weight of the shelf and links.

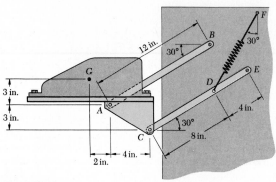

Fig. P16.19

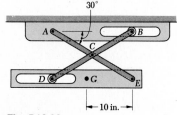

Fig. P16.20

16.20 Members ACE and DCB are each 20 in. long and are connected by a pin at C. The mass center of the 10-lb member DE is located at G. Determine (a) the acceleration of DE immediately after the system has been released from rest in the position shown, (b) the corresponding force exerted by roller D on member DE. Neglect the weight of members ACE and DCB.

16.21 The control rod AC is guided by two pins which slide freely in parallel curved slots of radius 200 mm. The rod has a mass of 10 kg, and its mass center is located at point G. Knowing that for the position shown the *vertical* component of the velocity of C is 1.25 m/s upward and the *vertical* component of the acceleration of C is 5 m/s² upward, determine the magnitude of the force **P**.

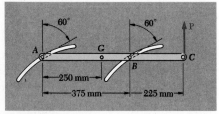

Fig. P16.21

*16.22 A 12-kg block is placed on a 3-kg platform BD which is held in the position shown by three wires. Determine the accelerations of the block and of the platform immediately after wire AB has been cut. Assume (a) that the block is rigidly attached to BD, (b) that μ = 0 between the block and BD.

*16.23 The coefficient of friction between the 12-kg block and the 3-kg platform BD is 0.50. Determine the accelerations of the block and of the platform immediately after wire AB has been cut.

*16.24 Draw the shear and bending moment diagrams for the vertical rod BC of Prob. 16.18.

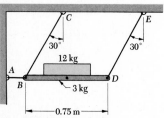

Fig. P16.22 and P16.23

∗16.25 Draw the shear and bending-moment diagrams for the rod *BC* of Prob. 16.17.

16.26 A turbine-generator unit is shut off when its rotor is rotating at 3600 rpm; it is observed that the rotor coasts to rest in 7.10 min. Knowing that the 1850-kg rotor has a radius of gyration of 234 mm, determine the average magnitude of the couple due to bearing friction.

16.27 An electric motor is rotating at 900 rpm when the load and power are cut off. The rotor weighs 200 lb and has a radius of gyration of 9 in. If the kinetic friction of the rotor produces a couple of moment 15 lb · in., how many revolutions will the rotor execute before stopping?

16.28 Disk *A* weighs 10 lb and is at rest when it is placed in contact with a conveyor belt moving at a constant speed. The link *AB* connecting the center of the disk to the support at *B* is of negligible weight. Knowing that $r = 5$ in. and $\mu = 0.40$, determine the angular acceleration of the disk while slipping occurs.

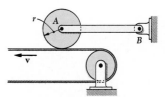

Fig. P16.28 and P16.29

16.29 The uniform disk *A* is at rest when it is placed in contact with a conveyor belt moving at a constant speed. Neglecting the weight of the link *AB*, derive an expression for the angular acceleration of the disk while slipping occurs.

16.30 Each of the double pulleys shown has a mass moment of inertia of 10 kg · m² and is initially at rest. The outside radius is 400 mm, and the inner radius is 200 mm. Determine (*a*) the angular acceleration of each pulley, (*b*) the angular velocity of each pulley at $t = 2$ s, (*c*) the angular velocity of each pulley after point *A* on the cord has moved 2 m.

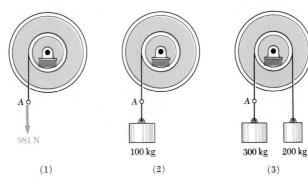

Fig. P16.30

16.31 Solve Prob. 12.16*a* assuming that each pulley is of 8-in. radius and has a centroidal mass moment of inertia 0.25 lb · ft · s².

Fig. P16.32 and P16.34

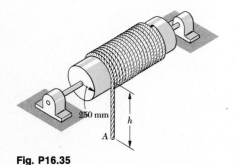

Fig. P16.35

16.32 The flywheel shown weighs 250 lb and has a radius of gyration of 15 in. A block A of weight 30 lb is attached to a wire wrapped around the rim of radius $r = 20$ in. The system is released from rest. Neglecting the effect of friction, determine (a) the acceleration of block A, (b) the speed of block A after it has moved 6 ft.

16.33 Solve Prob. 16.32 assuming that the weight of block A is 60 lb.

16.34 In order to determine the mass moment of inertia of a flywheel of radius $r = 600$ mm a block of mass 12 kg is attached to a cord which is wrapped around the rim of the flywheel. The block is released from rest and is observed to fall 3 m in 4.6 s. To eliminate bearing friction from the computation, a second block of mass 24 kg is used and is observed to fall 3 m in 3.1 s. Assuming that the moment of the couple due to friction is constant, determine the mass moment of inertia of the flywheel.

16.35 A rope of total mass 10 kg and total length 20 m is wrapped around the drum of a hoist as shown. The mass of the drum and shaft is 18 kg, and they have a combined radius of gyration of 200 mm. Knowing that the system is released from rest when a length $h = 5$ m hangs from the drum, determine the initial angular acceleration of the drum.

16.36 The 15-in.-radius brake drum is attached to a larger flywheel which is not shown. The total mass moment of inertia of the flywheel and drum is 15 lb·ft·s². Knowing that the initial angular velocity is 150 rpm clockwise, determine the force which must be exerted by the hydraulic cylinder if the system is to stop in 10 revolutions.

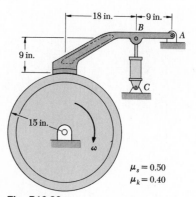

Fig. P16.36

16.37 Solve Prob. 16.36, assuming that the initial angular velocity of the flywheel is 150 rpm counterclockwise.

16.38 A cylinder of radius r and mass m is placed with no initial velocity on a belt as shown. Denoting by μ the coefficient of friction at A and at B and assuming that $\mu < 1$, determine the angular acceleration $\boldsymbol{\alpha}$ of the cylinder.

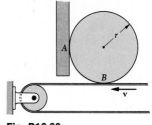

Fig. P16.38

16.39 Shaft A and friction disk B have a combined mass of 15 kg and a combined radius of gyration of 150 mm. Shaft D and friction wheel C rotate with a constant angular velocity of 1000 rpm. Disk B is at rest when it is brought into contact with the rotating wheel. Knowing that disk B accelerates uniformly for 12 s before acquiring its final angular velocity, determine the magnitude of the friction force between the disk and the wheel.

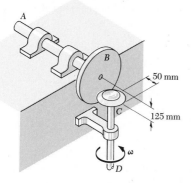

Fig. P16.39

16.40 Gear A weighs 30 lb and has a radius of gyration of 6 in.; gear B weighs 9 lb and has a radius of gyration of 2.5 in. A couple of moment $M = 15$ lb · in. is applied to gear B. Determine (a) the angular acceleration of gear A, (b) the time required for the angular velocity of gear A to increase from 100 to 500 rpm.

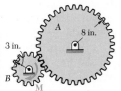

Fig. P16.40

16.41 Disk A is of mass 5 kg and has an initial angular velocity of 300 rpm clockwise. Disk B is of mass 1.8 kg and is at rest when it is placed in contact with disk A. Knowing that $\mu = 0.30$ between the disks and neglecting bearing friction, determine (a) the angular acceleration of each disk, (b) the reaction at the support C.

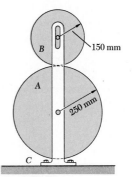

Fig. P16.41

16.42 In Prob. 16.41, (a) determine the final angular velocity of each disk, (b) show that the final angular velocities are independent of μ.

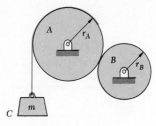

Fig. P16.43

16.43 Assuming that no slipping occurs, derive a formula for the acceleration of block C in terms of its mass m, the radii r_A and r_B, and the centroidal mass moments of inertia $\bar{I}_A$ and $\bar{I}_B$ of the two disks.

16.44 A coder C, used to record in digital form the rotation of a shaft S, is connected to the shaft by means of the gear train shown, which consists of four gears of the same thickness and of the same material. Two of the gears have a radius r and the other two a radius nr. Let I_R denote the ratio M/α of the moment M of the couple applied to the shaft S and of the resulting angular acceleration α of S. (I_R is sometimes called the "reflected moment of inertia" of the coder and gear train.) Determine I_R in terms of the gear ratio n, the moment of inertia I_0 of the first gear, and the moment of inertia I_C of the coder. Neglect the moments of inertia of the shafts.

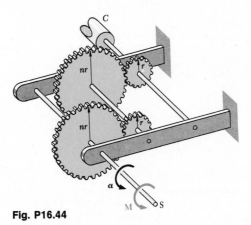

Fig. P16.44

16.45 A cylinder of radius r and mass m rests on two small casters A and B as shown. Initially, the cylinder is at rest and is set in motion by rotating caster B clockwise at high speed so that slipping occurs between the cylinder and caster B. Denoting by μ the coefficient of friction and neglecting the moment of inertia of the free caster A, derive an expression for the angular acceleration of the cylinder.

Fig. P16.45

16.46 In Prob. 16.45, assume that no slipping can occur between caster B and the cylinder (such a case would exist if the cylinder and caster had gear teeth along their rims). Derive an expression for the maximum counterclockwise acceleration α of the cylinder if it is not to lose contact with the caster at A.

16.47 Show that the system of the effective forces for a rigid slab in plane motion reduces to a single vector, and express the distance from the mass center G of the slab to the line of action of this vector in terms of the centroidal radius of gyration $\bar{k}$ of the slab, the magnitude $\bar{a}$ of the acceleration of G, and the angular acceleration α.

16.48   The uniform slender rod AB weighs 8 lb and is at rest on a frictionless horizontal surface. A force **P** of magnitude 2 lb is applied at A in a horizontal direction perpendicular to the rod. Determine (a) the angular acceleration of the rod, (b) the acceleration of the center of the rod, (c) the point of the rod which has no acceleration.

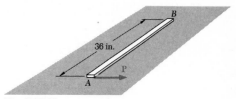

16.49   In Prob. 16.48, determine the point of the rod AB at which the force **P** should be applied if the acceleration of point B is to be zero. Knowing that the magnitude of **P** is 2 lb, determine the corresponding angular acceleration of the rod and the acceleration of the center of the rod.

Fig. P16.48

16.50   Shortly after being fired, the missile shown has a mass of 9000 kg and is moving upward with an acceleration of 16 m/s². If, at this instant, the rocket engine A fails, while rocket engine B continues to operate, determine (a) the acceleration of the mass center of the missile, (b) the angular acceleration of the missile. Assume the missile to be a uniform slender rod 15 m long.

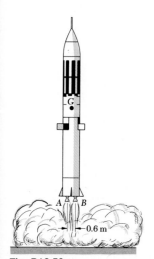

Fig. P16.50

16.51   A 50-kg space satellite has a radius of gyration of 450 mm with respect to the y axis, and is symmetrical with respect to the zx plane. The orientation of the satellite is changed by firing four small rockets A, B, C, and D which are equally spaced around the perimeter of the satellite. While being fired, each rocket produces a thrust **T** of magnitude 10 N directed as shown. Determine the angular acceleration of the satellite and the acceleration of its mass center G (a) when all four rockets are fired, (b) when all rockets except rocket D are fired.

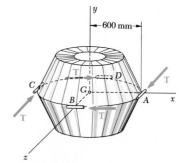

16.52   Solve Prob. 16.51 assuming that only rocket A is fired.

Fig. P16.51

16.53 An 18-ft beam weighing 400 lb is lowered from a considerable height by means of two cables unwinding from overhead cranes. As the beam approaches the ground, the crane operators apply brakes to slow the unwinding motion. The deceleration of cable A is 21 ft/s², while that of cable B is 3 ft/s². Determine the tension in each cable.

16.54 An 18-ft beam weighing 400 lb is lowered from a considerable height by means of two cables unwinding from overhead cranes. As the beam approaches the ground, the crane operators apply brakes to slow the unwinding motion. Determine the acceleration of each cable at that instant, knowing that $T_A = 325$ lb and $T_B = 275$ lb.

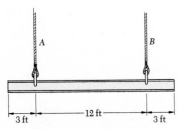

Fig. P16.53 and P16.54

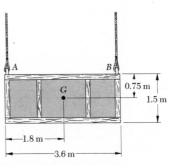

Fig. P16.55 and P16.56

16.55 The 180-kg crate shown is being lowered by means of two overhead cranes. Knowing that at the instant shown the deceleration of cable A is 7 m/s², while that of cable B is 1 m/s², determine the tension in each cable.

16.56 The 180-kg crate is being lowered by means of two overhead cranes. As the crate approaches the ground, the crane operators apply brakes to slow the motion. Determine the acceleration of each cable at that instant, knowing that $T_A = 1450$ N and $T_B = 1200$ N.

16.57 Solve Sample Prob. 16.4, assuming that the disk rests flat on a frictionless horizontal surface and that the cord is pulled horizontally with a force of magnitude 180 N.

16.58 A turbine disk and shaft have a combined mass of 100 kg and a centroidal radius of gyration of 50 mm. The unit is lifted by two ropes looped around the shaft as shown. Knowing that for each rope $T_A = 270$ N and $T_B = 320$ N, determine (a) the angular acceleration of the unit, (b) the acceleration of its mass center.

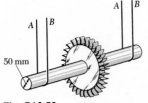

Fig. P16.58

16.59 A uniform slender bar of mass m is suspended from two *springs* as shown. If spring BD breaks, determine at that instant (*a*) the angular acceleration of the bar, (*b*) the acceleration of point A, (*c*) the acceleration of point B.

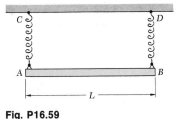

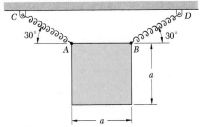

Fig. P16.59

Fig. P16.60

16.60 Solve Prob. 16.59, assuming that the bar is replaced by the uniform square plate shown.

16.61 A sphere of mass m and radius r is projected along a rough horizontal surface with a linear velocity $\mathbf{v}_0$ and with $\boldsymbol{\omega}_0 = 0$. The sphere will decelerate and then reach a uniform motion. Denoting by μ the coefficient of friction, determine (*a*) the linear and angular acceleration of the sphere before it reaches a uniform motion, (*b*) the time required for the motion to become uniform, (*c*) the distance traveled before the motion becomes uniform, (*d*) the final linear and angular velocities of the sphere.

Fig. P16.61

16.62 Solve Prob. 16.61, assuming that the sphere is replaced by a uniform disk of radius r and mass m.

16.63 A sphere of mass m and radius r is dropped with no initial velocity on a belt which moves with a constant velocity $\mathbf{v}_0$. At first the sphere will both slide and roll on the belt. Denoting by μ the coefficient of friction between the sphere and the belt, determine the distance the sphere will move before it starts rolling without sliding.

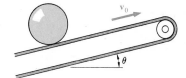

Fig. P16.63

16.64 Solve Prob. 16.63, assuming $\theta = 10°$, $\mu = 0.30$, and $v_0 = 3 \text{ m/s}$.

16.5. Constrained Plane Motion. Most engineering applications deal with rigid bodies which are moving under given constraints. Cranks, for example, are constrained to rotate about a fixed axis, wheels roll without sliding, connecting rods must describe certain prescribed motions. In all such cases, definite relations exist between the components of the acceleration $\bar{\mathbf{a}}$ of the mass center G of the body considered and its angular acceleration $\boldsymbol{\alpha}$; the corresponding motion is said to be a *constrained motion*.

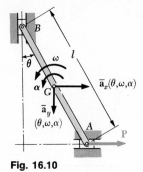

Fig. 16.10

The solution of a problem involving a constrained plane motion calls first for a *kinematic analysis* of the problem. Consider, for example, a slender rod AB of length l and mass m whose extremities are connected to blocks of negligible mass which slide along horizontal and vertical frictionless tracks. The rod is pulled by a force $\mathbf{P}$ applied at A (Fig. 16.10). We know from Sec. 15.8 that the acceleration $\bar{\mathbf{a}}$ of the mass center G of the rod may be determined at any given instant from the position of the rod, its angular velocity, and its angular acceleration at that instant. Suppose, for instance, that the values of θ, ω, and α are known at a given instant and that we wish to determine the corresponding value of the force $\mathbf{P}$, as well as the reactions at A and B. We should first *determine the components $\bar{a}_x$ and $\bar{a}_y$ of the acceleration of the mass center G* by the method of Sec. 15.8. We then carry the values obtained into Fig. 16.11

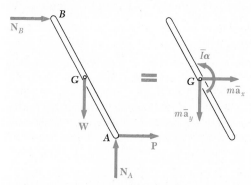

Fig. 16.11

or Fig. 16.12, depending upon whether D'Alembert's principle is applied directly or the method of dynamic equilibrium is used. The unknown forces $\mathbf{P}$, $\mathbf{N}_A$, and $\mathbf{N}_B$ may then be determined by writing and solving the appropriate equations.

Suppose now that the applied force $\mathbf{P}$, the angle θ, and the angular velocity ω of the rod are known at a given instant and that we wish to find the angular acceleration α of the rod and the components $\bar{a}_x$ and $\bar{a}_y$ of the acceleration of its mass center at that instant, as well as the reactions at A and B. The preliminary kinematic study of the problem will have for its object *to express the components $\bar{a}_x$ and $\bar{a}_y$ of the acceleration of G in terms of the angular acceleration α of the rod*. This will be done by first expressing the acceleration of a suitable reference point such as A in terms of the angular acceleration α. The components $\bar{a}_x$ and $\bar{a}_y$ of the acceleration of G may then be determined in terms of α, and the expressions obtained carried into Fig. 16.11 or Fig. 16.12. Three equations may then be derived in terms of α, N_A, and N_B, and solved for the three unknowns (see Sample Prob. 16.10).

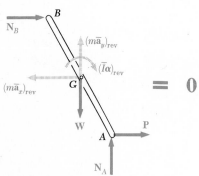

Fig. 16.12

When a mechanism consists of *several moving parts*, the method just described may be applied to each part of the mecha-

nism. The procedure required to determine the various unknowns is then similar to the procedure followed in the case
of the equilibrium of a system of connected rigid bodies (Sec.
6.12).

We have analyzed earlier two particular cases of constrained
plane motion, the translation of a rigid body, in which the
angular acceleration of the body is constrained to be zero, and
the centroidal rotation, in which the acceleration $\bar{a}$ of the mass
center of the body is constrained to be zero. Two other particular cases of constrained plane motion are of special interest, the
noncentroidal rotation of a rigid body and the *rolling motion* of
a disk or wheel. These two cases should be analyzed by one
of the general methods described above. However, in view of
the range of their applications, they deserve a few special comments.

Noncentroidal Rotation.   This is the motion of a rigid body
constrained to rotate about a fixed axis which does not pass
through its mass center. Such a motion is called a *noncentroidal
rotation*. The mass center G of the body moves along a circle
of radius $\bar{r}$ centered at the point O, where the axis of rotation
intersects the plane of reference (Fig. 16.13). Denoting, respec-
tively, by $\boldsymbol{\omega}$ and $\boldsymbol{\alpha}$ the angular velocity and the angular acceleration of the line OG, we obtain the following expressions for
the tangential and normal components of the acceleration of G:

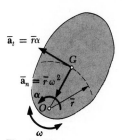

Fig. 16.13

$$\bar{a}_t = \bar{r}\alpha \qquad \bar{a}_n = \bar{r}\omega^2 \qquad (16.12)$$

Since line OG belongs to the body, its angular velocity $\boldsymbol{\omega}$ and
its angular acceleration $\boldsymbol{\alpha}$ also represent the angular velocity and
the angular acceleration of the body in its motion relative to
G. Equations (16.12) define, therefore, the kinematic relation
existing between the motion of the mass center G and the motion
of the body about G. They should be used to eliminate $\bar{a}_t$ and
$\bar{a}_n$ from the equations obtained by applying D'Alembert's principle (Fig. 16.14) or the method of dynamic equilibrium (Fig.
16.15).

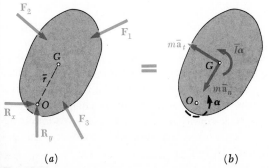

(a)   (b)

Fig. 16.14

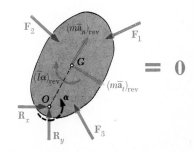

Fig. 16.15

An interesting relation may be obtained by equating the moments about the fixed point O of the forces and vectors shown respectively in parts a and b of Fig. 16.14. We write

$$+\!\!\uparrow \Sigma M_O = \bar{I}\alpha + (m\bar{r}\alpha)\bar{r} = (\bar{I} + m\bar{r}^2)\alpha$$

But, according to the parallel-axis theorem, we have $\bar{I} + m\bar{r}^2 = I_O$, where I_O denotes the moment of inertia of the rigid body about the fixed axis. We write, therefore,

$$\Sigma M_O = I_O\alpha \qquad (16.13)$$

While formula (16.13) expresses an important relation between the sum of the moments of the external forces about the fixed point O and the product $I_O\alpha$, it should be clearly understood that this formula *does not mean* that the system of the external forces is equivalent to a couple of moment $I_O\alpha$. The system of the effective forces, and thus the system of the external forces, reduces to a couple only when O coincides with G, that is, *only when the rotation is centroidal* (Sec. 16.2). In the more general case of noncentroidal rotation, the system of the external forces does not reduce to a couple.

A particular case of noncentroidal rotation is of special interest: the case of *uniform rotation*, in which the angular velocity ω is constant. Since α is zero, the inertia couple in Fig. 16.15 vanishes and the inertia vector reduces to its normal component. This component (also called *centrifugal force*) represents the tendency of the rigid body to break away from the axis of rotation.

Rolling Motion. Another important case of plane motion is the motion of a disk or wheel rolling on a plane surface. If the disk is constrained to roll without sliding, the acceleration $\bar{a}$ of its mass center G and its angular acceleration α are not independent. Assuming the disk to be balanced, so that its mass center and its geometric center coincide, we first write that the distance $\bar{x}$ traveled by G during a rotation θ of the disk is $\bar{x} = r\theta$, where r is the radius of the disk. Differentiating this relation twice, we write

$$\bar{a} = r\alpha \qquad (16.14)$$

Recalling that the system of the effective forces in plane motion reduces to a vector $m\bar{a}$ and a couple $\bar{I}\alpha$, we find that, in the particular case of the rolling motion of a balanced disk, the effective forces reduce to a vector of magnitude $mr\alpha$ attached at G and to a couple of magnitude $\bar{I}\alpha$. We may thus express that the external forces are equivalent to the vector and couple shown in Fig. 16.16.

When a disk *rolls without sliding,* there is no relative motion between the point of the disk which is in contact with the ground and the ground itself. As far as the computation of the friction force **F** is concerned, a rolling disk may thus be compared with a block at rest on a surface. The magnitude F of the friction force may have any value, as long as it does not exceed the maximum value $F_m = \mu_s N$, where μ_s is the coefficient of static friction and N the magnitude of the normal force. In the case of a rolling disk, the magnitude F of the friction force should therefore be determined independently of N by solving the equation obtained from Fig. 16.16.

When *sliding is impending,* the friction force reaches its maximum value $F_m = \mu_s N$ and may be obtained from N.

When the disk *rolls and slides* at the same time, a relative motion exists between the point of the disk which is in contact with the ground and the ground itself, and the force of friction has the magnitude $F_k = \mu_k N$, where μ_k is the coefficient of kinetic friction. In this case, however, the motion of the mass center G of the disk and the rotation of the disk about G are independent, and $\bar{a}$ is not equal to $r\alpha$.

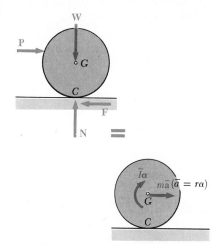

Fig. 16.16

These three different cases may be summarized as follows:

Rolling, no sliding:	$F \le \mu_s N$	$\bar{a} = r\alpha$
Rolling, sliding impending:	$F = \mu_s N$	$\bar{a} = r\alpha$
Rolling and sliding:	$F = \mu_k N$	$\bar{a}$ *and* α *independent*

When it is not known whether a disk slides or not, it should first be assumed that the disk rolls without sliding. If F is found smaller than, or equal to, $\mu_s N$, the assumption is proved correct. If F is found larger than $\mu_s N$, the assumption is incorrect and the problem should be started again, assuming rolling and sliding.

When a disk is *unbalanced,* i.e., when its mass center G does not coincide with its geometric center O, the relation (16.14) does not hold between $\bar{a}$ and α. A similar relation will hold, however, between the magnitude a_O of the acceleration of the geometric center and the angular acceleration α,

$$a_O = r\alpha \tag{16.15}$$

To determine $\bar{a}$ in terms of the angular acceleration α and the angular velocity ω of the disk, we may use the relative-acceleration formula,

$$\begin{aligned}\bar{\mathbf{a}} = \mathbf{a}_G &= \mathbf{a}_O + \mathbf{a}_{G/O} \\ &= \mathbf{a}_O + (\mathbf{a}_{G/O})_t + (\mathbf{a}_{G/O})_n\end{aligned} \tag{16.16}$$

where the three component accelerations obtained have the directions indicated in Fig. 16.17 and the magnitudes $a_O = r\alpha$, $(a_{G/O})_t = (OG)\alpha$, and $(a_{G/O})_n = (OG)\omega^2$.

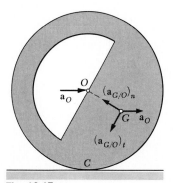

Fig. 16.17

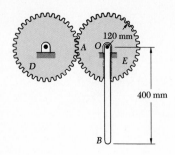

120 mm

400 mm

The portion AOB of a mechanism consists of a 400-mm steel rod OB welded to a gear E of radius 120 mm which may rotate about a horizontal shaft O. It is actuated by a gear D and, at the instant shown, has a clockwise angular velocity of 8 rad/s and a counterclockwise angular acceleration of 40 rad/s². Knowing that rod OB has a mass of 3 kg and gear E a mass of 4 kg and a radius of gyration of 85 mm, determine (a) the tangential force exerted by gear D on gear E, (b) the components of the reaction at shaft O.

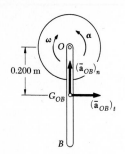

0.200 m

$(\bar{a}_{OB})_n$

G_{OB}

$(\bar{a}_{OB})_t$

B

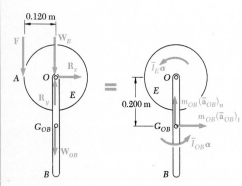

0.120 m

0.200 m

Solution. In determining the effective forces of the rigid body AOB we shall consider separately gear E and rod OB. Therefore, we shall first determine the components of the acceleration of the mass center G_{OB} of the rod:

$$(\bar{a}_{OB})_t = \bar{r}\alpha = (0.200 \text{ m})(40 \text{ rad/s}^2) = 8 \text{ m/s}^2$$
$$(\bar{a}_{OB})_n = \bar{r}\omega^2 = (0.200 \text{ m})(8 \text{ rad/s})^2 = 12.8 \text{ m/s}^2$$

Equations of Motion. Two sketches of the rigid body AOB have been drawn. The first shows the external forces consisting of the weight $\mathbf{W}_E$ of gear E, the weight $\mathbf{W}_{OB}$ of rod OB, the force $\mathbf{F}$ exerted by gear D, and the components $\mathbf{R}_x$ and $\mathbf{R}_y$ of the reaction at O. The magnitudes of the weights are, respectively,

$$W_E = m_E g = (4 \text{ kg})(9.81 \text{ m/s}^2) = 39.2 \text{ N}$$
$$W_{OB} = m_{OB} g = (3 \text{ kg})(9.81 \text{ m/s}^2) = 29.4 \text{ N}$$

The second sketch shows the effective forces, which consist of a couple $\bar{I}_E \alpha$ (since gear E is in centroidal rotation) and of a couple and two vector components at the mass center of OB. Since the accelerations are known, we compute the magnitudes of these components and couples:

$$\bar{I}_E \alpha = m_E \bar{k}_E^2 \alpha = (4 \text{ kg})(0.085 \text{ m})^2(40 \text{ rad/s}^2) = 1.156 \text{ N} \cdot \text{m}$$
$$m_{OB}(\bar{a}_{OB})_t = (3 \text{ kg})(8 \text{ m/s}^2) = 24.0 \text{ N}$$
$$m_{OB}(\bar{a}_{OB})_n = (3 \text{ kg})(12.8 \text{ m/s}^2) = 38.4 \text{ N}$$
$$\bar{I}_{OB} \alpha = (\tfrac{1}{12} m_{OB} L^2)\alpha = \tfrac{1}{12}(3 \text{ kg})(0.400 \text{ m})^2(40 \text{ rad/s}^2) = 1.600 \text{ N} \cdot \text{m}$$

Expressing that the system of the external forces is equivalent to the system of the effective forces, we write the following equations:

$+\!\!\uparrow \Sigma M_O = \Sigma(M_O)_{\text{eff}}$:

$$F(0.120 \text{ m}) = \bar{I}_E \alpha + m_{OB}(\bar{a}_{OB})_t(0.200 \text{ m}) + \bar{I}_{OB}\alpha$$
$$F(0.120 \text{ m}) = 1.156 \text{ N} \cdot \text{m} + (24.0 \text{ N})(0.200 \text{ m}) + 1.600 \text{ N} \cdot \text{m}$$

$$F = 63.0 \text{ N} \qquad \qquad \mathbf{F} = 63.0 \text{ N} \downarrow \quad \blacktriangleleft$$

$\xrightarrow{+} \Sigma F_x = \Sigma(F_x)_{\text{eff}}$: $\qquad R_x = m_{OB}(\bar{a}_{OB})_t$

$$R_x = 24.0 \text{ N} \qquad \qquad \mathbf{R}_x = 24.0 \text{ N} \rightarrow \quad \blacktriangleleft$$

$+\!\!\uparrow \Sigma F_y = \Sigma(F_y)_{\text{eff}}$: $\qquad R_y - F - W_E - W_{OB} = m_{OB}(\bar{a}_{OB})_n$
$$R_y - 63.0 \text{ N} - 39.2 \text{ N} - 29.4 \text{ N} = 38.4 \text{ N}$$

$$R_y = 170.0 \text{ N} \qquad \qquad \mathbf{R}_y = 170.0 \text{ N} \uparrow \quad \blacktriangleleft$$

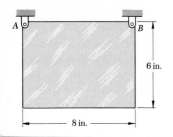

A rectangular plate, 6 by 8 in., weighs 60 lb and is suspended from two pins A and B. If pin B is suddenly removed, determine (a) the angular acceleration of the plate, (b) the components of the reactions at pin A, immediately after pin B has been removed.

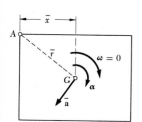

a. Angular Acceleration. We observe that as the plate rotates about point A, its mass center G describes a circle of radius $\bar{r}$ with center at A.

Since the plate is released from rest ($\omega = 0$), the normal component of the acceleration of G is zero. The magnitude of the acceleration $\bar{\mathbf{a}}$ of the mass center G is thus $\bar{a} = \bar{r}\alpha$. We draw the diagram shown to express that the external forces are equivalent to the effective forces:

$$+\!\!\downarrow \Sigma M_A = \Sigma(M_A)_{\text{eff}}: \qquad W\bar{x} = (m\bar{a})\bar{r} + \bar{I}\alpha$$

Since $\bar{a} = \bar{r}\alpha$, we have

$$W\bar{x} = m(\bar{r}\alpha)\bar{r} + \bar{I}\alpha \qquad \alpha = \dfrac{W\bar{x}}{\dfrac{W}{g}\bar{r}^2 + \bar{I}} \qquad (1)$$

The centroidal moment of inertia of the plate is

$$\bar{I} = \dfrac{m}{12}(a^2 + b^2) = \dfrac{60 \text{ lb}}{12(32.2 \text{ ft/s}^2)}[(\tfrac{8}{12} \text{ ft})^2 + (\tfrac{6}{12} \text{ ft})^2]$$

$$= 0.1078 \text{ lb} \cdot \text{ft} \cdot \text{s}^2$$

Substituting this value of $\bar{I}$ together with $W = 60$ lb, $\bar{r} = \tfrac{5}{12}$ ft, and $\bar{x} = \tfrac{4}{12}$ ft into Eq. (1), we obtain

$$\alpha = +46.4 \text{ rad/s}^2 \qquad \alpha = 46.4 \text{ rad/s}^2 \,\downarrow \ \blacktriangleleft$$

b. Reaction at A. Using the computed value of α, we determine the magnitude of the vector $m\bar{\mathbf{a}}$ attached at G,

$$m\bar{a} = m\bar{r}\alpha = \dfrac{60 \text{ lb}}{32.2 \text{ ft/s}^2}(\tfrac{5}{12} \text{ ft})(46.4 \text{ rad/s}^2) = 36.0 \text{ lb}$$

Showing this result on the diagram, we write the equations of motion

$$\overset{+}{\rightarrow} \Sigma F_x = \Sigma(F_x)_{\text{eff}}: \qquad A_x = -\tfrac{3}{5}(36 \text{ lb})$$
$$A_x = -21.6 \text{ lb} \qquad \mathbf{A}_x = 21.6 \text{ lb} \leftarrow \ \blacktriangleleft$$

$$+\!\!\uparrow \Sigma F_y = \Sigma(F_y)_{\text{eff}}: \qquad A_y - 60 \text{ lb} = -\tfrac{4}{5}(36 \text{ lb})$$
$$A_y = +31.2 \text{ lb} \qquad \mathbf{A}_y = 31.2 \text{ lb} \uparrow \ \blacktriangleleft$$

The couple $\bar{I}\alpha$ is not involved in the last two equations; nevertheless, it should be indicated on the diagram.

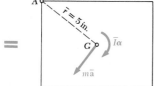

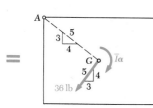

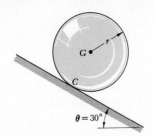

A sphere of radius r and weight W is released with no initial velocity on the incline and rolls without slipping. Determine (a) the minimum value of the coefficient of friction compatible with the rolling motion, (b) the velocity of the center G of the sphere after the sphere has rolled 10 ft, (c) the velocity of G if the sphere were to move 10 ft down a frictionless 30° incline.

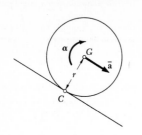

a. Minimum μ for Rolling Motion. The external forces **W**, **N**, and **F** form a system equivalent to the system of effective forces represented by the vector $m\bar{\mathbf{a}}$ and the couple $\bar{I}\alpha$. Since the sphere rolls without sliding, we have $\bar{a} = r\alpha$.

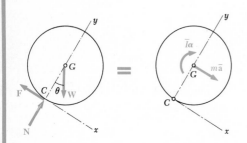

$$+\!\!\downarrow\ \Sigma M_C = \Sigma(M_C)_{\text{eff}}: \qquad (W\sin\theta)r = (m\bar{a})r + \bar{I}\alpha$$
$$(W\sin\theta)r = (mr\alpha)r + \bar{I}\alpha$$

Noting that $m = W/g$ and $\bar{I} = \frac{2}{5}mr^2$, we write

$$(W\sin\theta)r = \left(\frac{W}{g}r\alpha\right)r + \frac{2}{5}\frac{W}{g}r^2\alpha \qquad \alpha = +\frac{5g\sin\theta}{7r}$$

$$\bar{a} = r\alpha = \frac{5g\sin\theta}{7} = \frac{5(32.2\text{ ft/s}^2)\sin 30°}{7} = 11.50\text{ ft/s}^2$$

$$+\!\!\searrow\ \Sigma F_x = \Sigma(F_x)_{\text{eff}}: \qquad W\sin\theta - F = m\bar{a}$$

$$W\sin\theta - F = \frac{W}{g}\frac{5g\sin\theta}{7}$$

$$F = +\tfrac{2}{7}W\sin\theta = \tfrac{2}{7}W\sin 30° \qquad \mathbf{F} = 0.143W \ \searrow\ 30°$$

$$+\!\!\nearrow\ \Sigma F_y = \Sigma(F_y)_{\text{eff}}: \qquad N - W\cos\theta = 0$$
$$N = W\cos\theta = 0.866W \qquad \mathbf{N} = 0.866W \ \measuredangle\ 60°$$

$$\mu_{\min} = \frac{F}{N} = \frac{0.143W}{0.866W} \qquad \mu_{\min} = 0.165 \ \blacktriangleleft$$

b. Velocity of Rolling Sphere. We have uniformly accelerated motion,

$$\bar{v}_0 = 0 \qquad \bar{a} = 11.50\text{ ft/s}^2 \qquad \bar{x} = 10\text{ ft} \qquad \bar{x}_0 = 0$$
$$\bar{v}^2 = \bar{v}_0^2 + 2\bar{a}(\bar{x} - \bar{x}_0) \qquad \bar{v}^2 = 0 + 2(11.50\text{ ft/s}^2)(10\text{ ft})$$
$$\bar{v} = 15.17\text{ ft/s} \qquad \bar{\mathbf{v}} = 15.17\text{ ft/s} \ \searrow\ 30° \ \blacktriangleleft$$

c. Velocity of Sliding Sphere. Assuming now no friction, we have $F = 0$ and obtain

$$+\!\!\downarrow\ \Sigma M_G = \Sigma(M_G)_{\text{eff}}: \qquad 0 = \bar{I}\alpha \qquad \alpha = 0$$

$$+\!\!\searrow\ \Sigma F_x = \Sigma(F_x)_{\text{eff}}: \qquad W\sin 30° = m\bar{a} \qquad 0.50W = \frac{W}{g}\bar{a}$$

$$\bar{a} = +16.1\text{ ft/s}^2 \qquad \bar{\mathbf{a}} = 16.1\text{ ft/s}^2 \ \searrow\ 30°$$

Substituting $\bar{a} = 16.1\text{ ft/s}^2$ into the equations for uniformly accelerated motion, we obtain

$$\bar{v}^2 = \bar{v}_0^2 + 2\bar{a}(\bar{x} - \bar{x}_0) \qquad \bar{v}^2 = 0 + 2(16.1\text{ ft/s}^2)(10\text{ ft})$$
$$\bar{v} = 17.94\text{ ft/s} \qquad \bar{\mathbf{v}} = 17.94\text{ ft/s} \ \searrow\ 30° \ \blacktriangleleft$$

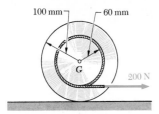

100 mm — 60 mm

G

200 N

A cord is wrapped around the inner drum of a wheel and pulled horizontally with a force of 200 N. The wheel has a mass of 50 kg and a radius of gyration of 70 mm. Knowing that $\mu_s = 0.20$ and $\mu_k = 0.15$, determine the acceleration of G and the angular acceleration of the wheel.

a. **Assume Rolling without Sliding.** In this case, we have

$$\bar{a} = r\alpha = (0.100 \text{ m})\alpha$$

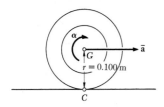

α

G $\bar{a}$

r = 0.100 m

C

By comparing the friction force obtained with the maximum available friction force, we shall determine whether this assumption is justified. The moment of inertia of the wheel is

$$\bar{I} = m\bar{k}^2 = (50 \text{ kg})(0.070 \text{ m})^2 = 0.245 \text{ kg} \cdot \text{m}^2$$

Equations of Motion

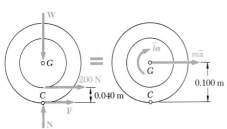

W

G

=

200 N

C 0.040 m

F

N

$\bar{I}\alpha$

G

$m\bar{a}$

0.100 m

C

$+\!\!\downarrow \Sigma M_C = \Sigma(M_C)_{\text{eff}}$: $(200 \text{ N})(0.040 \text{ m}) = m\bar{a}(0.100 \text{ m}) + \bar{I}\alpha$
$8.00 \text{ N} = (50 \text{ kg})(0.100 \text{ m})\alpha(0.100 \text{ m}) + (0.245 \text{ kg} \cdot \text{m}^2)\alpha$
$\alpha = +10.74 \text{ rad/s}^2$
$\bar{a} = r\alpha = (0.100 \text{ m})(10.74 \text{ rad/s}^2) = 1.074 \text{ m/s}^2$

$\xrightarrow{+} \Sigma F_x = \Sigma(F_x)_{\text{eff}}$: $F + 200 \text{ N} = m\bar{a}$
$F + 200 \text{ N} = (50 \text{ kg})(1.074 \text{ m/s}^2)$
$F = -146.3 \text{ N}$ $\mathbf{F} = 146.3 \text{ N} \leftarrow$

$+\!\uparrow \Sigma F_y = \Sigma(F_y)_{\text{eff}}$:
$N - W = 0$ $N = W = mg = (50 \text{ kg})(9.81 \text{ m/s}^2) = 490.5 \text{ N}$
$\mathbf{N} = 490.5 \text{ N} \uparrow$

Maximum Available Friction Force

$$F_{\text{max}} = \mu_s N = 0.20\,(490.5 \text{ N}) = 98.1 \text{ N}$$

Since $F > F_{\text{max}}$, the assumed motion is impossible.

b. **Rolling and Sliding.** Since the wheel must roll and slide at the same time, we draw a new diagram, where $\bar{a}$ and $\boldsymbol{\alpha}$ are independent and where

$$F = F_k = \mu_k N = 0.15\,(490.5 \text{ N}) = 73.6 \text{ N}$$

From the computation of part *a*, it appears that $\mathbf{F}$ should be directed to the left. We write the following equations of motion:

W

0.060 m

G

=

200 N

C

F = 73.6 N

N

$\bar{I}\alpha$

G

$m\bar{a}$

0.100 m

C

$\xrightarrow{+} \Sigma F_x = \Sigma(F_x)_{\text{eff}}$: $200 \text{ N} - 73.6 \text{ N} = (50 \text{ kg})\bar{a}$
$\bar{a} = +2.53 \text{ m/s}^2$ $\bar{a} = 2.53 \text{ m/s}^2 \rightarrow$ ◄

$+\!\!\downarrow \Sigma M_G = \Sigma(M_G)_{\text{eff}}$:
$(73.6 \text{ N})(0.100 \text{ m}) - (200 \text{ N})(0.060 \text{ m}) = (0.245 \text{ kg} \cdot \text{m}^2)\alpha$
$\alpha = -18.94 \text{ rad/s}^2$ $\alpha = 18.94 \text{ rad/s}^2 \,\rotatebox{0}{$\curvearrowright$}$ ◄

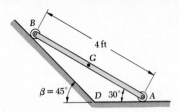

B

4 ft

G

$\beta = 45°$ D 30° A

The extremities of a 4-ft rod, weighing 50 lb, may move freely and with no friction along two straight tracks as shown. If the rod is released with no velocity from the position shown, determine (a) the angular acceleration of the rod, (b) the reactions at A and B.

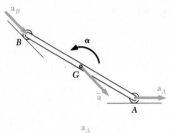

Kinematics of Motion. Since the motion is constrained, the acceleration of G must be related to the angular acceleration $\boldsymbol{\alpha}$. To obtain this relation, we shall first determine the magnitude of the acceleration $\mathbf{a}_A$ of point A in terms of α; assuming $\boldsymbol{\alpha}$ directed counterclockwise and noting that $a_{B/A} = 4\alpha$, we write

$$\mathbf{a}_B = \mathbf{a}_A + \mathbf{a}_{B/A}$$
$$[a_B \nwarrow 45°] = [a_A \rightarrow] + [4\alpha \nearrow 60°]$$

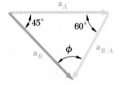

Noting that $\phi = 75°$ and using the law of sines, we obtain

$$a_A = 5.46\alpha \qquad a_B = 4.90\alpha$$

The acceleration of G is now obtained by writing

$$\bar{\mathbf{a}} = \mathbf{a}_G = \mathbf{a}_A + \mathbf{a}_{G/A}$$
$$\bar{\mathbf{a}} = [5.46\alpha \rightarrow] + [2\alpha \nearrow 60°]$$

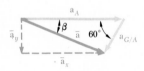

Resolving $\bar{\mathbf{a}}$ into x and y components, we obtain

$$\bar{a}_x = 5.46\alpha - 2\alpha \cos 60° = 4.46\alpha \qquad \bar{a}_x = 4.46\alpha \rightarrow$$
$$\bar{a}_y = -2\alpha \sin 60° = -1.732\alpha \qquad \bar{a}_y = 1.732\alpha \downarrow$$

Kinetics of Motion. We draw the two sketches shown to express that the system of external forces is equivalent to the system of effective forces represented by the vector of components $m\bar{a}_x$ and $m\bar{a}_y$ attached at G and the couple $\bar{I}\alpha$. We compute the following magnitudes:

$$\bar{I} = \frac{1}{12}ml^2 = \frac{1}{12}\frac{50\text{ lb}}{32.2\text{ ft/s}^2}(4\text{ ft})^2 = 2.07\text{ lb}\cdot\text{ft}\cdot\text{s}^2 \qquad \bar{I}\alpha = 2.07\alpha$$

$$m\bar{a}_x = \frac{50}{32.2}(4.46\alpha) = 6.93\alpha \qquad m\bar{a}_y = -\frac{50}{32.2}(1.732\alpha) = -2.69\alpha$$

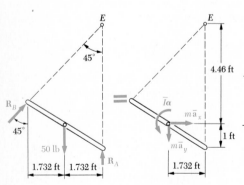

Equations of Motion

$$+\uparrow\,\Sigma M_E = \Sigma(M_E)_{\text{eff}}:$$
$$(50)(1.732) = (6.93\alpha)(4.46) + (2.69\alpha)(1.732) + 2.07\alpha$$
$$\alpha = +2.30\text{ rad/s}^2 \qquad \boldsymbol{\alpha} = 2.30\text{ rad/s}^2 \,\downarrow \quad \blacktriangleleft$$

$$\xrightarrow{+}\Sigma F_x = \Sigma(F_x)_{\text{eff}}: \qquad R_B \sin 45° = (6.93)(2.30) = 15.94$$
$$R_B = 22.5\text{ lb} \qquad \mathbf{R}_B = 22.5\text{ lb} \,\measuredangle\, 45° \quad \blacktriangleleft$$

$$+\uparrow\,\Sigma F_y = \Sigma(F_y)_{\text{eff}}: \qquad R_A + R_B \cos 45° - 50 = -(2.69)(2.30)$$
$$R_A = -6.19 - 15.94 + 50 = 27.9\text{ lb} \qquad \mathbf{R}_A = 27.9\text{ lb} \,\uparrow \quad \blacktriangleleft$$

PROBLEMS

16.65   Show that the couple $\bar{I}\alpha$ of Fig. 16.14 may be eliminated by attaching the vectors $m\bar{\mathbf{a}}_t$ and $m\bar{\mathbf{a}}_n$ at a point P called the *center of percussion,* located on line OG at a distance $GP = \bar{k}^2/\bar{r}$ from the mass center of the body.

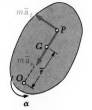

Fig. P16.65

16.66   A uniform slender rod, of length $L = 900$ mm and mass $m = 4$ kg, is supported as shown. A horizontal force **P** of magnitude 75 N is applied at end B. For $\bar{r} = \frac{1}{4}L = 225$ mm, determine (*a*) the angular acceleration of the rod, (*b*) the components of the reaction at C.

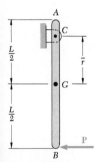

Fig. P16.66

16.67   In Prob. 16.66, determine (*a*) the distance $\bar{r}$ for which the horizontal component of the reaction at C is zero, (*b*) the corresponding angular acceleration of the rod.

16.68   A uniform slender rod, of length L and weight W, hangs freely from a hinge at A. A horizontal force **P** is applied as shown. For $h = L$, determine (*a*) the angular acceleration of the rod, (*b*) the components of the reaction at A.

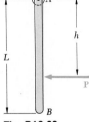

Fig. P16.68

16.69   In Prob. 16.68, determine (*a*) the distance h for which the horizontal component of the reaction at A is zero, (*b*) the corresponding angular acceleration of the rod.

16.70   A turbine disk of mass 75 kg rotates at a constant speed of 9600 rpm; the mass center of the disk coincides with the center of rotation O. Determine the reaction at O after a single vane at A, of mass 45 g, becomes loose and is thrown off.

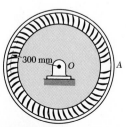

Fig. P16.70

16.71 A uniform slender rod of length l and mass m rotates about a vertical axis AA' at a constant angular velocity ω. Determine the tension in the rod at a distance x from the axis of rotation.

Fig. P16.71

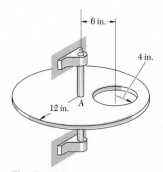

Fig. P16.72

16.72 An 8-in.-diameter hole is cut as shown in a thin disk of diameter 24 in. The disk rotates in a horizontal plane about its geometric center A at a constant angular velocity of 480 rpm. Knowing that the disk weighs 100 lb after the hole has been cut, determine the horizontal component of the force exerted by the shaft on the disk at A.

16.73 The rim of a flywheel has a mass of 1800 kg and a mean radius of 600 mm. As the flywheel rotates at a constant angular velocity of 360 rpm, radial forces are exerted on the rim by the spokes and internal forces are developed within the rim. Neglecting the weight of the spokes, determine (*a*) the internal forces in the rim, assuming the radial forces exerted by the spokes to be zero, (*b*) the radial force exerted by each spoke, assuming the tangential forces in the rim to be zero.

Fig. P16.73

16.74 and 16.75 A uniform beam of length L and weight W is supported as shown. If the cable attached at B suddenly breaks, determine (*a*) the reaction at the pin support, (*b*) the acceleration of point B.

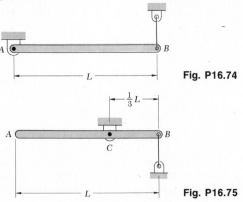

Fig. P16.74

Fig. P16.75

16.76 A 12-ft plank, weighing 80 lb, rests on two horizontal pipes *AB* and *CD* of a scaffolding. The pipes are 10 ft apart, and the plank overhangs 1 ft at each end. A man weighing 160 lb is standing on the plank when pipe *CD* suddenly breaks. Determine the initial acceleration of the man, knowing that he was standing (*a*) in the middle of the plank, (*b*) just above pipe *CD*.

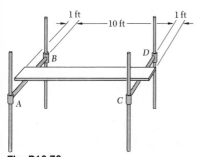

Fig. P16.76

16.77 A 2-kg slender rod is riveted to a 4-kg uniform disk as shown. The assembly swings freely in a vertical plane and, in the position shown, has an angular velocity of 4 rad/s clockwise. Determine (*a*) the angular acceleration of the assembly, (*b*) the components of the reaction at *A*.

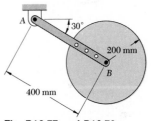

16.78 A 2-kg slender rod is riveted to a 4-kg uniform disk as shown. The assembly rotates in a vertical plane under the combined effect of gravity and a couple **M** which is applied to rod *AB*. Knowing that at the instant shown the assembly has an angular velocity of 6 rad/s and an angular acceleration of 10 rad/s² both counterclockwise, determine (*a*) the moment of the couple **M**, (*b*) the components of the reaction at *A*.

Fig. P16.77 and P16.78

16.79 After being released, the plate of Sample Prob. 16.7 is allowed to swing through 90°. Knowing that at that instant the angular velocity of the plate is 4.82 rad/s, determine (*a*) the angular acceleration of the plate, (*b*) the reaction at *A*.

16.80 Two uniform rods, *AB* of weight 6 lb and *CD* of weight 8 lb, are welded together to form the T-shaped assembly shown. The assembly rotates in a vertical plane about a horizontal shaft at *A*. Knowing that at the instant shown the assembly has a clockwise angular velocity of 8 rad/s and a counterclockwise angular acceleration of 24 rad/s², determine (*a*) the magnitude of the horizontal force **P**, (*b*) the components of the reaction at *A*.

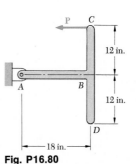

Fig. P16.80

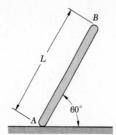

Fig. P16.81

16.81 A uniform rod of weight W is released from rest in the position shown. Assuming that the friction between end A and the surface is large enough to prevent sliding, determine (a) the angular acceleration of the rod just after release, (b) the normal reaction and the friction force at A, (c) the minimum value of μ compatible with the described motion.

16.82 Derive the equation $\Sigma M_C = I_C \alpha$ for the rolling disk of Fig. 16.16, where ΣM_C represents the sum of the moments of the external forces about the instantaneous center C and I_C the moment of inertia of the disk about C.

16.83 Show that, in the case of an unbalanced disk, the equation derived in Prob. 16.82 is valid only when the mass center G, the geometric center O, and the instantaneous center C happen to lie in a straight line.

16.84 A homogeneous cylinder C and a section of pipe P are in contact when they are released from rest. Knowing that both the cylinder and the pipe roll without slipping, determine the clear distance between them after 3 s.

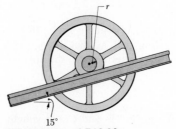

Fig. P16.84

16.85 A flywheel is rigidly attached to a shaft of 40-mm radius which may roll along parallel rails as shown. When released from rest, the system rolls a distance of 3 m in 30 s. Determine the centroidal radius of gyration of the system.

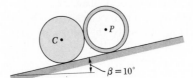

Fig. P16.85 and P16.86

16.86 A flywheel of centroidal radius of gyration $\bar{k} = 600$ mm is rigidly attached to a shaft of radius $r = 30$ mm which may roll along parallel rails. Knowing that the system is released from rest, determine the distance it will roll in 20 s.

16.87 through 16.90 A drum of 80-mm radius is attached to a disk of 160-mm radius. The disk and drum have a total mass of 5 kg and a radius of gyration of 120 mm. A cord is attached as shown and pulled with a force **P** of magnitude 20 N. Knowing that the disk rolls without sliding, determine (*a*) the angular acceleration of the disk and the acceleration of *G*, (*b*) the minimum value of the coefficient of friction compatible with this motion.

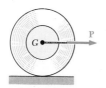

Fig. P16.87 and P16.91   **Fig. P16.88 and P16.92**   **Fig. P16.89 and P16.93**   **Fig. P16.90 and P16.94**

16.91 through 16.94 A drum of 4-in. radius is attached to a disk of 8-in. radius. The disk and drum have a total weight of 10 lb and a radius of gyration of 6 in. A cord is attached as shown and pulled with a force **P** of magnitude 5 lb. Knowing that $\mu = 0.20$, determine (*a*) whether or not the disk slides, (*b*) the angular acceleration of the disk and the acceleration of *G*.

16.95 The cord attached to the drum of Prob. 16.90 is pulled to the right in a direction forming an angle β with the horizontal and with a force **P** of magnitude 20 N. Determine the range of values of β for which the disk (*a*) rolls to the right, (*b*) rolls to the left. Assume that μ is large enough to prevent sliding.

16.96 and 16.97 The 15-lb carriage is supported as shown by two uniform disks each of weight 10 lb and radius 3 in. Knowing that the disks roll without sliding, determine the acceleration of the carriage when a force of 5 lb is applied to it.

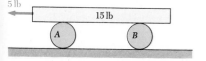

Fig. P16.96

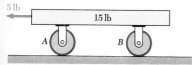

Fig. P16.97

16.98 A block B of mass m is attached to a cord wrapped around a cylinder of the same mass m and of radius r. The cylinder rolls without sliding on a horizontal surface. Determine the components of the accelerations of the center A of the cylinder and of the block B immediately after the system has been released from rest if (a) the block hangs freely, (b) the motion of the block is guided by a rigid member DAE, frictionless and of negligible mass, which is hinged to the cylinder at A.

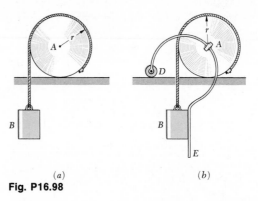

(a)   (b)

Fig. P16.98

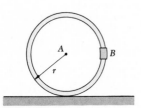

Fig. P16.99

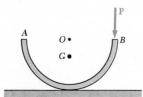

Fig. P16.100

16.99 A small block of mass m is attached at B to a hoop of mass m and radius r. Knowing that when the system is released from rest it starts to roll without sliding, determine (a) the angular acceleration of the hoop, (b) the acceleration of B.

16.100 A half section of pipe of mass m and radius r rests on a rough horizontal surface. A vertical force $\mathbf{P}$ is applied as shown. Assuming that the section rolls without sliding, derive an expression (a) for its angular acceleration, (b) for the minimum value of μ compatible with this motion. [*Hint.* Note that $OG = 2r/\pi$ and that, by the parallel-axis theorem, $\bar{I} = mr^2 - m(OG)^2$.]

16.101 Solve Prob. 16.100, assuming that the force $\mathbf{P}$ is applied at B and is directed horizontally to the right.

16.102 The mass center G of a 4-lb wheel of radius $R = 6$ in. is located at a distance $r = 2$ in. from its geometric center C. The centroidal radius of gyration is $\bar{k} = 3$ in. As the wheel rolls without sliding, its angular velocity varies and it is observed that $\omega = 12$ rad/s in the position shown. Determine the corresponding angular acceleration of the wheel.

Fig. P16.102

16.103   The motion of a 4-kg slender rod is guided by pins at A and B which slide freely in the slots shown. A vertical force **P** is applied at B, causing the rod to start from rest with a counter-clockwise angular acceleration of 12 rad/s². Determine (a) the magnitude and direction of **P**, (b) the reactions at A and B.

16.104   The motion of a 4-kg slender rod is guided by pins at A and B which slide freely in the slots shown. A couple **M** is applied to the rod which is initially at rest. Knowing that the initial acceleration of point B is 6 m/s² upward, determine (a) the moment of the couple, (b) the reactions at A and B.

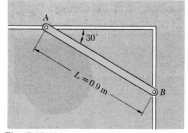

Fig. P16.103 and P16.104

16.105   End A of the 100-lb beam AB moves along the frictionless floor, while end B is supported by a 4-ft cable. Knowing that at the instant shown end A is moving to the left with a constant velocity of 8 ft/s, determine (a) the magnitude of the force **P**, (b) the corresponding tension in the cable.

16.106   Solve Prob. 16.105, assuming that at the instant shown end A moves with a velocity of 8 ft/s and an acceleration of 5 ft/s², both directed to the left.

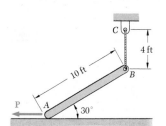

Fig. P16.105

16.107   The rod of Prob. 16.103 is released from rest in the position shown. Determine (a) the angular acceleration of the rod, (b) the reactions at A and B.

16.108   The 50-kg uniform rod AB is released from rest in the position shown. Knowing that end A may slide freely on the frictionless floor, determine (a) the angular acceleration of the rod, (b) the tension in wire BC, (c) the reaction at A.

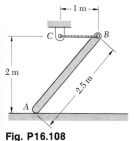

Fig. P16.108   **Fig. P16.109**

16.109   A 3- by 6-ft crate weighing 200 lb is released from rest in the position shown. Assuming corners A and B slide without friction, determine (a) the angular acceleration of the crate, (b) the reactions at A and B.

16.110   Two uniform rods AB and BC, each of weight 10 lb, are welded together to form a single 20-lb rigid body which is held by the three wires shown. Determine the tension in wires BD and CF immediately after wire CE has been cut.

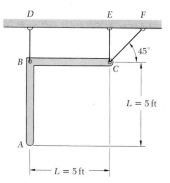

Fig. P16.110

16.111 Show that, for a rigid slab in plane motion, the equation $\Sigma M_A = I_A \alpha$, where ΣM_A represents the sum of the moments of the external forces about point A and I_A the moment of inertia of the slab about the same point A, is verified *if and only if* one of the following conditions is satisfied: (*a*) A is the mass center of the slab, (*b*) A has zero acceleration, (*c*) the acceleration of A is directed along a line joining point A and the mass center G.

16.112 The 10-lb sliding block is attached to the rotating crank BC by the uniform connecting rod AB which weighs 5 lb. Knowing that crank BC has a constant angular velocity of 240 rpm clockwise, determine the forces exerted on the connecting rod at A and B when $\beta = 0$.

16.113 Solve Prob. 16.112 when $\beta = 180°$.

16.114 Each of the bars shown is 600 mm long and has a mass of 4 kg. A horizontal and variable force $\mathbf{P}$ is applied at C, causing point C to move to the left with a constant speed of 10 m/s. Determine the force $\mathbf{P}$ for the position shown.

16.115 The two bars AB and BC are released from rest in the position shown. Each bar is 600 mm long and has a mass of 4 kg. Determine (*a*) the angular acceleration of each bar, (*b*) the reactions at A and C.

16.116 and 16.117 Two rods AB and BC, of mass m per unit length, are connected as shown to a disk which is made to rotate in a vertical plane at a constant angular velocity ω_0. For the position shown, determine the components of the forces exerted at A and B on rod AB.

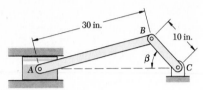

Fig. P16.112

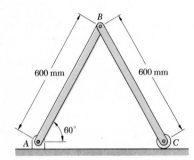

Fig. P16.114 and P16.115

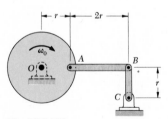

Fig. P16.116

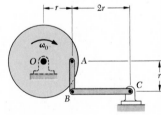

Fig. P16.117

16.118 A section of pipe rests on a plate. The plate is then given a constant acceleration **a** directed to the right. Assuming that the pipe rolls on the plate, determine (*a*) the acceleration of the pipe, (*b*) the distance through which the plate will move before the pipe reaches end A.

Fig. P16.118 and P16.119

16.119 A section of pipe rests on a plate. Knowing that $\mu = 0.40$ between the plate and the pipe, determine the largest acceleration **a** of the plate for which the pipe rolls without sliding.

16.120 and 16.121 Gear C weighs 6 lb and has a centroidal radius of gyration of 3 in. The uniform bar AB weighs 5 lb and gear D is stationary. If the system is released from rest in the position shown, determine (a) the angular acceleration of gear C, (b) the acceleration of point B.

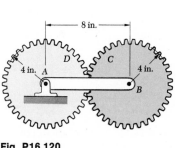

Fig. P16.120

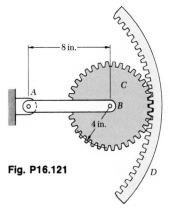

Fig. P16.121

16.122 The rectangular plate of Sample Prob. 16.7, which weighs 60 lb, is suspended from two inextensible wires as shown. If the wire at B is cut, determine at that instant (a) the angular acceleration of the plate, (b) the tension in the wire at A.

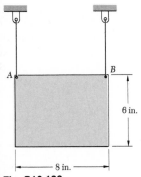

Fig. P16.122

*16.123 A uniform slender rod of length L and mass m is released from rest in the position shown. Derive an expression for (a) the angular acceleration of the rod, (b) the acceleration of end A, (c) the reaction at A, immediately after release. Neglect the mass and friction of the roller at A.

*16.124 A uniform rod AB, of mass 3 kg and length $L = 1.2$ m, is released from rest in the position shown. Knowing that $\beta = 30°$, determine the values immediately after release of (a) the angular acceleration of the rod, (b) the acceleration of end A, (c) the reaction at A. Neglect the mass and friction of the roller at A.

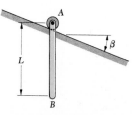

Fig. P16.123 and P16.124

*16.125   Each of the bars AB and BC is of length $L = 2$ ft and weight 5 lb. A couple $\mathbf{M}_0$ of moment 6 lb·ft is applied to bar BC. Determine the angular acceleration of each bar.

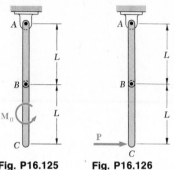

Fig. P16.125   **Fig. P16.126**

*16.126   Each of the bars AB and BC is of length $L = 2$ ft and weight 5 lb. A horizontal force $\mathbf{P}$ of magnitude 3 lb is applied at C. Determine the angular acceleration of each bar.

*16.127   Two uniform slender rods AB and CD, each of mass m, are connected by two wires as shown. If the wire BD is cut, determine at that instant the acceleration (a) of point D, (b) of point B.

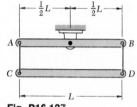

Fig. P16.127

*16.128   The slender bar AB is of length L and mass m. It is held in equilibrium by two counterweights, each of mass $\frac{1}{2}m$. If the wire at B is cut, determine at that instant the acceleration of (a) point A, (b) point B.

*16.129   Two uniform rods, each of weight W, are released from rest in the position shown. Determine the reaction at A immediately after release.

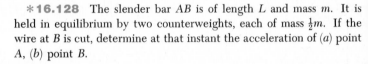

Fig. P16.128

*16.130   Solve Prob. 16.129, assuming that the connection at B is welded and that the rods move as a single rigid body. Also, determine the corresponding bending moment at B.

*16.131   In Prob. 16.125 the pin at B is severely rusted and the bars rotate as a single rigid body. Determine the bending moment which occurs at B.

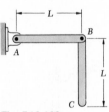

Fig. P16.129

*16.132   (a) Determine the magnitude and the location of the maximum bending moment in the rod of Prob. 16.68. (b) Show that the answer to part a is independent of the weight W of the rod.

*16.133   Draw the shear and bending-moment diagrams for the beam of Prob. 16.74 immediately after the cable at B breaks.

*16.134   Draw the shear and bending-moment diagrams for the bar of Prob. 16.128 immediately after the wire at B has been cut.

16.6. Rotation of a Three-dimensional Body about a Fixed Axis.

The rigid bodies considered in the preceding sections were assumed to be symmetrical with respect to the reference plane used to describe their motion. While such an assumption considerably simplifies the derivations involved, it also limits the range of application of the results obtained. One of the most important types of motion which *cannot* be analyzed directly by the methods considered so far is the rotation about a fixed axis of an unsymmetrical three-dimensional body.

We may, however, examine the particular case when the rotating body consists of several parallel slabs rigidly connected to a perpendicular shaft which is supported in bearings at A and B (Fig. 16.18). Since each slab rotates in its own plane, we may apply the method of Sec. 16.2 and reduce the effective forces of each slab to a force-couple system. By D'Alembert's principle the system of the external forces shown in part a of Fig. 16.18 must be equivalent to the system of the effective forces represented by the various force-couple systems shown in part b of the same figure. Equating components along, and moments about, each of the three coordinate axes, we write the following six equations:

$$\Sigma F_x = \Sigma(F_x)_{\text{eff}} \qquad \Sigma F_y = \Sigma(F_y)_{\text{eff}} \qquad \Sigma F_z = \Sigma(F_z)_{\text{eff}}$$
$$\Sigma M_x = \Sigma(M_x)_{\text{eff}} \qquad \Sigma M_y = \Sigma(M_y)_{\text{eff}} \qquad \Sigma M_z = \Sigma(M_z)_{\text{eff}} \qquad (16.17)$$

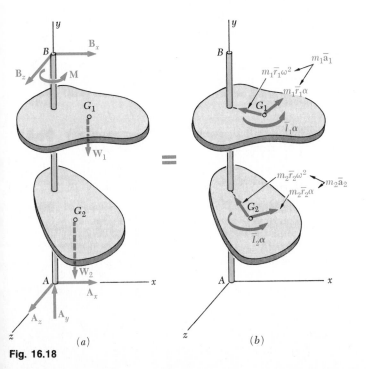

(a)   (b)

Fig. 16.18

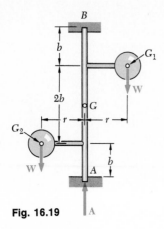

Fig. 16.19

It should be noted that this method may also be used when the bodies attached to the shaft are three-dimensional, *as long as each body is symmetrical with respect to a plane perpendicular to the shaft.*

As an example, we shall consider a shaft held in bearings at A and B and supporting two equal spheres of mass m and weight $W = mg$ (Fig. 16.19). We first observe that when the system is at rest, the shaft exerts no lateral thrust on its supports since the center of gravity G of the system is located directly above A. The system is said to be *statically balanced*. The reaction **A,** often referred to as a *static reaction,* is vertical, and its magnitude is $A = 2W = 2mg$. Let us now assume that the system rotates with a constant angular velocity ω. We shall apply D'Alembert's principle and express the fact that the system of the external forces, consisting of the weights and bearing reactions, is equivalent to the system of the effective forces. Since each sphere is symmetrical with respect to a plane perpendicular to the shaft and containing its own center, we may apply the method of Sec. 16.2 and replace the effective forces of each sphere by a force-couple system at its mass center. But in the present case the angular acceleration α is zero and the force-couple systems reduce respectively to the vectors $m\bar{a}_1$ and $m\bar{a}_2$ directed toward the axis of rotation. Since these two vectors, shown in part b of Fig. 16.20, have the same magnitude $mr\omega^2$ and are located at a distance $2b$ from each other, we conclude that the system of the effective forces is equivalent to a counterclockwise couple of moment $2bmr\omega^2$. Additional bearing reactions $\mathbf{A}_x$ and $\mathbf{B}_x$, therefore, are required in part a of the figure. These reactions, called *dynamic reactions,* must also form a counterclockwise couple of moment $2bmr\omega^2$. They must be

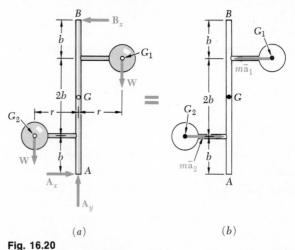

(*a*)   (*b*)

Fig. 16.20

directed as shown and have a magnitude $A_x = B_x = \frac{1}{2}mr\omega^2$. Although the system is statically balanced, it is clearly *not dynamically balanced*. A system which is not dynamically balanced will have a tendency to tear away from its bearings when rotating at high speed, as indicated by the values obtained for the bearing reactions. Moreover, since the bearing reactions are contained in the plane defined by the shaft and the centers of the spheres, they rotate with the shaft and cause the structure supporting it to vibrate. *Rotating systems should therefore be dynamically as well as statically balanced.* This may be done by rearranging the distribution of mass around the shaft, or by adding corrective masses, until the system of the effective forces becomes equivalent to zero.

The analysis of the rotating shaft of Figs. 16.19 and 16.20 could also be carried out by the method of dynamic equilibrium. The effective forces shown on the right-hand side of the equals sign in Fig. 16.20 would then be replaced by equal and opposite inertia vectors on the left-hand side of the equals sign (Fig. 16.21). The problem is thus reduced to a statics problem, in which the inertia vectors represent the "centrifugal forces" which tend to "pull" the spheres away from the shaft.

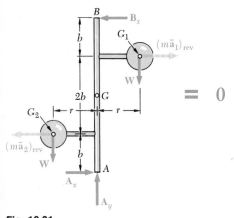

Fig. 16.21

In the general case of the rotation about a fixed axis of an unsymmetrical three-dimensional body, a typical element of mass should be isolated and its effective force should be determined. The sums of the components and moments of the effective forces may then be obtained by integration. While this approach is successfully used in Sample Prob. 16.12, it will lead, in general, to fairly involved computations. Other methods, beyond the scope of this chapter, may then be preferred. The approach indicated here shows clearly, however, that, even in as simple a motion as uniform rotation, the system of the effective forces *will not reduce to a single vector attached at the mass center of the rotating body* if the body does not possess a plane of symmetry perpendicular to the axis of rotation.

Consider, for example, an airplane making a turn of radius $\bar{r}$ in the banked position shown in Fig. 16.22. Since two symmetrical wing elements A and B are not at the same distance from

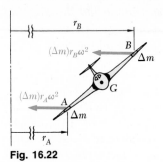

Fig. 16.22

the axis of rotation, the effective forces associated with A and B are not equal. The system of the effective forces corresponding to the various elements of the airplane, therefore, will *not* reduce to a single vector at the mass center G of the airplane. It will reduce to a force-couple system at G which, in turn, may be replaced by a single vector located above G (Fig. 16.23). However, since the difference between the magnitudes of the effective forces corresponding to A and B is relatively small, the

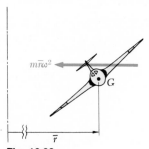

Fig. 16.23

distance from G to the resultant of the effective forces of the airplane is also relatively small. It is therefore customary in such problems, and in problems dealing with the motion of cars along banked highway curves, to neglect the effect of the nonsymmetry of the three-dimensional body on the position of the resultant of its effective forces.

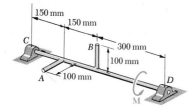

Two 100-mm rods A and B, each of mass 300 g, are welded to the shaft CD which is supported by bearings at C and D. If a couple $\mathbf{M}$ of moment equal to $6\,\mathrm{N}\cdot\mathrm{m}$ is applied to the shaft, determine the components of the dynamic reactions at C and D at the instant when the shaft has reached an angular velocity of 1200 rpm. Neglect the moment of inertia of the shaft itself.

Solution. The external forces consist of the dynamic reactions at C and D. (The weights and the corresponding static reactions are ignored since they cancel each other.) The system of the effective forces of each rod reduces to a vector $m\overline{a}$ at the mass center of the rod and a couple $\overline{I}\alpha$ of axis parallel to the shaft. Each vector $m\overline{a}$ is resolved into a tangential component of magnitude $m\overline{r}\alpha$ and a normal component of magnitude $m\overline{r}\omega^2$. The centroidal moment of inertia of each rod is

$$\overline{I} = \tfrac{1}{12}mL^2 = \tfrac{1}{12}(0.300\ \mathrm{kg})(0.100\ \mathrm{m})^2 = 2.5 \times 10^{-4}\ \mathrm{kg}\cdot\mathrm{m}^2$$

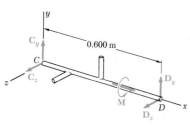

Angular Acceleration of Shaft. Equating the moments about the x axis of the external and effective forces, we first determine the angular acceleration of the shaft. The moments of the components of the vectors $m\overline{a}$, as well as the moments of the couples $\overline{I}\alpha$, must be taken into account.

$$+ \circlearrowleft\ \Sigma M_x = \Sigma(M_x)_{\mathrm{eff}}: \qquad M = 2(m\overline{r}\alpha)\overline{r} + 2\overline{I}\alpha$$
$$6\ \mathrm{N}\cdot\mathrm{m} = 2(0.300\ \mathrm{kg})(0.050\ \mathrm{m})^2\alpha + 2(2.5 \times 10^{-4}\ \mathrm{kg}\cdot\mathrm{m}^2)\alpha$$
$$\alpha = 3000\ \mathrm{rad/s^2}$$

Dynamic Reactions. Using the value obtained for α and noting that $\omega = 1200\ \mathrm{rpm} = 125.7\ \mathrm{rad/s}$, we compute the components of the vectors $m\overline{a}$:

$$m\overline{r}\alpha = (0.300\ \mathrm{kg})(0.050\ \mathrm{m})(3000\ \mathrm{rad/s^2}) = 45\ \mathrm{N}$$
$$m\overline{r}\omega^2 = (0.300\ \mathrm{kg})(0.050\ \mathrm{m})(125.7\ \mathrm{rad/s})^2 = 237\ \mathrm{N}$$

Equating successively the moments about and the components along the y and z axes of the external and effective forces, we write

$$+ \circlearrowleft\ \Sigma M_y = \Sigma(M_y)_{\mathrm{eff}}:$$
$$-(0.600\ \mathrm{m})D_z = (0.150\ \mathrm{m})(m\overline{r}\omega^2) - (0.300\ \mathrm{m})(m\overline{r}\alpha)$$
$$-(0.600\ \mathrm{m})D_z = (0.150\ \mathrm{m})(237\ \mathrm{N}) - (0.300\ \mathrm{m})(45\ \mathrm{N})$$
$$D_z = -36.8\ \mathrm{N} \quad \blacktriangleleft$$

$$+ \circlearrowleft\ \Sigma M_z = \Sigma(M_z)_{\mathrm{eff}}:$$
$$(0.600\ \mathrm{m})D_y = -(0.150\ \mathrm{m})(m\overline{r}\alpha) - (0.300\ \mathrm{m})(m\overline{r}\omega^2)$$
$$(0.600\ \mathrm{m})D_y = -(0.150\ \mathrm{m})(45\ \mathrm{N}) - (0.300\ \mathrm{m})(237\ \mathrm{N})$$
$$D_y = -129.8\ \mathrm{N} \quad \blacktriangleleft$$

$$\Sigma F_y = \Sigma(F_y)_{\mathrm{eff}}: \qquad C_y + D_y = -m\overline{r}\alpha - m\overline{r}\omega^2$$
$$C_y - 129.8\ \mathrm{N} = -45\ \mathrm{N} - 237\ \mathrm{N} \quad C_y = -152.2\ \mathrm{N} \quad \blacktriangleleft$$

$$\Sigma F_z = \Sigma(F_z)_{\mathrm{eff}}: \qquad C_z + D_z = -m\overline{r}\omega^2 + m\overline{r}\alpha$$
$$C_z - 36.8\ \mathrm{N} = -237\ \mathrm{N} + 45\ \mathrm{N} \quad C_z = -155.2\ \mathrm{N} \quad \blacktriangleleft$$

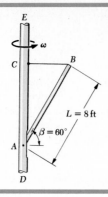

SAMPLE PROBLEM 16.12

A slender rod AB of length $L = 8$ ft and weight $W = 40$ lb is pinned at A to a vertical axle DE which rotates with a constant angular velocity $\omega = 15$ rad/s. The rod is maintained in position by means of a wire BC attached to the axle and to the end B of the rod. Determine the tension in the wire BC.

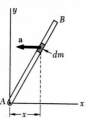

Kinematics of Motion. Choosing a coordinate system with origin at point A, we observe that each element dm of the rod moves in a horizontal circle of radius x. Each element is thus in plane motion and has an acceleration $\mathbf{a}$ of magnitude $a = x\omega^2$ directed horizontally toward the y axis.

Kinetics of Motion. By D'Alembert's principle, the system of the external forces, which consists of the weight $\mathbf{W}$, the force $\mathbf{T}$ exerted by the wire at B and the components $\mathbf{A}_x$ and $\mathbf{A}_y$ of the reaction at A, must be equivalent to the system of the effective forces $\mathbf{a}\,dm$ of the various elements. We write

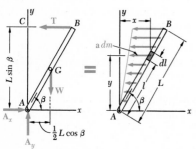

$$+\!\!\uparrow \Sigma M_A = \Sigma(M_A)_{\text{eff}}: \qquad T(L \sin \beta) - W(\tfrac{1}{2}L \cos \beta) = \int y(a\,dm) \qquad (1)$$

Denoting by l the distance from A to the element dm, we have

$$y = l \sin \beta \qquad a = x\omega^2 = (l \cos \beta)\omega^2 \qquad dm = \frac{m}{L}\,dl$$

Substituting these expressions into the integral and integrating from $l = 0$ to $l = L$, we obtain

$$\int y(a\,dm) = \int_0^L (l \sin \beta)(l \cos \beta)\omega^2 \left(\frac{m}{L}\,dl\right) = \tfrac{1}{3}mL^2\omega^2 \sin \beta \cos \beta \qquad (2)$$

Substituting this expression into (1), we have

$$T(L \sin \beta) - W(\tfrac{1}{2}L \cos \beta) = \tfrac{1}{3}mL^2\omega^2 \sin \beta \cos \beta$$

$$T = \tfrac{1}{2}W \cot \beta + \tfrac{1}{3}mL\omega^2 \cos \beta \qquad (3)$$

Substituting $W = 40$ lb, $m = 1.242$ lb $\cdot$ s^2/ft, $L = 8$ ft, $\beta = 60°$, and $\omega = 15$ rad/s into (3), we obtain

$$T = 384 \text{ lb} \quad \blacktriangleleft$$

Note. The resultant of the effective forces $\mathbf{a}\,dm$ does not pass through the mass center of the rod. To show this, we first determine the magnitude of the resultant:

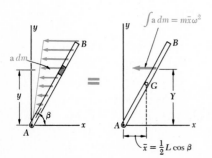

$$\int a\,dm = \int_0^L (l \cos \beta)\omega^2 \left(\frac{m}{L}\,dl\right) = \tfrac{1}{2}mL\omega^2 \cos \beta = m\bar{x}\omega^2 \qquad (4)$$

Since the resultant must be equivalent to the effective forces,

$$+\!\!\downarrow \Sigma M_A = \Sigma M_A: \qquad \int y(a\,dm) = Y\!\int a\,dm \qquad (5)$$

Substituting into (5) the expressions found in (2) and (4), we obtain

$$\tfrac{1}{3}mL^2\omega^2 \sin \beta \cos \beta = Y(\tfrac{1}{2}mL\omega^2 \cos \beta) \qquad Y = \tfrac{2}{3}L \sin \beta$$

PROBLEMS

16.135   The mass center of a loaded freight car is located 8 ft above the rails. Knowing that the gage is 4 ft $8\frac{1}{2}$ in. (distance between rails), determine the speed at which the car will overturn while rounding an unbanked curve of 2000-ft radius.

16.136   Solve Prob. 16.135 assuming that the rails are super-elevated so that the outer rail is $2\frac{1}{2}$ in. higher than the inner rail.

16.137   The center of gravity of a 1200-kg automobile is 0.5 m above the road, and the transverse distance between the wheels is 1.4 m. If the automobile takes a turn of 70-m radius on an unbanked highway, determine (*a*) the maximum safe velocity with regard to overturning, (*b*) the value of μ between the tires and pavement if the maximum safe velocity with regard to skidding is equal to that found for overturning.

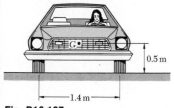

Fig. P16.137

16.138   The automobile described in Prob. 16.137 takes a turn of 70-m radius on a highway banked at 20°. Determine the maximum safe velocity with regard to overturning.

16.139   A stunt driver drives a small automobile of mass 600 kg (driver included) on the vertical wall of a cylindrical pit of radius 10 m at a speed of 50 km/h. Knowing that this is the slowest speed at which he can perform this stunt, determine (*a*) the coefficient of friction between tires and wall, (*b*) the normal component of the reaction of the wall at each wheel.

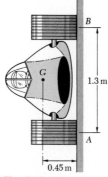

Fig. P16.139

16.140   If the coefficient of friction between the wall and the tires of the automobile of Prob. 16.139 is 0.60, determine (*a*) the lowest safe speed at which the stunt can be performed, (*b*) the normal component of the reaction of the wall at each wheel.

16.141 A stunt driver rides a motorcycle at 45 mph inside a vertical cylindrical wall. Assuming that the mass center of the driver and motorcycle travels in a circle of radius 38 ft, determine (*a*) the angle β at which the motorcycle is inclined to the horizontal, (*b*) the minimum safe value of μ.

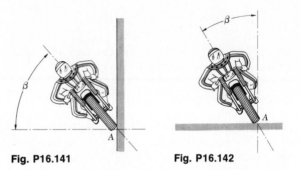

Fig. P16.141   **Fig. P16.142**

16.142 A motorcyclist drives around an unbanked curve at a speed of 40 mi/h. Assuming that the mass center of the driver and motorcycle travels in a circle of radius 160 ft, determine (*a*) the angle between the plane of the motorcycle and the vertical, (*b*) the minimum safe value of μ.

16.143 It has been found that the flywheel shown will be in perfect balance (statically and dynamically) if a mass of 350 g is attached at point *A* located 20 mm from the inside edge of the flywheel. However, when the 350-g mass is permanently attached to the flywheel, it is mistakenly attached at point *B* located 20 mm from the outside edge. Determine the resulting dynamic bearing reactions at *C* and *D* for a speed of 1200 rpm.

16.144 In Prob. 16.143, determine the dynamic bearing reactions at *C* and *D* for a speed of 1200 rpm before the 350-g mass was attached to the flywheel.

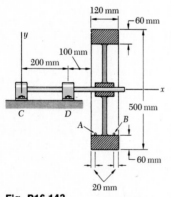

Fig. P16.143

16.145 Each element of the crankshaft shown is a homogeneous rod of weight *w* per unit length. Knowing that the crankshaft rotates with a constant angular velocity ω, determine the dynamic reactions at *A* and *B*.

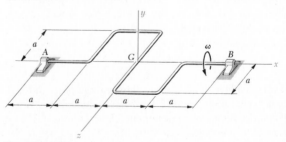

Fig. P16.145

16.146 Two L-shaped arms, each weighing 6 lb, are welded at the third points of the 3-ft shaft AB. Knowing that shaft AB rotates with a constant angular velocity ω of 300 rpm, determine the dynamic reactions at A and B.

16.147 The shaft of Prob. 16.145 is initially at rest ($\omega = 0$) and is accelerated at the rate $\alpha = 100 \text{ rad/s}^2$. Knowing that $w = 4 \text{ lb/ft}$ and $a = 3 \text{ in.}$, determine (a) the moment of the couple required to cause the acceleration, (b) the corresponding dynamic reactions at A and B.

16.148 The shaft of Prob. 16.146 is initially at rest ($\omega = 0$) when a couple $\mathbf{M}$ of moment 15 lb · ft is applied to the shaft AB. Determine (a) the angular acceleration of the shaft, (b) the corresponding dynamic reactions at A and B.

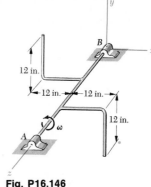

Fig. P16.146

16.149 Two uniform rods CD and DE, each of mass 2 kg, are welded to the shaft AB, which is at rest. If a couple $\mathbf{M}$ of moment 10 N · m is applied to the shaft, determine the dynamic reactions at A and B.

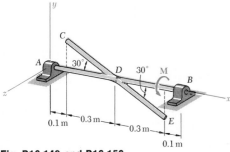

Fig. P16.149 and P16.150

16.150 Two uniform rods CD and DE, each of mass 2 kg, are welded to the shaft AB. At the instant shown the angular velocity of the shaft is 15 rad/s and the angular acceleration is 100 rad/s², both counterclockwise when viewed from the positive x axis. Determine (a) the moment of the couple $\mathbf{M}$ which must be applied to the shaft, (b) the corresponding dynamic reactions at A and B.

16.151 A door is made of a solid homogeneous panel 4 by 8 ft, weighing 75 lb. A man pushes on the knob D with a 10-lb force directed perpendicularly to the door. Determine (a) the angular acceleration of the door, (b) the horizontal components of the reactions at the hinges.

16.152 Locate the point on the face of the door of Prob. 16.151 at which a 10-lb force may be applied perpendicularly to the door without affecting the horizontal components of the reactions at the hinges.

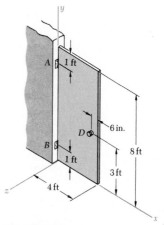

Fig. P16.151

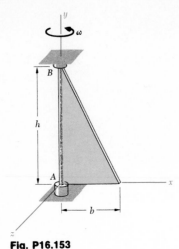

Fig. P16.153

16.153 A thin homogeneous triangular plate of mass m, base b, and height h is welded to a light axle which can rotate freely in bearings at A and B. Knowing that the plate rotates with a constant angular velocity ω, determine the dynamic reactions at A and B. (*Hint:* Consider the effective forces of elements consisting of horizontal strips.)

16.154 A thin homogeneous square plate of mass m and side a is welded to a vertical shaft AB with which it forms an angle of $45°$. Knowing that the shaft rotates with a constant angular velocity ω, determine the force-couple system representing the dynamic reaction at A. (*Hint:* Consider the effective forces of elements consisting of horizontal strips.)

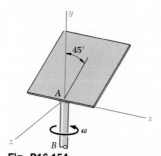

Fig. P16.154

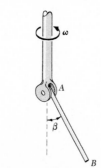

Fig. P16.155 and P16.156

16.155 A 2-ft uniform rod AB is attached at A to the pin of a clevis which rotates with a constant angular velocity ω. Determine (*a*) the angle β that the rod forms with the vertical when $\omega = 10$ rad/s, (*b*) the maximum value of ω for which the rod will remain vertical.

16.156 A uniform rod AB of length l and mass m is attached to the pin of a clevis which rotates with a constant angular velocity ω. Derive an expression (*a*) for the constant angle β that the rod forms with the vertical, (*b*) for the maximum value of ω for which the rod will remain vertical ($\beta = 0$).

16.157 A thin ring of radius b is attached by a collar at A to a vertical shaft which rotates with a constant angular velocity ω. Derive an expression (*a*) for the constant angle β that the plane of the ring forms with the vertical, (*b*) for the maximum value of ω for which the ring will remain vertical ($\beta = 0$).

∗16.158 The square plate of Prob. 16.154 is at rest ($\omega = 0$) when a couple of moment M_0 is applied to the shaft. Determine (*a*) the angular acceleration of the plate, (*b*) the force-couple system representing the dynamic reaction at A at that instant.

∗16.159 The triangular plate of Prob. 16.153 is initially at rest ($\omega = 0$) and has an angular acceleration $\boldsymbol{\alpha}$. Determine (*a*) the moment M of the couple required to cause the acceleration, (*b*) the corresponding dynamic reactions at A and B.

Fig. P16.157

REVIEW PROBLEMS

16.160 A 15-lb uniform disk is suspended from a link AB of negligible weight. If a 10-lb force is applied at B, determine the acceleration of B (*a*) if the connection at B is a frictionless pin, (*b*) if the connection at B is "frozen" and the system rotates about A as a rigid body.

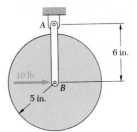

Fig. P16.160

16.161 and 16.162 A uniform plate of mass m is suspended in each of the ways shown. For each case determine the acceleration of the center of the plate immediately after the connection at B has been released.

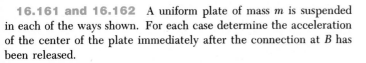

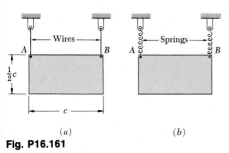
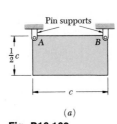

(*a*) (*b*) (*a*) (*b*)
Fig. P16.161 **Fig. P16.162**

16.163 A 100-lb uniform plate is suspended by two wires. Knowing that the plate is released from rest in the position shown, determine at that instant (*a*) the acceleration of the center of the plate, (*b*) the tension in each wire.

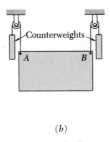

16.164 The flanged wheel shown rolls to the right with a constant velocity of 1.5 m/s. The rod AB is 1.2 m long and has a mass of 5 kg. Knowing that point A slides without friction on the horizontal surface, determine the reaction at A (*a*) when $\beta = 0$, (*b*) when $\beta = 180°$.

Fig. P16.163

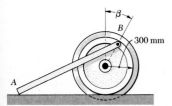

Fig. P16.164

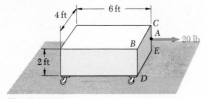

Fig. P16.165

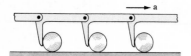

Fig. P16.167

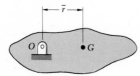

Fig. P16.168

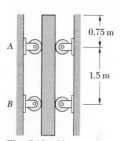

Fig. P16.169

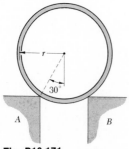

Fig. P16.171

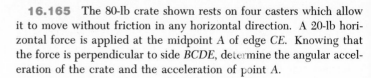

16.165 The 80-lb crate shown rests on four casters which allow it to move without friction in any horizontal direction. A 20-lb horizontal force is applied at the midpoint A of edge CE. Knowing that the force is perpendicular to side $BCDE$, determine the angular acceleration of the crate and the acceleration of point A.

16.166 Solve Prob. 16.165, assuming that the 20-lb horizontal force applied at A is parallel to the edge BC.

16.167 Identical cylinders of mass m and radius r are pushed by a series of moving arms. Assuming the coefficient of friction between all surfaces to be $\mu < 1$, and denoting by a the magnitude of the acceleration of the arms, derive an expression for (a) the maximum allowable value of a if each cylinder is to roll without sliding, (b) the minimum allowable value of a if each cylinder is to move to the right without rotating.

16.168 A rigid slab of weight W and centroidal radius of gyration $\bar{k}$ may rotate freely about a shaft O located at a distance $\bar{r}$ from its mass center G. If the slab is released from rest in the position shown, determine (a) the distance $\bar{r}$ for which the angular acceleration of the slab is maximum, (b) the corresponding vertical component of the reaction at O.

16.169 A 6-kg bar is held between four disks as shown. Each disk has a mass of 3 kg and a diameter of 200 mm. The disks may rotate freely, and the normal reaction exerted by each disk on the bar is sufficient to prevent slipping. If the bar is released from rest, determine (a) its acceleration immediately after release, (b) its velocity after it has dropped 0.75 m

16.170 In Prob. 16.169, determine (a) the acceleration of the bar after it has dropped 1.25 m, (b) the velocity of the bar after it has dropped 2.25 m.

16.171 A section of pipe, of mass 50 kg and radius 250 mm, rests on two corners as shown. Assuming that μ between the corners and the pipe is sufficient to prevent sliding, determine (a) the angular acceleration of the pipe just after corner B is removed, (b) the corresponding magnitude of the reaction at A.

Chapter
17

Kinetics of Rigid Bodies: Work and Energy

17.1. Principle of Work and Energy for a Rigid Body. In this chapter, the principle of work and energy will be used to analyze the motion of rigid bodies and of systems of rigid bodies. As was pointed out in Chap. 13, the method of work and energy is particularly well adapted to the solution of problems involving velocities and displacements. Its main advantage resides in the fact that the work of forces and the kinetic energy of particles are scalar quantities.

Consider a rigid body of mass m, and let P be a particle of the body, of mass Δm. The kinetic energy of this particle is $\Delta T = \frac{1}{2}(\Delta m)v^2$, where v is the speed of the particle. Consider now a displacement of the rigid body during which the particle P moves from a position P_1 to a position P_2, and denote by $\Delta U_{1\to 2}$ the work of all the forces acting on P during the displacement. We recall from Sec. 13.3 that the principle of work and energy for a particle states that

$$\Delta T_1 + \Delta U_{1\to 2} = \Delta T_2$$

where ΔT_1 = kinetic energy of the particle at P_1
ΔT_2 = kinetic energy of the particle at P_2
Similar relations may be written for all the particles forming the body. Adding the kinetic energy of all the particles, and

the work of all the forces involved, we write

$$T_1 + U_{1\to2} = T_2 \tag{17.1}$$

where T_1, T_2 = initial and final values of total kinetic energy of
the particles forming the rigid body

$U_{1\to2}$ = work of all forces acting on the various particles
of the body

As indicated in Chap. 13, work and energy are expressed in joules
(J) if SI units are used, and in ft · lb or in · lb if U.S. customary
units are used.

The total kinetic energy

$$T = \Sigma\tfrac{1}{2}(\Delta m)v^2 \tag{17.2}$$

is obtained by adding positive scalar quantities and is itself a
positive scalar quantity. We shall see later how T may be determined for various types of motion of a rigid body.

The expression $U_{1\to2}$ in (17.1) represents the work of all the
forces acting on the various particles of the body, whether these
forces are internal or external. However, as we shall see presently, the total work of the internal forces holding together the
particles of a rigid body is zero. Consider two particles A and
B of a rigid body and the two equal and opposite forces $\mathbf{F}$ and
$\mathbf{F}'$ they exert on each other (Fig. 17.1). While, in general, the

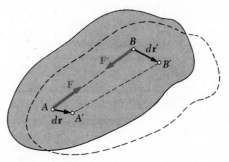

Fig. 17.1

displacements $d\mathbf{r}$ and $d\mathbf{r}'$ of the two particles are different, the
components of these displacements along AB must be equal;
otherwise, the particles would not remain at the same distance
from each other, and the body would not be rigid. Therefore,
the work of $\mathbf{F}$ is equal in magnitude and opposite in sign to
the work of $\mathbf{F}'$, and their sum is zero. Thus, the total work of
the internal forces acting on the particles of a rigid body is zero,
and *the expression $U_{1\to2}$ in Eq. (17.1) reduces to the work of the
external forces* acting on the body during the displacement considered.

17.2. Work of Forces Acting on a Rigid Body. We saw in Sec. 13.2 that the work of a force $\mathbf{F}$ during a displacement of its point of application from A_1 $(s = s_1)$ to A_2 $(s = s_2)$ is

$$U_{1 \to 2} = \int_{s_1}^{s_2} (F \cos \alpha)\, ds \qquad (17.3)$$

where F is the magnitude of the force $\mathbf{F}$, and α the angle that $\mathbf{F}$ forms at any given instant with the direction of motion of its point of application.

In computing the work of the external forces acting on a rigid body, it is often convenient to determine the work of a couple without considering separately the work of each of the two forces forming the couple. Consider the two forces $\mathbf{F}$ and $\mathbf{F}'$ forming a couple of moment $M = Fr$ and acting on a rigid body (Fig. 17.2). Any small displacement of the rigid body bringing A and B, respectively, into A' and B'' may be divided into two parts, one in which points A and B undergo equal displacements $d\mathbf{r}_1$, the other in which A' remains fixed while B' moves into B'' through a displacement $d\mathbf{r}_2$ of magnitude $ds_2 = r\, d\theta$. In the

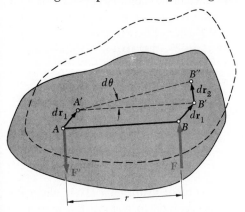

Fig. 17.2

first part of the motion, the work of $\mathbf{F}$ is equal in magnitude and opposite in sign to the work of $\mathbf{F}'$, and their sum is zero. In the second part of the motion, only force $\mathbf{F}$ works, and its work is $dU = F\, ds_2 = Fr\, d\theta = M\, d\theta$. Thus, the work of a couple of moment M acting on a rigid body is

$$dU = M\, d\theta \qquad (17.4)$$

where $d\theta$ is the small angle expressed in radians through which the body rotates. We again note that work should be expressed in units obtained by multiplying units of force by units of length. The work of the couple during a finite rotation of the rigid body is obtained by integrating both members of (17.4) from the initial value θ_1 of the angle θ to its final value θ_2. We write

$$U_{1 \to 2} = \int_{\theta_1}^{\theta_2} M \, d\theta \qquad (17.5)$$

When the moment M of the couple is constant, formula (17.5) reduces to

$$U_{1 \to 2} = M(\theta_2 - \theta_1) \qquad (17.6)$$

It was pointed out in Sec. 13.2 that a number of forces encountered in problems of kinetics *do no work.* They are forces applied to fixed points or acting in a direction perpendicular to the displacement of their point of application. Among the forces which do no work the following have been listed: the reaction at a frictionless pin when the body supported rotates about the pin, the reaction at a frictionless surface when the body in contact moves along the surface, the weight of a body when its center of gravity moves horizontally. We should also indicate now that, *when a rigid body rolls without sliding on a fixed surface, the friction force* **F** *at the point of contact C does no work.* The velocity v_C of the point of contact C is zero, and the work of the friction force **F** during a small displacement of the rigid body is $dU = F \, ds_C = F(v_C \, dt) = 0$.

17.3. Kinetic Energy in Translation. Consider a rigid body in translation. Since, at a given instant, all the particles of the body have the same velocity as the mass center, we may substitute $v = \bar{v}$ in (17.2). We thus obtain the following expression for the kinetic energy of a rigid body in translation:

$$T = \Sigma \tfrac{1}{2}(\Delta m)\bar{v}^2 = \tfrac{1}{2}(\Sigma \, \Delta m)\bar{v}^2$$

$$T = \tfrac{1}{2}m\bar{v}^2 \qquad (17.7)$$

This expression should be entered in formula (17.1) when the method of work and energy is used to analyze the motion of a rigid body in translation (either rectilinear or curvilinear).

We note that the expression obtained in (17.7) represents the kinetic energy of a particle of mass m moving with the velocity $\bar{\mathbf{v}}$ of the mass center. The approach used in Chap. 13, where bodies as large as automobiles or trains were considered as single particles, is therefore correct as long as the bodies considered are in translation.

17.4. Kinetic Energy in Rotation. Consider a rigid body rotating about a fixed axis intersecting the reference plane at O (Fig. 17.3). If the angular velocity of the body at a given instant is ω (rad/s), the speed of a particle P located at a distance r from the axis of rotation is $v = r\omega$ at that instant. Substituting

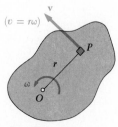

Fig. 17.3

for v into (17.2), we obtain the following expression for the kinetic energy of the body:

$$T = \Sigma \tfrac{1}{2}(\Delta m)(r\omega)^2 = \tfrac{1}{2}\omega^2 \Sigma r^2\, \Delta m$$

Observing that the sum $\Sigma r^2\, \Delta m$ represents the moment of inertia I_0 of the rigid body about the axis of rotation, we write

$$T = \tfrac{1}{2}I_0\omega^2 \qquad\qquad (17.8)$$

This expression should be entered in formula (17.1) when the method of work and energy is used to analyze the motion of a rigid body rotating about a fixed axis.

We note that formula (17.8) is valid with respect to any fixed axis of rotation, whether or not this axis passes through the mass center of the body. Moreover, the result obtained is not limited to the rotation of plane slabs or to the rotation of bodies which are symmetrical with respect to the reference plane. Formula (17.8) may be applied to the rotation of any rigid body about a fixed axis, regardless of the shape of the body or of the location of the axis of rotation.

17.5. Systems of Rigid Bodies. When a problem involves several rigid bodies, each rigid body may be considered separately and the principle of work and energy may be applied to each body. Adding the kinetic energies of all the particles and considering the work of all the forces involved, we may also write the equation of work and energy for the entire system. We have

$$T_1 + U_{1 \to 2} = T_2 \qquad\qquad (17.9)$$

where T represents the arithmetic sum of the kinetic energies of the rigid bodies forming the system (all terms are positive) and $U_{1 \to 2}$ the work of all the forces acting on the various bodies, whether these forces are *internal* or *external* from the point of view of the system as a whole.

The method of work and energy is particularly useful in solving problems involving pin-connected members, or blocks and pulleys connected by inextensible cords, or meshed gears. In all these cases, the internal forces occur by pairs of equal and opposite forces, and the points of application of the forces in each pair *move through equal distances* during a small displacement of the system. As a result, the work of the internal forces is zero, and $U_{1 \to 2}$ reduces to the work of the *forces external to the system.*

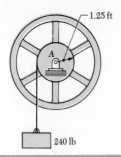

1.25 ft

A 240-lb block is suspended from an inextensible cable which is wrapped around a drum of 1.25-ft radius rigidly attached to a flywheel. The drum and flywheel have a combined centroidal moment of inertia $\bar{I} = 10.5$ lb $\cdot$ ft $\cdot$ s^2. At the instant shown, the velocity of the block is 6 ft/s directed downward. Knowing that the bearing at A is poorly lubricated and that the bearing friction is equivalent to a couple $\mathbf{M}$ of moment 60 lb $\cdot$ ft, determine the velocity of the block after it has moved 4 ft downward.

240 lb

Solution. We consider the system formed by the flywheel and the block. Since the cable is inextensible, the work done by the internal forces exerted by the cable cancels. The initial and final positions of the system and the external forces acting on the system are as shown.

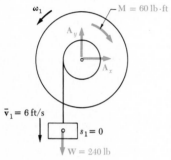

ω_1

A_y

A_x

M = 60 lb·ft

$\bar{v}_1 = 6$ ft/s

$s_1 = 0$

W = 240 lb

Kinetic Energy. *Position 1.* We have

$$\bar{v}_1 = 6 \text{ ft/s} \qquad \omega_1 = \frac{\bar{v}_1}{r} = \frac{6 \text{ ft/s}}{1.25 \text{ ft}} = 4.80 \text{ rad/s}$$

$$T_1 = \tfrac{1}{2}m\bar{v}_1^2 + \tfrac{1}{2}\bar{I}\omega_1^2$$

$$= \frac{1}{2}\frac{240 \text{ lb}}{32.2 \text{ ft/s}^2}(6 \text{ ft/s})^2 + \tfrac{1}{2}(10.5 \text{ lb} \cdot \text{ft} \cdot \text{s}^2)(4.80 \text{ rad/s})^2$$

$$= 255 \text{ ft} \cdot \text{lb}$$

Position 2. Noting that $\omega_2 = \bar{v}_2/1.25$, we write

$$T_2 = \tfrac{1}{2}m\bar{v}_2^2 + \tfrac{1}{2}\bar{I}\omega_2^2$$

$$= \frac{1}{2}\frac{240}{32.2}(\bar{v}_2)^2 + (\tfrac{1}{2})(10.5)\left(\frac{\bar{v}_2}{1.25}\right)^2 = 7.09\bar{v}_2^2$$

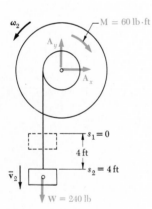

ω_2

A_y

A_x

M = 60 lb·ft

$s_1 = 0$

4 ft

$s_2 = 4$ ft

$\bar{v}_2$

W = 240 lb

Work. During the motion, only the weight $\mathbf{W}$ of the block and the friction couple $\mathbf{M}$ do work. Noting that $\mathbf{W}$ does positive work and that the friction couple $\mathbf{M}$ does negative work, we write

$$s_1 = 0 \qquad s_2 = 4 \text{ ft}$$

$$\theta_1 = 0 \qquad \theta_2 = \frac{s_2}{r} = \frac{4 \text{ ft}}{1.25 \text{ ft}} = 3.20 \text{ rad}$$

$$U_{1\rightarrow 2} = W(s_2 - s_1) - M(\theta_2 - \theta_1)$$

$$= (240 \text{ lb})(4 \text{ ft}) - (60 \text{ lb} \cdot \text{ft})(3.20 \text{ rad})$$

$$= 768 \text{ ft} \cdot \text{lb}$$

Principle of Work and Energy

$$T_1 + U_{1\rightarrow 2} = T_2$$

$$255 \text{ ft} \cdot \text{lb} + 768 \text{ ft} \cdot \text{lb} = 7.09\bar{v}_2^2$$

$$\bar{v}_2 = 12.01 \text{ ft/s} \qquad \bar{v}_2 = 12.01 \text{ ft/s} \downarrow \quad \blacktriangleleft$$

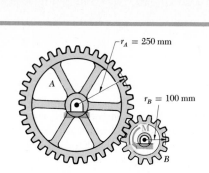

- $r_A = 250$ mm
- $r_B = 100$ mm

Gear A has a mass of 10 kg and a radius of gyration of 200 mm, while gear B has a mass of 3 kg and a radius of gyration of 80 mm. The system is at rest when a couple $\mathbf{M}$ of moment 6 N·m is applied to gear B. Neglecting friction, determine (a) the number of revolutions executed by gear B before its angular velocity reaches 600 rpm, (b) the tangential force which gear B exerts on gear A.

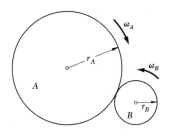

Motion of Entire System. Noting that the peripheral speeds of the gears are equal, we write

$$r_A \omega_A = r_B \omega_B \qquad \omega_A = \omega_B \frac{r_B}{r_A} = \omega_B \frac{100 \text{ mm}}{250 \text{ mm}} = 0.40 \omega_B$$

For $\omega_B = 600$ rpm, we have

$$\omega_B = 62.8 \text{ rad/s} \qquad \omega_A = 0.40 \omega_B = 25.1 \text{ rad/s}$$
$$\bar{I}_A = m_A \bar{k}_A^2 = (10 \text{ kg})(0.200 \text{ m})^2 = 0.400 \text{ kg·m}^2$$
$$\bar{I}_B = m_B \bar{k}_B^2 = (3 \text{ kg})(0.080 \text{ m})^2 = 0.0192 \text{ kg·m}^2$$

Kinetic Energy. Since the system is initially at rest, $T_1 = 0$. Adding the kinetic energies of the two gears when $\omega_B = 600$ rpm, we obtain

$$T_2 = \tfrac{1}{2}\bar{I}_A \omega_A^2 + \tfrac{1}{2}\bar{I}_B \omega_B^2$$
$$= \tfrac{1}{2}(0.400 \text{ kg·m}^2)(25.1 \text{ rad/s})^2 + \tfrac{1}{2}(0.0192 \text{ kg·m}^2)(62.8 \text{ rad/s})^2$$
$$= 163.9 \text{ J}$$

Work. Denoting by θ_B the angular displacement of gear B, we have

$$U_{1 \to 2} = M\theta_B = (6 \text{ N·m})(\theta_B \text{ rad}) = (6\,\theta_B) \text{ J}$$

Principle of Work and Energy

$$T_1 + U_{1 \to 2} = T_2: \qquad 0 + (6\,\theta_B) \text{ J} = 163.9 \text{ J}$$
$$\theta_B = 27.32 \text{ rad} \qquad \theta_B = 4.35 \text{ rev} \quad \blacktriangleleft$$

Motion of Gear A. *Kinetic Energy.* Initially, gear A is at rest, $T_1 = 0$. When $\omega_B = 600$ rpm, the kinetic energy of gear A is

$$T_2 = \tfrac{1}{2}\bar{I}_A \omega_A^2 = \tfrac{1}{2}(0.400 \text{ kg·m}^2)(25.1 \text{ rad/s})^2 = 126.0 \text{ J}$$

Work. The forces acting on gear A are as shown. The tangential force $\mathbf{F}$ does work equal to the product of its magnitude and of the length $\theta_A r_A$ of the arc described by the point of contact. Since $\theta_A r_A = \theta_B r_B$, we have

$$U_{1 \to 2} = F(\theta_B r_B) = F(27.3 \text{ rad})(0.100 \text{ m}) = F(2.73 \text{ m})$$

Principle of Work and Energy

$$T_1 + U_{1 \to 2} = T_2: \qquad 0 + F(2.73 \text{ m}) = 126.0 \text{ J}$$
$$F = +46.1 \text{ N} \qquad \mathbf{F} = 46.1 \text{ N} \nearrow \quad \blacktriangleleft$$

PROBLEMS

17.1 The rotor of a generator has an angular velocity of 3600 rpm when the generator is taken off line. The 150-kg rotor, which has a centroidal radius of gyration of 250 mm, then coasts to rest. Knowing that the kinetic friction of the rotor produces a couple of moment 2 N · m, determine the number of revolutions that the rotor executes before coming to rest.

17.2 A large flywheel of mass 1800 kg has a radius of gyration of 0.75 m. It is observed that 2500 revolutions are required for the flywheel to coast from an angular velocity of 450 rpm to rest. Determine the average moment of the couple due to kinetic friction in the bearings.

17.3 Three disks of the same thickness and same material are attached to a shaft as shown. Disks A and B each have a radius r; disk C has a radius nr. A couple **M** of constant moment is applied when the system is at rest and is removed after the system has executed one revolution. Determine the radius of disk C which results in the largest final speed of a point on the rim of disk C.

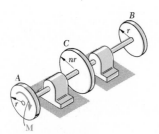

Fig. P17.3 and P17.4

17.4 Three disks of the same thickness and same material are attached to a shaft as shown. Disks A and B weigh 15 lb each and are of radius $r = 10$ in. A couple **M** of moment 30 lb · ft is applied to disk A when the system is at rest. Determine the radius nr of disk C if the angular velocity of the system is to be 600 rpm after 10 revolutions.

17.5 The flywheel of a punching machine weighs 900 lb and has a radius of gyration of 30 in. Each punching operation requires 2000 ft · lb of work. (a) Knowing that the speed of the flywheel is 200 rpm just before a punching, determine the speed immediately after the punching. (b) If a constant 25-lb · ft couple is applied to the shaft of the flywheel, determine the number of revolutions executed before the speed is again 200 rpm.

17.6 The flywheel of a small punch rotates at 300 rpm. It is known that 750 ft · lb of work must be done each time a hole is punched. It is desired that the speed of the flywheel after one punching be not less than 90 percent of the original speed of 300 rpm. (a) Determine the required moment of inertia of the flywheel. (b) If a constant 30-lb · ft couple is applied to the shaft of the flywheel, determine the number of revolutions which must occur between each punching, knowing that the initial velocity is to be 300 rpm at the start of each punching.

17.7 Two cylinders are attached by cords to an 18-kg double pulley which has a radius of gyration of 300 mm. When the system is at rest and in equilibrium, a 3-kg collar is added to the 12-kg cylinder. Neglecting friction, determine the velocity of each cylinder after the pulley has completed one revolution.

17.8 Solve Prob. 17.7, assuming that the 3-kg collar is added to the 6-kg cylinder.

17.9 Using the principle of work and energy, solve Prob. 16.32*b*.

17.10 Using the principle of work and energy, solve Prob. 16.30*c*.

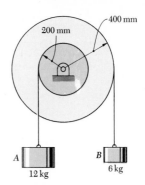

Fig. P17.7

17.11 A disk of constant thickness and initially at rest is placed in contact with the belt, which moves with a constant velocity **v**. Denoting by μ the coefficient of friction between the disk and the belt, derive an expression for the number of revolutions executed by the disk before it reaches a constant angular velocity.

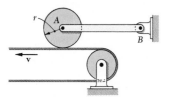

Fig. P17.11 and P17.12

17.12 Disk *A*, of weight 8 lb and radius $r = 5$ in., is at rest when it is placed in contact with the belt, which moves with a constant speed $v = 60$ ft/s. Knowing that $\mu = 0.15$ between the disk and the belt, determine the number of revolutions executed by the disk before it reaches a constant angular velocity.

17.13 Shafts *AB* and *CD* are connected by a friction disk and wheel as shown. The mass moment of inertia of the friction disk and its shaft *AB* is 0.360 kg · m², and the mass moment of inertia of the wheel and its shaft *CD* is 0.216 kg · m². The system is initially at rest when a couple **M** of constant moment 15 N · m is applied to shaft *AB* for 20 revolutions. Assuming no slipping occurs, determine the resulting angular velocity of each shaft when $x = 37.5$ mm.

17.14 The mechanism of Prob. 17.13 is set so that $x = 75$ mm; the initial angular velocity of shaft *CD* is 500 rpm. It is desired to increase its velocity to 600 rpm in 3 revolutions of shaft *CD*. Determine the moment of the couple **M** which must be applied to shaft *AB*.

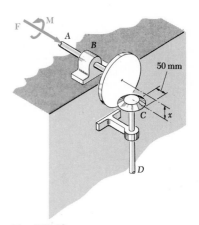

Fig. P17.13

17.15 The 15 in.-radius brake drum is attached to a larger fly-wheel which is not shown. The total mass moment of inertia of the flywheel and drum is 15 lb · ft · s². Knowing that the initial angular velocity is 150 rpm clockwise, determine the force which must be exerted by the hydraulic cylinder if the system is to come to rest in 10 revolutions.

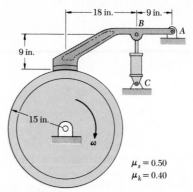

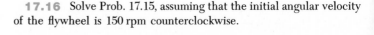

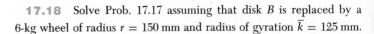

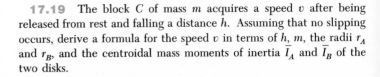

Fig. P17.15

17.16 Solve Prob. 17.15, assuming that the initial angular velocity of the flywheel is 150 rpm counterclockwise.

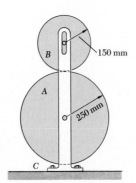

17.17 Disk A is attached to a motor and is made to rotate with a constant angular velocity of 300 rpm clockwise. Disk B has a mass of 5 kg and is at rest when it is placed in contact with disk A. Knowing that $\mu = 0.25$ and neglecting bearing friction, determine the number of revolutions executed by disk B before it reaches a constant angular velocity.

Fig. P17.17

17.18 Solve Prob. 17.17 assuming that disk B is replaced by a 6-kg wheel of radius $r = 150$ mm and radius of gyration $\bar{k} = 125$ mm.

17.19 The block C of mass m acquires a speed v after being released from rest and falling a distance h. Assuming that no slipping occurs, derive a formula for the speed v in terms of h, m, the radii r_A and r_B, and the centroidal mass moments of inertia $\bar{I}_A$ and $\bar{I}_B$ of the two disks.

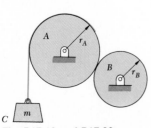

17.20 Disk A weighs 8 lb, and its radius is $r_A = 10$ in.; disk B weighs 3 lb and its radius is $r_B = 6$ in. Knowing that the system is released from rest, determine the velocity of the 2-lb block C after it has fallen 5 ft.

Fig. P17.19 and P17.20

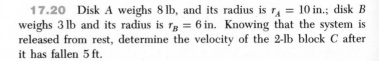

17.21 Each of the gears A and B has a mass of 2 kg and a radius of gyration of 70 mm, while gear C has a mass of 10 kg and a radius of gyration of 175 mm. A couple $\mathbf{M}$ of constant moment 12 N · m is applied to gear C. Determine (a) the number of revolutions required for the angular velocity of gear C to increase from 100 to 450 rpm, (b) the corresponding tangential force acting on gear A.

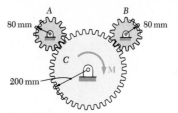

17.22 Solve Prob. 17.21 assuming that the 12-N · m couple is applied to gear B.

Fig. P17.21

17.23 Solve Sample Prob. 17.2 assuming that the 6-N · m couple is applied to gear A.

17.6. Kinetic Energy in Plane Motion.

Consider a rigid body in plane motion. We saw in Sec. 15.7 that at any given instant the velocities of the various particles of the body are the same as if the body were rotating about an axis perpendicular to the reference plane, called the *instantaneous axis of rotation*. We may therefore use the results obtained in Sec. 17.4 to express at any instant the kinetic energy of a rigid body in plane motion. Recalling that the instantaneous axis of rotation intersects the reference plane at a point C called the *instantaneous center of rotation* (Fig. 17.4), we denote by I_C the moment of inertia of the body about the instantaneous axis. If the body has an angular velocity ω at the instant considered, its kinetic energy is therefore

$$T = \tfrac{1}{2}I_C\omega^2 \tag{17.10}$$

But we know from the parallel-axis theorem (Sec. 9.11) that $I_C = m\bar{r}^2 + \bar{I}$, where $\bar{I}$ is the moment of inertia of the body about a centroidal axis perpendicular to the reference plane and $\bar{r}$ the distance from the instantaneous center C to the mass center G. Substituting for I_C into (17.10), we write

$$T = \tfrac{1}{2}(m\bar{r}^2 + \bar{I})\omega^2 = \tfrac{1}{2}m(\bar{r}\omega)^2 + \tfrac{1}{2}\bar{I}\omega^2$$

or, since $\bar{r}\omega$ represents the magnitude $\bar{v}$ of the velocity of the mass center,

$$T = \tfrac{1}{2}m\bar{v}^2 + \tfrac{1}{2}\bar{I}\omega^2 \tag{17.11}$$

Either of the expressions (17.10) and (17.11) may be entered in formula (17.1) when the method of work and energy is used to analyze the plane motion of a rigid body.

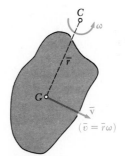

Fig. 17.4

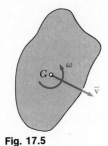

Fig. 17.5

We note that the expression given for the kinetic energy in (17.11) is separated into two distinct parts: (1) the kinetic energy $\frac{1}{2}m\bar{v}^2$ corresponding to the motion of the mass center G, and (2) the kinetic energy $\frac{1}{2}\bar{I}\omega^2$ corresponding to the rotation of the body about G (Fig. 17.5). This formula thus conforms to the unified approach we have used so far, and we shall apply it in preference to (17.10) in the Sample Problems of this chapter.

We also note that the results obtained in this section are not limited to the motion of plane slabs or to the motion of bodies which are symmetrical with respect to the reference plane. They may be applied to the study of the plane motion of any rigid body, regardless of its shape.

17.7. Conservation of Energy. We saw in Sec. 13.6 that the work of conservative forces, such as the weight of a body or the force exerted by a spring, may be expressed as a change in potential energy. When a rigid body, or a system of rigid bodies, moves under the action of conservative forces, the principle of work and energy stated in Sec. 17.1 may be expressed in a modified form. Substituting for $U_{1 \rightarrow 2}$ from (13.17) into (17.1), we write

$$T_1 + V_1 = T_2 + V_2 \tag{17.12}$$

Formula (17.12) indicates that, when a rigid body, or a system of rigid bodies, moves under the action of conservative forces, *the sum of the kinetic energy and of the potential energy of the system remains constant.* It should be noted that, in the case of the plane motion of a rigid body, the kinetic energy of the body should include both the *translational* term $\frac{1}{2}m\bar{v}^2$ and the *rotational* term $\frac{1}{2}\bar{I}\omega^2$.

As an example of application of the principle of conservation of energy, we shall consider a slender rod AB, of length l and mass m, whose extremities are connected to blocks of negligible mass sliding along horizontal and vertical tracks. We assume that the rod is released with no initial velocity from a horizontal position (Fig. 17.6a), and we wish to determine its angular velocity after it has rotated through an angle θ (Fig. 17.6b).

Since the initial velocity is zero, we have $T_1 = 0$. Measuring the potential energy from the level of the horizontal track, we write $V_1 = 0$. After the rod has rotated through θ, the center of gravity G of the rod is at a distance $\frac{1}{2}l \sin \theta$ below the reference level and we have

$$V_2 = -\tfrac{1}{2}Wl \sin \theta = -\tfrac{1}{2}mgl \sin \theta$$

Observing that, in this position, the instantaneous center of the rod is located at C, and that $CG = \frac{1}{2}l$, we write $\bar{v}_2 = \frac{1}{2}l\omega$ and

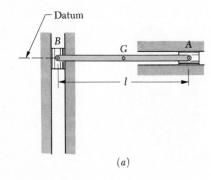

(a)

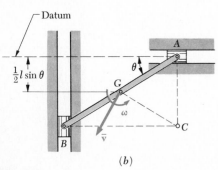

(b)

Fig. 17.6

obtain

$$T_2 = \tfrac{1}{2}m\bar{v}_2^2 + \tfrac{1}{2}\bar{I}\omega_2^2 = \tfrac{1}{2}m(\tfrac{1}{2}l\omega)^2 + \tfrac{1}{2}(\tfrac{1}{12}ml^2)\omega^2$$
$$= \frac{1}{2}\frac{ml^2}{3}\omega^2$$

Applying the principle of conservation of energy, we write

$$T_1 + V_1 = T_2 + V_2$$
$$0 = \frac{1}{2}\frac{ml^2}{3}\omega^2 - \tfrac{1}{2}mgl\sin\theta$$
$$\omega = \left(\frac{3g}{l}\sin\theta\right)^{1/2}$$

We recall that the advantages of the method of work and energy, as well as its shortcomings, were indicated in Sec. 13.4. In this connection, we wish to mention that the method of work and energy must be supplemented by the method of dynamic equilibrium when reactions at fixed axles, at rollers, or at sliding blocks are to be determined. For example, in order to compute the reactions at the extremities A and B of the rod of Fig. 17.6b, a diagram should be drawn to express that the system of the external forces applied to the rod is equivalent to the vector $m\bar{\mathbf{a}}$ and the couple $\bar{I}\boldsymbol{\alpha}$. The angular velocity ω of the rod, however, is determined by the method of work and energy before the equations of motion are solved for the reactions. The complete analysis of the motion of the rod and of the forces exerted on the rod requires, therefore, the combined use of the method of work and energy and of the principle of equivalence of the external and effective forces.

17.8. Power. *Power* was defined in Sec. 13.8 as the time rate at which work is done. In the case of a body acted upon by a force $\mathbf{F}$, and moving with a velocity $\mathbf{v}$, the power was expressed as follows [Eq. (13.20)]:

$$\text{Power} = \frac{dU}{dt} = \frac{(F\cos\alpha)\,ds}{dt} = (F\cos\alpha)v$$

where α is the angle between $\mathbf{F}$ and $\mathbf{v}$. In the case of a rigid body acted upon by a couple $\mathbf{M}$ and rotating at an angular velocity $\boldsymbol{\omega}$, the power may be expressed in a similar way in terms of M and ω,

$$\text{Power} = \frac{dU}{dt} = \frac{M\,d\theta}{dt} = M\omega \qquad (17.13)$$

The various units used to measure power, such as the watt and the horsepower, were defined in Sec. 13.8.

A sphere, a cylinder, and a hoop, each having the same mass and the same radius, are released from rest on an incline. Determine the velocity of each body after it has rolled through a distance corresponding to a change in elevation h.

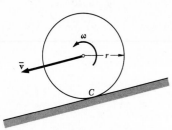

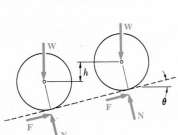

Solution. We shall first solve the problem in general terms and then find particular results for each body. We denote the mass by m, the centroidal moment of inertia by $\bar{I}$, the weight by W, and the radius by r.

Since each body rolls, the instantaneous center of rotation is located at C and we write

$$\omega = \frac{\bar{v}}{r}$$

Kinetic Energy

$$T_1 = 0$$
$$T_2 = \tfrac{1}{2}m\bar{v}^2 + \tfrac{1}{2}\bar{I}\omega^2$$
$$= \tfrac{1}{2}m\bar{v}^2 + \tfrac{1}{2}\bar{I}\left(\frac{\bar{v}}{r}\right)^2 = \tfrac{1}{2}\left(m + \frac{\bar{I}}{r^2}\right)\bar{v}^2$$

Work. Since the friction force $\mathbf{F}$ in rolling motion does no work,

$$U_{1\to2} = Wh$$

Principle of Work and Energy

$$T_1 + U_{1\to2} = T_2$$
$$0 + Wh = \tfrac{1}{2}\left(m + \frac{\bar{I}}{r^2}\right)\bar{v}^2 \qquad \bar{v}^2 = \frac{2Wh}{m + \bar{I}/r^2}$$

Noting that $W = mg$, we rearrange the result and obtain

$$\bar{v}^2 = \frac{2gh}{1 + \bar{I}/mr^2}$$

Velocities of Sphere, Cylinder, and Hoop. Introducing successively the particular expressions for $\bar{I}$, we obtain

Sphere:	$\bar{I} = \tfrac{2}{5}mr^2$	$\bar{v} = 0.845\sqrt{2gh}$ ◄
Cylinder:	$\bar{I} = \tfrac{1}{2}mr^2$	$\bar{v} = 0.816\sqrt{2gh}$ ◄
Hoop:	$\bar{I} = mr^2$	$\bar{v} = 0.707\sqrt{2gh}$ ◄

Remark. We may compare the results with the velocity attained by a frictionless block sliding through the same distance. The solution is identical to the above solution except that $\omega = 0$; we find $\bar{v} = \sqrt{2gh}$.

Comparing the results, we note that the velocity of the body is independent of both its mass and radius. However, the velocity does depend upon the quotient $\bar{I}/mr^2 = \bar{k}^2/r^2$, which measures the ratio of the rotational kinetic energy to the translational kinetic energy. Thus the hoop, which has the largest $\bar{k}$ for a given radius r, attains the smallest velocity, while the sliding block, which does not rotate, attains the largest velocity.

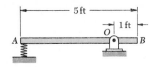

A 30-lb slender rod AB is 5 ft long and is pivoted about a point O which is 1 ft from end B. The other end is pressed against a spring of constant $k = 1800$ lb/in. until the spring is compressed 1 in. The rod is then in a horizontal position. If the rod is released from this position, determine its angular velocity and the reaction at the pivot O as the rod passes through a vertical position.

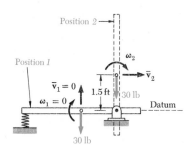

Position 1. Potential Energy. Since the spring is compressed 1 in., we have $x_1 = 1$ in.

$$V_e = \tfrac{1}{2}kx_1^2 = \tfrac{1}{2}(1800 \text{ lb/in.})(1 \text{ in.})^2 = 900 \text{ in} \cdot \text{lb}$$

Choosing the datum as shown, we have $V_g = 0$; therefore,

$$V_1 = V_e + V_g = 900 \text{ in} \cdot \text{lb} = 75 \text{ ft} \cdot \text{lb}$$

Kinetic Energy. Since the velocity in position 1 is zero, we have $T_1 = 0$.

Position 2. Potential Energy. The elongation of the spring is zero, and we have $V_e = 0$. Since the center of gravity of the rod is now 1.5 ft above the datum,

$$V_g = (30 \text{ lb})(+1.5 \text{ ft}) = 45 \text{ ft} \cdot \text{lb}$$
$$V_2 = V_e + V_g = 45 \text{ ft} \cdot \text{lb}$$

Kinetic Energy. Denoting by ω_2 the angular velocity of the rod in position 2, we note that the rod rotates about O and write $\bar{v}_2 = \bar{r}\omega_2 = 1.5\omega_2$.

$$\bar{I} = \tfrac{1}{12}ml^2 = \frac{1}{12}\frac{30 \text{ lb}}{32.2 \text{ ft/s}^2}(5 \text{ ft})^2 = 1.941 \text{ lb} \cdot \text{ft} \cdot \text{s}^2$$

$$T_2 = \tfrac{1}{2}m\bar{v}_2^2 + \tfrac{1}{2}\bar{I}\omega_2^2 = \frac{1}{2}\frac{30}{32.2}(1.5\omega_2)^2 + \tfrac{1}{2}(1.941)\omega_2^2 = 2.019\omega_2^2$$

Conservation of Energy

$$T_1 + V_1 = T_2 + V_2: \qquad 0 + 75 \text{ ft} \cdot \text{lb} = 2.019\omega_2^2 + 45 \text{ ft} \cdot \text{lb}$$
$$\omega_2 = 3.86 \text{ rad/s} \; \downarrow \quad \blacktriangleleft$$

Reaction in Position 2. Since $\omega_2 = 3.86$ rad/s, the components of the acceleration of G as the rod passes through position 2 are

$$\bar{a}_n = \bar{r}\omega_2^2 = (1.5 \text{ ft})(3.86 \text{ rad/s})^2 = 22.3 \text{ ft/s}^2 \qquad \bar{a}_n = 22.3 \text{ ft/s}^2 \downarrow$$
$$\bar{a}_t = \bar{r}\alpha \qquad \bar{a}_t = \bar{r}\alpha \rightarrow$$

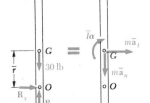

We express that the system of external forces is equivalent to the system of effective forces represented by the vector of components $m\bar{a}_t$ and $m\bar{a}_n$ attached at G and the couple $\bar{I}\alpha$.

$+\!\downarrow \Sigma M_O = \Sigma(M_O)_{\text{eff}}:$ $\qquad 0 = \bar{I}\alpha + m(\bar{r}\alpha)\bar{r} \qquad \alpha = 0$

$\xrightarrow{} \Sigma F_x = \Sigma(F_x)_{\text{eff}}:$ $\qquad R_x = m(\bar{r}\alpha) \qquad R_x = 0$

$+\!\uparrow \Sigma F_y = \Sigma(F_y)_{\text{eff}}:$ $\qquad R_y - 30 \text{ lb} = -m\bar{a}_n$

$$R_y - 30 \text{ lb} = -\frac{30 \text{ lb}}{32.2 \text{ ft/s}^2}(22.3 \text{ ft/s}^2)$$

$$R_y = +9.22 \text{ lb} \qquad R = 9.22 \text{ lb} \uparrow \quad \blacktriangleleft$$

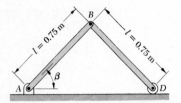

Each of the two slender rods shown is 0.75 m long and has a mass of 6 kg. If the system is released from rest when $\beta = 60°$, determine (a) the angular velocity of rod AB when $\beta = 20°$, (b) the velocity of point D at the same instant.

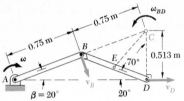

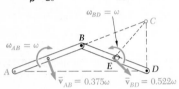

Kinematics of Motion When $\beta = 20°$. Since $\mathbf{v}_B$ is perpendicular to the rod AB and $\mathbf{v}_D$ is horizontal, the instantaneous center of rotation of rod BD is located at C. Considering the geometry of the figure, we obtain

$$BC = 0.75 \text{ m} \qquad CD = 2(0.75 \text{ m}) \sin 20° = 0.513 \text{ m}$$

Applying the law of cosines to triangle CDE, where E is located at the mass center of rod BD, we find $EC = 0.522$ m. Denoting by ω the angular velocity of rod AB, we have

$$\bar{v}_{AB} = (0.375 \text{ m})\omega \qquad \bar{\mathbf{v}}_{AB} = 0.375\omega \searrow$$
$$v_B = (0.75 \text{ m})\omega \qquad \mathbf{v}_B = 0.75\omega \searrow$$

Since rod BD seems to rotate about point C, we may write

$$v_B = (BC)\omega_{BD} \qquad (0.75 \text{ m})\omega = (0.75 \text{ m})\omega_{BD} \qquad \omega_{BD} = \omega \;\circlearrowleft$$
$$\bar{v}_{BD} = (EC)\omega_{BD} \qquad (0.522 \text{ m})\omega \qquad \bar{\mathbf{v}}_{BD} = 0.522\omega \searrow$$

Position 1. Potential Energy. Choosing the datum as shown, and observing that $W = (6 \text{ kg})(9.81 \text{ m/s}^2) = 58.9$ N, we have

$$V_1 = 2W\bar{y}_1 = 2(58.9 \text{ N})(0.325 \text{ m}) = 38.3 \text{ J}$$

Kinetic Energy. Since the system is at rest, $T_1 = 0$.

Position 2. Potential Energy

$$V_2 = 2W\bar{y}_2 = 2(58.9 \text{ N})(0.128 \text{ m}) = 15.1 \text{ J}$$

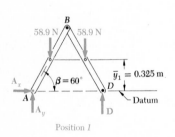

Position 1

Kinetic Energy

$$\bar{I}_{AB} = \bar{I}_{BD} = \tfrac{1}{12}ml^2 = \tfrac{1}{12}(6 \text{ kg})(0.75 \text{ m})^2 = 0.281 \text{ kg}\cdot\text{m}^2$$
$$T_2 = \tfrac{1}{2}m\bar{v}_{AB}^2 + \tfrac{1}{2}\bar{I}_{AB}\omega_{AB}^2 + \tfrac{1}{2}m\bar{v}_{BD}^2 + \tfrac{1}{2}\bar{I}_{BD}\omega_{BD}^2$$
$$= \tfrac{1}{2}(6)(0.375\omega)^2 + \tfrac{1}{2}(0.281)\omega^2 + \tfrac{1}{2}(6)(0.522\omega)^2 + \tfrac{1}{2}(0.281)\omega^2$$
$$= 1.520\omega^2$$

Conservation of Energy

$$T_1 + V_1 = T_2 + V_2: \qquad 0 + 38.3 \text{ J} = 1.520\omega^2 + 15.1 \text{ J}$$
$$\omega = 3.91 \text{ rad/s} \qquad \omega_{AB} = 3.91 \text{ rad/s} \;\circlearrowleft \quad \blacktriangleleft$$

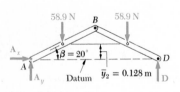

Position 2

Velocity of Point D

$$v_D = (CD)\omega = (0.513 \text{ m})(3.91 \text{ rad/s}) = 2.01 \text{ m/s}$$
$$\mathbf{v}_D = 2.01 \text{ m/s} \rightarrow \quad \blacktriangleleft$$

PROBLEMS

17.24 A cord is wrapped around a cylinder of radius r and mass m as shown. If the cylinder is released from rest, determine the velocity of the center of the cylinder after it has moved through a distance h.

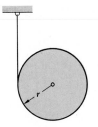

Fig. P17.24

17.25 Two 10-kg disks, each of radius $r = 0.3$ m, are connected by a cord. At the instant shown, the angular velocity of disk B is 20 rad/s clockwise. Determine how far disk A will rise before the angular velocity of disk B is reduced to 4 rad/s.

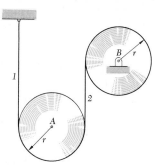

Fig. P17.25 and P17.26

17.26 Two identical disks, each of weight W and radius r, are connected by a cord as shown. If the system is released from rest, determine (a) the velocity of the center of disk A after it has moved through a distance h, (b) the tension in portion 2 of the cord.

17.27 A half cylinder of weight W and radius r is released from rest in the position shown. Assuming that the half cylinder rolls without sliding, determine (a) its angular velocity after it has rolled through 90°, (b) the normal reaction at the surface at the same instant. [*Hint*. Note that $GO = 4r/3\pi$ and that, by the parallel-axis theorem, $\bar{I} = \frac{1}{2}mr^2 - m(GO)^2$.]

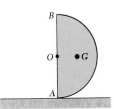

Fig. P17.27

17.28 A sphere of weight W and radius r rolls without slipping inside a curved surface of radius R. Knowing that the sphere is released from rest in the position shown, derive an expression (a) for the linear velocity of the sphere as it passes through B, (b) for the magnitude of the vertical reaction at that instant.

17.29 Solve Prob. 17.28, assuming that the sphere is replaced by a uniform cylinder of weight W and radius r.

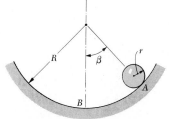

Fig. P17.28

17.30 A slender rod of length l and mass m is pivoted at one end as shown. It is released from rest in a horizontal position and swings freely. (*a*) Determine the angular velocity of the rod as it passes through a vertical position and the corresponding reaction at the pivot. (*b*) Solve part *a* for $m = 1.5$ kg and $l = 0.9$ m.

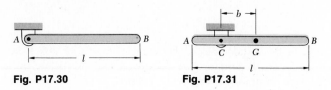

Fig. P17.30   Fig. P17.31

17.31 A uniform rod of length l is pivoted about a point C located at a distance b from its center G. The rod is released from rest in a horizontal position. Determine (*a*) the distance b so that the angular velocity of the rod as it passes through a vertical position is maximum, (*b*) the value of the maximum angular velocity.

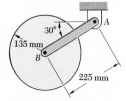

Fig. P17.32

17.32 A 15-kg disk is attached to a 6-kg rod AB as shown. The system is released from rest in the position shown. Determine the velocity of point B as it passes through its lowest position, (*a*) assuming that the disk is welded to the rod, (*b*) assuming that the disk is attached to the rod by a frictionless pin at B.

17.33 and 17.34 Gear C weighs 6 lb and has a centroidal radius of gyration of 3 in. The uniform bar AB weighs 5 lb, and gear D is stationary. If the system is released from rest in the position shown, determine the velocity of point B after bar AB has rotated through $90°$.

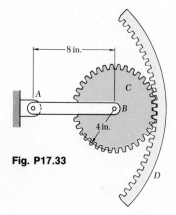

Fig. P17.33

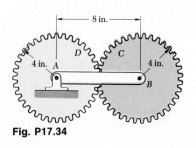

Fig. P17.34

17.35 The mass center G of a 1.5-kg wheel of radius $R = 150$ mm is located at a distance $r = 50$ mm from its geometric center C. The centroidal radius of gyration of the wheel is $\bar{k} = 75$ mm. As the wheel rolls without sliding, its angular velocity is observed to vary. Knowing that in position 1 the angular velocity is 10 rad/s, determine the angular velocity of the wheel (a) in position 2, (b) in position 3.

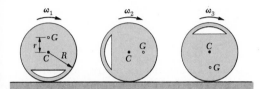

Fig. P17.35 and P17.36

17.36 The mass center G of a wheel of radius R is located at a distance r from its geometric center C. The centroidal radius of gyration of the wheel is denoted by $\bar{k}$. As the wheel rolls freely and without sliding on a horizontal plane, its angular velocity is observed to vary. Denoting by ω_1, ω_2, and ω_3, respectively, the angular velocity of the wheel when G is directly above C, level with C, and directly below C, show that ω_1, ω_2, and ω_3 satisfy the relation

$$\frac{\omega_2^2 - \omega_1^2}{\omega_3^2 - \omega_2^2} = \frac{g/R + \omega_1^2}{g/R + \omega_3^2}$$

17.37 The motion of a 0.6-m slender rod is guided by pins at A and B which slide freely in the slots shown. Knowing that the rod is released from rest when $\theta = 0$ and that end A is given a slight push to the right, determine (a) the angle θ for which the speed of end A is maximum, (b) the corresponding maximum speed of A.

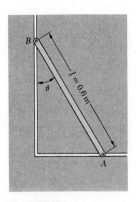

17.38 In Prob. 17.37, determine the velocity of ends A and B (a) when $\theta = 30°$, (b) when $\theta = 90°$.

17.39 and 17.40 The 15-lb carriage is supported as shown by two uniform disks, each of weight 10 lb and radius 3 in. Knowing that the system is initially at rest, determine the velocity of the carriage after it has moved 2 ft. Assume that the disks roll without sliding.

Fig. P17.37

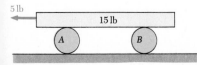

Fig. P17.39

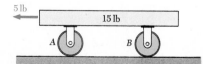

Fig. P17.40

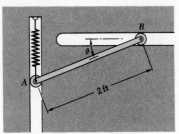

Fig. P17.41 and P17.42

17.41 The ends of a 25-lb rod AB are constrained to move along the slots shown. A spring of constant 3 lb/in. is attached to end A. Knowing that the rod is released from rest when $\theta = 0$ and that the initial tension in the spring is zero, determine the maximum distance through which end A will move.

17.42 The ends of a 25-lb rod AB are constrained to move along the slots shown. A spring of constant 10 lb/in. is attached to end A in such a way that its tension is zero when $\theta = 0$. If the rod is released from rest when $\theta = 60°$, determine the angular velocity of the rod when $\theta = 30°$.

17.43 The uniform rods AB and BC are of mass 4.5 and 1.5 kg respectively. If the system is released from rest in the position shown, determine the angular velocity of rod BC as it passes through a vertical position.

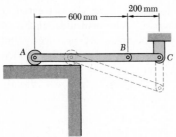

Fig. P17.43

17.44 Each of the two rods shown is of length $L = 3$ ft and weight $W = 5$ lb. Point D is constrained to move along a vertical line. If the system is released from rest when rod AB is horizontal, determine the velocity of points B and D as rod BD passes through a horizontal position.

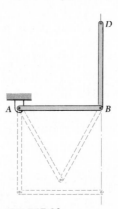

Fig. P17.44

17.45 In Prob. 17.44, determine the velocity of points B and D when points A and D are at the same elevation.

17.46   Determine the velocity of pin B as the rods of Sample Prob. 17.5 strike the horizontal surface.

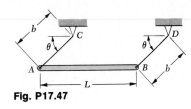

Fig. P17.47

17.47   A uniform rod of length L and weight W is attached to two wires, each of length b. The rod is released from rest when $\theta = 0$ and swings to the position $\theta = 90°$, at which time wire BD suddenly breaks. Determine the tension in wire AC (a) immediately before wire BD breaks, (b) immediately after wire BD breaks.

∗17.48   A small matchbox is placed on top of the rod AB. End B of the rod is given a slight horizontal push, causing it to slide on the horizontal floor. Assuming no friction and neglecting the weight of the matchbox, determine the angle θ through which the rod will have rotated when the matchbox loses contact with the rod.

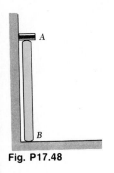

Fig. P17.48

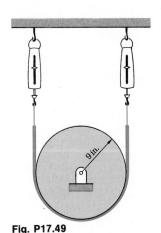

Fig. P17.49

17.49   The experimental setup shown is used to measure the power output of a small turbine. When the turbine is operating at 200 rpm, the readings of the two spring scales are 10 and 22 lb, respectively. Determine the power being developed by the turbine.

17.50   In Sample Prob. 17.2 determine the power being delivered to gear B at the instant when (a) the gear starts rotating, (b) the gear attains an angular velocity $\omega_B = 300$ rpm.

17.51   Knowing that the maximum allowable couple which can be applied to a shaft is $12\ \text{kN} \cdot \text{m}$, determine the maximum power which can be transmitted by the shaft (a) at 100 rpm, (b) at 1000 rpm.

17.52   Determine the moment of the couple which must be exerted by a motor to develop $\frac{1}{4}$ hp at a speed of (a) 3600 rpm, (b) 720 rpm.

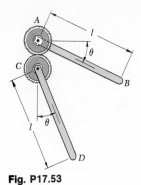

Fig. P17.53

REVIEW PROBLEMS

17.53 Two uniform rods, each of mass m, are attached to two gears of the same radius as shown. The rods are released from rest in the position $\theta = 0$. Neglecting the mass of the gears, determine the angular velocity of the rods when (*a*) $\theta = 30°$, (*b*) $\theta = 45°$, (*c*) $\theta = 60°$.

17.54 The motion of a 16-kg sliding panel is guided by rollers at B and C. The counterweight A has a mass of 12 kg and is attached to a cable as shown. If the system is released from rest, determine for each case shown the velocity of the counterweight as it strikes the ground. Neglect the effect of friction.

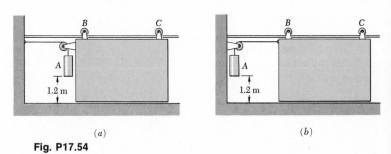

(*a*)   (*b*)

Fig. P17.54

17.55 The gear train shown consists of four gears of the same thickness and of the same material; two gears are of radius r, and the other two are of radius nr. The system is at rest when the couple M_0 is applied to shaft C. Denoting by I_0 the moment of inertia of a gear of radius r, determine the angular velocity of shaft A if the couple M_0 is applied for one revolution of shaft C.

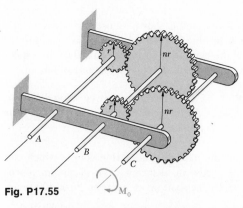

Fig. P17.55

17.56 Solve Prob. 17.55, assuming that the couple M_0 is applied to shaft A for one revolution of shaft A.

17.57   A 200-lb bin is suspended from two rods AB and CD of negligible weight. The bin is released from rest in the position $\theta = 0$. As the bin passes through its lowest position, determine (a) the velocity of the bin, (b) the reactions at A and C. (c) Show that the answer to part b is independent of the length l of the rods.

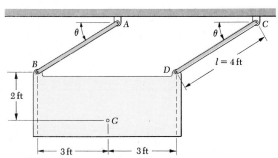

Fig. P17.57

17.58   Solve Prob. 17.57, assuming that each rod weighs 50 lb.

17.59   A flywheel of centroidal radius of gyration $\bar{k} = 600$ mm is rigidly attached to a shaft of radius $r = 30$ mm which may roll along parallel rails. Knowing that the system is released from rest, determine the velocity of the center of the shaft after it has moved 4 m.

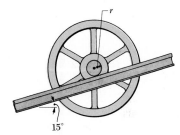

Fig. P17.59

17.60   A small collar of mass m is attached at B to the rim of a hoop of mass m and radius r. The hoop rolls without sliding on a horizontal plane. Find the angular velocity ω_1 of the hoop when B is directly above the center A in terms of g and r, knowing that the angular velocity of the hoop is $3\omega_1$ when B is directly below A.

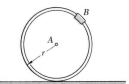

Fig. P17.60

17.61   A small disk A is driven at a constant angular velocity of 1200 rpm and is pressed against disk B, which is initially at rest. The normal force between disks is 10 lb, and $\mu_k = 0.20$. Knowing that disk B weighs 50 lb, determine the number of revolutions executed by disk B before its speed reaches 120 rpm.

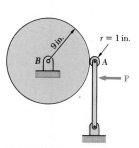

Fig. P17.61

17.62 Two uniform rods, AB of weight 4 lb and BC of weight 2 lb, are released from rest in the position shown. Neglecting both friction and the weight of the collar, determine the velocity of point B when $\theta = 90°$.

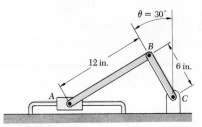

Fig. P17.62

17.63 Using the methods of Chap. 17, solve Prob. 16.169*b*.

17.64 Using the methods of Chap. 17, solve Prob. 16.170*b*.

Chapter
18

Kinetics of Rigid Bodies: Impulse and Momentum

18.1. Principle of Impulse and Momentum for a Rigid Body. In this chapter, the principle of impulse and momentum will be used to analyze the motion of rigid bodies and of systems of rigid bodies. As was pointed out in Chap. 14, the method of impulse and momentum is particularly well adapted to the solution of problems involving time and velocities. Moreover, the principle of impulse and momentum provides the only practicable method for the solution of problems of impact.

Since a rigid body of mass m may be considered as made of a large number of particles of mass Δm, Eq. (14.36), which was derived in Sec. 14.12 for a system of particles, may now be used to analyze the motion of a rigid body. We write

$$\text{Syst Momenta}_1 + \text{Syst Ext Imp}_{1 \to 2} = \text{Syst Momenta}_2 \quad (18.1)$$

This equation relates the three systems of vectors shown in Fig. 18.1. It expresses that the system formed by the momenta of the particles of the rigid body at time t_1 and the system of the impulses of the external forces applied to the body from t_1 to t_2 are together equivalent to the system formed by the momenta of the particles at time t_2.

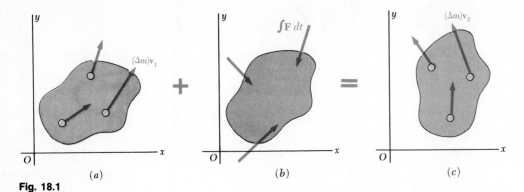

Fig. 18.1

We recall from Sec. 14.12 that, in the case of particles moving in the xy plane, three scalar equations are required to express the same relationship. The first two equations may be obtained by considering respectively the x and y components of the vectors shown in Fig. 18.1; they will relate the linear momenta of the particles and the linear impulses of the given forces in the x and y directions respectively. The third equation, which is obtained by computing the moments about O of the same vectors, will relate the angular momenta of the particles and the angular impulses of the forces.

Before we may conveniently apply the principle of impulse and momentum to the study of the plane motion of a rigid body, we must learn to express more simply the systems of vectors representing the momenta of the particles forming the rigid body. This is the object of the next section.

18.2. Momentum of a Rigid Body in Plane Motion. We shall see in this section that the system of the momenta of the particles forming a rigid body may be replaced at any given instant by an equivalent system consisting of a *momentum vector* attached at the mass center of the body and a *momentum couple*. This reduction of the system of momenta to a vector and a couple is similar to the reduction of the system of effective forces described in Chap. 16. While the method involved is quite general, our present analysis will be limited to the plane motion of rigid slabs and of rigid bodies which are symmetrical with respect to the reference plane.†

Translation. Consider first the case of a rigid body in translation. Since each particle, at any given instant, has the same velocity $\bar{\mathbf{v}}$ as the mass center G of the body, the momentum of a particle of mass Δm is represented by the vector $(\Delta m)\bar{\mathbf{v}}$.

---

† Or, more generally, to the motion of rigid bodies which have a principal axis of inertia perpendicular to the reference plane.

The resultant of the momenta of the various particles of the body is thus

$$\Sigma(\Delta m)\bar{\mathbf{v}} = (\Sigma\,\Delta m)\bar{\mathbf{v}} = m\bar{\mathbf{v}}$$

where m is the mass of the body. On the other hand, a reasoning similar to that used in Sec. 16.2 for the effective forces of a rigid body in translation shows that the sum of the moments about G of the momenta $(\Delta m)\bar{\mathbf{v}}$ is zero. We thus conclude that, in the case of a rigid body in translation, *the system of the momenta of the particles forming the body is equivalent to a single momentum vector $m\bar{\mathbf{v}}$ attached at the mass center G* (Fig. 18.2). This vector is the *linear momentum* of the body.

Centroidal Rotation. When a rigid body rotates about a fixed centroidal axis perpendicular to the reference plane, the velocity $\bar{\mathbf{v}}$ of its mass center G is zero, since G is fixed, and it follows from Eqs. (14.10) of Sec. 14.3 that the sums of the x and y components of the momenta of the various particles of the body are zero:

$$\Sigma(\Delta m)v_x = (\Sigma\,\Delta m)\bar{v}_x = 0 \qquad \Sigma(\Delta m)v_y = (\Sigma\,\Delta m)\bar{v}_y = 0$$

Therefore, in the case of a rigid body in centroidal rotation, *the system of the momenta of the particles forming the body is equivalent to a single momentum couple.* If, as we have assumed in this section, the body is symmetrical with respect to the reference plane, the axis of the momentum couple coincides with the axis of rotation.

The moment of the momentum couple is obtained by summing the moments about G of the momenta $(\Delta m)\mathbf{v}$ of the various particles. This sum is denoted by H_G and referred to as the *angular momentum* of the body about the axis through G. Observing that the momentum $(\Delta m)\mathbf{v}$ of a particle P is perpendicular to the line GP (Fig. 18.3a) and that $v = r\omega$, where r is the distance from G to P and ω the angular velocity of the body at the instant considered, we write

$$+\!\curvearrowright H_G = \Sigma r(\Delta m)v = \Sigma r(\Delta m)r\omega = \omega\Sigma r^2\,\Delta m$$

Since $\Sigma r^2\,\Delta m$ represents the centroidal moment of inertia $\bar{I}$ of the body, we have

$$H_G = \bar{I}\omega \qquad (18.2)$$

Recalling from Sec. 15.3 that the angular velocity of a rigid body rotating about a fixed axis may be represented by a vector $\boldsymbol{\omega}$ of magnitude ω directed along that axis, we note that the momentum couple may be represented by a couple vector $\bar{I}\boldsymbol{\omega}$ of magnitude $\bar{I}\omega$ directed along the axis of rotation. We there-

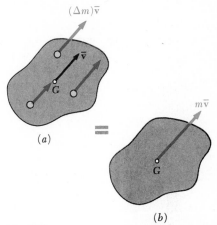

Fig. 18.2

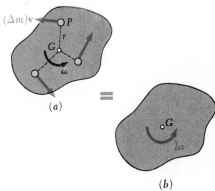

Fig. 18.3

fore conclude that, in the case of the centroidal rotation of a
rigid body symmetrical with respect to the reference plane, *the
system of the momenta of the particles of the body is equivalent
to a momentum couple* $\bar{I}\omega$, where ω denotes the angular velocity
of the body (Fig. 18.3*b*).

General Plane Motion. In the general case of the plane
motion of a rigid body, the velocity of any given particle P of
the body may be resolved into the velocity $\bar{v}$ of the mass center
G and the relative velocity v' of P with respect to G. The

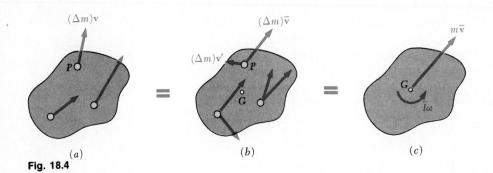

Fig. 18.4

momentum $(\Delta m)v$ of each particle (Fig. 18.4*a*) may thus be
resolved into the vectors $(\Delta m)\bar{v}$ and $(\Delta m)v'$ shown in Fig. 18.4*b*.
Recalling the results obtained above in the particular cases of
translation and centroidal rotation, we find that the vectors
$(\Delta m)\bar{v}$ may be replaced by their resultant $m\bar{v}$ attached at G and
that the vectors $(\Delta m)v'$ reduce to the couple $\bar{I}\omega$. While this
vector and this couple may in turn be replaced by a single
vector,† it is generally found more convenient to represent the
system of momenta by the momentum vector $m\bar{v}$ attached at
G and the momentum couple $\bar{I}\omega$ (Fig. 18.4*c*). The momentum
vector is associated with the translation of the body with G and
represents the *linear momentum* of the body. The momentum
couple corresponds to the rotation of the body about G and its
moment $\bar{I}\omega$ represents the *angular momentum* H_G *of the body
about an axis through* G. Recalling the definition given in Sec.
14.12 for the angular momentum of a system of particles, we
note that the angular momentum H_A of the body about an axis
through an arbitrary point A may be obtained by adding algebraically the moment about A of the momentum vector $m\bar{v}$ and
the moment $\bar{I}\omega$ of the momentum couple.‡

† See Probs. 18.13 and 18.14.
‡ Note that, in general, H_A is *not* equal to $I_A\omega$ (see Prob. 18.15).

18.3. Application of the Principle of Impulse and Momentum to the Analysis of the Plane Motion of a Rigid Body.

We saw in the preceding section that, in the most general case of the plane motion of a rigid body, the system of the momenta of the particles of the body reduces to a vector $m\overline{\mathbf{v}}$ attached at G and a couple $\overline{I}\omega$. Replacing the system of momenta in parts a and c of Fig. 18.1 by the equivalent momentum vector and momentum couple, we obtain the three diagrams shown in Fig. 18.5. This figure expresses graphically, in the case

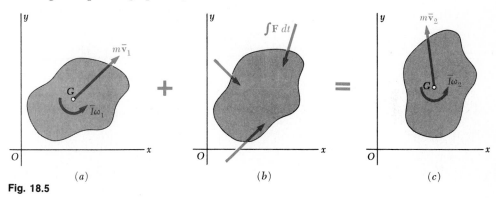

Fig. 18.5

of the plane motion of a rigid body symmetrical with respect to the reference plane, the fundamental relation

$$\text{Syst Momenta}_1 + \text{Syst Ext Imp}_{1\rightarrow2} = \text{Syst Momenta}_2 \quad (18.1)$$

Three equations of motion may be derived from Fig. 18.5. Two equations are obtained by summing and equating the x and y *components* of the momenta and impulses, and the third by summing and equating the *moments* of these vectors *about any given point*. The coordinate axes may be chosen fixed in space, or they may be allowed to move with the mass center of the body while maintaining a fixed direction. In either case, the point about which moments are taken should keep the same position relative to the coordinate axes during the interval of time considered.

In deriving the three equations of motion for a rigid body, care should be taken not to add indiscriminately linear and angular momenta. Confusion will be avoided if it is kept in mind that $m\overline{v}_x$ and $m\overline{v}_y$ represent the *components of a vector*, namely, the linear momentum vector $m\overline{\mathbf{v}}$, while $\overline{I}\omega$ represents the *moment of a couple*, namely, the momentum couple $\overline{I}\omega$. Thus the angular momentum $\overline{I}\omega$ should be added only to the *moment* of the linear momentum $m\overline{\mathbf{v}}$, never to this vector itself nor to its components. All quantities involved will then be expressed in the same units, namely, $N \cdot m \cdot s$ or $lb \cdot ft \cdot s$.

As an example, we shall consider the motion of a wheel of radius r, mass m, and centroidal moment of inertia $\bar{I}$, which rolls down an incline without sliding. Assuming that the initial angular velocity ω_1 of the wheel is known, we propose to determine its angular velocity ω_2, after a time interval t, and the values of the components F and N of the reaction of the incline.

The system of momenta at times t_1 and t_2 are shown, respectively, in parts a and c of Fig. 18.6. They are represented in both cases by a vector $m\bar{v}$ attached at the center G of the wheel and a couple $\bar{I}\omega$. Choosing coordinate axes centered at G with

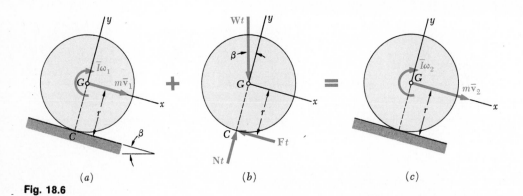

(a)   (b)   (c)

Fig. 18.6

the x and y axes parallel and perpendicular, respectively, to the incline, we note that the lines of action of the forces $\mathbf{W}$, $\mathbf{F}$, and $\mathbf{N}$ remain fixed with respect to the coordinate axes. Since these forces are also constant in magnitude, their impulses will be represented by the vectors $\mathbf{W}t$, $\mathbf{F}t$, and $\mathbf{N}t$ shown in part b of Fig. 18.6. Denoting by β the angle the incline forms with the horizontal, we add and equate successively the x components, y components, and moments about C (positive clockwise) of the momenta and impulses shown in Fig. 18.6. We write

$+\searrow$ x components:
$$m\bar{v}_1 - Ft + Wt\sin\beta = m\bar{v}_2 \tag{18.3}$$

$+\nearrow$ y components:
$$Nt - Wt\cos\beta = 0 \tag{18.4}$$

$+\downarrow$ moments about C:
$$\bar{I}\omega_1 + m\bar{v}_1 r + (Wt\sin\beta)r = \bar{I}\omega_2 + m\bar{v}_2 r \tag{18.5}$$

We note that all quantities involved in Eqs. (18.3) and (18.4) are expressed in the same units, namely, N·s or lb·s. Since, in Eq. (18.5), the magnitudes of the linear impulse $Wt\sin\beta$ and of the linear momenta $m\bar{v}_1$ and $m\bar{v}_2$ are multiplied by the radius

r, all quantities involved in that equation are also expressed in the same units, namely, $N \cdot m \cdot s$ or $lb \cdot ft \cdot s$. Observing that $\bar{v} = r\omega$, we write

$$mr\omega_1 - Ft + Wt \sin \beta = mr\omega_2 \qquad (18.3')$$
$$Nt - Wt \cos \beta = 0 \qquad (18.4')$$
$$(\bar{I} + mr^2)\omega_1 + (Wt \sin \beta)r = (\bar{I} + mr^2)\omega_2 \qquad (18.5')$$

Equation (18.5') may be solved for ω_2 and Eq. (18.4') for N; Eq. (18.3') may be solved for F after substituting for ω_2.

Noncentroidal Rotation.   In this particular case of plane motion, the magnitude of the velocity of the mass center of the body is $\bar{v} = \bar{r}\omega$, where $\bar{r}$ represents the distance from the mass center to the fixed axis of rotation and ω the angular velocity of the body at the instant considered; the magnitude of the momentum vector attached at G is thus $m\bar{v} = m\bar{r}\omega$. Summing the moments about O of the momentum vector and momentum couple (Fig. 18.7) and using the parallel-axis theorem for moments of inertia, we find that the angular momentum of the body about O is

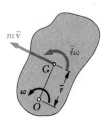

Fig. 18.7

$$\bar{I}\omega + (m\bar{r}\omega)\bar{r} = (\bar{I} + m\bar{r}^2)\omega = I_O\omega \qquad (18.6)$$

Equating the moments about O of the momenta and impulses in (18.1), we write

$$I_O\omega_1 + \sum \int_{t_1}^{t_2} M_O \, dt = I_O\omega_2 \qquad (18.7)$$

In the general case of plane motion of a rigid body symmetrical with respect to the reference plane, Eq. (18.7) may be used with respect to the instantaneous axis of rotation under certain conditions. It is recommended, however, that all problems of plane motion be solved by the general method described earlier in this section.

18.4. Systems of Rigid Bodies.   The motion of several rigid bodies may be analyzed by applying the principle of impulse and momentum to each body separately (Sample Prob. 18.1).

However, in solving problems involving no more than three unknowns (including the impulses of unknown reactions), it is often found convenient to apply the principle of impulse and momentum to the system as a whole. The momentum and impulse diagrams are drawn for the entire system of bodies. The diagrams of momenta should include a momentum vector, a momentum couple, or both, for each moving part of the system. Impulses of forces internal to the system may be omitted from the impulse diagram since they occur in pairs of equal and opposite vectors. Summing and equating successively the x components, y components, and moments of all vectors involved, one obtains three relations which express that the momenta at time t_1 and the impulses of the external forces form a system equipollent to the system of the momenta at time t_2.† Again, care should be taken not to add indiscriminately linear and angular momenta; each equation should be checked to make sure that consistent units have been used. This approach has been used in Sample Prob. 18.3 and, further on, in Sample Probs. 18.4 and 18.5.

18.5. Conservation of Angular Momentum. When no external force acts on a rigid body or a system of rigid bodies, the impulses of the external forces are zero and the system of the momenta at time t_1 is equipollent to the system of the momenta at time t_2. Summing and equating successively the x components, y components, and moments of the momenta at times t_1 and t_2, we conclude that the total linear momentum of the system is conserved in any direction and that its total angular momentum is conserved about any point.

There are many engineering applications, however, in which *the linear momentum is not conserved*, yet in which *the angular momentum H_O of the system about a given point O is conserved*:

$$(H_O)_1 = (H_O)_2 \tag{18.8}$$

Such cases occur when the lines of action of all external forces pass through O or, more generally, when the sum of the angular impulses of the external forces about O is zero.

Problems involving *conservation of angular momentum* about a point O may be solved by the general method of impulse and momentum, i.e., by drawing momentum and impulse diagrams as described in Secs. 18.3 and 18.4. Equation (18.8) is then obtained by summing and equating moments about O (Sample Prob. 18.3). As we shall see later in Sample Prob. 18.4, two additional equations may be written by summing and equating x and y components; these equations may be used to determine two unknown linear impulses, such as the impulses of the reaction components at a fixed point.

† Note that, as in Sec. 16.4, we cannot speak of *equivalent* systems since we are not dealing with a single rigid body.

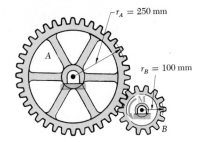

$r_A = 250$ mm

$r_B = 100$ mm

A

M

B

Gear A has a mass of 10 kg and a radius of gyration of 200 mm, while gear B has a mass of 3 kg and a radius of gyration of 80 mm. The system is at rest when a couple **M** of moment 6 N $\cdot$ m is applied to gear B. Neglecting friction, determine (*a*) the time required for the angular velocity of gear B to reach 600 rpm, (*b*) the tangential force which gear B exerts on gear A. These gears have been previously considered in Sample Prob. 17.2.

Solution. We apply the principle of impulse and momentum to each gear separately. Since all forces and the couple are constant, their impulses are obtained by multiplying them by the unknown time t. We recall from Sample Prob. 17.2 that the centroidal moments of inertia and the final angular velocities are

$$\bar{I}_A = 0.400 \text{ kg} \cdot \text{m}^2 \qquad \bar{I}_B = 0.0192 \text{ kg} \cdot \text{m}^2$$
$$(\omega_A)_2 = 25.1 \text{ rad/s} \qquad (\omega_B)_2 = 62.8 \text{ rad/s}$$

Principle of Impulse and Momentum for Gear A. The systems of initial momenta, impulses, and final momenta are shown in three separate sketches.

$\bar{I}_A(\omega_A)_1 = 0$ — A

$A_x t$, $A_y t$, r_A, Ft — A

$\bar{I}_A(\omega_A)_2$ — A

Syst Momenta₁ + Syst Ext Imp₁→₂ = Syst Momenta₂

$+\circlearrowleft$ moments about A: $\qquad 0 - Ftr_A = -\bar{I}_A(\omega_A)_2$

$$Ft(0.250 \text{ m}) = (0.400 \text{ kg} \cdot \text{m}^2)(25.1 \text{ rad/s})$$
$$Ft = 40.2 \text{ N} \cdot \text{s}$$

Principle of Impulse and Momentum for Gear B.

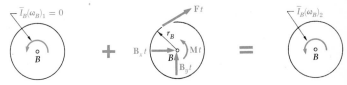

$\bar{I}_B(\omega_B)_1 = 0$ — B

$B_x t$, r_B, Mt, $B_y t$, Ft — B

$\bar{I}_B(\omega_B)_2$ — B

Syst Momenta₁ + Syst Ext Imp₁→₂ = Syst Momenta₂

$+\circlearrowleft$ moments about B: $\qquad 0 + Mt - Ftr_B = \bar{I}_B(\omega_B)_2$

$$+(6 \text{ N} \cdot \text{m})t - (40.2 \text{ N} \cdot \text{s})(0.100 \text{ m}) = (0.0192 \text{ kg} \cdot \text{m}^2)(62.8 \text{ rad/s})$$

$$t = 0.871 \text{ s} \blacktriangleleft$$

Recalling that $Ft = 40.2$ N $\cdot$ s, we write

$$F(0.871 \text{ s}) = 40.2 \text{ N} \cdot \text{s} \qquad F = +46.2 \text{ N}$$

Thus, the force exerted by gear B on gear A is $\quad \text{F} = 46.2 \text{ N} \swarrow \quad \blacktriangleleft$

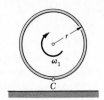

SAMPLE PROBLEM 18.2

A hoop of radius r and mass m is placed on a horizontal surface with no linear velocity but with a clockwise angular velocity ω_1. Denoting by μ the coefficient of friction between the hoop and the surface, determine (a) the time t_2 at which the hoop will start rolling without sliding, (b) the linear and angular velocities of the hoop at time t_2.

Solution. Since the entire mass m is located at a distance r from the center of the hoop, we have $\bar{I} = mr^2$. While the hoop is sliding relative to the surface, it is acted upon by the normal force $\mathbf{N}$, the friction force $\mathbf{F}$, and its weight $\mathbf{W}$ of magnitude $W = mg$.

Principle of Impulse and Momentum. We apply the principle of impulse and momentum to the hoop from the time t_1 when it is placed on the surface until the time t_2 when it starts rolling without sliding.

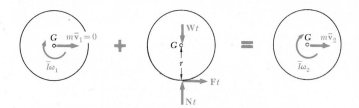

$$\text{Syst Momenta}_1 + \text{Syst Ext Imp}_{1\to2} = \text{Syst Momenta}_2$$

$+\uparrow y$ components: $\qquad\qquad Nt - Wt = 0 \qquad\qquad (1)$

$\xrightarrow{+} x$ components: $\qquad\qquad\qquad Ft = m\bar{v}_2 \qquad\qquad (2)$

$+\circlearrowleft$ moments about G: $\qquad -\bar{I}\omega_1 + Ftr = -\bar{I}\omega_2 \qquad (3)$

From (1) we obtain $N = W = mg$. For $t < t_2$, sliding occurs at point C and we have $F = \mu N = \mu mg$. Substituting for F into (2), we write

$$\mu mgt = m\bar{v}_2 \qquad \bar{v}_2 = \mu gt \qquad (4)$$

Substituting $F = \mu mg$ and $\bar{I} = mr^2$ into (3),

$$-mr^2\omega_1 + \mu mgtr = -mr^2\omega_2 \qquad \omega_2 = \omega_1 - \frac{\mu g}{r}t \qquad (5)$$

The hoop will start rolling without sliding when the velocity $\mathbf{v}_C$ of the point of contact is zero. At that time, $t = t_2$, point C becomes the instantaneous center of rotation, and we have $\bar{v}_2 = r\omega_2$. Substituting from (4) and (5), we write

$$\bar{v}_2 = r\omega_2 \qquad \mu gt_2 = r\left(\omega_1 - \frac{\mu g}{r}t_2\right) \qquad t_2 = \frac{r\omega_1}{2\mu g} \quad \blacktriangleleft$$

Substituting this expression for t_2 into (4),

$$\bar{v}_2 = \mu gt_2 = \mu g\frac{r\omega_1}{2\mu g} \qquad \bar{v}_2 = \tfrac{1}{2}r\omega_1 \qquad \mathbf{\bar{v}}_2 = \tfrac{1}{2}r\omega_1 \rightarrow \quad \blacktriangleleft$$

$$\omega_2 = \frac{\bar{v}_2}{r} \qquad \omega_2 = \tfrac{1}{2}\omega_1 \qquad \omega_2 = \tfrac{1}{2}\omega_1 \,\circlearrowright \quad \blacktriangleleft$$

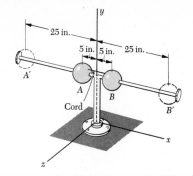

25 in.

5 in. 5 in. 25 in.

A'

A

B

Cord

B'

x

y

z

Two solid spheres of radius 3 in., weighing 2 lb each, are mounted at A and B on the horizontal rod $A'B'$, which rotates freely about the vertical with a counterclockwise angular velocity of 6 rad/s. The spheres are held in position by a cord which is suddenly cut. Knowing that the centroidal moment of inertia of the rod and pivot is $\bar{I}_R = 0.25$ lb · ft · s², determine (a) the angular velocity of the rod after the spheres have moved to positions A' and B', (b) the energy lost due to the plastic impact of the spheres and the stops at A' and B'.

a. Principle of Impulse and Momentum. In order to determine the final angular velocity of the rod, we shall express that the initial momenta of the various parts of the system and the impulses of the external forces are together equipollent to the final momenta of the system.

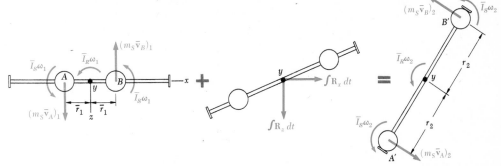

Observing that the external forces consist of the weights and the reaction at the pivot, which have no moment about the y axis, and noting that $\bar{v}_A = \bar{v}_B = \bar{r}\omega$, we write

$+\uparrow$ moments about y axis:

$$2(m_S\bar{r}_1\omega_1)\bar{r}_1 + 2\bar{I}_S\omega_1 + \bar{I}_R\omega_1 = 2(m_S\bar{r}_2\omega_2)\bar{r}_2 + 2\bar{I}_S\omega_2 + \bar{I}_R\omega_2$$

$$(2m_S\bar{r}_1^2 + 2\bar{I}_S + \bar{I}_R)\omega_1 = (2m_S\bar{r}_2^2 + 2\bar{I}_S + \bar{I}_R)\omega_2 \qquad (1)$$

which expresses that *the angular momentum of the system about the y axis is conserved.* We now compute

$$\bar{I}_S = \tfrac{2}{5}m_S a^2 = \frac{2}{5}\left(\frac{2\text{ lb}}{32.2\text{ ft/s}^2}\right)(\tfrac{3}{12}\text{ ft})^2 = 0.00155\text{ lb · ft · s}^2$$

$$m_S\bar{r}_1^2 = \frac{2}{32.2}\left(\frac{5}{12}\right)^2 = 0.0108 \qquad m_S\bar{r}_2^2 = \frac{2}{32.2}\left(\frac{25}{12}\right)^2 = 0.2696$$

Substituting these values and $\bar{I}_R = 0.25$, $\omega_1 = 6$ rad/s into (1):

$$0.275(6\text{ rad/s}) = 0.792\omega_2 \qquad \omega_2 = 2.08\text{ rad/s} \uparrow \quad \blacktriangleleft$$

b. Energy Lost. The kinetic energy of the system at any instant is

$$T = 2(\tfrac{1}{2}m_S\bar{v}^2 + \tfrac{1}{2}\bar{I}_S\omega^2) + \tfrac{1}{2}\bar{I}_R\omega^2 = \tfrac{1}{2}(2m_S\bar{r}^2 + 2\bar{I}_S + \bar{I}_R)\omega^2$$

Recalling the numerical values found above, we have

$$T_1 = \tfrac{1}{2}(0.275)(6)^2 = 4.95\text{ ft · lb} \qquad T_2 = \tfrac{1}{2}(0.792)(2.08)^2 = 1.713\text{ ft · lb}$$

$$\Delta T = T_2 - T_1 = 1.71 - 4.95 \qquad \Delta T = -3.24\text{ ft · lb} \quad \blacktriangleleft$$

PROBLEMS

18.1 The rotor of a steam turbine has an angular velocity of 5400 rpm when the steam supply is suddenly cut off. The rotor weighs 200 lb and has a radius of gyration of 6 in. If kinetic friction produces a couple of magnitude 24 lb · in., determine the time required for the rotor to coast to rest.

18.2 A large flywheel weighs 3 tons and has a radius of gyration of 42 in. It is observed that 9 min 18 s is required for the flywheel to coast to rest from an angular velocity of 300 rpm. Determine the average moment of the couple due to kinetic friction in the bearings of the flywheel.

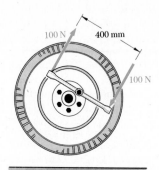

18.3 A bolt located 50 mm from the center of an automobile wheel is tightened by applying the couple shown for 0.1 s. Assuming that the wheel is free to rotate and is initially at rest, determine the resulting angular velocity of the wheel. The 20-kg wheel has a radius of gyration of 250 mm.

Fig. P18.3

18.4 Solve Prob. 17.3, assuming that the couple $\mathbf{M}$ is applied for a time t_0 and then removed.

18.5 A cylinder of radius r and mass m is placed in a corner with an initial counterclockwise angular velocity $\boldsymbol{\omega}_0$. Denoting by μ the coefficient of friction at A and B, derive an expression for the time required for the cylinder to come to rest.

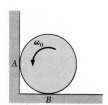

Fig. P18.5

18.6 Solve Prob. 18.5, assuming that the surface at A is frictionless.

18.7 Using the principle of impulse and momentum, solve Prob. 16.39.

18.8 Using the principle of impulse and momentum, solve Prob. 16.30b.

18.9 Disks A and B are of mass 5 and 1.8 kg, respectively. The disks are initially at rest and the coefficient of friction between them is 0.20. A couple $\mathbf{M}$ of moment 4 N · m is applied to disk A for 1.50 s and then removed. Determine (a) whether slipping occurs between the disks, (b) the final angular velocity of each disk.

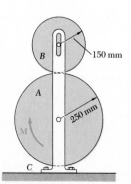

Fig. P18.9

18.10 In Prob. 18.9, determine (a) the largest couple $\mathbf{M}$ for which no slipping occurs, (b) the corresponding final angular velocity of each disk.

18.11   The flywheel of a small hoisting engine has a radius of gyration of 2 ft and weighs 750 lb. If the power is cut off when the angular velocity of the flywheel is 100 rpm clockwise, determine the time required for the system to come to rest.

18.12   In Prob. 18.11, determine the time required for the angular velocity of the flywheel to be reduced to 50 rpm.

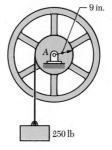

250 lb

Fig. P18.11

18.13   Show that the system of momenta for a rigid slab in plane motion reduces to a single vector, and express the distance from the mass center G to the line of action of this vector in terms of the centroidal radius of gyration $\bar{k}$ of the slab, the magnitude $\bar{v}$ of the velocity of G, and the angular velocity ω.

18.14   Show that, when a rigid slab rotates about a fixed axis through O perpendicular to the slab, the system of momenta of its particles is equivalent to a single vector of magnitude $m\bar{r}\omega$, perpendicular to the line OG, and applied to a point P on this line, called the *center of percussion*, at a distance $GP = \bar{k}^2/\bar{r}$ from the mass center of the slab.

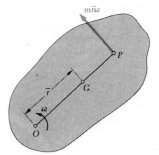

Fig. P18.14

18.15   Show that the sum H_A of the moments about a point A of the momenta of the particles of a rigid slab in plane motion is equal to $I_A\omega$, where ω is the angular velocity of the slab at the instant considered and I_A the moment of inertia of the slab about A, *if and only if* one of the following conditions is satisfied: (*a*) A is the mass center of the slab, (*b*) A is the instantaneous center of rotation, (*c*) the velocity of A is directed along a line joining point A and the mass center G.

18.16   Consider a rigid slab initially at rest and subjected to an impulsive force $\mathbf{F}$ contained in the plane of the slab. We define the *center of percussion* P as the point of intersection of the line of action of $\mathbf{F}$ with the perpendicular drawn from G. (*a*) Show that the instantaneous center of rotation C of the slab is located on line GP at a distance $GC = \bar{k}^2/GP$ on the opposite side of G. (*b*) Show that, if the center of percussion were located at C, the instantaneous center of rotation would be located at P.

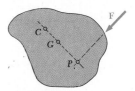

Fig. P18.16

18.17   A cord is wrapped around a solid cylinder of radius r and mass m as shown. If the cylinder is released from rest at time $t = 0$, determine the velocity of the center of the cylinder at a time t.

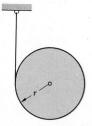

Fig. P18.17

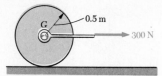

Fig. P18.18

18.18 A 100-kg cylindrical roller is initially at rest and is acted upon by a 300-N force as shown. Assuming that the body rolls without slipping, determine (a) the velocity of the center G after 6 s, (b) the friction force required to prevent slipping.

18.19 A section of thin-walled pipe of radius r is released from rest at time $t = 0$. Assuming that the pipe rolls without slipping, determine (a) the velocity of the center at time t, (b) the coefficient of friction required to prevent slipping.

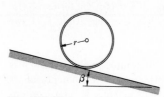

Fig. P18.19

18.20 and 18.21 The 15-lb carriage is supported as shown by two uniform disks, each of weight 10 lb and radius 3 in. Knowing that the carriage is initially at rest, determine the velocity of the carriage 3 s after the 5-lb force is applied. Assume that the disks roll without sliding.

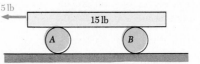

Fig. P18.20

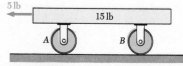

Fig. P18.21

Fig. P18.22 and P18.23

18.22 A sphere of mass m and radius r is projected along a rough horizontal surface with a linear velocity $\bar{v}_0$ but with no angular velocity $(\omega_0 = 0)$. Determine (a) the final velocity of the sphere, (b) the time at which the velocity of the sphere becomes constant in terms of $\bar{v}_0$ and μ.

18.23 A sphere of mass m and radius r is projected along a rough horizontal surface with the initial velocities indicated. If the final velocity of the sphere is to be zero, express (a) the required ω_0 in terms of $\bar{v}_0$ and r, (b) the time required for the sphere to come to rest in terms of $\bar{v}_0$ and μ.

18.24 Solve Sample Prob. 18.3, assuming that, after the cord is cut, sphere B moves to position B' but that an obstruction prevents sphere A from moving.

18.25 A 4-kg bar AB is attached by a pin at D to a 5-kg circular plate which may rotate freely about a vertical axis. Knowing that when the bar is vertical the angular velocity of the plate is 90 rpm, determine (a) the angular velocity of the plate after the bar has swung into a horizontal position and has come to rest against pin C, (b) the energy lost due to the plastic impact at C.

18.26 Disks A and B are made of the same material and are of the same thickness; they may rotate freely about the vertical shaft. Disk B is at rest when it is dropped onto disk A which is rotating with an angular velocity of 400 rpm. Knowing that the mass of disk A is 4 kg, determine (a) the final angular velocity of the disks, (b) the change in kinetic energy of the system.

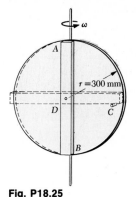

Fig. P18.25

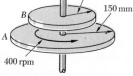

Fig. P18.26

18.27 Solve Prob. 18.26 assuming that disk B has an initial angular velocity of 800 rpm counterclockwise when viewed from above.

18.28 A small 250-g ball may slide in a slender tube of length 1 m and of mass 1 kg which rotates freely about a vertical axis passing through its center C. If the angular velocity of the tube is 10 rad/s as the ball passes through C, determine the angular velocity of the tube (a) just before the ball leaves the tube, (b) just after the ball has left the tube.

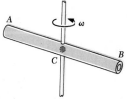

Fig. P18.28

18.29 The rod AB is of mass m and slides freely inside the tube CD which is also of mass m. The angular velocity of the assembly was ω_1 when the rod was entirely inside the tube ($x = 0$). Neglecting the effect of friction, determine the angular velocity of the assembly when $x = \frac{1}{2}L$.

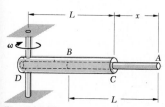

Fig. P18.29

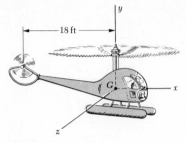

Fig. P18.30

18.30 In the helicopter shown, a vertical tail propeller is used to prevent rotation of the cab as the speed of the main blades is changed. Assuming that the tail propeller is not operating, determine the final angular velocity of the cab after the speed of the main blades has been changed from 200 to 300 rpm. The speed of the main blades is measured relative to the cab, which has a centroidal moment of inertia of 800 lb · ft · s². Each of the four main blades is assumed to be a 15-ft slender rod weighing 60 lb.

18.31 Assuming that the tail propeller in Prob. 18.30 is operating and that the angular velocity of the cab remains zero, determine the final horizontal velocity of the cab when the speed of the main blades is changed from 200 to 300 rpm. The cab weighs 1500 lb and is initially at rest. Also determine the force exerted by the tail propeller if this change in speed takes places uniformly in 12 s.

18.32 Two 4-kg disks and a small motor are mounted on a 6-kg rectangular platform which is free to rotate about a central vertical spindle. The normal operating speed of the motor is 180 rpm. If the motor is started when the system is at rest, determine the angular velocity of all elements of the system after the motor has attained its normal operating speed. Neglect the weight of the motor, and assume that the system is perfectly lubricated.

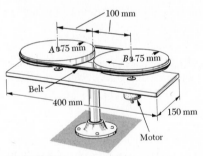

Fig. P18.32

18.33 Solve Prob. 18.32, assuming (*a*) that the belt is removed, (*b*) that the belt is looped around the disks in a figure 8.

18.34 Knowing that in Prob. 18.28 the speed of the ball is 1.2 m/s as it passes through C, determine the radial and transverse components of the velocity of the ball as it leaves the tube at B.

18.35 In Prob. 18.29, determine the velocity of the rod relative to the tube when $x = \frac{1}{2}L$.

18.6. Impulsive Motion. We saw in Chap. 14 that the method of impulse and momentum is the only practicable method for the solution of problems involving impulsive motion. Now we shall also find that, compared with the various problems considered in the preceding sections, problems involving impulsive motion are particularly well adapted to a solution by the method of impulse and momentum. The computation of linear impulses and angular impulses is quite simple, since, the time interval considered being very short, the bodies involved may be assumed to occupy the same position during that time interval.

18.7. Eccentric Impact. In Secs. 14.5 to 14.7, we learned to solve problems of *central impact*, i.e., problems in which the mass centers of the two colliding bodies are located on the line of impact. We shall now analyze the *eccentric impact* of two rigid bodies. Consider two bodies which collide, and denote by $\mathbf{v}_A$ and $\mathbf{v}_B$ the velocities before impact *of the two points of contact A and B* (Fig. 18.8a). Under the impact, the two bodies will *deform* and, at the end of the period of deformation, the velocities $\mathbf{u}_A$ and $\mathbf{u}_B$ of A and B will have equal components along the line of impact nn (Fig. 18.8b). A period of *restitution* will then take place, at the end of which A and B will have velocities $\mathbf{v}_A'$ and $\mathbf{v}_B'$ (Fig. 18.8c). Assuming the bodies frictionless, we find that the forces they exert on each other are directed along the line of impact. Denoting, respectively, by $\int P\,dt$ and $\int R\,dt$ the magnitude of the impulse of one of these forces during the period of deformation and during the period of restitution, we recall that the coefficient of restitution e is defined as the ratio

$$e = \frac{\int R\,dt}{\int P\,dt} \tag{18.9}$$

We propose to show that the relation established in Sec. 14.6 between the relative velocities of two particles before and after impact also holds between the components along the line of impact of the relative velocities of the two points of contact A and B. We propose to show, therefore, that

$$(v_B')_n - (v_A')_n = e[(v_A)_n - (v_B)_n] \tag{18.10}$$

We shall first assume that the motion of each of the two colliding bodies of Fig. 18.8 is unconstrained. Thus the only impulsive forces exerted on the bodies during the impact are applied at A and B respectively. Consider the body to which point A belongs and draw the three momentum and impulse

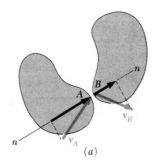
(a)

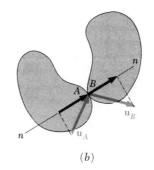
(b)

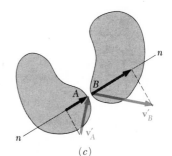
(c)

Fig. 18.8

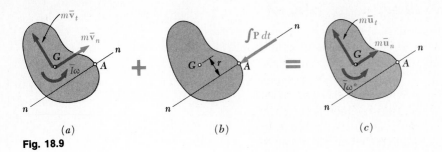

Fig. 18.9

diagrams corresponding to the period of deformation (Fig. 18.9). We denote by $\bar{\mathbf{v}}$ and $\bar{\mathbf{u}}$, respectively, the velocity of the mass center at the beginning and at the end of the period of deformation, and by $\boldsymbol{\omega}$ and $\boldsymbol{\omega}^*$ the angular velocity of the body at the same instants. Summing and equating the components of the momenta and impulses along the line of impact nn, we write

$$m\bar{v}_n - \int P\,dt = m\bar{u}_n \tag{18.11}$$

Summing and equating the moments about G of the momenta and impulses, we also write

$$\bar{I}\omega - r\int P\,dt = \bar{I}\omega^* \tag{18.12}$$

where r represents the perpendicular distance from G to the line of impact. Considering now the period of restitution, we obtain in a similar way

$$m\bar{u}_n - \int R\,dt = m\bar{v}'_n \tag{18.13}$$
$$\bar{I}\omega^* - r\int R\,dt = \bar{I}\omega' \tag{18.14}$$

where $\bar{\mathbf{v}}'$ and $\boldsymbol{\omega}'$ represent, respectively, the velocity of the mass center and the angular velocity of the body after impact. Solving (18.11) and (18.13) for the two impulses and substituting into (18.9), and then solving (18.12) and (18.14) for the same two impulses and substituting again into (18.9), we obtain the following two alternate expressions for the coefficient of restitution:

$$e = \frac{\bar{u}_n - \bar{v}'_n}{\bar{v}_n - \bar{u}_n} \qquad\qquad e = \frac{\omega^* - \omega'}{\omega - \omega^*} \tag{18.15}$$

Multiplying by r the numerator and denominator of the second expression obtained for e, and adding respectively to the numerator and denominator of the first expression, we have

$$e = \frac{\bar{u}_n + r\omega^* - (\bar{v}'_n + r\omega')}{\bar{v}_n + r\omega - (\bar{u}_n + r\omega^*)} \tag{18.16}$$

Observing that $\bar{v}_n + r\omega$ represents the component $(v_A)_n$ along nn of the velocity of the point of contact A and that, similarly,

$\bar{u}_n + r\omega^*$ and $\bar{v}'_n + r\omega'$ represent, respectively, the components $(u_A)_n$ and $(v'_A)_n$, we write

$$e = \frac{(u_A)_n - (v'_A)_n}{(v_A)_n - (u_A)_n} \qquad (18.17)$$

The analysis of the motion of the second body leads to a similar expression for e in terms of the components along nn of the successive velocities of point B. Recalling that $(u_A)_n = (u_B)_n$, and eliminating these two velocity components by a manipulation similar to the one used in Sec. 14.6, we obtain relation (18.10).

If one or both of the colliding bodies is constrained to rotate about a fixed point O, as in the case of a compound pendulum (Fig. 18.10a), an impulsive reaction will be exerted at O (Fig. 18.10b). We shall verify that, while their derivation must be

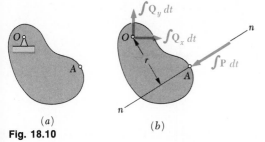

(a)

(b)

Fig. 18.10

modified, Eqs. (18.17) and (18.10) remain valid. Applying formula (18.7) to the period of deformation and to the period of restitution, we write

$$I_O\omega - r\!\int\! P \, dt = I_O\omega^* \qquad (18.18)$$
$$I_O\omega^* - r\!\int\! R \, dt = I_O\omega' \qquad (18.19)$$

where r represents the perpendicular distance from the fixed point O to the line of impact. Solving (18.18) and (18.19) for the two impulses and substituting into (18.9), and then observing that $r\omega$, $r\omega^*$, and $r\omega'$ represent the components along nn of the successive velocities of point A, we write

$$e = \frac{\omega^* - \omega'}{\omega - \omega^*} = \frac{r\omega^* - r\omega'}{r\omega - r\omega^*} = \frac{(u_A)_n - (v'_A)_n}{(v_A)_n - (u_A)_n}$$

and check that Eq. (18.17) still holds. Thus Eq. (18.10) remains valid when one or both of the colliding bodies is constrained to rotate about a fixed point O.

In order to determine the velocities of the two colliding bodies after impact, relation (18.10) should be used in conjunction with one or several other equations obtained by applying the principle of impulse and momentum (Sample Prob. 18.5).

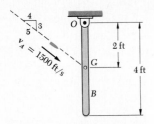

SAMPLE PROBLEM 18.4

A 0.125-lb bullet A is fired with an initial velocity of 1500 ft/s into a 50-lb wooden beam B which is suspended from a hinge at O. Knowing that the beam is initially at rest, determine (a) the angular velocity of the beam immediately after the bullet becomes embedded in the beam, (b) the impulsive reaction at the hinge, assuming that the bullet becomes embedded in 0.0002 s.

Solution. *Principle of Impulse and Momentum.* We consider the bullet and the beam as a single system and express that the initial momenta of the bullet and beam and the impulses of the external forces are together equipollent to the final momenta of the system. Since the time interval $\Delta t = 0.0002$ s is very short, we neglect all nonimpulsive forces and consider only the external impulses $\mathbf{R}_x \, \Delta t$ and $\mathbf{R}_y \, \Delta t$.

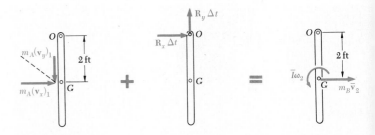

$+\circlearrowleft$ moments about O: $m_A(v_x)_1(2 \text{ ft}) + 0 = \bar{I}\omega_2 + m_B\bar{v}_2(2 \text{ ft})$ (1)

$\xrightarrow{+} x$ components: $m_A(v_x)_1 + R_x \, \Delta t = m_B\bar{v}_2$ (2)

$+\uparrow y$ components: $-m_A(v_y)_1 + R_y \, \Delta t = 0$ (3)

The components of the velocity of the bullet and the centroidal moment of inertia of the beam are

$$(v_x)_1 = \tfrac{4}{5}(1500 \text{ ft/s}) = 1200 \text{ ft/s} \qquad (v_y)_1 = \tfrac{3}{5}(1500 \text{ ft/s}) = 900 \text{ ft/s}$$

$$\bar{I} = \tfrac{1}{12}ml^2 = \frac{1}{12}\frac{50 \text{ lb}}{32.2 \text{ ft/s}^2}(4 \text{ ft})^2 = 2.07 \text{ lb} \cdot \text{ft} \cdot \text{s}^2$$

Substituting these values into (1) and noting that $\bar{v}_2 = (2 \text{ ft})\omega_2$:

$$(0.125/32.2)(1200)(2) = 2.07\omega_2 + (50/32.2)(2\omega_2)(2)$$
$$\omega_2 = 1.125 \text{ rad/s} \qquad \omega_2 = 1.125 \text{ rad/s} \circlearrowleft \quad \blacktriangleleft$$

Substituting $\bar{v}_2 = (2 \text{ ft})(1.125 \text{ rad/s}) = 2.25 \text{ ft/s}$ into (2), we solve Eqs. (2) and (3) for R_x and R_y, respectively.

$$(0.125/32.2)(1200) + R_x(0.0002) = (50/32.2)(2.25)$$
$$R_x = -5820 \text{ lb} \qquad \mathbf{R}_x = 5820 \text{ lb} \leftarrow \quad \blacktriangleleft$$
$$-(0.125/32.2)(900) + R_y(0.0002) = 0$$
$$R_y = +17,470 \text{ lb} \qquad \mathbf{R}_y = 17,470 \text{ lb} \uparrow \quad \blacktriangleleft$$

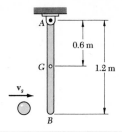

A 2-kg sphere moving horizontally to the right with an initial velocity of 5 m/s strikes the lower end of an 8-kg rigid rod AB. The rod is suspended from a hinge at A and is initially at rest. Knowing that the coefficient of restitution between the rod and sphere is 0.80, determine the angular velocity of the rod and the velocity of the sphere immediately after the impact.

Principle of Impulse and Momentum. We consider the rod and sphere as a single system and express that the initial momenta of the rod and sphere and the impulses of the external forces are together equipollent to the final momenta of the system. We note that the only impulsive force external to the system is the impulsive reaction at A.

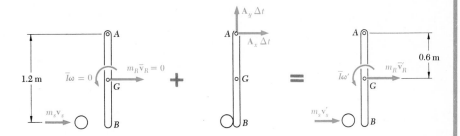

$+\uparrow$ moments about A:
$$m_s v_s(1.2\text{ m}) = m_s v_s'(1.2\text{ m}) + m_R \bar{v}_R'(0.6\text{ m}) + \bar{I}\omega' \quad (1)$$

Since the rod rotates about A, we have $\bar{v}_R' = \bar{r}\omega' = (0.6\text{ m})\omega'$. Also,

$$\bar{I} = \tfrac{1}{12}mL^2 = \tfrac{1}{12}(8\text{ kg})(1.2\text{ m})^2 = 0.96\text{ kg}\cdot\text{m}^2$$

Substituting these values and the given data into Eq. (1), we have

$(2\text{ kg})(5\text{ m/s})(1.2\text{ m})$
$$= (2\text{ kg})v_s'(1.2\text{ m}) + (8\text{ kg})(0.6\text{ m})\omega'(0.6\text{ m}) + (0.96\text{ kg}\cdot\text{m}^2)\omega'$$
$$12 = 2.4v_s' + 3.84\omega' \quad (2)$$

Relative Velocities. Choosing positive to the right, we write

$$v_B' - v_s' = e(v_s - v_B)$$

Substituting $v_s = 5$ m/s, $v_B = 0$, and $e = 0.80$, we obtain

$$v_B' - v_s' = 0.80(5\text{ m/s}) \quad (3)$$

Again noting that the rod rotates about A, we write

$$v_B' = (1.2\text{ m})\omega' \quad (4)$$

Solving Eqs. (2) to (4) simultaneously, we obtain

$$\omega' = +3.21\text{ rad/s} \qquad \omega' = 3.21\text{ rad/s} \,\downarrow \quad \blacktriangleleft$$
$$v_s' = -0.143\text{ m/s} \qquad \mathbf{v}_s' = 0.143\text{ m/s} \leftarrow \quad \blacktriangleleft$$

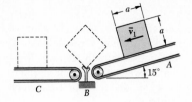

A square package of side a and mass m moves down a conveyor belt A with a constant velocity $\bar{\mathbf{v}}_1$. At the end of the conveyor belt, the corner of the package strikes a rigid support at B. Assuming that the impact at B is perfectly plastic, derive an expression for the smallest magnitude of the velocity $\bar{\mathbf{v}}_1$ for which the package will rotate about B and reach conveyor belt C.

Principle of Impulse and Momentum. Since the impact between the package and the support is perfectly plastic, the package rotates about B during the impact. We apply the principle of impulse and momentum to the package and note that the only impulsive force external to the package is the impulsive reaction at B.

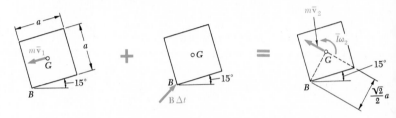

Syst Momenta$_1$ + Syst Ext Imp$_{1\to2}$ = Syst Momenta$_2$

$+\gamma$ moments about B: $\qquad (m\bar{v}_1)(\tfrac{1}{2}a) + 0 = (m\bar{v}_2)(\tfrac{1}{2}\sqrt{2}a) + \bar{I}\omega_2 \qquad$ (1)

Since the package rotates about B, we have $\bar{v}_2 = (\overline{GB})\omega_2 = \tfrac{1}{2}\sqrt{2}a\omega_2$. We substitute this expression, together with $\bar{I} = \tfrac{1}{6}ma^2$, into Eq. (1):

$$(m\bar{v}_1)(\tfrac{1}{2}a) = m(\tfrac{1}{2}\sqrt{2}a\omega_2)(\tfrac{1}{2}\sqrt{2}a) + \tfrac{1}{6}ma^2\omega_2 \qquad \bar{v}_1 = \tfrac{4}{3}a\omega_2 \qquad (2)$$

Principle of Conservation of Energy. We apply the principle of conservation of energy between position 2 and position 3.

Position 2. $V_2 = Wh_2$. Recalling that $\bar{v}_2 = \tfrac{1}{2}\sqrt{2}a\omega_2$, we write

$$T_2 = \tfrac{1}{2}m\bar{v}_2^2 + \tfrac{1}{2}\bar{I}\omega_2^2 = \tfrac{1}{2}m(\tfrac{1}{2}\sqrt{2}a\omega_2)^2 + \tfrac{1}{2}(\tfrac{1}{6}ma^2)\omega_2^2 = \tfrac{1}{3}ma^2\omega_2^2$$

Position 3. Since the package must reach conveyor belt B, it must pass through position 3 where G is directly above B. Also, since we wish to determine the smallest velocity for which the package will reach this position, we choose $\bar{v}_3 = \omega_3 = 0$. Therefore $T_3 = 0$ and $V_3 = Wh_3$.

Conservation of Energy

$$T_2 + V_2 = T_3 + V_3$$
$$\tfrac{1}{3}ma^2\omega_2^2 + Wh_2 = 0 + Wh_3$$
$$\omega_2^2 = \frac{3W}{ma^2}(h_3 - h_2) = \frac{3g}{a^2}(h_3 - h_2) \qquad (3)$$

Substituting the computed values of h_2 and h_3 into Eq. (3), we obtain

$$\omega_2^2 = \frac{3g}{a^2}(0.707a - 0.612a) = \frac{3g}{a^2}(0.095a) \qquad \omega_2 = \sqrt{0.285g/a}$$
$$\bar{v}_1 = \tfrac{4}{3}a\omega_2 = \tfrac{4}{3}a\sqrt{0.285g/a} \qquad v_1 = 0.712\sqrt{ga} \qquad \blacktriangleleft$$

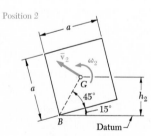

Position 2

$\overline{GB} = \tfrac{1}{2}\sqrt{2}a = 0.707a$

$h_2 = \overline{GB}\sin(45° + 15°)$

$\quad = 0.612a$

Position 3

$h_3 = \overline{GB} = 0.707a$

PROBLEMS

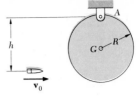

Fig. P18.36

18.36 A 45-g bullet is fired with a horizontal velocity of 400 m/s into a 9-kg wooden disk of radius $R = 200$ mm. Knowing that $h = 200$ mm and that the disk is initially at rest, determine (a) the velocity of the center of the disk immediately after the bullet becomes embedded, (b) the impulsive reaction at A, assuming that the bullet becomes embedded in 0.001 s.

18.37 In Prob. 18.36, determine (a) the required distance h if the impulsive reaction at A is to be zero, (b) the corresponding velocity of the center of the disk after the bullet becomes embedded.

18.38 A bullet weighing 0.08 lb is fired with a horizontal velocity of 1500 ft/s into the 15-lb wooden rod AB of length $L = 24$ in. The rod, which is initially at rest, is suspended by a cord of length $L = 24$ in. Knowing that $h = 6$ in., determine the velocity of each end of the rod immediately after the bullet becomes embedded.

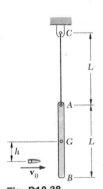

Fig. P18.38

18.39 In Prob. 18.38, determine the distance h for which, immediately after the bullet becomes embedded, the instantaneous center of rotation of the rod is point C.

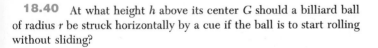

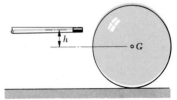

Fig. P18.40

18.40 At what height h above its center G should a billiard ball of radius r be struck horizontally by a cue if the ball is to start rolling without sliding?

18.41 A uniform slender rod AB is equipped at both ends with the hooks shown and is supported by a frictionless horizontal table. Initially the rod is hooked at A to a fixed pin C about which it rotates with the constant angular velocity ω_1. Suddenly end B of the rod hits and gets hooked to the pin D, causing end A to be released. Determine the magnitude of the angular velocity ω_2 of the rod in its subsequent rotation about D.

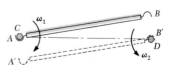

Fig. P18.41

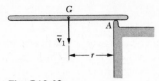

Fig. P18.42

18.42 A slender rigid rod of mass m and with a centroidal radius of gyration $\bar{k}$ strikes a rigid knob at A with a vertical velocity of magnitude $\bar{v}_1$ and no angular velocity. Assuming that the impact is perfectly plastic, show that the magnitude of the velocity of G after impact is

$$\bar{v}_2 = \bar{v}_1 \frac{r^2}{r^2 + \bar{k}^2}$$

18.43 Assuming that in Prob. 18.42 the impact with knob A is perfectly elastic ($e = 1$), (a) show that the magnitude of the velocity of G after impact is

$$\bar{v}_2 = \bar{v}_1 \frac{r^2 - \bar{k}^2}{r^2 + \bar{k}^2}$$

(b) derive an expression for the angular velocity of the rod after impact.

18.44 A slender rod of mass m and length L is released from rest in the position shown. The coefficient of friction is sufficient to prevent sliding at A. Assuming perfectly elastic impact ($e = 1$) at D, determine the distance d for which the rod will rebound with no angular velocity.

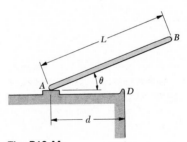

Fig. P18.44

18.45 Solve Prob. 18.44, assuming $e = \frac{1}{2}$.

18.46 A slender rod of length L is falling with a velocity $\bar{v}_1$ at the instant when the cords simultaneously become taut. Assuming that the impacts are perfectly plastic, determine the angular velocity of the rod and the velocity of its mass center immediately after the cords become taut.

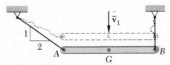

Fig. P18.46

18.47 A square block of mass m is falling with a velocity $\bar{v}_1$ when it strikes a small obstruction at B. Assuming that the impact between corner A and the obstruction B is perfectly plastic, determine the angular velocity of the block and the velocity of its mass center G immediately after impact.

18.48 Solve Prob. 18.47, assuming that the impact between corner A and the obstruction B is perfectly elastic.

18.49 A uniformly loaded rectangular crate is released from rest in the position shown. Assuming that the floor is sufficiently rough to prevent slipping and that the impact at B is perfectly plastic, determine the largest value of the ratio b/a for which corner A will remain in contact with the floor.

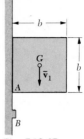

Fig. P18.47

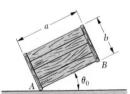

Fig. P18.49

18.50 A uniform sphere of radius r rolls without slipping down the incline shown. It hits the horizontal surface and, after slipping for a while, starts rolling again. Assuming that the sphere does not bounce as it hits the horizontal surface, determine its angular velocity and the velocity of its mass center after it has resumed rolling.

Fig. P18.50

18.51 Solve Prob. 18.50, assuming $r = 4$ in., $\beta = 30°$, and $v_1 = 5$ ft/s.

18.52 A sphere A of mass m and radius r rolls without slipping with a velocity v_0 on a horizontal plane. It hits squarely an identical sphere B which is at rest. Denoting by μ the coefficient of friction between the spheres and the plane, neglecting the friction between the spheres, and assuming perfectly elastic impact $(e = 1)$, determine (a) the linear and angular velocity of each sphere immediately after impact, (b) the velocity of each sphere after it has started rolling uniformly. (c) Discuss the special case when $\mu = 0$.

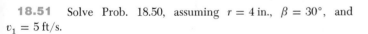

Fig. P18.52

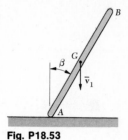

Fig. P18.53

18.53 A slender rod of length l strikes a frictionless floor at A with a vertical velocity $\bar{v}_1$ and no angular velocity. Assuming that the impact at A is perfectly elastic, derive an expression for the angular velocity of the rod immediately after impact.

18.54 Solve Prob. 18.53, assuming that the impact at A is perfectly plastic.

18.55 A slender rod of mass m and length l is held in the position shown. Roller B is given a slight push to the right and moves along the horizontal plane, while roller A is constrained to move vertically. Determine the magnitudes of the impulses exerted on the rollers A and B as roller A strikes the ground. Assume perfectly plastic impact.

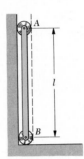

Fig. P18.55

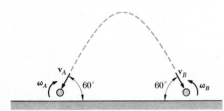

Fig. P18.57

18.56 In Prob. 18.47, determine the line of action of the impulsive force exerted on the block by the obstruction at B.

18.57 A small rubber ball of radius r is thrown against a rough floor with a velocity $\mathbf{v}_A$ of magnitude v_0 and a "backspin" $\boldsymbol{\omega}_A$ of magnitude ω_0. It is observed that the ball bounces from A to B, then from B to A, then from A to B, etc. Assuming perfectly elastic impact, determine (a) the required magnitude ω_0 of the "backspin" in terms of v_0 and r, (b) the minimum required value of the coefficient of friction.

18.58 Two identical rods AB and CD, each of length L, may move freely on a frictionless horizontal surface. Rod AB is rotating about its mass center with an angular velocity $\boldsymbol{\omega}_0$ when end B strikes end C of rod CD, which is at rest. Knowing that at the instant of impact the rods are parallel and assuming perfectly elastic impact ($e = 1$), determine the angular velocity of each rod and the velocity of its mass center immediately after impact.

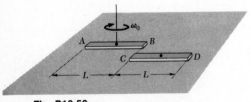

Fig. P18.58

18.59 Solve Prob. 18.58, assuming that the impact is perfectly plastic ($e = 0$).

*18.8. Gyroscopes.   Our study of kinetics is limited to
the plane motion of rigid bodies. The analysis of the general
motion of a rigid body in space is indeed beyond the scope of
this text. Among the problems which we shall *not* attempt to
solve with the means at our disposal are those involving the
general motion of a top or of a gyroscope. However, we shall
consider in this section a particular case of gyroscopic motion
which requires only a very limited knowledge of the motion of
a rigid body in space and yet has many useful engineering applications.

A *gyroscope* consists of a body of revolution mounted on an
axle which coincides with its geometric axis. Consider a gyroscope which spins at a very large rate ω about an axle AB
directed along the z axis (Fig. 18.11). This motion may be represented by a vector $\boldsymbol{\omega}$ of magnitude equal to the *rate of spin* ω
of the gyroscope and directed along the *spin axis* AB in such
a way that an observer located at the tip of the vector $\boldsymbol{\omega}$ will
see the gyroscope rotate counterclockwise.

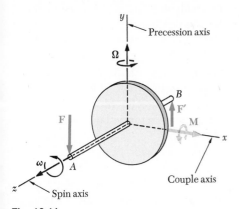

Fig. 18.11

If we attempt to rotate the axle of the gyroscope about the
x axis by applying two vertical forces **F** and **F'** at A and B,
forming a couple of moment M, we find that the axle will resist
the motion we try to impose and that *it will rotate about the*
y axis with an angular velocity Ω. This motion of the gyroscope
is called *precession;* it may be represented by a vector $\boldsymbol{\Omega}$ of
magnitude equal to the *rate of precession* Ω of the gyroscope
and directed along the y axis, or *precession axis,* in such a way
that a man located at the tip of the vector will observe this
motion as counterclockwise.

In order to explain this apparently odd behavior, we shall
consider the gyroscope in two different positions and apply the
principle of impulse and momentum. We first consider the gyro-

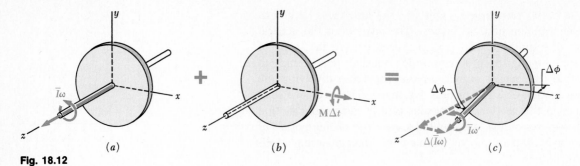

Fig. 18.12

scope in its initial position at $t_1 = 0$. Denoting by $\overline{I}$ its centroidal moment of inertia, we find that the system of the momenta of its particles reduces to a couple of moment $\overline{I}\omega$ (Sec. 18.2). This momentum couple is contained in the xy plane and may therefore be represented by the couple vector $\overline{I}\omega$ directed along the z axis (Fig. 18.12a). Next we consider the gyroscope at time $t_2 = \Delta t$, after its axle has rotated about the y axis through an angle $\Delta\phi$ as a result of our miscalculated effort. We may neglect the relatively small angular momentum of the gyroscope about the y axis and consider only the large momentum couple vector $\overline{I}\omega'$ associated with the rotation of the gyroscope about its axle (Fig. 18.12c). The change in momentum during the time interval Δt is represented by the couple vector $\Delta(\overline{I}\omega)$. This couple vector has the same direction as the couple vector $\mathbf{M}\,\Delta t$, which represents the angular impulse of the couple formed by the two forces applied to the axle (Fig. 18.12b).

Far from being strange, the behavior of the gyroscope is therefore entirely predictable. Noting that the couple vector $\Delta(\overline{I}\omega)$ has a magnitude equal to $\overline{I}\omega\,\Delta\phi$, we equate the angular impulse and the change in angular momentum and write

$$M\,\Delta t = \overline{I}\omega\,\Delta\phi$$

Dividing both members by Δt and letting Δt approach zero, we find that the moment M of the *gyroscopic couple* is equal to

$$M = \overline{I}\omega\,\frac{d\phi}{dt}$$

$$M = I\omega\Omega \tag{18.20}$$

where ω = rate of spin
$\quad\Omega = d\phi/dt$ = rate of precession of the axis of the gyroscope

We note that the spin axis, the couple axis (about which the force couple is applied), and the precession axis are at right angles to each other. The relative position of these axes may easily be found if we remember that the extremity of the vector couple $\bar{I}\omega$ representing the momentum of the gyroscope tends to move in the direction defined by the vector **M** representing the gyroscopic couple.

If the motion just described is maintained for an appreciable length of time, the gyroscope is said to be in *steady precession*. Steady precession may be obtained only if the axis of the applied couple is maintained perpendicular to the spin axis and to the precession axis.

We shall now consider another case of steady precession, in which the axis of the gyroscope forms with the precession axis an angle β different from 90° (Fig. 18.13). Neglecting again the angular momentum associated with the motion of precession, we observe that the change in momentum during a time interval

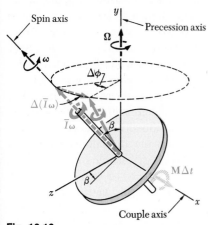

Fig. 18.13

Δt is represented by the vector $\Delta(\bar{I}\omega)$ of magnitude $\bar{I}\omega \sin \beta \, \Delta\phi$. Equating angular impulse and change in angular momentum, we write

$$M \, \Delta t = \bar{I}\omega \sin \beta \, \Delta\phi$$

and obtain

$$M = \bar{I}\omega\Omega \sin \beta \tag{18.21}$$

where Ω represents the rate of precession $d\phi/dt$. Thus, the motion described will be obtained if a couple of moment M satisfying Eq. (18.21) is applied about an axis which is maintained perpendicular to the spin axis and to the precession axis.

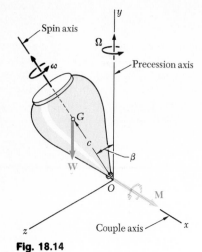

Fig. 18.14

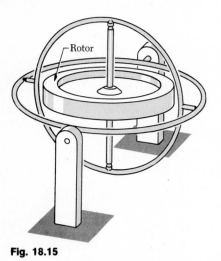

Fig. 18.15

An interesting example of gyroscopic motion is provided by a top supported at a fixed point O (Fig. 18.14). The weight **W** of the top and the normal component of the reaction at O form a couple of moment $M = Wc \sin \beta$. Assuming that the angle β the axis of the top forms with the vertical remains constant, we substitute for M in formula (18.21) and write

$$Wc = \bar{I}\omega\Omega \qquad (18.22)$$

We should note that the result obtained is valid only for a very large rate of spin and that the top will just fall if it does not spin. Besides, the top will move in *steady* precession, at a constant rate Ω and with its axis forming a constant angle β with the vertical, only if it is given initially a rate of spin ω and a rate of precession Ω which satisfy Eq. (18.22). Otherwise, *the precession will be unsteady* and will be accompanied by variations in β, referred to as *nutation*. The study of the general case of motion of the top is a difficult one which requires a full knowledge of the motion of rigid bodies in space.†

A gyroscope mounted in a Cardan's suspension (Fig. 18.15) is free from any applied couple and thus will maintain a fixed direction with respect to a newtonian frame of reference. This was one of the facts used by the French physicist Jean Foucault (1819–1868) to prove that the earth rotates about its axis. The *gyrocompass* used to indicate the true north consists essentially of a gyroscope whose axis is maintained in a horizontal plane and oscillates symmetrically about the direction north-south. Other sensitive instruments based on gyroscopic motion include the inertial guidance systems of rockets, as well as the turn indicator, the artificial horizon, and the automatic pilot used on airplanes.

Because of the relatively large couples required to change the orientation of their axles, gyroscopes are used as stabilizers in torpedoes and ships. Spinning bullets and shells remain tangent to their trajectory because of gyroscopic action. And a bicycle is easier to keep balanced at high speeds because of the stabilizing effect of its spinning wheels. However, gyroscopic action is not always welcome and must be taken into account in the design of bearings supporting rotating shafts subjected to forced precession. The reactions exerted by its propellers on an airplane which changes its direction of flight must also be taken into consideration and compensated for whenever possible.

† Cf. Synge and Griffith, "Principles of Mechanics," McGraw-Hill Book Company, New York, sec. 14.2, or Becker, "Introduction to Theoretical Mechanics," McGraw-Hill Book Company, New York, sec. 12.14.

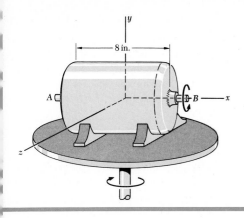

The rotor of an electric motor weighs 6 lb and has a radius of gyration of 2 in. The angular velocity of the rotor is 3600 rpm counterclockwise as viewed from the positive x axis. Determine the reactions exerted by the bearings on the axle AB when the motor is rotated about the y axis clockwise as viewed from above and at a rate of 6 rpm.

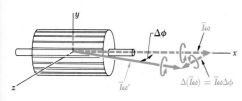

Solution. During the time interval Δt, the momentum couple vector $\bar{I}\omega$ maintains a constant magnitude $\bar{I}\omega$ but is rotated in a horizontal plane through an angle $\Delta\phi$ into a new position $\bar{I}\omega'$. Thus, the change in angular momentum is represented by the vector $\Delta(\bar{I}\omega)$ parallel to the z axis and of magnitude $\bar{I}\omega\,\Delta\phi$. Denoting by $\mathbf{M}_z$ the couple applied about the z axis, we write that the angular impulse $\mathbf{M}_z\,\Delta t$ must equal the change in angular momentum.

$$M_z\,\Delta t = \bar{I}\omega\,\Delta\phi \qquad M_z = \bar{I}\omega\,\frac{\Delta\phi}{\Delta t}$$

Denoting by $\boldsymbol{\Omega}$ the angular velocity about the vertical y axis, we write $\Omega = \Delta\phi/\Delta t$ and obtain

$$M_z = \bar{I}\omega\Omega \qquad\qquad (1)$$

Using the given data, we have

$$\omega = 3600 \text{ rpm} = 377 \text{ rad/s}$$
$$\Omega = 6 \text{ rpm} = 0.628 \text{ rad/s}$$

$$\bar{I} = m\bar{k}^2 = \frac{6 \text{ lb}}{32.2 \text{ ft/s}^2}(\tfrac{2}{12} \text{ ft})^2 = 0.00518 \text{ lb} \cdot \text{ft} \cdot \text{s}^2$$

Substituting these values into (1), we obtain the moment of the gyroscopic couple.

$$M_z = (0.00518 \text{ lb} \cdot \text{ft} \cdot \text{s}^2)(377 \text{ rad/s})(0.628 \text{ rad/s})$$
$$= 1.226 \text{ lb} \cdot \text{ft}$$

Reactions at A and B. The gyroscopic reactions $\mathbf{A}_g$ and $\mathbf{B}_g$ must be *equivalent* to the gyroscopic couple $\mathbf{M}_z$.

$$M_z = A_g(\tfrac{8}{12} \text{ ft})$$
$$1.226 \text{ lb} \cdot \text{ft} = A_g(\tfrac{8}{12} \text{ ft})$$

$$A_g = 1.84 \text{ lb} \qquad \mathbf{A}_g = 1.84 \text{ lb} \downarrow$$
$$B_g = 1.84 \text{ lb} \qquad \mathbf{B}_g = 1.84 \text{ lb} \uparrow$$

Since the static reactions are $\mathbf{A}_s = 3 \text{ lb} \uparrow$ and $\mathbf{B}_s = 3 \text{ lb} \uparrow$, the total reactions are

$$\mathbf{A} = 1.16 \text{ lb} \uparrow \qquad \mathbf{B} = 4.84 \text{ lb} \uparrow \quad \blacktriangleleft$$

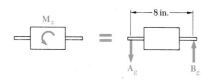

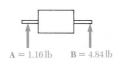

A = 1.16 lb   B = 4.84 lb

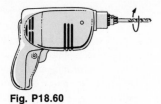

Fig. P18.60

PROBLEMS

18.60 The main rotating parts of a hand drill rotate clockwise when viewed from behind. In what direction will the front end of the drill tend to move if the man holding the drill (a) rotates it counterclockwise in a horizontal plane, (b) tips it downward while maintaining it in the same vertical plane?

18.61 The major rotating elements of a given propeller-driven airplane rotate clockwise when viewed from behind. If the pilot turns the airplane to his right, determine in what direction the nose of the airplane will tend to move under the action of the gyroscopic couple exerted by the rotating elements.

18.62 A three-bladed airplane propeller has a mass of 150 kg and a radius of gyration of 0.9 m. Knowing that the propeller rotates at 1500 rpm, determine the moment of the couple applied by the propeller to its shaft when the airplane travels in a circular path of 400-m radius at 600 km/h.

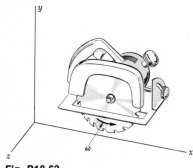

Fig. P18.63

18.63 The blade of a portable saw and the rotor of its motor have a combined mass of 1.2 kg and a radius of gyration of 35 mm. Determine the couple that a man must exert on the handle to rotate the saw about the y axis with a constant angular velocity of 3 rad/s clockwise, as viewed from above, when the blade rotates at the rate $\omega = 1800$ rpm as shown.

18.64 The flywheel of an automobile engine, which is mounted on the crankshaft, is equivalent to a 16-in.-diameter steel plate of $\frac{15}{16}$-in. thickness. At a time when the flywheel is rotating at 4000 rpm the automobile is traveling around a curve of 600-ft radius at a speed of 60 mi/h. Determine, at that time, the magnitude of the gyroscopic couple exerted by the flywheel on the horizontal crankshaft. (Specific weight of steel = 490 lb/ft³.)

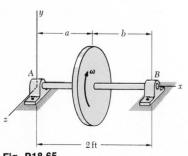

Fig. P18.65

18.65 The rotor of a given turbine may be approximated by a 50-lb disk of 12-in. radius. Knowing that the turbine rotates clockwise at 10,000 rpm as viewed from the positive x axis, determine the components due to gyroscopic action of the forces exerted by the bearings on axle AB if the instantaneous angular velocity of the turbine housing is 2 rad/s clockwise as viewed from (a) the positive y axis, (b) the positive x axis.

18.66 Solve Prob. 18.65 at an instant when the angular velocity of the turbine housing is 2 rad/s clockwise as viewed from the positive z axis.

18.67 The essential structure of a certain type of aircraft turn indicator is shown. Springs AC and BD are initially stretched and exert equal vertical forces at A and B when the airplane is traveling in a straight path. Knowing that the disk weighs $\frac{1}{2}$ lb and spins at the rate of 10,000 rpm, determine the angle through which the yoke will rotate when the airplane executes a horizontal turn of radius 2000 ft at a speed of 400 mi/h. The constant of each spring is 3 lb/in.

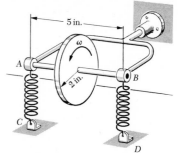

Fig. P18.67

18.68 The rotor shown spins at the rate of 12 000 rpm about the 150-mm shaft CD. At the same time the gimbal rotates about the vertical axis AB at the rate of 20 rpm. Knowing that the mass of the rotor is 30 g and that it has a centroidal radius of gyration of 20 mm, determine the magnitude of the reactions at C and D due to gyroscopic action.

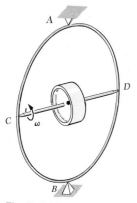

Fig. P18.68

18.69 A toy top has a radius of gyration of 20 mm about its geometric axis and is set spinning at the rate of 1500 rpm. Assuming steady precession, determine the rate of precession (a) for $\beta = 10°$, (b) for $\beta = 20°$.

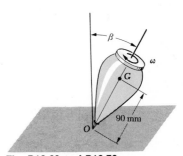

Fig. P18.69 and P18.70

18.70 A toy top having a radius of gyration of 20 mm with respect to its geometric axis is set in motion as shown. It is observed that the top precesses through two complete turns in 1 s. Assuming steady precession, determine the rate at which the top spins about its geometric axis.

18.71 A 5-lb disk of 9-in. diameter is attached to the end of a rod AB of negligible weight which is supported by a ball-and-socket joint at A. If the rate of steady precession Ω of the disk about the vertical is observed to be 24 rpm, determine the rate of spin ω of the disk about AB when $\beta = 30°$.

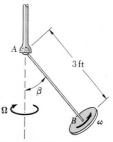

Fig. P18.71

18.72 Solve Prob. 18.71 assuming that the rod AB weighs 2 lb.

18.73 A homogeneous sphere of radius a and mass m is attached to a light rod of length $4a$. The rod forms an angle of $30°$ with the vertical and rotates about AC with a constant angular velocity Ω. (a) Assuming that the sphere does not spin about the rod ($\omega = 0$), determine the tension in the cord BC. (b) Determine the spin ω (magnitude and sense) which should be given to the sphere if the tension in the cord BC is to be zero.

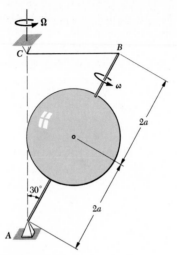

Fig. P18.73

REVIEW PROBLEMS

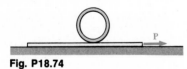

Fig. P18.74

18.74 A 300-mm-diameter pipe of mass 20 kg rests on a 5-kg plate. The pipe and plate are initially at rest when a force P of magnitude 150 N is applied for 0.50 s. Knowing that $\mu = 0.20$ between the plate and *both* the pipe and the floor, determine (a) whether the pipe slides with respect to the plate, (b) the resulting velocities of the pipe and of the plate.

18.75 Solve Prob. 18.74, assuming that the magnitude of the force P is 100 N.

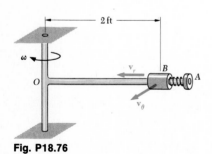

Fig. P18.76

18.76 Collar B weighs 3 lb and may slide freely on rod OA which in turn may rotate freely in the horizontal plane. The assembly is rotating with an angular velocity $\omega = 1.5$ rad/s when a spring located between A and B is released, projecting the collar along the rod with an initial relative speed $v_r = 5$ ft/s. Knowing that the moment of inertia about O of the rod and spring is 0.15 lb · ft · s^2, determine (a) the minimum distance between the collar and point O in the ensuing motion, (b) the corresponding angular velocity of the assembly.

18.77   In Prob. 18.76, determine the required magnitude of the initial relative velocity $\mathbf{v}_r$ if during the ensuing motion the minimum distance between collar B and point O is to be 1 ft.

18.78   A uniform slender rod of length L is dropped onto rigid supports at A and B. Immediately before striking A the velocity of the rod is $\bar{\mathbf{v}}_1$. Since support B is slightly lower than support A, the rod strikes A before it strikes B. Assuming perfectly elastic impact at both A and B, determine the angular velocity of the rod and the velocity of its mass center immediately after the rod (*a*) strikes support A, (*b*) strikes support B, (*c*) again strikes support A.

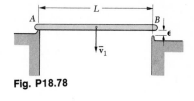
Fig. P18.78

18.79   A rigid rod of length l and mass m is released in the position shown. It is observed that the rod rebounds to a horizontal position after striking the inclined surface. (*a*) Determine the coefficient of restitution between the knob K and the surface. (*b*) Show that the same rebound may be expected for any position of the knob K.

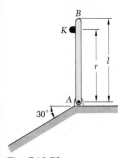
Fig. P18.79

18.80   The plank CDE of mass m_P rests on top of a small pivot at D. A gymnast A of mass m stands on the plank at end C; a second gymnast B of the same mass m jumps from a height h and strikes the plank at E. Assuming perfectly plastic impact, determine the height to which gymnast A will rise. (Assume that gymnast A stands completely rigid.)

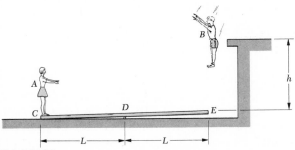
Fig. P18.80

18.81   Solve Prob. 18.80, assuming that the impact between gymnast B and the plank is perfectly elastic.

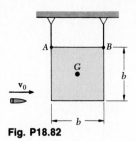

Fig. P18.82

18.82 A bullet of mass m is fired with a horizontal velocity $\mathbf{v}_0$ into the lower corner of a square panel of much larger mass M. The panel is held by two vertical wires as shown. Determine the velocity of the center G of the panel immediately after the bullet becomes embedded.

18.83 A sphere of mass m_S is dropped from a height h and strikes at A the uniform slender plank AB of mass m which is held by two inextensible cords. Knowing that the impact is perfectly plastic and that the sphere remains attached to the plank, determine the velocity of the sphere immediately after impact.

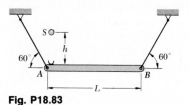

Fig. P18.83

18.84 Solve Prob. 18.82, assuming that the panel is held as shown by a single hinge located at the center of its top edge.

18.85 Solve Prob. 18.83 when $m = 3\,\text{kg}$, $m_S = 1\,\text{kg}$, $L = 1.2\,\text{m}$, and $h = 1.5\,\text{m}$.

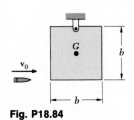

Fig. P18.84

Chapter
19

Mechanical Vibrations

19.1. Introduction. A *mechanical vibration* is the motion of a particle or a body which oscillates about a position of equilibrium. Most vibrations in machines and structures are undesirable because of the increased stresses and energy losses which accompany them. They should therefore be eliminated or reduced as much as possible by appropriate design. The analysis of vibrations has become increasingly important in recent years owing to the current trend toward higher-speed machines and lighter structures. There is every reason to expect that this trend will continue and that an even greater need for vibration analysis will develop in the future.

The analysis of vibrations is a very extensive subject to which entire texts have been devoted. We shall therefore limit our present study to the simpler types of vibrations, namely, the vibrations of a body or a system of bodies with one degree of freedom.

A mechanical vibration generally results when a system is displaced from a position of stable equilibrium. The system tends to return to this position under the action of restoring forces (either elastic forces, as in the case of a mass attached to a spring, or gravitational forces, as in the case of a pendulum). But the system generally reaches its original position with a certain acquired velocity which carries it beyond that position. Since the process can be repeated indefinitely, the system keeps moving back and forth across its position of equilibrium. The time interval required for the system to complete a full cycle of motion is called the *period* of the vibration. The number of cycles per unit time defines the *frequency*, and the maximum displacement of the system from its position of equilibrium is called the *amplitude* of the vibration.

When the motion is maintained by the restoring forces only, the vibration is said to be a *free vibration* (Secs. 19.2 to 19.6). When a periodic force is applied to the system, the resulting motion is described as a *forced vibration* (Sec. 19.7). When the effects of friction may be neglected, the vibrations are said to be *undamped*. However, all vibrations are actually *damped* to some degree. If a free vibration is only slightly damped, its amplitude slowly decreases until, after a certain time, the motion comes to a stop. But damping may be large enough to prevent any true vibration; the system then slowly regains its original position (Sec. 19.8). A damped forced vibration is maintained as long as the periodic force which produces the vibration is applied. The amplitude of the vibration, however, is affected by the magnitude of the damping forces (Sec. 19.9).

VIBRATIONS WITHOUT DAMPING

19.2. Free Vibrations of Particles. Simple Harmonic Motion. Consider a body of mass m attached to a spring of constant k (Fig. 19.1a). Since, at the present time, we are concerned only with the motion of its mass center, we shall refer to this body as a particle. When the particle is in static equilibrium, the forces acting on it are its weight $\mathbf{W}$ and the force $\mathbf{T}$ exerted by the spring, of magnitude $T = k\delta_{st}$, where δ_{st} denotes the elongation of the spring. We have, therefore,

$$W = k\delta_{st} \qquad (19.1)$$

Suppose now that the particle is displaced through a distance x_m from its equilibrium position and released with no initial velocity. If x_m has been chosen smaller than δ_{st}, the particle will move back and forth through its equilibrium position; a vibration of amplitude x_m has been generated. Note that the vibration may also be produced by imparting a certain initial velocity to the particle when it is in its equilibrium position $x = 0$ or, more generally, by starting the particle from any given position $x = x_0$ with a given initial velocity v_0.

To analyze the vibration, we shall consider the particle in a position P at some arbitrary time t (Fig. 19.1b). Denoting by x the displacement OP measured from the equilibrium position

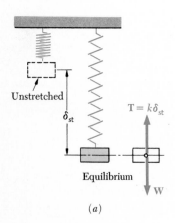

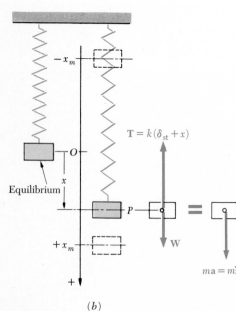

Fig. 19.1

O (positive downward), we note that the forces acting on the particle are its weight $\mathbf{W}$ and the force $\mathbf{T}$ exerted by the spring which, in this position, has a magnitude $T = k(\delta_{st} + x)$. Recalling (19.1), we find that the magnitude of the resultant $\mathbf{F}$ of the two forces (positive downward) is

$$F = W - k(\delta_{st} + x) = -kx \qquad (19.2)$$

Thus the *resultant* of the forces exerted on the particle is proportional to the displacement *OP measured from the equilibrium position*. Recalling the sign convention, we note that $\mathbf{F}$ is always directed *toward* the equilibrium position O. Substituting for F into the fundamental equation $F = ma$ and recalling that a is the second derivative $\ddot{x}$ of x with respect to t, we write

$$m\ddot{x} + kx = 0 \qquad (19.3)$$

Note that the same sign convention should be used for the acceleration $\ddot{x}$ and for the displacement x, namely, positive downward.

Equation (19.3) is a linear differential equation of the second order. Setting

$$p^2 = \frac{k}{m} \qquad (19.4)$$

we write (19.3) in the form

$$\ddot{x} + p^2 x = 0 \qquad (19.5)$$

The motion defined by Eq. (19.5) is called *simple harmonic motion*. It is characterized by the fact that *the acceleration is proportional to the displacement and of opposite direction*. We note that each of the functions $x_1 = \sin pt$ and $x_2 = \cos pt$ satisfies (19.5). These functions, therefore, constitute two *particular solutions* of the differential equation (19.5). As we shall see presently, the *general solution* of (19.5) may be obtained by multiplying the two particular solutions by arbitrary constants A and B and adding. We write

$$x = Ax_1 + Bx_2 = A \sin pt + B \cos pt \qquad (19.6)$$

Differentiating, we obtain successively the velocity and acceleration at time t,

$$v = \dot{x} = Ap \cos pt - Bp \sin pt \qquad (19.7)$$

$$a = \ddot{x} = -Ap^2 \sin pt - Bp^2 \cos pt \qquad (19.8)$$

Substituting from (19.6) and (19.8) into (19.5), we verify that the expression (19.6) provides a solution of the differential equation

(19.5). Since this expression contains two arbitrary constants A and B, the solution obtained is the general solution of the differential equation. The values of the constants A and B depend upon the *initial conditions* of the motion. For example, we have $A = 0$ if the particle is displaced from its equilibrium position and released at $t = 0$ with no initial velocity, and we have $B = 0$ if P is started from O at $t = 0$ with a certain initial velocity. In general, substituting $t = 0$ and the initial values x_0 and v_0 of the displacement and velocity into (19.6) and (19.7), we find $A = v_0/p$ and $B = x_0$.

The expressions obtained for the displacement, velocity, and acceleration of a particle may be written in a more compact form if we observe that (19.6) expresses that the displacement $x = OP$ is the sum of the x components of two vectors **A** and **B**, respectively of magnitude A and B, directed as shown in Fig. 19.2a. As t varies, both vectors rotate clockwise; we also note that the magnitude of their resultant $\overrightarrow{OQ}$ is equal to the maxi-

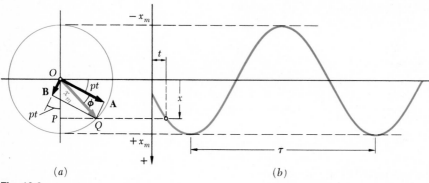

(a)   (b)

Fig. 19.2

mum displacement x_m. The simple harmonic motion of P along the x axis may thus be obtained by projecting on this axis the motion of a point Q describing an *auxiliary circle* of radius x_m *with a constant angular velocity p.* Denoting by ϕ the angle formed by the vectors $\overrightarrow{OQ}$ and **A**, we write

$$OP = OQ \sin{(pt + \phi)} \qquad (19.9)$$

which leads to new expressions for the displacement, velocity, and acceleration of P,

$$\boxed{x = x_m \sin{(pt + \phi)}} \qquad (19.10)$$

$$v = \dot{x} = x_m p \cos{(pt + \phi)} \qquad (19.11)$$

$$a = \ddot{x} = -x_m p^2 \sin{(pt + \phi)} \qquad (19.12)$$

The displacement-time curve is represented by a sine curve (Fig. 19.2b), and the maximum value x_m of the displacement is called the *amplitude* of the vibration. The angular velocity p of the point Q which describes the auxiliary circle is known as the *circular frequency* of the vibration and is measured in rad/s, while the angle ϕ which defines the initial position of Q on the circle is called the *phase angle*. We note from Fig. 19.2 that a full *cycle* has been described after the angle pt has increased by 2π rad. The corresponding value of t, denoted by τ, is called the *period* of the vibration and is measured in seconds. We have

$$\text{Period} = \tau = \frac{2\pi}{p} \qquad (19.13)$$

The number of cycles described per unit of time is denoted by f and is known as the *frequency* of the vibration. We write

$$\text{Frequency} = f = \frac{1}{\tau} = \frac{p}{2\pi} \qquad (19.14)$$

The unit of frequency is a frequency of 1 cycle per second, corresponding to a period of 1 s. In terms of base units the unit of frequency is thus 1/s or s^{-1}. It is called a *hertz* (Hz) in the SI system of units. It also follows from Eq. (19.14) that a frequency of $1\,s^{-1}$ or 1 Hz corresponds to a circular frequency of 2π rad/s. In problems involving angular velocities expressed in revolutions per minute (rpm), we have $1\,\text{rpm} = \frac{1}{60}\,s^{-1} = \frac{1}{60}\,\text{Hz}$, or $1\,\text{rpm} = (2\pi/60)$ rad/s.

Recalling that p was defined in (19.4) in terms of the constant k of the spring and the mass m of the particle, we observe that the period and the frequency are independent of the initial conditions and of the amplitude of the vibration. Note that τ and f depend on the *mass* rather than on the *weight* of the particle and thus are independent of the value of g.

The velocity-time and acceleration-time curves may be represented by sine curves of the same period as the displacement-time curve, but with different phase angles. From (19.11) and (19.12), we note that the maximum values of the magnitudes of the velocity and acceleration are

$$v_m = x_m p \qquad a_m = x_m p^2 \qquad (19.15)$$

Since the point Q describes the auxiliary circle, of radius x_m, at the constant angular velocity p, its velocity and acceleration are equal, respectively, to the expressions (19.15). Recalling Eqs. (19.11) and (19.12), we find, therefore, that the velocity and acceleration of P may be obtained at any instant by projecting on the x axis vectors of magnitudes $v_m = x_m p$ and $a_m = x_m p^2$ representing respectively the velocity and acceleration of Q at the same instant (Fig. 19.3).

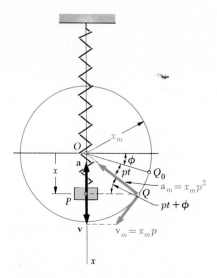

Fig. 19.3

The results obtained are not limited to the solution of the problem of a mass attached to a spring. They may be used to analyze the rectilinear motion of a particle *whenever the resultant* **F** *of the forces acting on the particle is proportional to the displacement x and directed toward O.* The fundamental equation of motion $F = ma$ may then be written in the form (19.5), which characterizes simple harmonic motion. Observing that the coefficient of x in (19.5) represents the square of the circular frequency p of the vibration, we easily obtain p and, after substitution into (19.13) and (19.14), the period τ and the frequency f of the vibration.

19.3. Simple Pendulum (Approximate Solution). Most of the vibrations encountered in engineering applications may be represented by a simple harmonic motion. Many others, although of a different type, may be *approximated* by a simple harmonic motion, provided that their amplitude remains small. Consider for example a *simple pendulum,* consisting of a bob of mass m attached to a cord of length l, which may oscillate in a vertical plane (Fig. 19.4a). At a given time t, the cord forms an angle θ with the vertical. The forces acting on the bob are its weight **W** and the force **T** exerted by the cord (Fig. 19.4b). Resolving the vector $m\mathbf{a}$ into tangential and normal components,

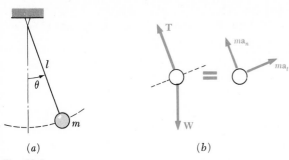

(a)   (b)

Fig. 19.4

with ma_t directed to the right, i.e., in the direction corresponding to increasing values of θ, and observing that $a_t = l\alpha = l\ddot{\theta}$, we write

$\Sigma F_t = ma_t:$ $-W \sin \theta = ml\ddot{\theta}$

Noting that $W = mg$ and dividing through by ml, we obtain

$$\ddot{\theta} + \frac{g}{l} \sin \theta = 0 \qquad (19.16)$$

For oscillations of small amplitude, we may replace $\sin \theta$ by θ, expressed in radians, and write

$$\ddot{\theta} + \frac{g}{l}\theta = 0 \qquad (19.17)$$

Comparison with (19.5) shows that the equation obtained is that of a simple harmonic motion and that the circular frequency p of the oscillations is equal to $(g/l)^{1/2}$. Substitution into (19.13) yields the period of the small oscillations of a pendulum of length l,

$$\tau = \frac{2\pi}{p} = 2\pi \sqrt{\frac{l}{g}} \qquad (19.18)$$

∗19.4. Simple Pendulum (Exact Solution). Formula (19.18) is only approximate. To obtain an exact expression for the period of the oscillations of a simple pendulum, we must return to (19.16). Multiplying both terms by $2\dot{\theta}$ and integrating from an initial position corresponding to the maximum deflection, that is, $\theta = \theta_m$ and $\dot{\theta} = 0$, we write

$$\dot{\theta}^2 = \frac{2g}{l}(\cos \theta - \cos \theta_m)$$

or

$$\left(\frac{d\theta}{dt}\right)^2 = \frac{2g}{l}(\cos \theta - \cos \theta_m)$$

Replacing $\cos \theta$ by $1 - 2 \sin^2 (\theta/2)$ and $\cos \theta_m$ by a similar expression, solving for dt, and integrating over a quarter period from $t = 0$, $\theta = 0$ to $t = \tau/4$, $\theta = \theta_m$, we have

$$\tau = 2 \sqrt{\frac{l}{g}} \int_0^{\theta_m} \frac{d\theta}{\sqrt{\sin^2 (\theta_m/2) - \sin^2 (\theta/2)}}$$

The integral in the right-hand member is known as an *elliptic integral;* it cannot be expressed in terms of the usual algebraic or trigonometric functions. However, setting

$$\sin (\theta/2) = \sin (\theta_m/2) \sin \phi$$

we may write

$$\tau = 4 \sqrt{\frac{l}{g}} \int_0^{\pi/2} \frac{d\phi}{\sqrt{1 - \sin^2 (\theta_m/2) \sin^2 \phi}} \qquad (19.19)$$

where the integral obtained, commonly denoted by K, may be found in *tables of elliptic integrals* for various values of $\theta_m/2$.†
In order to compare the result just obtained with that of the preceding section, we write (19.19) in the form

$$\tau = \frac{2K}{\pi} \left(2\pi \sqrt{\frac{l}{g}} \right) \qquad (19.20)$$

Formula (19.20) shows that the actual value of the period of a simple pendulum may be obtained by multiplying the approximate value (19.18) by the correction factor $2K/\pi$. Values of the correction factor are given in Table 19.1 for various values of

Table 19.1 Correction Factor for the Period of a
Simple Pendulum

θ_m	0°	10°	20°	30°	60°	90°	120°	150°	180°
K	1.571	1.574	1.583	1.598	1.686	1.854	2.157	2.768	∞
$2K/\pi$	1.000	1.002	1.008	1.017	1.073	1.180	1.373	1.762	∞

the amplitude θ_m. We note that for ordinary engineering computations the correction factor may be omitted as long as the amplitude does not exceed 10°.

† See, for example, Dwight, "Table of Integrals and Other Mathematical Data," The Macmillan Company, or Peirce, "A Short Table of Integrals," Ginn and Company.

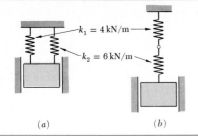

(a)   (b)

A 50-kg block moves between vertical guides as shown. The block is pulled 40 mm down from its equilibrium position and released. For each spring arrangement, determine the period of the vibration, the maximum velocity of the block, and the maximum acceleration of the block.

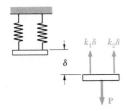

a. Springs Attached in Parallel. We first determine the constant k of a single spring equivalent to the two springs *by finding the magnitude of the force* **P** required to cause a given deflection δ. Since for a deflection δ the magnitudes of the forces exerted by the springs are, respectively, $k_1\delta$ and $k_2\delta$, we have

$$P = k_1\delta + k_2\delta = (k_1 + k_2)\delta$$

The constant k of the single equivalent spring is

$$k = \frac{P}{\delta} = k_1 + k_2 = 4\,\text{kN/m} + 6\,\text{kN/m} = 10\,\text{kN/m} = 10^4\,\text{N/m}$$

Period of Vibration: Since $m = 50$ kg, Eq. (19.4) yields

$$p^2 = \frac{k}{m} = \frac{10^4\,\text{N/m}}{50\,\text{kg}} \qquad p = 14.14\,\text{rad/s}$$

$$\tau = 2\pi/p \qquad\qquad \tau = 0.444\,\text{s} \quad \blacktriangleleft$$

Maximum Velocity: $\quad v_m = x_m p = (0.040\,\text{m})(14.14\,\text{rad/s})$

$$v_m = 0.566\,\text{m/s} \qquad \text{v}_m = 0.566\,\text{m/s}\,\updownarrow \quad \blacktriangleleft$$

Maximum Acceleration: $\quad a_m = x_m p^2 = (0.040\,\text{m})(14.14\,\text{rad/s})^2$

$$a_m = 8.00\,\text{m/s}^2 \qquad \text{a}_m = 8.00\,\text{m/s}^2\,\updownarrow \quad \blacktriangleleft$$

b. Springs Attached in Series. We first determine the constant k of a single spring equivalent to the two springs *by finding the total elongation* δ of the springs under a given static load **P**. To facilitate the computation, a static load of magnitude $P = 12$ kN is used.

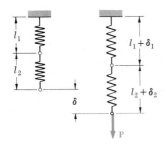

$$\delta = \delta_1 + \delta_2 = \frac{P}{k_1} + \frac{P}{k_2} = \frac{12\,\text{kN}}{4\,\text{kN/m}} + \frac{12\,\text{kN}}{6\,\text{kN/m}} = 5\,\text{m}$$

$$k = \frac{P}{\delta} = \frac{12\,\text{kN}}{5\,\text{m}} = 2.4\,\text{kN/m} = 2400\,\text{N/m}$$

Period of Vibration: $\quad p^2 = \frac{k}{m} = \frac{2400\,\text{N/m}}{50\,\text{kg}} \qquad p = 6.93\,\text{rad/s}$

$$\tau = \frac{2\pi}{p} \qquad\qquad \tau = 0.907\,\text{s} \quad \blacktriangleleft$$

Maximum Velocity: $\quad v_m = x_m p = (0.040\,\text{m})(6.93\,\text{rad/s})$

$$v_m = 0.277\,\text{m/s} \qquad \text{v}_m = 0.277\,\text{m/s}\,\updownarrow \quad \blacktriangleleft$$

Maximum Acceleration: $\quad a_m = x_m p^2 = (0.040\,\text{m})(6.93\,\text{rad/s})^2$

$$a_m = 1.920\,\text{m/s}^2 \qquad \text{a}_m = 1.920\,\text{m/s}^2\,\updownarrow \quad \blacktriangleleft$$

PROBLEMS

19.1 A particle moves in simple harmonic motion with an amplitude of 3 in. and a period of 0.60 s. Find the maximum velocity and the maximum acceleration.

19.2 The analysis of the motion of a particle shows a maximum acceleration of 30 m/s² and a frequency of 120 cycles per minute. Assuming that the motion is simple harmonic, determine (*a*) the amplitude, (*b*) the maximum velocity.

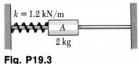
$k = 1.2$ kN/m

A

2 kg

Fig. P19.3

19.3 Collar A is attached to the spring shown and may slide without friction on the horizontal rod. If the collar is moved 75 mm from its equilibrium position and released, determine the period, the maximum velocity, and the maximum acceleration of the resulting motion.

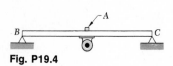

Fig. P19.4

19.4 A variable-speed motor is rigidly attached to the beam BC. The rotor is slightly unbalanced and causes the beam to vibrate with a circular frequency equal to the motor speed. When the speed of the motor is less than 600 rpm or more than 1200 rpm, a small object placed at A is observed to remain in contact with the beam. For speeds between 600 and 1200 rpm the object is observed to "dance" and actually to lose contact with the beam. Determine the amplitude of the motion of A when the speed of the motor is (*a*) 600 rpm, (*b*) 1200 rpm. Give answers in both SI and U.S. customary units.

A m

k

Fig. P19.5, P19.6, and P19.8

19.5 The 10-lb collar rests on, but is not attached to, the spring shown. The collar is depressed 1.5 in. and released. If the ensuing motion is to be simple harmonic, determine (*a*) the largest permissible value of the spring constant k, (*b*) the corresponding frequency of the motion.

19.6 The 5-kg collar is attached to a spring of constant $k = 800$ N/m as shown. If the collar is given a displacement of 50 mm from its equilibrium position and released, determine for the ensuing motion (*a*) the period, (*b*) the maximum velocity of the collar, (*c*) the maximum acceleration of the collar.

19.7 In Prob. 19.6, determine the position, velocity, and acceleration of the collar 0.20 s after it has been released.

19.8 An 8-lb collar is attached to a spring of constant $k = 5$ lb/in. as shown. If the collar is given a displacement of 2 in. downward from its equilibrium position and released, determine (*a*) the time required for the collar to move 3 in. upward, (*b*) the corresponding velocity and acceleration of the collar.

19.9 and 19.10 A 35-kg block is supported by the spring arrangement shown. If the block is moved vertically downward from its equilibrium position and released, determine (a) the period and frequency of the resulting motion, (b) the maximum velocity and acceleration of the block if the amplitude of the motion is 20 mm.

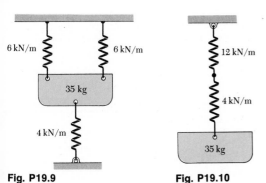

6 kN/m   6 kN/m

35 kg

4 kN/m

Fig. P19.9

12 kN/m

4 kN/m

35 kg

Fig. P19.10

19.11 Denoting by δ_{st} the static deflection of a beam under a given load, show that the frequency of vibration of the load is

$$f = \frac{1}{2\pi}\sqrt{\frac{g}{\delta_{st}}}$$

Neglect the mass of the beam, and assume that the load remains in contact with the beam.

δ_{st}

Fig. P19.11

19.12 The frequency of vibration of the system shown is observed to be 2.5 cycles per second. After cylinder B is removed, the frequency is observed to be 3 cycles per second. Knowing that cylinder B weighs 2 lb, determine the weight of cylinder A.

19.13 A simple pendulum of length l is suspended in an elevator. A mass m is attached to a spring of constant k and is carried in the same elevator. Determine the period of vibration of both the pendulum and the mass if the elevator has an upward acceleration **a**.

A

B

Fig. P19.12

19.14 Determine (a) the required length l of a simple pendulum if the period of small oscillations is to be 2 s, (b) the required amplitude of this pendulum if the maximum velocity of the bob is to be 200 mm/s.

19.15 and 19.16 An 80-lb block is attached to the rod AB and to the two springs shown. Knowing that $h = 30$ in., determine the period of small oscillations of the block. Neglect the weight of the rod and assume that each spring can act in either tension or compression.

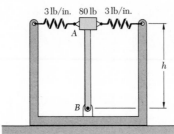

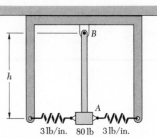

Fig. P19.15   **Fig. P19.16**

19.17 In Prob. 19.15, determine the length h of rod AB for which the period of small oscillations is (*a*) 2 s, (*b*) infinite.

***19.18** A particle is placed with no initial velocity on a frictionless *plane* tangent to the surface of the earth. (*a*) Show that the particle will theoretically execute simple harmonic motion with a period of oscillation equal to that of a simple pendulum of length equal to the radius of the earth. (*b*) Compute the theoretical period of oscillation and show that it is equal to the periodic time of an earth satellite describing a low-altitude circular orbit. [*Hint.* See Eq. (12.35).]

19.19 A small block of mass m rests on a frictionless horizontal surface and is attached to a taut string. Denoting by T the tension in the string, determine the period and frequency of small oscillations of the block in a direction perpendicular to the string. Show that the longest period occurs when $a = b = \frac{1}{2}l$.

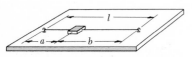

Fig. P19.19 and P19.20

19.20 A 3-lb block rests on a frictionless horizontal surface and is attached to a taut string. Knowing that the tension in the string is 8 lb, determine the frequency of small oscillations of the block when $a = 12$ in. and $b = 18$ in.

***19.21** Expanding the integrand in (19.19) into a series of even powers of $\sin \phi$ and integrating, show that the period of a simple pendulum of length l may be approximated by the formula

$$\tau = 2\pi \sqrt{\frac{l}{g}} \left(1 + \tfrac{1}{4} \sin^2 \frac{\theta_m}{2} \right)$$

where θ_m is the amplitude of the oscillations.

*19.22   Using the data of Table 19.1, determine the period of a simple pendulum of length 800 mm (*a*) for small oscillations, (*b*) for oscillations of amplitude $\theta_m = 30°$, (*c*) for oscillations of amplitude $\theta_m = 90°$.

*19.23   Using the formula given in Prob. 19.21, determine the amplitude θ_m for which the period of a simple pendulum is $\frac{1}{2}$ percent longer than the period of the same pendulum for small oscillations.

*19.24   Using a table of elliptic integrals, determine the period of a simple pendulum of length $l = 800$ mm if the amplitude of the oscillations is $\theta_m = 40°$.

19.5. Free Vibrations of Rigid Bodies.   The analysis of the vibrations of a rigid body or of a system of rigid bodies possessing a single degree of freedom is similar to the analysis of the vibrations of a particle. An appropriate variable, such as a distance x or an angle θ, is chosen to define the position of the body or system of bodies, and an equation relating this variable and its second derivative with respect to t is written. If the equation obtained is of the same form as (19.5), i.e., if we have

$$\ddot{x} + p^2x = 0 \qquad \text{or} \qquad \ddot{\theta} + p^2\theta = 0 \qquad (19.21)$$

the vibration considered is a simple harmonic motion. The period and frequency of the vibration may then be obtained by identifying p and substituting into (19.13) and (19.14).

In general, a simple way to obtain one of Eqs. (19.21) is to express that the system of the external forces is equivalent to the system of the effective forces by drawing a diagram of the body for an arbitrary value of the variable and writing the appropriate equation of motion. We recall that our goal should be *the determination of the coefficient* of the variable x or θ, *not* the determination of the variable itself or of the derivatives $\ddot{x}$ or $\ddot{\theta}$. Setting this coefficient equal to p^2, we obtain the circular frequency p, from which τ and f may be determined.

The method we have outlined may be used to analyze vibrations which are truly represented by a simple harmonic motion, or vibrations of small amplitude which can be *approximated* by a simple harmonic motion. As an example, we shall determine the period of the small oscillations of a square plate of side $2b$

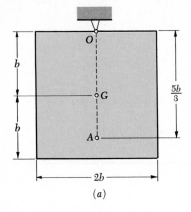

(a)

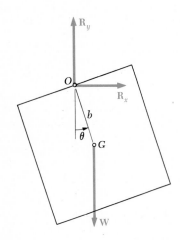

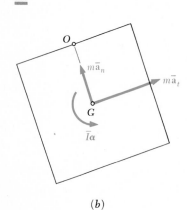

(b)

Fig. 19.5

which is suspended from the midpoint O of one side (Fig. 19.5a). We consider the plate in an arbitrary position defined by the angle θ that the line OG forms with the vertical and draw a diagram to express that the weight $\mathbf{W}$ of the plate and the components $\mathbf{R}_x$ and $\mathbf{R}_y$ of the reaction at O are equivalent to the vectors $m\overline{\mathbf{a}}_t$ and $m\overline{\mathbf{a}}_n$ and to the couple $\overline{I}\boldsymbol{\alpha}$ (Fig. 19.5b). Since the angular velocity and angular acceleration of the plate are equal, respectively, to $\dot{\theta}$ and $\ddot{\theta}$, the magnitudes of the two vectors are, respectively, $mb\ddot{\theta}$ and $mb\dot{\theta}^2$, while the moment of the couple is $\overline{I}\ddot{\theta}$. In previous applications of this method (Chap. 16), we tried whenever possible to assume the correct sense for the acceleration. Here, however, we must assume the same positive sense for θ and $\ddot{\theta}$ in order to obtain an equation of the form (19.21). Consequently, the angular acceleration $\ddot{\theta}$ will be assumed positive counterclockwise, even though this assumption is obviously unrealistic. Equating moments about O, we write

$$+\!\!\downarrow \qquad -W(b\sin\theta) = (mb\ddot{\theta})b + \overline{I}\ddot{\theta}$$

Noting that $\overline{I} = \frac{1}{12}m[(2b)^2 + (2b)^2] = \frac{2}{3}mb^2$ and $W = mg$, we obtain

$$\ddot{\theta} + \frac{3}{5}\frac{g}{b}\sin\theta = 0 \qquad (19.22)$$

For oscillations of small amplitude, we may replace $\sin\theta$ by θ, expressed in radians, and write

$$\ddot{\theta} + \frac{3}{5}\frac{g}{b}\theta = 0 \qquad (19.23)$$

Comparison with (19.21) shows that the equation obtained is that of a simple harmonic motion and that the circular frequency p of the oscillations is equal to $(3g/5b)^{1/2}$. Substituting into (19.13), we find that the period of the oscillations is

$$\tau = \frac{2\pi}{p} = 2\pi\sqrt{\frac{5b}{3g}} \qquad (19.24)$$

The result obtained is valid only for oscillations of small amplitude. A more accurate description of the motion of the plate is obtained by comparing Eqs. (19.16) and (19.22). We note that the two equations are identical if we choose l equal to $5b/3$. This means that the plate will oscillate as a simple pendulum of length $l = 5b/3$, and the results of Sec. 19.4 may be used to correct the value of the period given in (19.24). The point A of the plate located on line OG at a distance $l = 5b/3$ from O is defined as the *center of oscillation* corresponding to O (Fig. 19.5a).

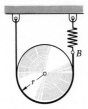

A cylinder of weight W and radius r is suspended from a looped cord as shown. One end of the cord is attached directly to a rigid support, while the other end is attached to a spring of constant k. Determine the period and frequency of vibration of the cylinder.

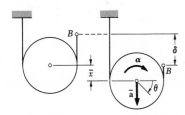

Kinematics of Motion. We express the linear displacement and the acceleration of the cylinder in terms of the angular displacement θ. Choosing the positive sense clockwise and measuring the displacements from the equilibrium position, we write

$$\bar{x} = r\theta \qquad \delta = 2\bar{x} = 2r\theta$$
$$\alpha = \ddot{\theta}\,\rangle \qquad \bar{a} = r\alpha = r\ddot{\theta} \qquad \bar{\mathbf{a}} = r\ddot{\theta} \downarrow \qquad (1)$$

Equations of Motion. The system of external forces acting on the cylinder consists of the weight $\mathbf{W}$ and of the forces $\mathbf{T}_1$ and $\mathbf{T}_2$ exerted by the cord. We express that this system is equivalent to the system of effective forces represented by the vector $m\bar{\mathbf{a}}$ attached at G and the couple $\bar{I}\alpha$.

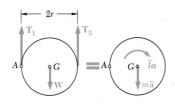

$$+\,\rangle\ \Sigma M_A = \Sigma(M_A)_{\text{eff}}\text{:} \qquad Wr - T_2(2r) = m\bar{a}r + \bar{I}\alpha \qquad (2)$$

When the cylinder is in its position of equilibrium, the tension in the cord is $T_0 = \frac{1}{2}W$. We note that, for an angular displacement θ, the magnitude of $\mathbf{T}_2$ is

$$T_2 = T_0 + k\delta = \tfrac{1}{2}W + k\delta = \tfrac{1}{2}W + k(2r\theta) \qquad (3)$$

Substituting from (1) and (3) into (2), and recalling that $\bar{I} = \frac{1}{2}mr^2$, we write

$$Wr - (\tfrac{1}{2}W + 2kr\theta)(2r) = m(r\ddot{\theta})r + \tfrac{1}{2}mr^2\ddot{\theta}$$

$$\ddot{\theta} + \frac{8}{3}\frac{k}{m}\theta = 0$$

The motion is seen to be simple harmonic, and we have

$$p^2 = \frac{8}{3}\frac{k}{m} \qquad p = \sqrt{\frac{8}{3}\frac{k}{m}}$$

$$\tau = \frac{2\pi}{p} \qquad\qquad \tau = 2\pi\sqrt{\frac{3}{8}\frac{m}{k}} \qquad \blacktriangleleft$$

$$f = \frac{p}{2\pi} \qquad\qquad f = \frac{1}{2\pi}\sqrt{\frac{8}{3}\frac{k}{m}} \qquad \blacktriangleleft$$

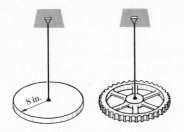

A circular disk, weighing 20 lb and of radius 8 in., is suspended from a wire as shown. The disk is rotated (thus twisting the wire) and then released; the period of the torsional vibration is observed to be 1.13 s. A gear is then suspended from the same wire, and the period of torsional vibration for the gear is observed to be 1.93 s. Assuming that the moment of the couple exerted by the wire is proportional to the angle of twist, determine (a) the torsional spring constant of the wire, (b) the centroidal moment of inertia of the gear, (c) the maximum angular velocity reached by the gear if it is rotated through 90° and released.

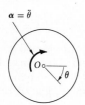

$\alpha = \ddot{\theta}$

a. **Vibration of Disk.** Denoting by θ the angular displacement of the disk, we express that the magnitude of the couple exerted by the wire is $M = K\theta$, where K is the torsional spring constant of the wire. Since this couple must be equivalent to the couple $\bar{I}\alpha$ representing the effective forces of the disk, we write

$$+\circlearrowleft \Sigma M_O = \Sigma(M_O)_{\text{eff}}: \qquad +K\theta = -\bar{I}\ddot{\theta}$$

$$\ddot{\theta} + \frac{K}{\bar{I}}\theta = 0$$

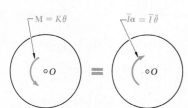

$-M = K\theta \qquad -\bar{I}\alpha = \bar{I}\ddot{\theta}$

The motion is seen to be simple harmonic, and we have

$$p^2 = \frac{K}{\bar{I}} \qquad \tau = \frac{2\pi}{p} \qquad \tau = 2\pi\sqrt{\frac{\bar{I}}{K}} \qquad (1)$$

For the disk, we have

$$\tau = 1.13\,\text{s} \qquad \bar{I} = \tfrac{1}{2}mr^2 = \frac{1}{2}\left(\frac{20\,\text{lb}}{32.2\,\text{ft/s}^2}\right)(\tfrac{8}{12}\,\text{ft})^2 = 0.138\,\text{lb}\cdot\text{ft}\cdot\text{s}^2$$

Substituting into (1), we obtain

$$1.13 = 2\pi\sqrt{\frac{0.138}{K}} \qquad K = 4.27\,\text{lb}\cdot\text{ft/rad} \quad \blacktriangleleft$$

b. **Vibration of Gear.** Since the period of vibration of the gear is 1.93 s and $K = 4.27$ lb · ft/rad, Eq. (1) yields

$$1.93 = 2\pi\sqrt{\frac{\bar{I}}{4.27}} \qquad \bar{I}_{\text{gear}} = 0.403\,\text{lb}\cdot\text{ft}\cdot\text{s}^2 \quad \blacktriangleleft$$

c. **Maximum Angular Velocity of the Gear.** Since the motion is simple harmonic, we have

$$\theta = \theta_m \sin pt \qquad \omega = \theta_m p \cos pt \qquad \omega_m = \theta_m p$$

Recalling that $\theta_m = 90° = 1.571$ rad and $\tau = 1.93$ s, we write

$$\omega_m = \theta_m p = \theta_m(2\pi/\tau) = (1.571\,\text{rad})(2\pi/1.93\,\text{s})$$

$$\omega_m = 5.11\,\text{rad/s} \quad \blacktriangleleft$$

PROBLEMS

19.25 and 19.26 The uniform rod shown weighs 10 lb and is attached to a spring of constant $k = 3$ lb/in. If end A of the rod is depressed 1.5 in. and released, determine (a) the period of vibration, (b) the maximum velocity of end A.

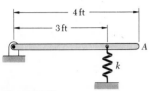

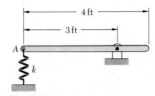

Fig. P19.25   **Fig. P19.26**

19.27 A belt is placed over the rim of a 15-kg disk as shown and then attached to a 5-kg cylinder and to a spring of constant $k = 600$ N/m. If the cylinder is moved 50 mm down from its equilibrium position and released, determine (a) the period of vibration, (b) the maximum velocity of the cylinder. Assume friction is sufficient to prevent the belt from slipping on the rim.

19.28 In Prob. 19.27, determine (a) the frequency of vibration, (b) the maximum tension which occurs in the belt at B and at C.

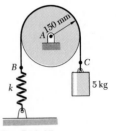

Fig. P19.27

19.29 A disk is rigidly attached to a gear of radius r which may roll on the gear rack shown. The disk and gear have a total mass m and a combined radius of gyration $\bar{k}$. A spring of constant K is attached to the center of the gear. If the system is displaced slightly to the right and released, determine the period of the resulting motion.

19.30 A uniform square plate of mass m is supported in a horizontal plane by a vertical pin at B and is attached at A to a spring of constant k. If corner A is given a small displacement and released, determine the period of the resulting motion.

Fig. P19.29

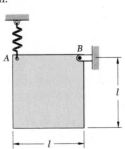

Fig. P19.30

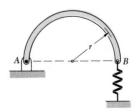

Fig. P19.31

19.31 A semicircular rod of mass m is supported in a horizontal plane by a vertical pin at A and is attached at B to a spring of constant k. If end B is given a small displacement and released, determine the frequency of the resulting motion.

19.32 A square plate of mass m is held by eight springs, each of constant k. Knowing that each spring can act in either tension or compression, determine the frequency of the resulting vibration (*a*) if the plate is given a small vertical displacement and released, (*b*) if the plate is rotated through a small angle about G and released.

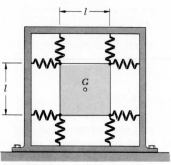

Fig. P19.32

19.33 A *compound pendulum* is defined as a rigid slab which oscillates about a fixed point O, called the center of suspension. Show that the period of oscillation of a compound pendulum is equal to the period of a simple pendulum of length OA, where the distance from A to the mass center G is $GA = \bar{k}^2/\bar{r}$. Point A is defined as the center of oscillation and coincides with the center of percussion defined in Prob. 18.14.

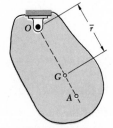

Fig. P19.33 and P19.35

19.34 Show that, if the compound pendulum of Prob. 19.33 is suspended from A instead of O, the period of oscillation is the same as before and the new center of oscillation is located at O.

19.35 A rigid slab oscillates about a fixed point O. Show that the smallest period of oscillation occurs when the distance $\bar{r}$ from point O to the mass center G is equal to $\bar{k}$.

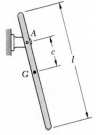

19.36 A uniform bar of length l may oscillate about a hinge at A located a distance c from its mass center G. (*a*) Determine the frequency of small oscillations if $c = \frac{1}{2}l$. (*b*) Determine a second value of c for which the frequency of small oscillations is the same as that found in part *a*.

Fig. P19.36

19.37 For the rod of Prob. 19.36, determine (*a*) the distance c for which the frequency of oscillation is maximum, (*b*) the corresponding minimum period.

19.38 A homogeneous wire is bent to form a square of side l which is supported by a ball-and-socket joint at A. Determine the period of small oscillations of the square (a) in the plane of the square, (b) in a direction perpendicular to the square.

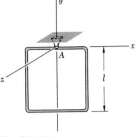

Fig. P19.38

19.39 A thin disk of radius r may oscillate about the axis AB located in the plane of the disk at a distance c from the mass center G. (a) Determine the period of small oscillations if $c = r$. (b) Determine a second value of c for which the period of oscillation is the same as that found in part a.

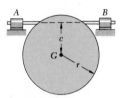

Fig. P19.39

19.40 Solve Prob. 19.39 for $r = 600$ mm.

19.41 A 75-mm-radius hole is cut in a 200-mm-radius uniform disk which is attached to a frictionless pin at its geometric center O. Determine (a) the period of small oscillations of the disk, (b) the length of a simple pendulum which has the same period.

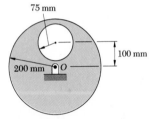

Fig. P19.41

19.42 The centroidal radii of gyration $\bar{k}_x$ and $\bar{k}_z$ of an airplane are determined by letting the airplane oscillate as a compound pendulum. In the arrangement shown, the distance from the mass center G to the point of suspension O is known to be 12 ft. Knowing that the observed periods of oscillation about axes through O parallel to the x and z axes are, respectively, 4.3 and 4.4 s, determine the centroidal radii of gyration $\bar{k}_x$ and $\bar{k}_z$.

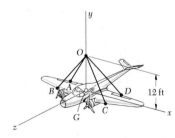

Fig. P19.42

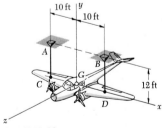

Fig. P19.43

19.43 The centroidal radius of gyration $\bar{k}_y$ of an airplane is determined by suspending the airplane by two 12-ft-long cables as shown. The airplane is rotated through a small angle about the vertical through G and then released. Knowing that the observed period of oscillation is 3.3 s, determine the centroidal radius of gyration $\bar{k}_y$.

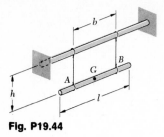

Fig. P19.44

19.44 A slender rod of length l is suspended from two vertical wires of length h, each located a distance $\frac{1}{2}b$ from the mass center G. Determine the period of oscillation when (a) the rod is rotated through a small angle about a vertical axis passing through G and released, (b) the rod is given a small horizontal translation along AB and released.

19.45 A slender rod of weight 6 lb is suspended from a steel wire which is known to have a torsional spring constant $K = 0.25 \text{ lb} \cdot \text{in./rad}$. If the rod is rotated through 180° about the vertical and then released, determine (a) the period of oscillation, (b) the maximum velocity of end A of the rod.

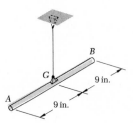

Fig. P19.45

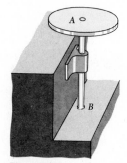

Fig. P19.46

19.46 A uniform disk of radius 200 mm and of mass 8 kg is attached to a vertical shaft which is rigidly held at B. It is known that the disk rotates through 3° when a 4 N · m static couple is applied to the disk. If the disk is rotated through 6° and then released, determine (a) the period of the resulting vibration, (b) the maximum velocity of a point on the rim of the disk.

19.47 A steel casting is rigidly bolted to the disk of Prob. 19.46. Knowing that the period of torsional vibration of the disk and casting is 0.650 s, determine the moment of inertia of the casting with respect to the shaft AB.

19.48 A period of 4.10 s is observed for the angular oscillations of a 1-lb gyroscope rotor suspended from a wire as shown. Knowing that a period of 6.20 s is obtained when a 2-in.-diameter steel sphere is suspended in the same fashion, determine the centroidal radius of gyration of the rotor. (Specific weight of steel = 490 lb/ft³.)

Fig. P19.48

19.6. Application of the Principle of Conservation of Energy.

We saw in Sec. 19.2 that, when a particle of mass m is in simple harmonic motion, the resultant $\mathbf{F}$ of the forces exerted on the particle has a magnitude proportional to the displacement x measured from the position of equilibrium O and is directed toward O; we write $F = -kx$. Referring to Sec. 13.6, we note that $\mathbf{F}$ is a *conservative force* and that the corresponding potential energy is $V = \frac{1}{2}kx^2$, where V is assumed equal to zero in the equilibrium position $x = 0$. Since the velocity of the particle is equal to $\dot{x}$, its kinetic energy is $T = \frac{1}{2}m\dot{x}^2$ and we may express that the total energy of the particle is conserved by writing

$$T + V = \text{constant} \qquad \tfrac{1}{2}m\dot{x}^2 + \tfrac{1}{2}kx^2 = \text{constant}$$

Setting $p^2 = k/m$ as in (19.4), where p is the circular frequency of the vibration, we have

$$\dot{x}^2 + p^2x^2 = \text{constant} \qquad\qquad (19.25)$$

Equation (19.25) is characteristic of simple harmonic motion; it may be obtained directly from (19.5) by multiplying both terms by $2\dot{x}$ and integrating.

The principle of conservation of energy provides a convenient way for determining the period of vibration of a rigid body or of a system of rigid bodies possessing a single degree of freedom, once it has been established that the motion of the system is a simple harmonic motion, or that it may be approximated by a simple harmonic motion. Choosing an appropriate variable, such as a distance x or an angle θ, we consider two particular positions of the system:

1. *The displacement of the system is maximum;* we have $T_1 = 0$, and V_1 may be expressed in terms of the amplitude x_m or θ_m (choosing $V = 0$ in the equilibrium position).
2. *The system passes through its equilibrium position;* we have $V_2 = 0$, and T_2 may be expressed in terms of the maximum velocity $\dot{x}_m$ or $\dot{\theta}_m$.

We then express that the total energy of the system is conserved and write $T_1 + V_1 = T_2 + V_2$. Recalling from (19.15) that for simple harmonic motion the maximum velocity is equal to the product of the amplitude and of the circular frequency p, we find that the equation obtained may be solved for p.

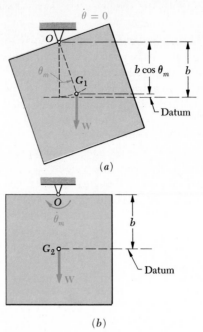

Fig. 19.6

As an example, we shall consider again the square plate of Sec. 19.5. In the position of maximum displacement (Fig. 19.6a), we have

$$T_1 = 0 \qquad V_1 = W(b - b \cos \theta_m) = Wb(1 - \cos \theta_m)$$

or, since $1 - \cos \theta_m = 2 \sin^2 (\theta_m/2) \approx 2(\theta_m/2)^2 = \theta_m^2/2$ for oscillations of small amplitude,

$$T_1 = 0 \qquad V_1 = \tfrac{1}{2}Wb\theta_m^2 \tag{19.26}$$

As the plate passes through its position of equilibrium (Fig. 19.6b), its velocity is maximum and we have

$$T_2 = \tfrac{1}{2}m\bar{v}_m^2 + \tfrac{1}{2}\bar{I}\omega_m^2 = \tfrac{1}{2}mb^2\dot{\theta}_m^2 + \tfrac{1}{2}\bar{I}\dot{\theta}_m^2 \qquad V_2 = 0$$

or, recalling from Sec. 19.5 that $\bar{I} = \tfrac{2}{3}mb^2$,

$$T_2 = \tfrac{1}{2}(\tfrac{5}{3}mb^2)\dot{\theta}_m^2 \qquad V_2 = 0 \tag{19.27}$$

Substituting from (19.26) and (19.27) into $T_1 + V_1 = T_2 + V_2$, and noting that the maximum velocity $\dot{\theta}_m$ is equal to the product $\theta_m p$, we write

$$\tfrac{1}{2}Wb\theta_m^2 = \tfrac{1}{2}(\tfrac{5}{3}mb^2)\theta_m^2 p^2 \tag{19.28}$$

which yields $p^2 = 3g/5b$ and

$$\tau = \frac{2\pi}{p} = 2\pi\sqrt{\frac{5b}{3g}} \tag{19.29}$$

as previously obtained.

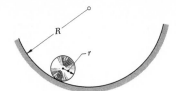

Determine the period of small oscillations of a cylinder of radius r which rolls without slipping inside a curved surface of radius R.

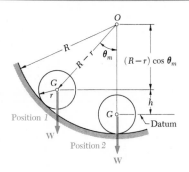

Position 1

Position 2

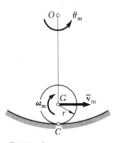

Position 2

Solution. We denote by θ the angle which line OG forms with the vertical. Since the cylinder rolls without slipping, we may apply the principle of conservation of energy between position 1, where $\theta = \theta_m$, and position 2, where $\theta = 0$.

Position 1. **Kinetic Energy.** Since the velocity of the cylinder is zero, we have $T_1 = 0$.

Potential Energy. Choosing a datum as shown and denoting by W the weight of the cylinder, we have

$$V_1 = Wh = W(R - r)(1 - \cos\theta)$$

Noting that for small oscillations $(1 - \cos\theta) = 2\sin^2(\theta/2) \approx \theta^2/2$, we have

$$V_1 = W(R - r)\frac{\theta_m^2}{2}$$

Position 2. Denoting by $\dot{\theta}_m$ the angular velocity of line OG as the cylinder passes through position 2, and observing that point C is the instantaneous center of rotation of the cylinder, we write

$$\bar{v}_m = (R - r)\dot{\theta}_m \qquad \omega_m = \frac{\bar{v}_m}{r} = \frac{R - r}{r}\dot{\theta}_m$$

Kinetic Energy:

$$T_2 = \tfrac{1}{2}m\bar{v}_m^2 + \tfrac{1}{2}\bar{I}\omega_m^2$$
$$= \tfrac{1}{2}m(R - r)^2\dot{\theta}_m^2 + \tfrac{1}{2}(\tfrac{1}{2}mr^2)\left(\frac{R - r}{r}\right)^2\dot{\theta}_m^2$$
$$= \tfrac{3}{4}m(R - r)^2\dot{\theta}_m^2$$

Potential Energy: $V_2 = 0$

Conservation of Energy

$$T_1 + V_1 = T_2 + V_2$$

$$0 + W(R - r)\frac{\theta_m^2}{2} = \tfrac{3}{4}m(R - r)^2\dot{\theta}_m^2 + 0$$

Since $\dot{\theta}_m = p\theta_m$ and $W = mg$, we write

$$mg(R - r)\frac{\theta_m^2}{2} = \tfrac{3}{4}m(R - r)^2(p\theta_m)^2 \qquad p^2 = \frac{2}{3}\frac{g}{R - r}$$

$$\tau = \frac{2\pi}{p} \qquad \tau = 2\pi\sqrt{\frac{3}{2}\frac{R - r}{g}} \quad \blacktriangleleft$$

Fig. P19.49

PROBLEMS

19.49 Using the method of Sec. 19.6, determine the period of a simple pendulum of length l.

19.50 Using the method of Sec. 19.6, solve Prob. 19.6.

19.51 Using the method of Sec. 19.6, solve Prob. 19.9.

19.52 Using the method of Sec. 19.6, solve Prob. 19.10.

19.53 Using the method of Sec. 19.6, solve Prob. 19.15.

19.54 It is observed that when a 15-lb weight is attached to the rim of a 6-ft-diameter flywheel, the period of small oscillations of the flywheel is 21 s. Neglecting axle friction, determine the centroidal moment of inertia of the flywheel.

Fig. P19.54

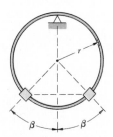

Fig. P19.55

19.55 Two collars, each of weight W, are attached as shown to a hoop of radius r and of negligible weight. (*a*) Show that for any value of β the period is $\tau = 2\pi\sqrt{2r/g}$. (*b*) Show that the same result is obtained if the weight of the hoop is not neglected.

Fig. P19.56

19.56 A thin homogeneous wire is bent into the shape of an equilateral triangle of side l. Determine the period of small oscillations if the wire figure is suspended as shown.

19.57 Solve Prob. 19.56, assuming that the wire triangle is suspended from a pin located at the midpoint of one side.

19.58 Using the method of Sec. 19.6, solve Prob. 19.44.

19.59 Using the method of Sec. 19.6, solve Prob. 19.41.

19.60 A section of uniform pipe is suspended from two vertical cables attached at A and B. Determine the period of oscillation in terms of l_1 and l_2 when point B is given a small horizontal displacement to the right and released.

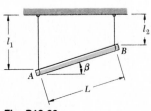

Fig. P19.60

19.61 The motion of the uniform rod AB is guided by the cord BC and by the small roller at A. Determine the frequency of oscillation when the end B of the rod is given a small horizontal displacement and released.

19.62 A uniform disk of weight W and radius r may roll on the horizontal surface shown. In case a, a weight W of negligible dimensions is bolted to the disk at a distance $\frac{1}{2}r$ from the geometric center of the disk. In case b, a vertical force of constant magnitude W is applied to a cord which is attached to the disk at a distance $\frac{1}{2}r$ from the center of the disk. Determine the ratio τ_b/τ_a of the two periods of oscillation.

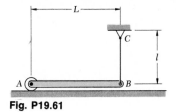

Fig. P19.61

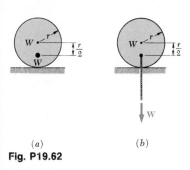

(a) (b)

Fig. P19.62

19.63 The 8-kg rod AB is bolted to the 12-kg disk. Knowing that the disk rolls without sliding, determine the period of small oscillations of the system.

19.64 Using the method of Sec. 19.6, solve Prob. 19.30.

19.65 Using the method of Sec. 19.6, solve Prob. 19.31.

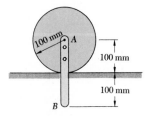

Fig. P19.63

19.66 The slender rod AB of mass m is attached to two collars of negligible mass. Knowing that the system lies in a horizontal plane and is in equilibrium in the position shown, determine the period of vibration if the collar A is given a small displacement and released.

19.67 Solve Prob. 19.66, assuming that rod AB is of mass m and that each collar is of mass m_c.

19.68 A slender rod AB of negligible mass connects two collars, each of mass m_c. Knowing that the system lies in a horizontal plane and is in equilibrium in the position shown, determine the period of vibration if the collar at A is given a small displacement and released.

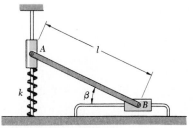

Fig. P19.66 and P19.68

19.69 Two uniform rods AB and CD, each of length l and weight W, are attached to two gears as shown. Neglecting the weight of the gears, determine for each of the positions shown the period of small oscillations of the system.

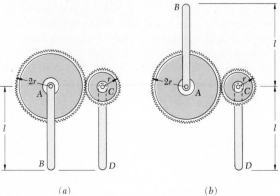

Fig. P19.69

19.70 Solve Prob. 16.69, assuming that $l = 30$ in., $W = 5$ lb, and $r = 4$ in.

19.71 A uniform rod of length L is supported by a ball-and-socket joint at A and by the vertical wire CD. Derive an expression for the period of oscillation of the rod if end B is given a small horizontal displacement and then released.

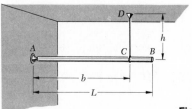

Fig. P19.71

19.72 Solve Prob. 19.71, assuming that $L = 10$ ft, $b = 8$ ft, and $h = 6$ ft.

Fig. P19.73

*19.73 As a submerged body moves through a fluid, the particles of the fluid flow around the body and thus acquire kinetic energy. In the case of a sphere moving in an ideal fluid, the total kinetic energy acquired by the fluid is $\frac{1}{4}\rho V v^2$, where ρ is the mass density of the fluid, V the volume of the sphere, and v the velocity of the sphere. Consider a 1-lb hollow spherical shell of radius 3 in. which is held submerged in a tank of water by a spring of constant 3 lb/in. (a) Neglecting fluid friction, determine the period of vibration of the shell when it is displaced vertically and then released. (b) Solve part a, assuming that the tank is accelerated upward at the constant rate of 10 ft/s^2.

19.7. Forced Vibrations. The most important vibrations from the point of view of engineering applications are the *forced vibrations* of a system. These vibrations occur when a system is subjected to a periodic force or when it is elastically connected to a support which has an alternating motion.

Consider first the case of a body of mass m suspended from a spring and subjected to a periodic force **P** of magnitude $P = P_m \sin \omega t$ (Fig. 19.7). This force may be an actual external force applied to the body, or it may be a centrifugal force produced by the rotation of some unbalanced part of the body. (See Sample Prob. 19.5.) Denoting by x the displacement of the body measured from its equilibrium position, we write the equation of motion,

$$+\downarrow \Sigma F = ma: \qquad P_m \sin \omega t + W - k(\delta_{st} + x) = m\ddot{x}$$

Recalling that $W = k\delta_{st}$, we have

$$m\ddot{x} + kx = P_m \sin \omega t \qquad (19.30)$$

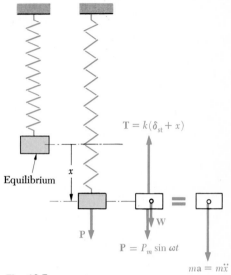

Fig. 19.7

Next we consider the case of a body of mass m suspended from a spring attached to a moving support whose displacement δ is equal to $\delta_m \sin \omega t$ (Fig. 19.8). Measuring the displacement x of the body from the position of static equilibrium corresponding to $\omega t = 0$, we find that the total elongation of the spring at time t is $\delta_{st} + x - \delta_m \sin \omega t$. The equation of motion is thus

$$+\downarrow \Sigma F = ma: \qquad W - k(\delta_{st} + x - \delta_m \sin \omega t) = m\ddot{x}$$

Recalling that $W = k\delta_{st}$, we have

$$m\ddot{x} + kx = k\delta_m \sin \omega t \qquad (19.31)$$

We note that Eqs. (19.30) and (19.31) are of the same form and that a solution of the first equation will satisfy the second if we set $P_m = k\delta_m$.

A differential equation like (19.30) or (19.31), possessing a right-hand member different from zero, is said to be *nonhomogeneous*. Its general solution is obtained by adding a particular solution of the given equation to the general solution of the corresponding *homogeneous* equation (with right-hand member equal to zero). A *particular solution* of (19.30) or (19.31) may be obtained by trying a solution of the form

$$x_{part} = x_m \sin \omega t \qquad (19.32)$$

Substituting x_{part} for x into (19.30), we find

$$-m\omega^2 x_m \sin \omega t + k x_m \sin \omega t = P_m \sin \omega t$$

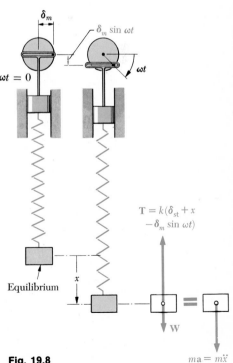

Fig. 19.8

which may be solved for the amplitude,

$$x_m = \frac{P_m}{k - m\omega^2}$$

Recalling from (19.4) that $k/m = p^2$, where p is the circular frequency of the free vibration of the body, we write

$$x_m = \frac{P_m/k}{1 - (\omega/p)^2} \tag{19.33}$$

Substituting from (19.32) into (19.31), we obtain in a similar way

$$x_m = \frac{\delta_m}{1 - (\omega/p)^2} \tag{19.33'}$$

The homogeneous equation corresponding to (19.30) or (19.31) is Eq. (19.3), defining the free vibration of the body. Its general solution, called the *complementary function*, was found in Sec. 19.2,

$$x_{\text{comp}} = A \sin pt + B \cos pt \tag{19.34}$$

Adding the particular solution (19.32) and the complementary function (19.34), we obtain the *general solution* of Eqs. (19.30) and (19.31),

$$x = A \sin pt + B \cos pt + x_m \sin \omega t \tag{19.35}$$

We note that the vibration obtained consists of two superposed vibrations. The first two terms in (19.35) represent a free vibration of the system. The frequency of this vibration, called the *natural frequency* of the system, depends only upon the constant k of the spring and the mass m of the body, and the constants A and B may be determined from the initial conditions. This free vibration is also called a *transient* vibration since, in actual practice, it will soon be damped out by friction forces (Sec. 19.9).

The last term in (19.35) represents the *steady-state* vibration produced and maintained by the impressed force or impressed support movement. Its frequency is the *forced frequency* imposed by this force or movement, and its amplitude x_m, defined by (19.33) or (19.33′), depends upon the *frequency ratio* ω/p. The ratio of the amplitude x_m of the steady-state vibration to the static deflection P_m/k caused by a force P_m, or to the amplitude δ_m of the support movement, is called the *magnification factor*. From (19.33) and (19.33′), we obtain

$$\text{Magnification factor} = \frac{x_m}{P_m/k} = \frac{x_m}{\delta_m} = \frac{1}{1 - (\omega/p)^2} \quad (19.36)$$

The magnification factor has been plotted in Fig. 19.9 against the frequency ratio ω/p. We note that, when $\omega = p$, the amplitude of the forced vibration becomes infinite. The impressed force or impressed support movement is said to be in *resonance* with the given system. Actually, the amplitude of the vibration remains finite because of damping forces (Sec. 19.9); nevertheless, such a situation should be avoided, and the forced frequency should not be chosen too close to the natural frequency of the system. We also note that for $\omega < p$ the coefficient of $\sin \omega t$ in (19.35) is positive, while for $\omega > p$ this coefficient is negative. In the first case the forced vibration is *in phase* with the impressed force or impressed support movement, while in the second case it is 180° *out of phase*.

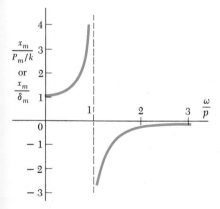

Fig. 19.9

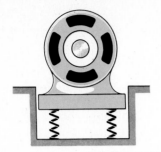

A motor weighing 350 lb is supported by four springs, each having a constant of 750 lb/in. The unbalance of the rotor is equivalent to a weight of 1 oz located 6 in. from the axis of rotation. Knowing that the motor is constrained to move vertically, determine (a) the speed in rpm at which resonance will occur, (b) the amplitude of the vibration of the motor at a speed of 1200 rpm.

a. Resonance Speed. The resonance speed is equal to the circular frequency (in rpm) of the free vibration of the motor. The mass of the motor and the equivalent constant of the supporting springs are

$$m = \frac{350 \text{ lb}}{32.2 \text{ ft/s}^2} = 10.87 \text{ lb} \cdot \text{s}^2/\text{ft}$$

$$k = 4(750 \text{ lb/in.}) = 3000 \text{ lb/in.} = 36{,}000 \text{ lb/ft}$$

$$p = \sqrt{\frac{k}{m}} = \sqrt{\frac{36{,}000}{10.87}} = 57.5 \text{ rad/s} = 549 \text{ rpm}$$

Resonance speed = 549 rpm ◄

b. Amplitude of Vibration at 1200 rpm. The angular velocity of the motor and the mass of the equivalent 1-oz weight are

$$\omega = 1200 \text{ rpm} = 125.7 \text{ rad/s}$$

$$m = (1 \text{ oz})\frac{1 \text{ lb}}{16 \text{ oz}}\frac{1}{32.2 \text{ ft/s}^2} = 0.00194 \text{ lb} \cdot \text{s}^2/\text{ft}$$

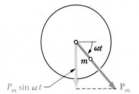

The magnitude of the centrifugal force due to the unbalance of the rotor is

$$P_m = ma_n = mr\omega^2 = (0.00194 \text{ lb} \cdot \text{s}^2/\text{ft})(\tfrac{6}{12} \text{ ft})(125.7 \text{ rad/s})^2 = 15.3 \text{ lb}$$

The static deflection that would be caused by a constant load P_m is

$$\frac{P_m}{k} = \frac{15.3 \text{ lb}}{3000 \text{ lb/in.}} = 0.00510 \text{ in.}$$

Substituting the value of P_m/k together with the known values of ω and p into Eq. (19.33), we obtain

$$x_m = \frac{P_m/k}{1 - (\omega/p)^2} = \frac{0.00510 \text{ in.}}{1 - (125.7/57.5)^2}$$

$$x_m = 0.00135 \text{ in.} ◄$$

Note. Since $\omega > p$, the vibration is 180° out of phase with the centrifugal force due to the unbalance of the rotor. For example, when the unbalanced mass is directly below the axis of rotation, the position of the motor is $x_m = 0.00135$ in. above the position of equilibrium.

PROBLEMS

19.74 A block of mass m is suspended from a spring of constant k and is acted upon by a vertical periodic force of magnitude $P = P_m \sin \omega t$. Determine the range of values of ω for which the amplitude of the vibration exceeds twice the static deflection caused by a constant force of magnitude P_m.

19.75 In Prob. 19.74, determine the range of values of ω for which the amplitude of the vibration is less than the static deflection caused by a constant force of magnitude P_m.

19.76 A simple pendulum of length l is suspended from a collar C which is forced to move horizontally according to the relation $x_C = \delta_m \sin \omega t$. Determine the range of values of ω for which the amplitude of the motion of the bob exceeds $2\delta_m$. (Assume δ_m is small compared to the length l of the pendulum.)

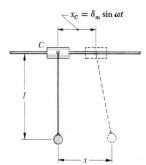

Fig. P19.76

19.77 In Prob. 19.76, determine the range of values of ω for which the amplitude of the motion of the bob is less than δ_m.

19.78 A 500-lb motor is supported by a light horizontal beam. The unbalance of the rotor is equivalent to a weight of 1 oz located 10 in. from the axis of rotation. Knowing that the static deflection of the beam due to the weight of the motor is 0.220 in., determine (*a*) the speed (in rpm) at which resonance will occur, (*b*) the amplitude of the steady-state vibration of the motor at a speed of 800 rpm.

Fig. P19.78

19.79 Solve Prob. 19.78, assuming that the 500-lb motor is supported by a nest of springs having a total constant of 400 lb/in.

19.80 A motor of mass 45 kg is supported by four springs, each of constant 100 kN/m. The motor is constrained to move vertically, and the amplitude of its movement is observed to be 0.5 mm at a speed of 1200 rpm. Knowing that the mass of the rotor is 14 kg, determine the distance between the mass center of the rotor and the axis of the shaft.

19.81 In Prob. 19.80, determine the amplitude of the vertical movement of the motor at a speed of (*a*) 200 rpm, (*b*) 1600 rpm, (*c*) 900 rpm.

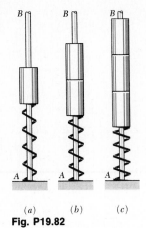

(a)   (b)   (c)

Fig. P19.82

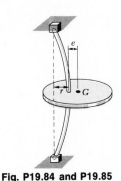

Fig. P19.84 and P19.85

19.82 Rod *AB* is rigidly attached to the frame of a motor running at a constant speed. When a collar of mass *m* is placed on the spring, it is observed to vibrate with an amplitude of 0.5 in. When two collars, each of mass *m*, are placed on the spring, the amplitude is observed to be 0.6 in. What amplitude of vibration should be expected when three collars, each of mass *m*, are placed on the spring? (Obtain two answers.)

19.83 Solve Prob. 19.82, assuming that the speed of the motor is changed and that one collar has an amplitude of 0.60 in. and two collars have an amplitude of 0.20 in.

19.84 A disk of mass *m* is attached to the midpoint of a vertical shaft which revolves at an angular velocity ω. Denoting by *k* the spring constant of the system for a horizontal movement of the disk and by *e* the eccentricity of the disk with respect to the shaft, show that the deflection of the center of the shaft may be written in the form

$$r = \frac{e(\omega/p)^2}{1 - (\omega/p)^2}$$

where $p = \sqrt{k/m}$.

19.85 A disk of mass 30 kg is attached to the midpoint of a shaft. Knowing that a static force of 200 N will deflect the shaft 0.6 mm, determine the speed of the shaft in rpm at which resonance will occur.

19.86 Knowing that the disk of Prob. 19.85 is attached to the shaft with an eccentricity *e* = 0.2 mm, determine the deflection *r* of the shaft at a speed of 900 rpm.

19.87 A variable-speed motor is rigidly attached to the beam *BC*. When the speed of the motor is less than 600 rpm or more than 1200 rpm, a small object placed at *A* is observed to remain in contact with the beam. For speeds between 600 and 1200 rpm the object is observed to "dance" and actually to lose contact with the beam. Determine the speed at which resonance will occur.

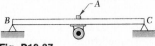

Fig. P19.87

19.88 As the speed of a spring-supported motor is slowly increased from 300 to 400 rpm, the amplitude of the vibration due to the unbalance of the rotor is observed to decrease continuously from 0.075 to 0.040 in. Determine the speed at which resonance will occur.

19.89 In Prob. 19.88, determine the speed for which the amplitude of the vibration is 0.100 in.

19.90 The amplitude of the motion of the pendulum bob shown is observed to be 60 mm when the amplitude of the motion of collar C is 15 mm. Knowing that the length of the pendulum is $l = 720\,\text{mm}$, determine the two possible values of the frequency of the horizontal movement of the collar C.

19.91 A small trailer of mass 200 kg with its load is supported by two springs, each of constant 20 kN/m. The trailer is pulled over a road, the surface of which may be approximated by a sine curve of amplitude 30 mm and of period 5 m (i.e., the distance between two successive crests is 5 m, and the vertical distance from a crest to a trough is 60 mm). Determine (a) the speed at which resonance will occur, (b) the amplitude of the vibration of the trailer at a speed of 60 km/h.

19.92 Knowing that the amplitude of the vibration of the trailer of Prob. 19.91 is not to exceed 15 mm, determine the smallest speed at which the trailer can be pulled over the road.

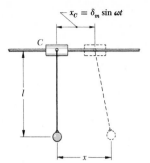

Fig. P19.90

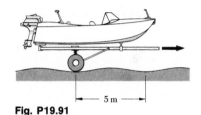

Fig. P19.91

DAMPED VIBRATIONS

*19.8. Damped Free Vibrations.

The vibrating systems considered in the first part of this chapter were assumed free of damping. Actually all vibrations are damped to some degree by friction forces. These forces may be caused by *dry friction*, or *Coulomb friction*, between rigid bodies, by *fluid friction* when a rigid body moves in a fluid, or by *internal friction* between the molecules of a seemingly elastic body.

A type of damping of special interest is the *viscous damping* caused by fluid friction at low and moderate speeds. Viscous damping is characterized by the fact that the friction force is *directly proportional to the speed* of the moving body. As an example, we shall consider again a body of mass m suspended from a spring of constant k, and we shall assume that the body is attached to the plunger of a dashpot (Fig. 19.10). The magnitude of the friction force exerted on the plunger by the surrounding fluid is equal to $c\dot{x}$, where the constant c, expressed in N·s/m or lb·s/ft and known as the *coefficient of viscous damping*, depends upon the physical properties of the fluid and the construction of the dashpot. The equation of motion is

$$+\!\downarrow \Sigma F = ma: \qquad W - k(\delta_{\text{st}} + x) - c\dot{x} = m\ddot{x}$$

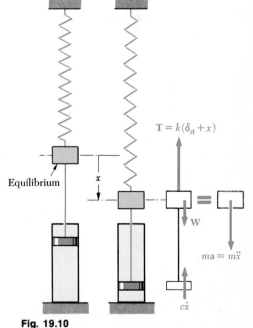

Fig. 19.10

Recalling that $W = k\delta_{st}$, we write

$$m\ddot{x} + c\dot{x} + kx = 0 \qquad (19.37)$$

Substituting $x = e^{\lambda t}$ into (19.37) and dividing through by $e^{\lambda t}$, we write the *characteristic equation*

$$m\lambda^2 + c\lambda + k = 0 \qquad (19.38)$$

and obtain the roots

$$\lambda = -\frac{c}{2m} \pm \sqrt{\left(\frac{c}{2m}\right)^2 - \frac{k}{m}} \qquad (19.39)$$

Defining the *critical damping coefficient* c_c as the value of c which makes the radical in (19.39) equal to zero, we write

$$\left(\frac{c_c}{2m}\right)^2 - \frac{k}{m} = 0 \qquad c_c = 2m\sqrt{\frac{k}{m}} = 2mp \qquad (19.40)$$

where p is the circular frequency of the system in the absence of damping. We may distinguish three different cases of damping, depending upon the value of the coefficient c.

1. *Heavy damping:* $c > c_c$. The roots λ_1 and λ_2 of the characteristic equation (19.38) are real and distinct, and the general solution of the differential equation (19.37) is

$$x = Ae^{\lambda_1 t} + Be^{\lambda_2 t} \qquad (19.41)$$

   This solution corresponds to a nonvibratory motion. Since λ_1 and λ_2 are both negative, x approaches zero as t increases indefinitely. However, the system actually regains its equilibrium position after a finite time.

2. *Critical damping:* $c = c_c$. The characteristic equation has a double root $\lambda = -c_c/2m = -p$, and the general solution of (19.37) is

$$x = (A + Bt)e^{-pt} \qquad (19.42)$$

   The motion obtained is again nonvibratory. Critically damped systems are of special interest in engineering applications since they regain their equilibrium position in the shortest possible time without oscillation.

3. *Light damping:* $c < c_c$. The roots of (19.38) are complex and conjugate, and the general solution of (19.37) is of the form

$$x = e^{-(c/2m)t}(A \sin qt + B \cos qt) \qquad (19.43)$$

where q is defined by the relation

$$q^2 = \frac{k}{m} - \left(\frac{c}{2m}\right)^2$$

Substituting $k/m = p^2$ and recalling (19.40), we write

$$q = p\sqrt{1 - \left(\frac{c}{c_c}\right)^2} \qquad (19.44)$$

where the constant c/c_c is known as the *damping factor*. A substitution similar to the one used in Sec. 19.2 enables us to write the general solution of (19.37) in the form

$$x = x_m e^{-(c/2m)t} \sin(qt + \phi) \qquad (19.45)$$

The motion defined by (19.45) is vibratory with diminishing amplitude (Fig. 19.11). Although this motion does not actually repeat itself, the time interval $\tau = 2\pi/q$, corresponding to two successive points where the curve (19.45) touches one of the limiting curves shown in Fig. 19.11, is commonly referred to as the period of the damped vibration. Recalling (19.44), we observe that τ is larger than the period of vibration of the corresponding undamped system.

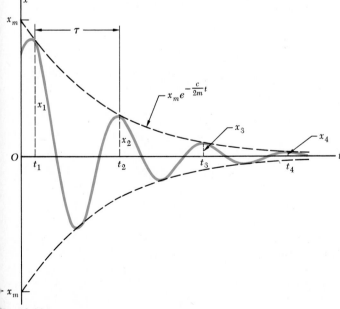

Fig. 19.11

∗19.9. Damped Forced Vibrations. If the system considered in the preceding section is subjected to a periodic force **P** of magnitude $P = P_m \sin \omega t$, the equation of motion becomes

$$m\ddot{x} + c\dot{x} + kx = P_m \sin \omega t \tag{19.46}$$

The general solution of (19.46) is obtained by adding a particular solution of (19.46) to the complementary function or general solution of the homogeneous equation (19.37). The complementary function is given by (19.41), (19.42), or (19.43), depending upon the type of damping considered. It represents a *transient* motion which is eventually damped out.

Our interest in this section is centered on the steady-state vibration represented by a particular solution of (19.46) of the form

$$x_{\text{part}} = x_m \sin{(\omega t - \varphi)} \tag{19.47}$$

Substituting x_{part} for x into (19.46), we obtain

$$-m\omega^2 x_m \sin{(\omega t - \varphi)} + c\omega x_m \cos{(\omega t - \varphi)}$$
$$+ kx_m \sin{(\omega t - \varphi)} = P_m \sin \omega t$$

Making $\omega t - \varphi$ successively equal to 0 and to $\pi/2$, we write

$$c\omega x_m = P_m \sin \varphi \tag{19.48}$$
$$(k - m\omega^2)x_m = P_m \cos \varphi \tag{19.49}$$

Squaring both members of (19.48) and (19.49) and adding, we have

$$[(k - m\omega^2)^2 + (c\omega)^2]x_m^2 = P_m^2 \tag{19.50}$$

Solving (19.50) for x_m and dividing (19.48) and (19.49) member by member, we obtain, respectively,

$$x_m = \frac{P_m}{\sqrt{(k - m\omega^2)^2 + (c\omega)^2}} \qquad \tan \varphi = \frac{c\omega}{k - m\omega^2} \tag{19.51}$$

Recalling from (19.4) that $k/m = p^2$, where p is the circular frequency of the undamped free vibration, and from (19.40) that $2mp = c_c$, where c_c is the critical damping coefficient of the system, we write

$$\frac{x_m}{P_m/k} = \frac{x_m}{\delta_m} = \frac{1}{\sqrt{[1 - (\omega/p)^2]^2 + [2(c/c_c)(\omega/p)]^2}} \tag{19.52}$$

$$\tan \varphi = \frac{2(c/c_c)(\omega/p)}{1 - (\omega/p)^2} \tag{19.53}$$

Formula (19.52) expresses the magnification factor in terms of the frequency ratio ω/p and damping factor c/c_c. It may be used to determine the amplitude of the steady-state vibration produced by an impressed force of magnitude $P = P_m \sin \omega t$ or by an impressed support movement $\delta = \delta_m \sin \omega t$. Formula (19.53) defines in terms of the same parameters the *phase difference* φ between the impressed force or impressed support movement and the resulting steady-state vibration of the damped system. The magnification factor has been plotted against the frequency ratio in Fig. 19.12 for various values of the damping factor. We observe that the amplitude of a forced vibration may be kept small by choosing a large coefficient of viscous damping c or by keeping the natural and forced frequencies far apart.

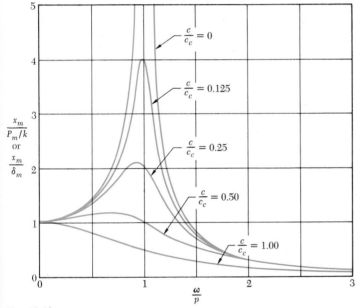

Fig. 19.12

＊19.10. Electrical Analogues. Oscillating electrical circuits are characterized by differential equations of the same type as those obtained in the preceding sections. Their analysis is therefore similar to that of a mechanical system, and the results obtained for a given vibrating system may be readily extended to the equivalent circuit. Conversely, any result obtained for an electrical circuit will also apply to the corresponding mechanical system.

Consider an electrical circuit consisting of an inductor of inductance L, a resistor of resistance R, and a capacitor of

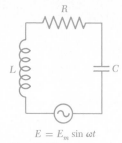

$E = E_m \sin \omega t$

Fig. 19.13

capacitance C, connected in series with a source of alternating voltage $E = E_m \sin \omega t$ (Fig. 19.13). It is recalled from elementary electromagnetic theory† that, if i denotes the current in the circuit and q the electric charge on the capacitor, the drop in potential is $L(di/dt)$ across the inductor, Ri across the resistor, and q/C across the capacitor. Expressing that the algebraic sum of the applied voltage and of the drops in potential around the circuit loop is zero, we write

$$E_m \sin \omega t - L \frac{di}{dt} - Ri - \frac{q}{C} = 0 \qquad (19.54)$$

Rearranging the terms and recalling that, at any instant, the current i is equal to the rate of change $\dot{q}$ of the charge q, we have

$$L\ddot{q} + R\dot{q} + \frac{1}{C}q = E_m \sin \omega t \qquad (19.55)$$

We verify that Eq. (19.55), which defines the oscillations of the electrical circuit of Fig. 19.13, is of the same type as Eq. (19.46), which characterizes the damped forced vibrations of the mechanical system of Fig. 19.10. By comparing the two equations, we may construct a table of the analogous mechanical and electrical expressions.

Table 19.2 may be used to extend to their electrical analogues the results obtained in the preceding sections for various mechanical systems. For instance, the amplitude i_m of the current in the circuit of Fig. 19.13 may be obtained by noting that it corresponds to the maximum value v_m of the velocity in the analogous mechanical system. Recalling that $v_m = \omega x_m$, substituting for x_m from Eq. (19.51), and replacing the constants of

Table 19.2 Characteristics of a Mechanical System and of Its Electrical Analogue

Mechanical System		Electrical Circuit	
m	mass	L	Inductance
c	Coefficient of viscous damping	R	Resistance
k	Spring constant	$1/C$	Reciprocal of capacitance
x	Displacement	q	Charge
v	Velocity	i	Current
P	Applied force	E	Applied voltage

† See Hammond, "Electrical Engineering," McGraw-Hill Book Company, or Smith, "Circuits, Devices, and Systems," John Wiley & Sons.

the mechanical system by the corresponding electrical expressions, we have

$$i_m = \frac{\omega E_m}{\sqrt{\left(\dfrac{1}{C} - L\omega^2\right)^2 + (R\omega)^2}}$$

$$i_m = \frac{E_m}{\sqrt{R^2 + \left(L\omega - \dfrac{1}{C\omega}\right)^2}} \tag{19.56}$$

The radical in the expression obtained is known as the *impedance* of the electrical circuit.

The analogy between mechanical systems and electrical circuits holds for transient as well as steady-state oscillations. The oscillations of the circuit shown in Fig. 19.14, for instance, are analogous to the damped free vibrations of the system of Fig. 19.10. As far as the initial conditions are concerned, we may note that closing the switch S when the charge on the capacitor is $q = q_0$ is equivalent to releasing the mass of the mechanical system with no initial velocity from the position $x = x_0$. We should also observe that, if a battery of constant voltage E is introduced in the electrical circuit of Fig. 19.14, closing the switch S will be equivalent to suddenly applying a force of constant magnitude P to the mass of the mechanical system of Fig. 19.10.

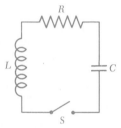

Fig. 19.14

The above discussion would be of questionable value if its only result were to make it possible for mechanics students to analyze electrical circuits without learning the elements of electromagnetism. It is hoped, rather, that this discussion will encourage the students to apply to the solution of problems in mechanical vibrations the mathematical techniques they may learn in later courses in electrical circuits theory. The chief value of the concept of electrical analogue, however, resides in its application to *experimental methods* for the determination of the characteristics of a given mechanical system. Indeed, an electrical circuit is much more easily constructed than a mechanical model, and the fact that its characteristics may be modified by varying the inductance, resistance, or capacitance of its various components makes the use of the electrical analogue particularly convenient.

To determine the electrical analogue of a given mechanical system, we shall focus our attention on each moving mass in the system and observe which springs, dashpots, or external forces are applied directly to it. An equivalent electrical loop may then be constructed to match each of the mechanical units

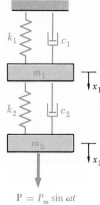

$P = P_m \sin \omega t$

Fig. 19.15

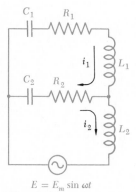

$E = E_m \sin \omega t$

Fig. 19.16

thus defined; the various loops obtained in that way will form together the desired circuit. Consider, for instance, the mechanical system of Fig. 19.15. We observe that the mass m_1 is acted upon by two springs of constants k_1 and k_2 and by two dashpots characterized by the coefficients of viscous damping c_1 and c_2. The electrical circuit should therefore include a loop consisting of an inductor of inductance L_1 proportional to m_1, of two capacitors of capacitance C_1 and C_2 inversely proportional to k_1 and k_2, respectively, and of two resistors of resistance R_1 and R_2, proportional to c_1 and c_2, respectively. Since the mass m_2 is acted upon by the spring k_2 and the dashpot c_2, as well as by the force $P = P_m \sin \omega t$, the circuit should also include a loop containing the capacitor C_2, the resistor R_2, the new inductor L_2, and the voltage source $E = E_m \sin \omega t$ (Fig. 19.16).

To check that the mechanical system of Fig. 19.15 and the electrical circuit of Fig. 19.16 actually satisfy the same differential equations, we shall first derive the equations of motion for m_1 and m_2. Denoting respectively by x_1 and x_2 the displacements of m_1 and m_2 from their equilibrium positions, we observe that the elongation of the spring k_1 (measured from the equilibrium position) is equal to x_1, while the elongation of the spring k_2 is equal to the relative displacement $x_2 - x_1$ of m_2 with respect to m_1. The equations of motion for m_1 and m_2 are therefore

$$m_1\ddot{x}_1 + c_1\dot{x}_1 + c_2(\dot{x}_1 - \dot{x}_2) + k_1x_1 + k_2(x_1 - x_2) = 0 \quad (19.57)$$

$$m_2\ddot{x}_2 + c_2(\dot{x}_2 - \dot{x}_1) + k_2(x_2 - x_1) = P_m \sin \omega t \quad (19.58)$$

Consider now the electrical circuit of Fig. 19.16; we denote respectively by i_1 and i_2 the current in the first and second loops, and by q_1 and q_2 the integrals $\int i_1\, dt$ and $\int i_2\, dt$. Noting that the charge on the capacitor C_1 is q_1, while the charge on C_2 is $q_1 - q_2$, we express that the sum of the potential differences in each loop is zero:

$$L_1\ddot{q}_1 + R_1\dot{q}_1 + R_2(\dot{q}_1 - \dot{q}_2) + \frac{q_1}{C_1} + \frac{q_1 - q_2}{C_2} = 0 \quad (19.59)$$

$$L_2\ddot{q}_2 + R_2(\dot{q}_2 - \dot{q}_1) + \frac{q_2 - q_1}{C_2} = E_m \sin \omega t \quad (19.60)$$

We easily check that Eqs. (19.59) and (19.60) reduce to (19.57) and (19.58), respectively, when the substitutions indicated in Table 19.2 are performed.

PROBLEMS

19.93   Show that, in the case of heavy damping $(c > c_c)$, a body never passes through its position of equilibrium O (a) if it is released with no initial velocity from an arbitrary position or (b) if it is started from O with an arbitrary initial velocity.

19.94   Show that, in the case of heavy damping $(c > c_c)$, a body released from an arbitrary position with an arbitrary initial velocity cannot pass more than once through its equilibrium position.

19.95   In the case of light damping, the displacements x_1, x_2, x_3, etc., shown in Fig. 19.11 may be assumed equal to the maximum displacements. Show that the ratio of any two successive maximum displacements x_n and x_{n+1} is a constant and that the natural logarithm of this ratio, called the *logarithmic decrement*, is

$$\ln \frac{x_n}{x_{n+1}} = \frac{2\pi(c/c_c)}{\sqrt{1 - (c/c_c)^2}}$$

19.96   In practice, it is often difficult to determine the logarithmic decrement defined in Prob. 19.95 by measuring two successive maximum displacements. Show that the logarithmic decrement may also be expressed as $(1/n) \ln (x_1/x_{n+1})$, where n is the number of cycles between readings of the maximum displacement.

19.97   Successive maximum displacements of a spring-mass-dashpot system similar to that shown in Fig. 19.10 are 75, 60, 48, and 38.4 mm. Knowing that $m = 20$ kg and $k = 800$ N/m, determine (a) the damping factor c/c_c, (b) the value of the coefficient of viscous damping c. (*Hint.* See Probs. 19.95 and 19.96.)

19.98   In a system with light damping $(c < c_c)$, the period of vibration is commonly defined as the time interval $\tau = 2\pi/q$ corresponding to two successive points where the displacement-time curve touches one of the limiting curves shown in Fig. 19.11. Show that the interval of time (a) between a maximum positive displacement and the following maximum negative displacement is $\frac{1}{2}\tau$, (b) between two successive zero displacements is $\frac{1}{2}\tau$, (c) between a maximum positive displacement and the following zero displacement is greater than $\frac{1}{4}\tau$.

19.99   The barrel of a field gun weighs 1400 lb and is returned into firing position after recoil by a recuperator of constant $k = 10,000$ lb/ft. Determine the value of the coefficient of damping of the recoil mechanism which causes the barrel to return into firing position in the shortest possible time without oscillation.

19.100 A critically damped system is released from rest at an arbitrary position x_0 when $t = 0$. (*a*) Determine the position of the system at any time *t*. (*b*) Apply the result obtained in part *a* to the barrel of the gun of Prob. 19.99, and determine the time at which the barrel is halfway back to its firing position.

19.101 Assuming that the barrel of the gun of Prob. 19.99 is modified, with a resulting increase in weight of 400 lb, determine the constant *k* of the recuperator which should be used if the recoil mechanism is to remain critically damped.

19.102 In the case of the forced vibration of a system with a given damping factor c/c_c, determine the frequency ratio ω/p for which the amplitude of the vibration is maximum.

19.103 Show that for a small value of the damping factor c/c_c, (*a*) the maximum amplitude of a forced vibration occurs when $\omega \approx p$, (*b*) the corresponding value of the magnification factor is approximately $\frac{1}{2}(c_c/c)$.

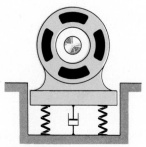

Fig. P19.104

19.104 A motor of mass 25 kg is supported by four springs, each having a constant of 200 kN/m. The unbalance of the rotor is equivalent to a mass of 30 g located 125 mm from the axis of rotation. Knowing that the motor is constrained to move vertically, determine the amplitude of the steady-state vibration of the motor at a speed of 1800 rpm, assuming (*a*) that no damping is present, (*b*) that the damping factor c/c_c is equal to 0.125.

19.105 Assume that the 25-kg motor of Prob. 19.104 is directly supported by a light horizontal beam. The static deflection of the beam due to the weight of the motor is observed to be 5.75 mm, and the amplitude of the vibration of the motor is 0.5 mm at a speed of 400 rpm. Determine (*a*) the damping factor c/c_c, (*b*) the coefficient of damping *c*.

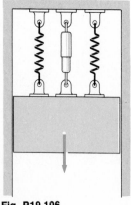

Fig. P19.106

19.106 A machine element weighing 800 lb is supported by two springs, each having a constant of 200 lb/in. A periodic force of maximum value 20 lb is applied to the element with a frequency of 2.5 cycles per second. Knowing that the coefficient of damping is 8 lb · s/in., determine the amplitude of the steady-state vibration of the element.

19.107 In Prob. 19.106, determine the required value of the coefficient of damping if the amplitude of the steady-state vibration of the element is to be 0.15 in.

19.108 A platform of mass 100 kg, supported by a set of springs equivalent to a single spring of constant $k = 80$ kN/m, is subjected to a periodic force of maximum magnitude 500 N. Knowing that the coefficient of damping is 2 kN · s/m, determine (a) the natural frequency in rpm of the platform *if* there were no damping, (b) the frequency in rpm of the periodic force corresponding to the maximum value of the magnification factor, assuming damping, (c) the amplitude of the *actual* motion of the platform for each of the frequencies found in parts a and b.

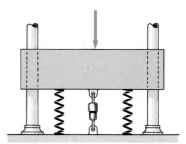

Fig. P19.108

*19.109 The suspension of an automobile may be approximated by the simplified spring-and-dashpot system shown. (a) Write the differential equation defining the absolute motion of the mass *m* when the system moves at a speed *v* over a road of sinusoidal cross section as shown. (b) Derive an expression for the amplitude of the absolute motion of *m*.

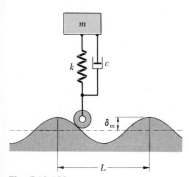

Fig. P19.109

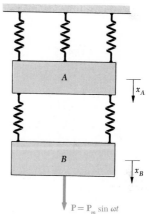

Fig. P19.110

*19.110 Two loads A and B, each of mass *m*, are suspended as shown by means of five springs of the same constant *k*. Load B is subjected to a force of magnitude $P = P_m \sin \omega t$. Write the differential equations defining the displacements x_A and x_B of the two loads from their equilibrium positions.

19.111 Determine the range of values of the resistance R for which oscillations will take place in the circuit shown when the switch S is closed.

19.112 Consider the circuit of Prob. 19.111 when the capacitance C is equal to zero. If the switch S is closed at time $t = 0$, determine (a) the final value of the current in the circuit, (b) the time t at which the current will have reached $(1 - 1/e)$ times its final value. (The desired value of t is known as the *time constant* of the circuit.)

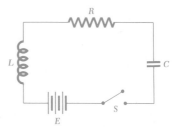

Fig. P19.111

19.113 through 19.116 Draw the electrical analogue of the mechanical system shown. (*Hint.* In Probs. 19.113 and 19.114, draw the loops corresponding to the free bodies m and A.)

19.117 and 19.118 Write the differential equations defining (*a*) the displacements of mass m and point A, (*b*) the currents in the corresponding loops of the electrical analogue.

19.119 and 19.120 Write the differential equations defining (*a*) the displacements of the masses m_1 and m_2, (*b*) the currents in the corresponding loops of the electrical analogue.

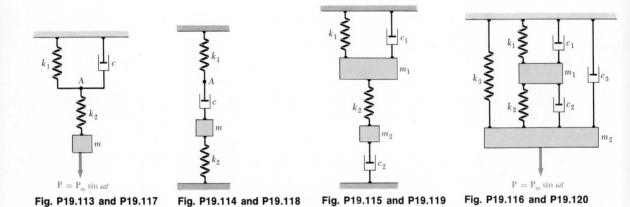

Fig. P19.113 and P19.117   **Fig. P19.114 and P19.118**   **Fig. P19.115 and P19.119**   **Fig. P19.116 and P19.120**

REVIEW PROBLEMS

19.121 A homogeneous wire of length $2l$ is bent as shown and allowed to oscillate about a smooth pin at B. Denoting by τ_0 the period of small oscillations when $\beta = 0$, determine the angle β for which the period of small oscillations is $2\tau_0$.

Fig. P19.121 and P19.122

19.122 Knowing that $l = 0.6\,\text{m}$ and $\beta = 30°$, determine the period of oscillation of the bent homogeneous wire shown.

19.123 An automobile wheel-and-tire assembly of total weight 47 lb is attached to a mounting plate of negligible weight which is suspended from a steel wire. The torsional spring constant of the wire is known to be $K = 0.40$ lb $\cdot$ in./rad. The wheel is rotated through 90° about the vertical and then released. Knowing that the period of oscillation is observed to be 30 s, determine the centroidal mass moment of inertia and the centroidal radius of gyration of the wheel-and-tire assembly.

19.124 The period of small oscillations about A of a connecting rod is observed to be 1.12 s. Knowing that the distance r_a is 7.50 in., determine the centroidal radius of gyration of the connecting rod.

Fig. P19.123

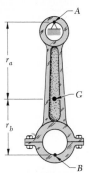

Fig. P19.124 and P19.125

19.125 A connecting rod is supported by a knife-edge at point A; the period of small oscillations is observed to be 0.945 s. The rod is then inverted and supported by a knife-edge at point B and the period of small oscillations is observed to be 0.850 s. Knowing that $r_a + r_b = 11.50$ in., determine (a) the location of the mass center G, (b) the centroidal radius of gyration $\bar{k}$.

19.126 A 150-kg electromagnet is at rest and is holding 100 kg of scrap steel when the current is turned off and the steel is dropped. Knowing that the cable and the supporting crane have a total stiffness equivalent to a spring of constant 200 kN/m, determine (a) the frequency, the amplitude, and the maximum velocity of the resulting motion, (b) the minimum tension which will occur in the cable during the motion, (c) the velocity of the magnet 0.03 s after the current is turned off.

Fig. P19.126

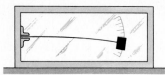

Fig. P19.127

19.127 A certain vibrometer used to measure vibration amplitudes consists essentially of a box containing a slender rod to which a mass m is attached; the natural frequency of the mass-rod system is known to be 5 Hz. When the box is rigidly attached to the casing of a motor rotating at 600 rpm, the mass is observed to vibrate with an amplitude of 1.6 mm relative to the box. Determine the amplitude of the vertical motion of the motor.

19.128 Solve Prob. 19.127, assuming that the speed of the motor is 150 rpm and that the other data are unchanged.

19.129 A rod of mass m and length L rests on the two pulleys A and B which rotate in the directions shown. Denoting by μ the coefficient of friction between the rod and the pulleys, determine the frequency of vibration if the rod is given a small displacement to the right and released.

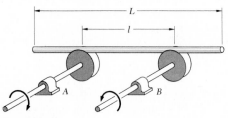

Fig. P19.129

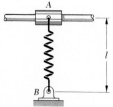

Fig. P19.130

19.130 A collar of mass m slides without friction on a horizontal rod and is attached to a spring AB of constant k. (a) If the unstretched length of the spring is just equal to l, show that the collar does not execute simple harmonic motion even when the amplitude of the oscillations is small. (b) If the unstretched length of the spring is less than l, show that the motion may be approximated by a simple harmonic motion for small oscillations.

19.131 The 30-lb beam is supported as shown by two uniform disks, each of weight 20 lb and radius 5 in. Knowing that the disks roll without sliding, determine the period of vibration of the system if the beam is given a small displacement to the right and released.

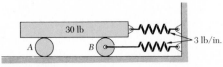

Fig. P19.131

19.132 Solve Prob. 19.131, assuming that the spring attached to the beam is removed.

APPENDIX
Moments of Inertia of Masses

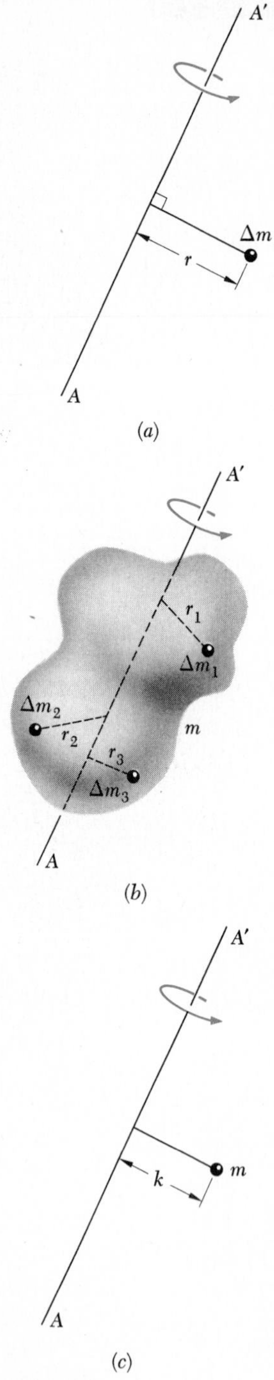

(a)

(b)

(c)

Fig. 9.20

MOMENTS OF INERTIA OF MASSES

9.10. Moment of Inertia of a Mass.* Consider a small mass Δm mounted on a rod of negligible mass which may rotate freely about an axis AA' (Fig. 9.20a). If a couple is applied to the system, the rod and mass, assumed initially at rest, will start rotating about AA'. The details of this motion will be studied later in dynamics. At present, we wish only to indicate that the time required for the system to reach a given speed of rotation is proportional to the mass Δm and to the square of the distance r. The product $r^2\,\Delta m$ provides, therefore, a measure of the *inertia* of the system, i.e., of the resistance the system offers when we try to set it in motion. For this reason, the product $r^2\,\Delta m$ is called the *moment of inertia* of the mass Δm with respect to the axis AA'.

Consider now a body of mass m which is to be rotated about an axis AA' (Fig. 9.20b). Dividing the body into elements of mass Δm_1, Δm_2, etc., we find that the resistance offered by the body is measured by the sum $r_1^2\,\Delta m_1 + r_2^2\,\Delta m_2 + \cdots$. This sum defines, therefore, the moment of inertia of the body with respect to the axis AA'. Increasing the number of elements, we find that the moment of inertia is equal, at the limit, to the integral

$$I = \int r^2\, dm \tag{9.28}$$

The *radius of gyration* k of the body with respect to the axis AA' is defined by the relation

$$I = k^2 m \qquad \text{or} \qquad k = \sqrt{\frac{I}{m}} \tag{9.29}$$

The radius of gyration k represents, therefore, the distance at which the entire mass of the body should be concentrated if its moment of inertia with respect to AA' is to remain unchanged (Fig. 9.20c). Whether it is kept in its original shape (Fig. 9.20b) or whether it is concentrated as shown in Fig. 9.20c, the mass m will react in the same way to a rotation, or *gyration*, about AA'.

If SI units are used, the radius of gyration k is expressed in meters and the mass m in kilograms. The moment of inertia of a mass, therefore, will be expressed in kg · m². If U.S. customary units are used, the radius of gyration is expressed in feet

---

° This repeats Secs. 9.10 through 9.14 of the volume on statics.

and the mass in slugs, i.e., in lb · s²/ft. The moment of inertia of a mass, then, will be expressed in lb · ft · s².†

The moment of inertia of a body with respect to a coordinate axis may easily be expressed in terms of the coordinates x, y, z of the element of mass dm (Fig. 9.21). Noting, for example, that the square of the distance r from the element dm to the y axis is $z^2 + x^2$, we express the moment of inertia of the body with respect to the y axis as

$$I_y = \int r^2 \, dm = \int (z^2 + x^2) \, dm$$

Similar expressions may be obtained for the moments of inertia with respect to the x and z axes. We write

$$
\begin{aligned}
I_x &= \int (y^2 + z^2) \, dm \\
I_y &= \int (z^2 + x^2) \, dm \\
I_z &= \int (x^2 + y^2) \, dm
\end{aligned}
\tag{9.30}
$$

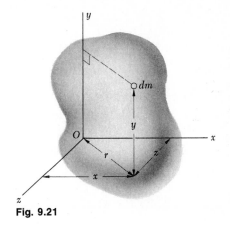

Fig. 9.21

9.11. Parallel-Axis Theorem. Consider a body of mass m. Let $Oxyz$ be a system of rectangular coordinates with origin at an arbitrary point O, and $Gx'y'z'$ a system of parallel *centroidal axes*, i.e., a system with origin at the center of gravity G of the body‡ and with axes x', y', z', respectively parallel to x, y, z (Fig. 9.22). Denoting by $\bar{x}$, $\bar{y}$, $\bar{z}$ the coordinates of G with respect to $Oxyz$, we write the following relations between the coordinates x, y, z of the element dm with respect to $Oxyz$ and its coordinates x', y', z' with respect to the centroidal axes $Gx'y'z'$:

$$x = x' + \bar{x} \qquad y = y' + \bar{y} \qquad z = z' + \bar{z} \tag{9.31}$$

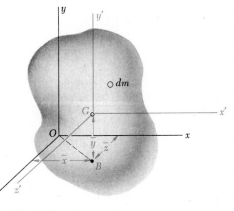

Fig. 9.22

---

†It should be kept in mind, when converting the moment of inertia of a mass from U.S. customary units to SI units, that the base unit pound used in the derived unit lb · ft · s² is a unit of force (*not* of mass) and, therefore, should be converted into newtons. We have

$$1 \text{ lb} \cdot \text{ft} \cdot \text{s}^2 = (4.45 \text{ N})(0.3048 \text{ m})(1 \text{ s})^2 = 1.356 \text{ N} \cdot \text{m} \cdot \text{s}^2$$

or, since $\text{N} = \text{kg} \cdot \text{m/s}^2$,

$$1 \text{ lb} \cdot \text{ft} \cdot \text{s}^2 = 1.356 \text{ kg} \cdot \text{m}^2$$

‡Note that the term centroidal is used to define an axis passing through the center of gravity G of the body, whether or not G coincides with the centroid of the volume of the body.

Referring to Eqs. (9.30), we may express the moment of inertia of the body with respect to the x axis as follows:

$$I_x = \int(y^2 + z^2)\,dm = \int[(y' + \bar{y})^2 + (z' + \bar{z})^2]\,dm$$
$$= \int(y'^2 + z'^2)\,dm + 2\bar{y}\int y'\,dm + 2\bar{z}\int z'\,dm + (\bar{y}^2 + \bar{z}^2)\int dm$$

The first integral in the expression obtained represents the moment of inertia $\bar{I}_{x'}$ of the body with respect to the centroidal axis x'; the second and third integrals represent the first moment of the body with respect to the $z'x'$ and $x'y'$ planes, respectively, and, since both planes contain G, the two integrals are zero; the last integral is equal to the total mass m of the body. We write, therefore,

$$I_x = \bar{I}_{x'} + m(\bar{y}^2 + \bar{z}^2) \tag{9.32}$$

and, similarly,

$$I_y = \bar{I}_{y'} + m(\bar{z}^2 + \bar{x}^2) \qquad I_z = \bar{I}_{z'} + m(\bar{x}^2 + \bar{y}^2)$$

$$(9.32')$$

We easily verify from Fig. 9.22 that the sum $\bar{z}^2 + \bar{x}^2$ represents the square of the distance OB between the y and y' axis. Similarly, $\bar{y}^2 + \bar{z}^2$ and $\bar{x}^2 + \bar{y}^2$ represent the squares of the distances between the x and x' axes, and the z and z' axes, respectively. Denoting by d the distance between an arbitrary axis AA' and a parallel centroidal axis BB' (Fig. 9.23), we may, therefore, write the following general relation between the moment of inertia I of the body with respect to AA' and its moment of inertia $\bar{I}$ with respect to BB':

$$I = \bar{I} + md^2 \tag{9.33}$$

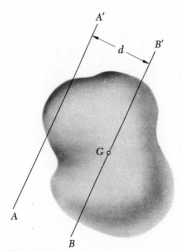

Fig. 9.23

Expressing the moments of inertia in terms of the corresponding radii of gyration, we may also write

$$k^2 = \bar{k}^2 + d^2 \tag{9.34}$$

where k and $\bar{k}$ represent the radii of gyration about AA' and BB', respectively.

9.12. Moments of Inertia of Thin Plates.

Consider a thin plate of uniform thickness t, made of a homogeneous material of density ρ (density = mass per unit volume). The mass moment of inertia of the plate with respect to an axis AA' *contained in the plane* of the plate (Fig. 9.24a) is

$$I_{AA',\text{mass}} = \int r^2 \, dm$$

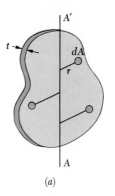

(a)

Since $dm = \rho t \, dA$, we write

$$I_{AA',\text{mass}} = \rho t \int r^2 \, dA$$

But r represents the distance of the element of area dA to the axis AA'; the integral is therefore equal to the moment of inertia of the area of the plate with respect to AA'. We have

$$I_{AA',\text{mass}} = \rho t I_{AA',\text{area}} \tag{9.35}$$

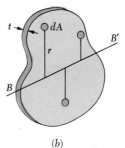

(b)

Similarly, we have with respect to an axis BB' perpendicular to AA' (Fig. 9.24b)

$$I_{BB',\text{mass}} = \rho t I_{BB',\text{area}} \tag{9.36}$$

Considering now the axis CC' *perpendicular* to the plate through the point of intersection C of AA' and BB' (Fig. 9.24c), we write

$$I_{CC',\text{mass}} = \rho t J_{C,\text{area}} \tag{9.37}$$

where J_C is the *polar* moment of inertia of the area of the plate with respect to point C.

Recalling the relation $J_C = I_{AA'} + I_{BB'}$ existing between polar and rectangular moments of inertia of an area, we write the following relation between the mass moments of inertia of a thin plate:

$$I_{CC'} = I_{AA'} + I_{BB'} \tag{9.38}$$

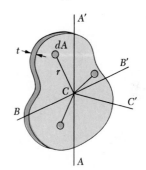

(c)   **Fig. 9.24**

Rectangular Plate. In the case of a rectangular plate of sides a and b (Fig. 9.25), we obtain the following mass moments of inertia with respect to axes through the center of gravity of the plate:

$$I_{AA',\text{mass}} = \rho t I_{AA',\text{area}} = \rho t (\tfrac{1}{12} a^3 b)$$
$$I_{BB',\text{mass}} = \rho t I_{BB',\text{area}} = \rho t (\tfrac{1}{12} a b^3)$$

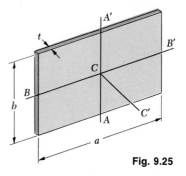

Fig. 9.25

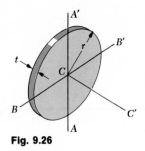

Fig. 9.26

Observing that the product ρabt is equal to the mass m of the plate, we write the mass moments of inertia of a thin rectangular plate as follows:

$$I_{AA'} = \tfrac{1}{12}ma^2 \qquad I_{BB'} = \tfrac{1}{12}mb^2 \qquad (9.39)$$
$$I_{CC'} = I_{AA'} + I_{BB'} = \tfrac{1}{12}m(a^2 + b^2) \qquad (9.40)$$

Circular Plate. In the case of a circular plate, or disk, of radius r (Fig. 9.26), we write

$$I_{AA',\text{mass}} = \rho t I_{AA',\text{area}} = \rho t(\tfrac{1}{4}\pi r^4)$$

Observing that the product $\rho\pi r^2 t$ is equal to the mass m of the plate and that $I_{AA'} = I_{BB'}$, we write the mass moments of inertia of a circular plate as follows:

$$I_{AA'} = I_{BB'} = \tfrac{1}{4}mr^2 \qquad (9.41)$$
$$I_{CC'} = I_{AA'} + I_{BB'} = \tfrac{1}{2}mr^2 \qquad (9.42)$$

9.13. Determination of the Moment of Inertia of a Three-dimensional Body by Integration. The moment of inertia of a three-dimensional body is obtained by computing the integral $I = \int r^2\, dm$. If the body is made of a homogeneous material of density ρ, we have $dm = \rho\, dV$ and write $I = \rho \int r^2\, dV$. This integral depends only upon the shape of the body. In order to compute it, it will generally be necessary to perform a triple, or at least a double, integration.

However, if the body possesses two planes of symmetry, it is usually possible to determine its moment of inertia through a single integration by choosing as an element of mass dm the mass of a thin slab perpendicular to the planes of symmetry. In the case of bodies of revolution, for example, the element of mass should be a thin disk (Fig. 9.27). Using formula (9.42), the moment of inertia of the disk with respect to the axis of revolution may be readily expressed as indicated in Fig. 9.27. Its moment of inertia with respect to each of the other two axes of coordinates will be obtained by using formula (9.41) and the parallel-axis theorem. Integration of the expressions obtained will yield the desired moments of inertia of the body of revolution.

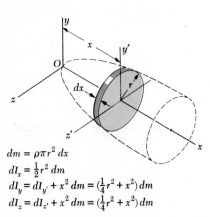

$dm = \rho\pi r^2\, dx$

$dI_x = \tfrac{1}{2}r^2\, dm$

$dI_y = dI_{y'} + x^2\, dm = (\tfrac{1}{4}r^2 + x^2)\, dm$

$dI_z = dI_{z'} + x^2\, dm = (\tfrac{1}{4}r^2 + x^2)\, dm$

Fig. 9.27 Determination of the moment of inertia of a body of revolution.

9.14. Moments of Inertia of Composite Bodies. The moments of inertia of a few common shapes are shown in Fig. 9.28. The moment of inertia with respect to a given axis of a body made of several of these simple shapes may be obtained by computing the moments of inertia of its component parts about the desired axis and adding them together. We should note, as we already have noted in the case of areas, that the radius of gyration of a composite body *cannot* be obtained by adding the radii of gyration of its component parts.

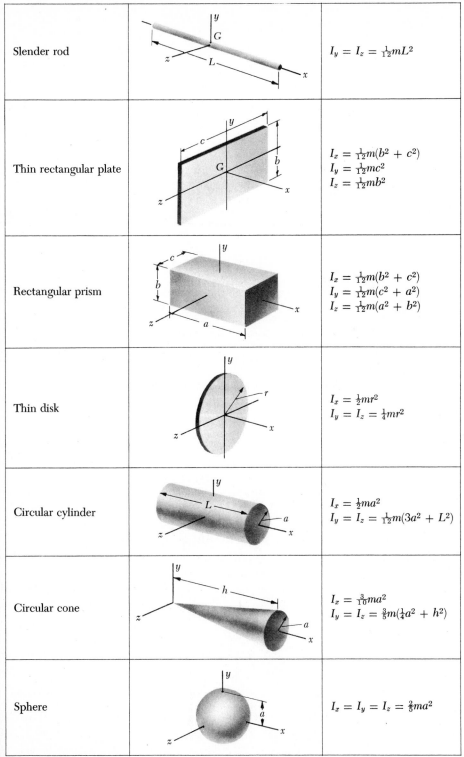

Slender rod	$I_y = I_z = \frac{1}{12}mL^2$
Thin rectangular plate	$I_x = \frac{1}{12}m(b^2 + c^2)$ $I_y = \frac{1}{12}mc^2$ $I_z = \frac{1}{12}mb^2$
Rectangular prism	$I_x = \frac{1}{12}m(b^2 + c^2)$ $I_y = \frac{1}{12}m(c^2 + a^2)$ $I_z = \frac{1}{12}m(a^2 + b^2)$
Thin disk	$I_x = \frac{1}{2}mr^2$ $I_y = I_z = \frac{1}{4}mr^2$
Circular cylinder	$I_x = \frac{1}{2}ma^2$ $I_y = I_z = \frac{1}{12}m(3a^2 + L^2)$
Circular cone	$I_x = \frac{3}{10}ma^2$ $I_y = I_z = \frac{3}{5}m(\frac{1}{4}a^2 + h^2)$
Sphere	$I_x = I_y = I_z = \frac{2}{5}ma^2$

Fig. 9.28 Mass moments of inertia of common geometric shapes

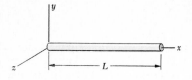

Determine the mass moment of inertia of a slender rod of length L and mass m with respect to an axis perpendicular to the rod and passing through one end of the rod.

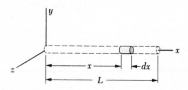

Solution. Choosing the differential element of mass shown, we write

$$dm = \frac{m}{L} dx$$

$$I_y = \int x^2 dm = \int_0^L x^2 \frac{m}{L} dx = \left[\frac{m}{L} \frac{x^3}{3} \right]_0^L \qquad I_y = \frac{mL^2}{3} \quad \blacktriangleleft$$

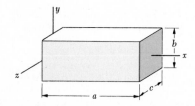

Determine the mass moment of inertia of the homogeneous rectangular prism shown with respect to the z axis.

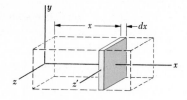

Solution. We choose as a differential element of mass the thin slab shown for which

$$dm = \rho bc \, dx$$

Referring to Sec. 9.12, we find that the moment of inertia of the element with respect to the z' axis is

$$dI_{z'} = \tfrac{1}{12}b^2 \, dm$$

Applying the parallel-axis theorem, we obtain the mass moment of inertia of the slab with respect to the z axis.

$$dI_z = dI_{z'} + x^2 \, dm = \tfrac{1}{12}b^2 \, dm + x^2 \, dm = (\tfrac{1}{12}b^2 + x^2) \, \rho bc \, dx$$

Integrating from $x = 0$ to $x = a$, we obtain

$$I_z = \int dI_z = \int_0^a (\tfrac{1}{12}b^2 + x^2) \, \rho bc \, dx = \rho abc \, (\tfrac{1}{12}b^2 + \tfrac{1}{3}a^2)$$

Since the total mass of the prism is $m = \rho abc$, we may write

$$I_z = m(\tfrac{1}{12}b^2 + \tfrac{1}{3}a^2) \qquad I_z = \tfrac{1}{12}m(4a^2 + b^2) \quad \blacktriangleleft$$

We note that if the prism is slender, b is small compared to a and the expression for I_z reduces to $ma^2/3$, which is the result obtained in Sample Prob. 9.9 when $L = a$.

SAMPLE PROBLEM 9.11

Determine the mass moment of inertia of a right circular cone with respect to (*a*) its longitudinal axis, (*b*) an axis through the apex of the cone and perpendicular to its longitudinal axis, (*c*) an axis through the centroid of the cone and perpendicular to its longitudinal axis.

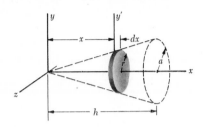

Solution. We choose the differential element of mass shown.

$$r = a\frac{x}{h} \qquad dm = \rho \pi r^2 \, dx = \rho \pi \frac{a^2}{h^2} x^2 \, dx$$

a. Moment of Inertia I_x. Using the expression derived in Sec. 9.12 for a thin disk, we compute the mass moment of inertia of the differential element with respect to the x axis.

$$dI_x = \tfrac{1}{2} r^2 \, dm = \tfrac{1}{2} \left(a\frac{x}{h}\right)^2 \left(\rho \pi \frac{a^2}{h^2} x^2 \, dx\right) = \tfrac{1}{2}\rho \pi \frac{a^4}{h^4} x^4 \, dx$$

Integrating from $x = 0$ to $x = h$, we obtain

$$I_x = \int dI_x = \int_0^h \tfrac{1}{2}\rho \pi \frac{a^4}{h^4} x^4 \, dx = \tfrac{1}{2}\rho \pi \frac{a^4}{h^4}\frac{h^5}{5} = \tfrac{1}{10}\rho \pi a^4 h$$

Since the total mass of the cone is $m = \tfrac{1}{3}\rho \pi a^2 h$, we may write

$$I_x = \tfrac{1}{10}\rho \pi a^4 h = \tfrac{3}{10}a^2(\tfrac{1}{3}\rho \pi a^2 h) = \tfrac{3}{10}ma^2 \qquad I_x = \tfrac{3}{10}ma^2 \quad \blacktriangleleft$$

b. Moment of Inertia I_y. The same differential element will be used. Applying the parallel-axis theorem and using the expression derived in Sec. 9.12 for a thin disk, we write

$$dI_y = dI_{y'} + x^2 \, dm = \tfrac{1}{4} r^2 \, dm + x^2 \, dm = (\tfrac{1}{4}r^2 + x^2) \, dm$$

Substituting the expressions for r and dm, we obtain

$$dI_y = \left(\frac{1}{4}\frac{a^2}{h^2}x^2 + x^2\right)\left(\rho \pi \frac{a^2}{h^2}x^2 \, dx\right) = \rho \pi \frac{a^2}{h^2}\left(\frac{a^2}{4h^2} + 1\right)x^4 \, dx$$

$$I_y = \int dI_y = \int_0^h \rho \pi \frac{a^2}{h^2}\left(\frac{a^2}{4h^2} + 1\right)x^4 \, dx = \rho \pi \frac{a^2}{h^2}\left(\frac{a^2}{4h^2} + 1\right)\frac{h^5}{5}$$

Introducing the total mass of the cone m, we rewrite I_y as follows:

$$I_y = \tfrac{3}{5}(\tfrac{1}{4}a^2 + h^2)\tfrac{1}{3}\rho \pi a^2 h \qquad I_y = \tfrac{3}{5}m(\tfrac{1}{4}a^2 + h^2) \quad \blacktriangleleft$$

c. Moment of Inertia $\bar{I}_{y''}$. We apply the parallel-axis theorem and write

$$I_y = \bar{I}_{y''} + m\bar{x}^2$$

Solving for $\bar{I}_{y''}$ and recalling that $\bar{x} = \tfrac{3}{4}h$, we have

$$\bar{I}_{y''} = I_y - m\bar{x}^2 = \tfrac{3}{5}m(\tfrac{1}{4}a^2 + h^2) - m(\tfrac{3}{4}h)^2$$

$$\bar{I}_{y''} = \tfrac{3}{20}m(a^2 + \tfrac{1}{4}h^2) \quad \blacktriangleleft$$

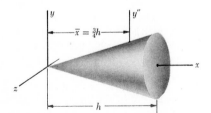

A steel forging consists of a rectangular prism 6 by 2 by 2 in. and of two cylinders of diameter 2 in. and length 3 in., as shown. Determine the mass moments of inertia with respect to the coordinate axes. (Specific weight of steel = 490 lb/ft³.)

Computation of Masses
Prism

$$V = 24 \text{ in}^3 \qquad W = \frac{(24 \text{ in}^3)(490 \text{ lb/ft}^3)}{1728 \text{ in}^3/\text{ft}^3} = 6.81 \text{ lb}$$

$$m = \frac{6.81 \text{ lb}}{32.2 \text{ ft/s}^2} = 0.211 \text{ lb} \cdot \text{s}^2/\text{ft}$$

Each Cylinder

$$V = \pi(1 \text{ in.})^2(3 \text{ in.}) = 9.42 \text{ in}^3$$

$$W = \frac{(9.42 \text{ in}^3)(490 \text{ lb/ft}^3)}{1728 \text{ in}^3/\text{ft}^3} = 2.67 \text{ lb}$$

$$m = \frac{2.67 \text{ lb}}{32.2 \text{ ft/s}^2} = 0.0829 \text{ lb} \cdot \text{s}^2/\text{ft}$$

Mass Moments of Inertia. The mass moments of inertia of each component are computed from Fig. 9.28, using the parallel-axis theorem when necessary. Note that all lengths should be expressed in feet.

Prism

$$I_x = I_z = \tfrac{1}{12}(0.211 \text{ lb} \cdot \text{s}^2/\text{ft})[(\tfrac{6}{12} \text{ ft})^2 + (\tfrac{2}{12} \text{ ft})^2] = 4.88 \times 10^{-3} \text{ lb} \cdot \text{ft} \cdot \text{s}^2$$
$$I_y = \tfrac{1}{12}(0.211 \text{ lb} \cdot \text{s}^2/\text{ft})[(\tfrac{2}{12} \text{ ft})^2 + (\tfrac{2}{12} \text{ ft})^2] = 0.977 \times 10^{-3} \text{ lb} \cdot \text{ft} \cdot \text{s}^2$$

Each cylinder

$$I_x = \tfrac{1}{2}ma^2 + m\bar{y}^2 = \tfrac{1}{2}(0.0829 \text{ lb} \cdot \text{s}^2/\text{ft})(\tfrac{1}{12} \text{ ft})^2$$
$$+ (0.0829 \text{ lb} \cdot \text{s}^2/\text{ft})(\tfrac{2}{12} \text{ ft})^2 = 2.59 \times 10^{-3} \text{ lb} \cdot \text{ft} \cdot \text{s}^2$$
$$I_y = \tfrac{1}{12}m(3a^2 + L^2) + m\bar{x}^2 = \tfrac{1}{12}(0.0829 \text{ lb} \cdot \text{s}^2/\text{ft})[3(\tfrac{1}{12} \text{ ft})^2 + (\tfrac{3}{12} \text{ ft})^2]$$
$$+ (0.0829 \text{ lb} \cdot \text{s}^2/\text{ft})(\tfrac{2.5}{12} \text{ ft})^2 = 4.17 \times 10^{-3} \text{ lb} \cdot \text{ft} \cdot \text{s}^2$$
$$I_z = \tfrac{1}{12}m(3a^2 + L^2) + m(\bar{x}^2 + \bar{y}^2) = \tfrac{1}{12}(0.0829)[3(\tfrac{1}{12})^2 + (\tfrac{3}{12})^2]$$
$$+ (0.0829)[(\tfrac{2.5}{12})^2 + (\tfrac{2}{12})^2] = 6.48 \times 10^{-3} \text{ lb} \cdot \text{ft} \cdot \text{s}^2$$

Entire Body. Adding the values obtained:

$$I_x = 4.88 \times 10^{-3} + 2(2.59 \times 10^{-3})$$
$$\qquad\qquad I_x = 10.06 \times 10^{-3} \text{ lb} \cdot \text{ft} \cdot \text{s}^2 \blacktriangleleft$$

$$I_y = 0.977 \times 10^{-3} + 2(4.17 \times 10^{-3})$$
$$\qquad\qquad I_y = 9.32 \times 10^{-3} \text{ lb} \cdot \text{ft} \cdot \text{s}^2 \blacktriangleleft$$

$$I_z = 4.88 \times 10^{-3} + 2(6.48 \times 10^{-3})$$
$$\qquad\qquad I_z = 17.84 \times 10^{-3} \text{ lb} \cdot \text{ft} \cdot \text{s}^2 \blacktriangleleft$$

Solve Sample Prob. 9.12 using SI units.

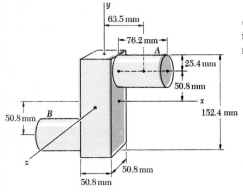

Solution. First, the dimensions are converted into millimeters (1 in. = 25.4 mm). Next, the density of steel ρ (mass per unit volume) is determined in SI units. Recalling that 1 ft = 0.3048 m and that the mass of a block weighing 1 lb is 0.454 kg, we have

$$\rho = (490 \text{ lb/ft}^3)\left(\frac{0.454 \text{ kg}}{1 \text{ lb}}\right)\left(\frac{1 \text{ ft}}{0.3048 \text{ m}}\right)^3 = 7850 \text{ kg/m}^3$$

Computation of Masses

Prism. $\quad V = (50.8 \text{ mm})^2(152.4 \text{ mm}) = 0.393 \times 10^6 \text{ mm}^3$

or, since $1 \text{ mm}^3 = (10^{-3} \text{ m})^3 = 10^{-9} \text{ m}^3$,

$$V = 0.393 \times 10^6 \times 10^{-9} \text{ m}^3 = 0.393 \times 10^{-3} \text{ m}^3$$
$$m = \rho V = (7.85 \times 10^3 \text{ kg/m}^3)(0.393 \times 10^{-3} \text{ m}^3) = 3.09 \text{ kg}$$

Each Cylinder

$$V = \pi r^2 h = \pi(25.4 \text{ mm})^2(76.2 \text{ mm}) = 0.1544 \times 10^6 \text{ mm}^3$$
$$= 0.1544 \times 10^{-3} \text{ m}^3$$
$$m = \rho V = (7.85 \times 10^3 \text{ kg/m}^3)(0.1544 \times 10^{-3} \text{ m}^3) = 1.212 \text{ kg}$$

Mass Moments of Inertia. The mass moments of inertia of each component are computed from Fig. 9.28, using the parallel-axis theorem when necessary. Note that all lengths should be expressed in millimeters.

Prism

$$I_x = I_z = \tfrac{1}{12}(3.09 \text{ kg})[(152.4 \text{ mm})^2 + (50.8 \text{ mm})^2] = 6640 \text{ kg} \cdot \text{mm}^2$$
$$I_y = \tfrac{1}{12}(3.09 \text{ kg})[(50.8 \text{ mm})^2 + (50.8 \text{ mm})^2] = 1329 \text{ kg} \cdot \text{mm}^2$$

Each Cylinder

$$I_x = \tfrac{1}{2}ma^2 + m\bar{y}^2 = \tfrac{1}{2}(1.212 \text{ kg})(25.4 \text{ mm})^2 + (1.212 \text{ kg})(50.8 \text{ mm})^2$$
$$= 3520 \text{ kg} \cdot \text{mm}^2$$
$$I_y = \tfrac{1}{12}m(3a^2 + L^2) + m\bar{x}^2 = \tfrac{1}{12}(1.212 \text{ kg})[3(25.4 \text{ mm})^2 + (76.2 \text{ mm})^2]$$
$$+ (1.212 \text{ kg})(63.5 \text{ mm})^2 = 5670 \text{ kg} \cdot \text{mm}^2$$
$$I_z = \tfrac{1}{12}m(3a^2 + L^2) + m(\bar{x}^2 + \bar{y}^2)$$
$$= \tfrac{1}{12}(1.212 \text{ kg})[3(25.4 \text{ mm})^2 + (76.2 \text{ mm})^2]$$
$$+ (1.212 \text{ kg})[(63.5 \text{ mm})^2 + (50.8 \text{ mm})^2] = 8800 \text{ kg} \cdot \text{mm}^2$$

Entire Body. Adding the values obtained, and observing that $1 \text{ mm}^2 = (10^{-3} \text{ m})^2 = 10^{-6} \text{ m}^2$, we have

$$I_x = 6640 \text{ kg} \cdot \text{mm}^2 + 2(3520 \text{ kg} \cdot \text{mm}^2) = 13.68 \times 10^3 \text{ kg} \cdot \text{mm}^2$$
$$I_x = 13.68 \times 10^{-3} \text{ kg} \cdot \text{m}^2 \quad \blacktriangleleft$$
$$I_y = 1329 \text{ kg} \cdot \text{mm}^2 + 2(5670 \text{ kg} \cdot \text{mm}^2) = 12.67 \times 10^3 \text{ kg} \cdot \text{mm}^2$$
$$I_y = 12.67 \times 10^{-3} \text{ kg} \cdot \text{m}^2 \quad \blacktriangleleft$$
$$I_z = 6640 \text{ kg} \cdot \text{mm}^2 + 2(8800 \text{ kg} \cdot \text{mm}^2) = 24.2 \times 10^3 \text{ kg} \cdot \text{mm}^2$$
$$I_z = 24.2 \times 10^{-3} \text{ kg} \cdot \text{m}^2 \quad \blacktriangleleft$$

Recalling that $1 \text{ lb} \cdot \text{ft} \cdot \text{s}^2 = 1.356 \text{ kg} \cdot \text{m}^2$ (see footnote, page 793) we may check these answers against the values obtained in Sample Prob. 9.12.

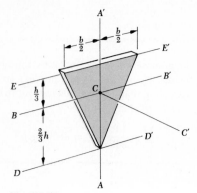

Fig. P9.72

PROBLEMS

9.72 A thin plate of mass m is cut in the shape of an isosceles triangle of width b and height h. Determine the mass moment of inertia of the plate with respect to (a) the centroidal axes AA' and BB' in the plane of the plate, (b) the centroidal axis CC' perpendicular to the plate.

9.73 Determine the mass moment of inertia of the plate of Prob. 9.72 with respect to (a) edge EE', (b) axis DD' which passes through a vertex and is parallel to edge EE'.

9.74 Determine the mass moment of inertia of a ring of mass m, cut from a thin uniform plate, with respect to (a) the diameter AA' of the ring, (b) the axis CC' perpendicular to the plane of the ring.

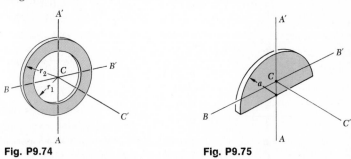

Fig. P9.74   **Fig. P9.75**

9.75 A thin semicircular plate has a radius a and a mass m. Determine the mass moment of inertia of the plate with respect to (a) the centroidal axis BB', (b) the centroidal axis CC'.

9.76 Determine by direct integration the mass moment of inertia of the homogeneous regular tetrahedron of mass m and side a with respect to an axis through A perpendicular to the base BCD. (*Hint.* Use the result of part b of Prob. 9.72.)

9.77 The area shown is revolved about the x axis to form a homogeneous solid of revolution of mass m. Express the mass moment of inertia of the solid with respect to the x axis in terms of m, a, and n. The expression obtained may be used to verify (a) the value given in Fig. 9.28 for a cone (with $n = 1$) and (b) the answer to Prob. 9.78 (with $n = \frac{1}{2}$).

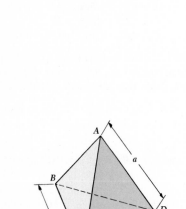

Fig. P9.76

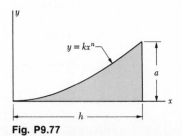

Fig. P9.77

9.78 Determine by direct integration the mass moment of inertia and the radius of gyration with respect to the x axis of the paraboloid shown, assuming a uniform density and a mass m.

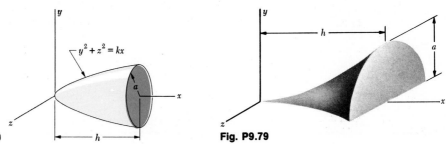

Fig. P9.78 and P9.80

Fig. P9.79

9.79 The homogeneous solid shown was obtained by rotating the area of Prob. 9.77, with $n = 2$, through 180° about the x axis. Determine the mass moment of inertia $\bar{I}_x$ in terms of m and a.

9.80 Determine by direct integration the mass moment of inertia and the radius of gyration with respect to the y axis of the paraboloid shown, assuming a uniform density and a mass m.

9.81 Determine by direct integration the mass moment of inertia and the radius of gyration of a circular cylinder of radius a, length L, and uniform density ρ, with respect to a diameter of its base.

9.82 Determine by direct integration the mass moment of inertia with respect to the y axis of the pyramid shown, assuming a uniform density and a mass m.

9.83 Determine by direct integration the mass moment of inertia and the radius of gyration of the right circular cone with respect to the z axis, assuming a uniform density and a mass m.

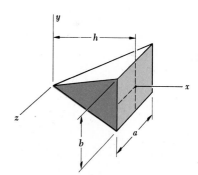

Fig. P9.82

9.84 Determine the mass moment of inertia of the frustum of a right circular cone of mass m with respect to its axis of symmetry.

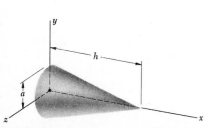

Fig. P9.83

Fig. P9.84

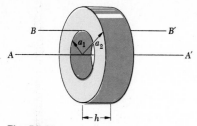

Fig. P9.85

9.85 Determine the mass moment of inertia of the homogeneous ring of mass m with respect to the axis AA'.

9.86 Determine the mass moment of inertia and the radius of gyration of the steel flywheel shown with respect to the axis of rotation. The web of the flywheel consists of a solid plate 25 mm thick. (Density of steel = 7850 kg/m³.)

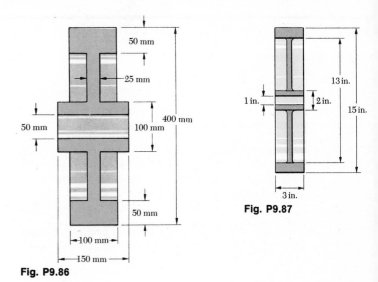

Fig. P9.86

Fig. P9.87

9.87 The cross section of a small flywheel is shown. The rim and hub are connected by eight spokes (two of which are shown in the cross section). Each spoke has a cross-sectional area of 0.1875 in². Determine the mass moment of inertia and radius of gyration of the flywheel with respect to the axis of rotation. (Specific weight of steel = 490 lb/ft³.)

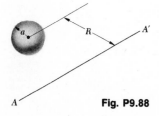

Fig. P9.88

9.88 In using the parallel-axis theorem, the error introduced by neglecting the centroidal moment of inertia is sometimes small. For a homogeneous sphere of radius a and mass m, (a) determine the mass moment of inertia with respect to an axis AA' at a distance R from the center of the sphere, (b) express as a function of a/R the relative error introduced by neglecting the centroidal moment of inertia, (c) determine the distance R in terms of a for which the relative error is 0.4 percent.

9.89 Two rods, *each* of length l and mass m, are welded together as shown. Knowing that rod CD is parallel to the x axis, determine the mass moment of inertia of the composite body with respect to (*a*) the x axis, (*b*) the y axis, (*c*) the z axis.

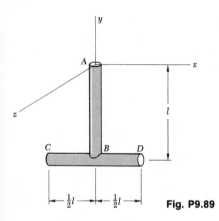

Fig. P9.89

9.90 In Prob. 9.89, determine the mass moments of inertia of the assembly with respect to centroidal axes parallel to the given x, y, and z axes.

9.91 For the homogeneous ring shown, which is of density ρ, determine (*a*) the mass moment of inertia with respect to the axis BB', (*b*) the value of a_1 for which, given a_2 and h, $I_{BB'}$ is maximum, (*c*) the corresponding value of $I_{BB'}$.

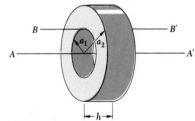

Fig. P9.91

9.92 and 9.93 Determine the mass moment of inertia and the radius of gyration of the steel link shown with respect to the x axis. (Specific weight of steel = 490 lb/ft³; density of steel = 7850 kg/m³.)

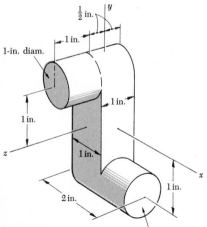

Fig. P9.92

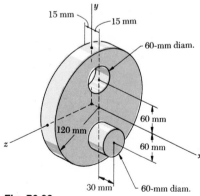

Fig. P9.93

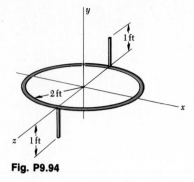

Fig. P9.94

9.94 A homogeneous wire, of weight 1.5 lb/ft, is used to form the figure shown. Determine the mass moment of inertia of the wire figure with respect to (*a*) the *x* axis, (*b*) the *y* axis, (*c*) the *z* axis.

9.95 Two holes, each of diameter 50 mm, are drilled through the steel block shown. Determine the mass moment of inertia of the body with respect to the axis of either of the holes. (Density of steel = 7850 kg/m³.)

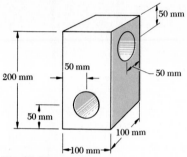

Fig. P9.95

INDEX

INDEX

Absolute acceleration, 610–613
Absolute motion of a particle, 434
Absolute system of units, 6, 456
Absolute velocity, 595–597, 605
Acceleration, 402, 431
  absolute, 610–613
  angular, 586, 588
  components of: normal, 442–444
    radial, 444–446
    rectangular, 431–432
    tangential, 442–444
    transverse, 444–446
  Coriolis, 621–623
  of gravity, 4, 411, 475
  in plane motion, 610–613
  relative: of a particle: 593–595
    in plane motion, 610–613
    with respect to axes in translation, 433
    with respect to rotating axes, 622
    of two particles, 414, 433–434
  in rotation, 586
Acceleration-time curve, 403, 420–421
Accuracy, numerical, 14–15
Action and reaction, 4, 185
Addition:
  of couples: in a plane, 68–69
    in space, 109–110
  of forces: concurrent: in a plane, 20, 29
    in space, 45
    nonconcurrent: in a plane, 75–76
    in space, 117–118
  of vectors, 18–20

Amplitude, 746, 749
Analogue, electrical, 781–784
Angle:
  of kinetic friction, 277
  lead, 292
  phase, 746
  of repose, 278
  of static friction, 277
Angular acceleration, 586, 588
Angular coordinate, 586
Angular impulse, 559
Angular momentum:
  conservation of, 556, 716
  of a particle, 556
  of a rigid body, 711
  of a system of particles, 558, 559
Angular velocity, 586, 587, 597
Apogee, 487
Archimedes, 2
Areal velocity, 474
Aristotle, 2
Associative property for sums of vectors, 20
Auxiliary circle, 748
Axioms of mechanics, 3–5
Axis:
  of precession, 735
  of rotation, instantaneous, 584, 603
  of spin, 735
  of symmetry, 142, 144
  of a wrench, 118
Axle friction, 300–301

Balancing of rotating shafts, 675
Ball-and-socket supports, 126
Ball supports, 125
Ballistic missiles, 488

Band brakes, 310–311
Banking of curves, 476
Beams, 240–257
  combined, 241
  loading of, 240–241
  span of, 241
  supports of, 240–241
  types of, 241
Bearings, 126, 300–303
  collar, 301
  end, 301–302
  journal, 300–301
  thrust, 301–303
Belt drive, 310–311
Belt friction, 309–311
Bending, 237
Bending moment, 237, 242–257
Bending-moment diagram, 244
Bernoulli, Jean, 369
Bow's notation, 188

Cables:
  with concentrated loads, 258–259
  with distributed loads, 259–260
  parabolic, 260–261
  reactions at, 125
  span of, 261, 269
Calculators:
  accuracy, 15
  use of, 22, 23, 27, 30, 36, 43, 46
Cardan's suspension, 738
Catenary, 267–269
Cathode-ray tube, 480
Center:
  of gravity, 138–139, 171–173, 462

Center:
  of oscillation, 758
  of percussion, 663, 721
  of pressure, 165, 338–339
  of rotation, instantaneous, 603–605
  of symmetry, 142, 144
  (*See also* Mass center)
Centimeter, 8
Central force, 473–474
Central impact, 543–547
Centrifugal force, 473, 656
Centroidal axes, principal, 343
Centroidal rotation:
  effective forces in, 635
  kinetic energy in, 688–689
  momentum in, 711
Centroids, 140–156, 171–175
  of areas and lines, 140–144
  of common shapes: of areas, 143–144
    of lines, 144
    of volumes, 173–175
  of composite areas and lines, 146
  of composite volumes, 175
  determination of, by integration, 153–154, 175
  of volumes, 171–175
Circular frequency, 749
Circular orbit, 487
Coefficient:
  of critical damping, 778
  of damping, 777
  of kinetic friction, 275–276
  of restitution, 544, 725
  of rolling resistance, 304
  of static friction, 275–276
  of viscous damping, 777
Collar bearings, 301
Complementary acceleration, 622
Complementary function, 772
Complete constraints, 86
Components:
  of acceleration (*see* Acceleration)

Components:
  of force, 21, 25–26, 40, 45
  of velocity (*see* Velocity)
Composite areas:
  centroids of, 146
  moments of inertia of, 330
Composite bodies:
  centroids of, 175
  moments of inertia of, 796
Composition of forces (*see* Addition, of forces)
Compound pendulum, 762
Compound trusses, 203
Compression, 59, 186, 249
Concurrent forces, 20–21
Conic section, 485
Conservation:
  of angular momentum, 716
  of energy: for particles, 516–517
    for rigid bodies, 696–697
    in vibrations, 765–766
  of momentum, 536–537, 716
Conservative force, 387, 516–517, 765
Constrained plane motion, 653
Constraining forces, 82
Constraints, 374
  complete, 86
  improper, 87–88
  partial, 86–88, 126
Coordinate:
  angular, 586
  position, 400
Coplanar forces, 20
Coplanar vectors, 20
Coriolis acceleration, 621–623
Coulomb friction, 274
Counters, 208
Couple vector, 108
Couples, 66–70, 108–110
  gyroscopic, 736
  inertia, 636
  momentum, 710
  in a plane, 66–70
    addition of, 68–69

Couples:
  in a plane: equivalent, 67–68
  in space, 108–110
    addition of, 109
    equivalent, 110
Critical damping, coefficient of, 778
Curvature, radius of, 443
Curvilinear motion of a particle:
  kinematics of, 428–446
  kinetics of, 459–489
Curvilinear translation, 583
Customary units, U.S., 8–13

D'Alembert, Jean, 2
D'Alemberts' principle, 632
Damped vibrations (*see* Vibrations)
Damping:
  coefficient of, 777
  critical, 778
  heavy, 778
  light, 778–779
  viscous, 777
Damping coefficient, 777
Damping factor, 779
Deceleration, 402
Decimeter, 8
Decrement, logarithmic, 785
Deformation, period of, 543–544, 725
Degrees of freedom, 387
Density, 140, 355
Dependent motions, 414–415
Determinate reactions, 87
Determinate structures, 213
Determinate trusses, 203
Diagram:
  acceleration-time, 403, 421
  bending-moment, 244
  displacement-time, 403, 421
  free-body, 34–35, 125
  shear, 242, 244
  velocity-displacement, 422
  velocity-time, 403, 421

Differential elements:
  for centroids: of areas, 153
    of volumes, 175
  for moments of inertia: of
    areas, 320–323
    of masses, 796
Direct central impact, 543–546
Direction cosines, 42
Direction of a force, 16
Disk clutches, 302
Disk friction, 301–303
Displacement, 369
  virtual, 372
Distance, 403
Distributed forces, 138, 320
Distributed loads, 163–164, 240
Dry friction, 274–275
Dynamic balancing, 675
Dynamic equilibrium:
  of a particle, 459
  of a rigid body: in noncen-
    troidal rotation, 655
    in plane motion, 636
Dynamics, definition of, 1, 399

Earth satellites, 485
Eccentric impact, 725–727
Eccentricity, 486, 776
Effective forces, 460
  for a rigid body: in plane mo-
    tion, 632–635
    in three dimensions, 673
Efficiency, 375–376, 525–526
Elastic impact, 546
Electrical analogue, 781–784
Elliptic orbit, 486
End bearings, 301
Energy:
  conservation of, 516–517, 696–
    697, 765–766
  kinetic: of a particle, 500–502
    of a rigid body: in plane
      motion, 695
      in rotation, 688
      in translation, 688

Energy:
  potential, 386–388, 513–516,
    696
  total mechanical, 516
Equations:
  of equilibrium: for a particle,
    33, 49
    for a rigid body: in a plane,
      81
      in space, 125
  of motion: for a particle, 459
    for a system of particles,
      461
    for a three-dimensional
      body, 673
    for a two-dimensional body:
      in noncentroidal rota-
        tion, 656
      in plane motion, 634, 636
Equilibrium:
  dynamic (see Dynamic equilib-
    rium)
  equations (see Equations)
  neutral, 389–390
  of a particle: in a plane,
    32–35
    in space, 49
  of a rigid body: in a plane,
    81–88, 99–100
    in space, 124–127
  stability of, 389–390
Equipollence of external forces
  and effective forces, 461
Equipollent systems of vectors,
  461, 462
Equivalence of external forces
  and effective forces for a
  rigid body, 632, 673
Equivalent forces, 58–59
Equivalent systems of forces:
  in a plane, 56–81
  in space, 106–124
Escape velocity, 411, 487
External forces, 56–57

Fan, 569–570

First moment:
  of area, 141
  of volume, 173
Fixed supports, 84, 126
Flexible cords (see Cables)
Fluid flow, 568
Fluid friction, 777
Focus of conic section, 485
Foot, 8
Force, 2
  central, 473–474
  centrifugal, 473, 656
  conservative, 387, 516, 765
  effective (see Effective forces)
  external, 56–57
  gravitational, 474–475, 500,
    513, 557
  impulsive, 534
  inertia, 459
  internal, 56–57, 460
  in a member, 188
  nonconservative, 517
  nonimpulsive, 535
  on a particle: in a plane,
    16–29
    in space, 40–45
  reversed effective (see Inertia
    vector)
  on a rigid body: in a plane,
    56–105
    in space, 106–137
Force-couple system, 70
Force systems, 56–81, 106–124
Forced frequency, 772–773
Forced vibrations:
  damped, 780–781
  undamped, 771–773
Frame of reference, 455
Frames, 210–213
Free-body diagram:
  of a particle, 34–35
  of a rigid body, 125
Free vibrations:
  damped, 777–779
  undamped, 746–777
Freedom, degrees of, 387, 415

Freely falling body, 413
Frequency, 746, 749
    circular, 749
    forced, 772–773
    natural, 772
Frequency ratio, 773
Friction, 274–319
    angles of, 277–278
    belt, 309–311
    circle of, 301
    coefficient of, 275–276
    Coulomb, 274
    dry, 274–280
    fluid, 777
    kinetic, 275–277
    laws of, 275–277
    static, 275–278
    wheel, 303–304
Frictionless surface, 83, 125

Gears:
    analysis of, 597, 613
    planetary, 600
Geneva mechanism, 624, 625
Geometric instability, 88
Gram, 6
Graphical methods:
    for analysis of trusses, 198–199
    for solution of rectilinear-motion
        problems, 420–421
Gravitation:
    constant of, 4, 474
    Newton's law of, 4–5, 474–475
Gravitational forces, 474–475,
        500, 513, 557
Gravitational potential energy,
        513
Gravitational system of units, 8
Gravity:
    acceleration of, 4, 411, 475
    center of, 138–139, 171–173
Guldinus, theorems of, 154–156
Gun, recoil, 539
Gyration, radius of, 324, 792
Gyrocompass, 738

Gyroscope, 735–738
Gyroscopic couple, 736

Hamilton, Sir William R., 2
Harmonic motion, simple, 746–
        747
Helicopter, 577, 724
Hertz (unit), 749
Hinges, 126
Horsepower, 526
Hydrostatic forces, 164–165
Hyperbolic trajectory, 486

Ideal machines, 375
Impact, 543
    central: direct, 543–546
        oblique, 546–547
    eccentric, 725–727
    elastic, 546
    line of, 543
    plastic, 545
Improper constraints, 87–88
Impulse:
    angular, 559
    linear, 533
    and momentum, principle (see
        Principle)
Impulsive force, 534
Impulsive motion, 534, 725
Inch, 10
Inclined axes, moments of iner-
        tia, 341
Indeterminate reactions, 86–88
Indeterminate structures, 213
Indeterminate trusses, 204
Inertia:
    moments of (see Moments of
        inertia)
    principal axes of, 341–343
    product of, 339–340
        parallel-axis theorem for,
        340
Inertia couple, 636
Inertia force, 459
Inertia vector:
    for a particle, 459

Inertia vector:
    for a rigid body in plane mo-
        tion, 636
Inertial system, 455
Initial conditions, 404
Input forces, 223
Input power, 526
Input work, 375
Instantaneous axis of rotation,
        584, 603
Instantaneous center of rotation,
        603–605
Internal forces, 56–57, 460
    in members, 188, 236–237
    in structures, 184–185
International system of units, 5–9

Jacks, 292
Jerk, 426
Jet engine, 569
Joints, method of, 187–192
Joule (unit), 370
Journal bearings, 300–301

Kepler, Johann, 489
Kepler's laws, 489
Kilogram, 6
Kilometer, 6
Kilonewton, 6
Kilopound, 10
Kilowatt, 526
Kinematics, 399
    of particles: in curvilinear mo-
        tion, 428–446
    in rectilinear motion, 400–422
    in relative motion, 414–415,
        621–623
    of rigid bodies: in plane mo-
        tion, 583–585
    in rotation, 586–588
    in translation, 585
Kinetic energy (see Energy)
Kinetic friction, 275–277
Kinetics, 399
    of particles, 454–582
    of rigid bodies, 631–744
Kip, 10

Lagrange, J. L., 2
Laws:
  of friction, 274–277
  Kepler's, 489
  Newton's (*see* Newton's law)
Lead of a screw, 292–293
Lead angle, 292
Line of action, 16, 58
Linear impulse, 533
Linear momentum:
  of a particle, 533
  of a rigid body, 711
Links, 83
Liter, 8
Loading of beams, 240–241
Logarithmic decrement, 785

Machines, 210, 223
  ideal, 375
  real, 375–376
Magnification factor, 773, 781
Magnitude of a force, 16
Mass, 2, 455
Mass center:
  of a rigid body, 635
  of a system of particles, 462
Mass moments of inertia, 792–797
Maxwell's diagram, 198–199
Mechanical efficiency, 375–376, 526
Mechanical energy, 516–517
Mechanics:
  definition of, 1
  Newtonian, 2–5
  principles of, 2–5
Megagram, 6
Meter, 6
Metric ton, 6
Metric units, 5–9
Mile, 10
Millimeter, 6
Mohr's circle, 344–345
Moment:
  bending, 237, 242–257
  of a couple, 66

Moment:
  first, 141, 173
  of a force, 60–61
  of momentum (*see* Angular momentum)
  second, 320–323
Moment-area method, 422
Moments of inertia, 320–357
  of areas, 320–343
    parallel-axis theorem for, 329–330
  of common geometric shapes, 331, 797
  of composite areas, 330
  of composite bodies, 796
  determination of, by integration, 322–323, 796
  inclined axes, 341
  of masses, 792–797
    parallel-axis theorem for, 793–794
  polar, 323
  principal, 343
  rectangular, 323, 793
  of thin plates, 795–796
Momentum:
  angular (*see* Angular momentum)
  conservation of, 536–537, 716
  linear (*see* Linear momentum)
  of a particle, 532–535
  of a rigid body, 709–715
Momentum couple, 710
Momentum vector, 710
Motion:
  absolute, 434
  under a central force, 473–474
  curvilinear, 428–446, 459–489
  equations of (*see* Equations)
  harmonic, simple, 746–747
  impulsive, 534
  about mass center, 635
  of mass center, 462, 463, 635
  Newton's laws of (*see* Newton's law)
  of a particle, 400–582
  plane (*see* Plane motion)

Motion:
  rectilinear, 400–422
  relative (*see* Relative motion)
  of a rigid body, 583–744
  rolling, 656–657
  of a system of particles, 460–463
Motion curves, 403, 420–422
Multiforce members, 210, 237

Natural frequency, 772
Neutral equilibrium, 389–390
Newton, Sir Isaac, 2
Newton (unit), 6
Newtonian frame of reference, 455
Newtonian mechanics, 2–5
Newton's law:
  of gravitation, 4–5, 474–475
  of motion: first, 3, 33–34
    second, 3, 454–456
    third, 4, 184, 185
Noncentroidal rotation:
  dynamic equilibrium in, 655
  effective forces in, 655
  equations of motion in, 655–656
  kinetic energy in, 688–689
  momentum in, 715
Nonconservative force, 517
Nonimpulsive force, 535
Nonrigid truss, 204
Normal component of acceleration, 442–444
Numerical accuracy, 14–15
Nutation, 738

Oblique central impact, 546–547
Orbit, 487
Oscillation, center of, 758
Oscillations:
  of a rigid body, 758, 765
  of a simple pendulum, 750–752
Output forces, 223
Output power, 526
Output work, 375
Overrigid trusses, 203

Pappus, theorems of, 154–156
Parabolic cable, 260–261
Parabolic trajectory, 433, 486
Parallel-axis theorem:
  for moments of inertia: of
    areas, 329–330
    of masses, 793–794
  for products of inertia, 340
Parallelogram law, 3, 18–19
Partial constraints, 86–88, 126
Particles, 3, 16
  equilibrium of: in a plane,
    32–35
    in space, 49
  free-body diagram of, 34–35
  kinematics of, 400–453
  kinetics of, 454–582
  moving on a slab, 621–623
  relative motion of, 414, 433,
    621–623
  systems of (see Systems)
  vibrations of (see Vibrations)
Pascal (unit), 165
Pendulum:
  compound, 762
  simple, 750–752
Percussion, center of, 663, 721
Perigee, 487
Period:
  of deformation, 543–544, 725
  of restitution, 543–544, 725
  of vibration, 746
    damped, 779
    undamped, 749
Periodic time, 488–489
Phase angle, 749
Phase difference, 781
Pile driver, 554
Pin-and-bracket supports, 126
Pins, 83, 187
Pitch:
  of a thread, 293
  of a wrench, 118
Plane of symmetry, 175
Plane motion, 583–585
  constrained, 653

Plane motion:
  dynamic equilibrium in, 636
  effective forces in, 632–635
  equations of motion in, 634,
    636
  kinematics of, 583–585
  kinetic energy in, 695
  momentum in, 709
Planetary gears, 600
Planetary motion, 489
Plastic impact, 545
Point of application of a force,
  16, 57
Polar coordinates, 444
Polar moment of inertia, 323
Pole, 323
Polygon rule, 20
Position coordinate, 400
  relative, 414
Position vector, 69, 428
  relative, 433
Potential energy, 386–388, 513–
  516, 696
Pound, 8
Pound force, 8
Pound mass, 12
Power, 525–526, 697
Precession, 735–738
  steady, 737
  unsteady, 738
Pressure, center of, 165,
  338–339
Principal axes of inertia, 341–
  343
Principal moments of inertia,
  343
Principle:
  of impulse and momentum: for
    a particle, 532–534
    for a rigid body, 709–710
    for a system of particles,
      535–536
  of transmissibility, 3, 58–59
  of virtual work, 372–375
  of work and energy: for a par-
    ticle, 500–504

Principle:
  of work and energy:
    for a rigid body, 685–686
    for a system of particles,
      504
Principles of mechanics, 2–5
Problem solution, method of,
  12–14
Product of inertia, 339–340
Projectile, 432–433, 472
Propeller, 570

Radial component:
  of acceleration, 444–446
  of velocity, 444–446
Radius:
  of curvature, 443
  of gyration, 324, 792
Rated speed, 476
Reactions at supports and con-
  nections, 83–84, 125–127
Real machines, 375–376
Rectangular components:
  of acceleration, 431–432
  of force, 25, 40
  of velocity, 429
Rectilinear motion of a particle,
  400–422
  uniform, 412
  uniformly accelerated, 412–
    413
Rectilinear-motion problems,
  solution of: analytical, 404–405
    graphical, 420–422
Rectilinear translation, 583
Reduction of a system of forces:
  in a plane, 75–76
  in space, 117–118
Redundant members, 203
Reference frame, 455
Relative acceleration (see Acceleration)
Relative motion:
  of a particle: with respect to
    axes in translation, 433

Relative motion:
  of a particle:
    with respect to rotating
      axes, 621–623
  of two particles, 414, 433–434
Relative position, 414, 433
Relative velocity:
  of a particle: in plane motion,
    595–597
    with respect to axes in
      translation, 433
    with respect to rotating
      axes, 621
  of two particles, 414, 433
Relativity, theory of, 2
Repose, angle of, 278
Resolution of a force:
  into components: in a plane,
    21, 25
    in space, 40–43
  into a force and a couple: in a
    plane, 69–70
    in space, 116
Resonance, 773
Restitution:
  coefficient of, 544, 725
  period of, 543–544, 725
Resultant of forces, 3, 17, 45, 117
  (*See also* Addition, of forces;
    Addition, of vectors)
Reversed effective force (*see* Inertia vector)
Revolution:
  body of, 155, 356
  surface of, 155
Right-hand rule, 108
Rigid body, 3, 56
  equilibrium of: in a plane,
    81–88, 99–100
    in space, 49
  free-body diagram of, 125
  kinematics of, 583–630
  kinetics of, 631–744
  vibrations of, 757–758
Rigid truss, 187
Rocket, 571

Rollers, 83, 125
Rolling motion, 656–657
Rolling resistance, 303–304
  coefficient of, 304
Rotating shafts, 675
Rotation, 584, 586–588, 673–676
  centroidal (*see* Centroidal rota-
    tion)
  dynamic equilibrium in, 655
  effective forces in, 635, 655
  equations of motion in, 655,
    673
  instantaneous axis of, 584, 603
  instantaneous center of, 603–
    605
  kinematics of, 586–588
  kinetic energy in, 688–689
  momentum in, 711–712
  noncentroidal (*see* Noncen-
    troidal rotation)
  of a three-dimensional body,
    673–676
  uniform, 587
  uniformly accelerated, 587–
    588
Rough surfaces, 83, 125–126

Sag, 261, 269
Satellites, 485
Scalar components, 26
Scalars, 17
Screws, 292–293
Second, 6
Second moment, 320–323
Sections, method of, 201–203
Self-locking screws, 293
Semimajor axis, 488–489
Semiminor axis, 488–489
Sense of a force, 16
Shafts, rotating, 675
Shear, 237, 242–244
Shear diagram, 244
SI units, 5–9
Significant figures, 14–15
Simple harmonic motion, 746–
  747

Simple pendulum, 750–752
Simple trusses, 187, 193
Slide rule:
  accuracy, 15
  use of, 22, 23, 28, 30, 36, 43,
    45, 46, 311
Slipstream, 569
Slug, 9
Space, 2
Space mechanics, 485–489
Space truss, 193
Space vehicles, 485, 571
Specific weight, 140, 172
Speed, 401, 429
  rated, 476
Spin, 735
Spring:
  force exerted by, 385–386,
    498–499
  potential energy, 387, 514–515
Spring constant, 385, 498
Square-threaded screws, 292–293
Stable equilibrium, 389–390
Static friction, 275–278
  angle of, 277
  coefficient of, 275–276
Statically determinate reactions,
  86–88
Statically determinate structures,
  213
Statically determinate trusses, 203
Statically indeterminate reactions,
  86–88
Statically indeterminate struc-
  tures, 213
Statically indeterminate trusses,
  204
Statics, definition of, 1, 399
Steady precession, 737–738
Steady-state vibrations, 772, 780
Stream of particles, 566–570
Structural shapes, properties of,
  332–333
Structures:
  analysis of, 184–223
  determinate, 213

Structures:
  indeterminate, 213
  internal forces in, 184
  two-dimensional, 60
Submerged surfaces, forces on, 164–165
Supports:
  ball, 125
  ball-and-socket, 126
  of beams, 240–241
  reactions at, 83–84, 125–127
Surface:
  frictionless, 83, 125
  of revolution, 155
  rough, 83, 125–126
  submerged, forces on, 164–165
Suspension bridges, 260
Symmetry:
  axis of, 142, 144
  center of, 142, 144
  plane of, 175
Systems:
  of forces, 56–81, 106–124
  of particles: angular momentum of, 558–559
    equations of motion for, 461
    impulse-momentum principle for, 535–536
    kinetic energy of, 504
    mass center of, 462
    variable, 566
    work-energy principle for, 504
  of units, 5–13

Tangential component of acceleration, 442–444
Tension, 59, 186, 236
Three-force body, 100
Thrust, 569
Thrust bearings, 301–303
Time, 2
Toggle vise, analysis of, 373, 375
Ton:
  metric, 6
  U.S., 10

Top, 738
Torsional vibrations, 760
Trajectory:
  of projectile, 433, 488
  of space vehicle, 486–488
Transfer formula (*see* Parallel-axis theorem)
Transient vibrations, 772, 780
Translation, 583, 585
  curvilinear, 583
  effective forces in, 634
  kinematics of, 585
  kinetic energy in, 688
  momentum in, 710–711
  rectilinear, 583
Transmissibility, principle of, 3, 58–59
Transverse component:
  of acceleration, 444–446
  of velocity, 444–446
Triangle rule, 19
Trusses, 185–193, 198–199, 201–204
  compound, 203
  determinate, 203
  indeterminate, 204
  overrigid, 203
  rigid, 187
  simple, 187, 193
  space, 193
  typical, 186
Two-dimensional structures, 60
Two-force body, 99–100

Unbalanced disk, 657
Uniform rectilinear motion, 412
Uniform rotation, 587
Uniformly accelerated rectilinear motion, 412–413
Uniformly accelerated rotation, 587–588
U.S. customary units, 8–13
Units, 5–13
  (*See also* specific systems of units)
Universal joints, 126

Unstable equilibrium, 389
Unstable rigid bodies, 87
Unsteady precession, 738

V belts, 311
Variable systems of particles, 566
Varignon's theorem, 62
Vector addition, 18–20
Vector components, 26
Vectors, 17–18
  bound, fixed, 17
  coplanar, 20
  couple, 108
  free, 18
  inertia, 459, 636
  momentum, 710
  sliding, 18, 58
Velocity, 401, 429
  absolute, 595–597, 605
  angular, 586, 588, 597
  areal, 474
  components of: radial, 444–446
    rectangular, 429
    transverse, 444–446
  escape, 411, 487
  in plane motion, 595–597
  relative (*see* Relative velocity)
  in rotation, 586
Velocity-displacement curve, 422
Velocity-time curve, 403, 420–421
Vibrations, 745–790
  damped: forced, 780–781
    free, 777–779
  forced, 771–773, 780–781
  free, 746–750, 757–758, 777–779
  frequency of, 746, 749
  period of, 746
  of rigid bodies, 757–758
  steady-state, 772, 780
  torsional, 760
  transient, 772, 780
  undamped: forced, 771–773
    free, 746–750, 757–758

Vibrometer, 790
Virtual displacement, 372
Virtual work, 372–375
    principle of, 372–373
Viscous damping, 777

Watt (unit), 526
Wedges, 291
Weight, 4, 475

Wheel friction, 303–304
Wheels, 125, 303, 657
Work:
    of a couple, 371, 384, 688
    and energy, principle of (*see*
        Principle, of work and en-
        ergy)
    of a force, 369, 384, 496–500
    of force exerted by spring,
        385–386

Work:
    of forces on a rigid body, 371,
        687–688
    of gravitational force, 500
    input and output, 375
    virtual, 372–375
    of a weight, 384–385, 498
Wrench, 118

Zero-force member, 191

Answers to Even-Numbered Problems

11.2 $t = 0$, $x = 15$ in., $a = -12$ in./s^2;
$t = 2$ s, $x = 7$ in., $a = 12$ in./s^2.
SI: $t = 0$; $x = 0.381$ m, $a = -0.305$ m/s^2;
$t = 2$ s, $x = 0.1778$ m, $a = 0.305$ m/s^2.

11.4 (a) 2 s, 4 s. (b) 8 m, 7.33 m.

11.6 -4 m/s; 12 m; 20 m.

11.8 (a) 3 ft/s^4. (b) $a = 3t^2$; $v = t^3 - 32$;
$x = \frac{1}{4}t^4 - 32t + 64$. SI: (a) 0.914 m/s^4.

11.10 25 s^{-2}.

11.12 (a) ± 3.34 ft/s. (b) 7.59 ft.
SI: (a) ± 1.018 m/s. (b) 2.31 m.

11.14 (a) 55.5 m. (b) Infinite.

11.16 142.7 ft/s. SI: 43.5 m/s.

11.20 $\sqrt{2gR}$.

11.22 (a) 6.90 m/s. (b) Infinite.

11.24 (a) 29.1 m. (b) 24.4 m/s.

11.26 (a) -3.20 ft/s^2. (b) 34.0 ft/s.
SI: (a) -0.975 m/s^2. (b) 10.36 m/s.

11.28 (b) 1.242 s; 24.8 ft. SI: (b) 1.242 s; 7.56 m.

11.30 1.200 m.

11.32 (a) 3 ft/s^2 ↑, 6 ft/s^2 ↓. (b) 12 ft/s^2 ↑.
SI: (a) 0.914 m/s^2 ↑, 1.829 m/s^2 ↓.
(b) 3.66 m/s^2 ↑.

11.34 (a) 45 mm/s^2 →; 30 mm/s^2 →.
(b) 90 mm/s →; 135 mm →.

11.36 (a) $v_A = -\frac{1}{2}(v_B + v_C)$. (b) 40.5 in. ↑;
27 in./s ↑. SI: (b) 1.029 m ↑; 0.686 m/s ↑.

11.38 $v_A = 60$ mm/s ↓; $v_B = 100$ mm/s ↑;
$v_C = 80$ mm/s ↓.

11.40 (b) 20 ft/s; 74 ft/s; 176 ft.
SI: (b) 6.10 m/s; 22.6 m/s; 53.6 m.

11.42 (a) 72 m. (b) 4 s and 15 s.

11.44 Accelerate for 14.94 s, 50.9 mi/h.
SI: 82.0 km/h.

11.46 (a) 270 m/s^2, -19.57 m/s^2. (b) 54 mm.
(c) 0.296 s.

11.48 (a) 16 s. (b) 310 m.

11.50 9.39 s.

11.52 22.5 ft/s^2. SI: 6.86 m/s^2.

11.54 (a) Local. (b) 122 s.

11.56 8 s.

11.58 (a) 123 ft/s. (b) 227 ft.
SI: (a) 37.5 m. (b) 69.2 m.

11.60 (a) 9.71 s. (b) 130 m.

11.62 (a) 90 ft/s; 180 ft. (b) 120 ft/s; 225 ft.
SI: (a) 27.4 m/s; 54.9 m.
(b) 36.6 m/s; 68.6 m.

11.64 104 ft. SI: 31.7 m.

11.66 (a) $v = 2.69$ m/s $\nearrow 21.8°$;
$a = 1.414$ m/s^2 $\searrow 45°$.
(b) $v = 1.803$ m/s $\measuredangle 33.7°$;
$a = 5.10$ m/s^2 $\measuredangle 11.3°$.

11.68 (a) $v = 2.83$ ft/s $\nwarrow 45°$;
$a = 6.32$ ft/s^2 $\measuredangle 71.6°$.
(b) $v = 4.01$ ft/s $\nwarrow 3.6°$;
$a = 2.03$ ft/s^2 $\measuredangle 10.6°$.
SI: (a) $v = 0.863$ m/s $\nwarrow 45°$;
$a = 1.926$ m/s^2 $\measuredangle 71.6°$.
(b) $v = 1.222$ m/s $\nwarrow 3.6°$;
$a = 0.619$ m/s^2 $\measuredangle 10.6°$.

11.70 (a) $v = 100\,\pi$ mm/s $\leftarrow$;
$a = 100\,\pi^2$ mm/s^2 $\downarrow$.

11.74 (a) 14.86 m/s. (b) 26.0 m.

11.76 14.54 ft from B. SI: 4.43 m from B.

11.78 (a) 4.79 m/s $\measuredangle 50°$. (b) 0.975 s.

11.80 6.1° or 32.5°.

11.82 (a) 30.09 m. (b) 181.9 m.

11.84 15° or 75°.

11.88 76.0 mi/h, 55.2° south of west.
SI: 122.3 km/h, 55.2° south of west.

11.90 (a) 187.4 km/h $\nwarrow 63.7°$.
(b) 208 m $\nwarrow 63.7°$. (c) 353 m.

11.92 (a) 8.62 knots, 55.1° east of north.
(b) 8.62 knots, 55.1° west of south.

11.94 623 m/s $\searrow 36.7°$; 9.81 m/s^2 $\downarrow$.

11.96 124.7 km/h.

11.98 89.7 ft/s. SI: 27.3 m/s.

11.100 9.18 ft/s^2. SI: 2.80 m/s^2.

11.102 4.56 m/s^2.

11.104 (a) 79.6 m. (b) 40.8 m.

11.106 3810 m.

11.108 592 mi. SI: 953 km.

11.110 2280 mi/h. SI: 3670 km/h.

11.112 (a) $v = 0$; $a = 120$ mm/s^2, $\gamma = 0$.
(b) $v = 170.8$ mm/s, $\beta = 69.4°$;
$a = 905$ mm/s^2, $\gamma = 135°$. Where β and
γ are measured counterclockwise from
the axis OA.

11.114 (a) $v = 2b\omega$; $a = 4b\omega^2$. (b) b. (Particle
describes a circle.)

11.116 $-(v_0/b) \sin^2\theta$.

11.118 $a = d \sec^2\theta \,(\ddot\theta + 2\,\dot\theta^2 \tan\theta)$.

11.120 4.28 m/s; 0.188 m.

11.122 (a) $v = \sqrt{2}\pi b$, $\beta = 135°$; $a = \frac{1}{2}\sqrt{5}\pi^2 b$,
$\gamma = 243.4°$. (b) $v = 0$; $a = \frac{1}{2}\pi^2 b$, $\gamma = 0$.
Where β and γ are measured
counterclockwise from the axis $\theta = 0$.

11.124 (a) 21.1°. (b) 1.02°.

11.126 78.8 m/s.

11.128 (a) 1286 ft. (b) 1159 ft.
SI: (a) 392 m. (b) 353 m.

11.130 (a) 120 mm/s $\uparrow$. (b) 120 mm/s $\downarrow$.

CHAPTER 12

12.2 At all points: mass $= m = 10$ kg.
Weight: 0°, 97.81 N; 45°, 98.07 N;
90°, 98.33 N.

12.4 3.22 ft/s^2 $\downarrow$; 20 lb.
SI: 0.981 m/s^2 $\downarrow$; 89 N.

12.6 (a) 4.55 s. (b) 0.45.

12.8 0.294 m/s^2.

12.10 23.9 N.

12.12 $F_{AB} = 500$ lb C; $F_{BC} = 500$ lb T.
SI: $F_{AB} = 2220$ N C; $F_{BC} = 2220$ N T.

12.14 (a) 29.2 kg. (b) Impossible, since a_B
cannot be greater than g.

12.16 (a) 16.1 ft/s^2 $\downarrow$; 21.5 ft/s^2 $\downarrow$; 0.767 ft/s^2 $\downarrow$.
(b) 64.4 ft/s $\downarrow$; 86.0 ft/s $\downarrow$; 3.07 ft/s $\downarrow$.
(c) 17.94 ft/s $\downarrow$; 20.7 ft/s $\downarrow$; 3.92 ft/s $\downarrow$.
SI: (a) 4.91 m/s^2 $\downarrow$; 6.55 m/s^2 $\downarrow$;
0.234 m/s^2 $\downarrow$. (b) 19.63 m/s $\downarrow$; 26.2 m/s $\downarrow$;
0.936 m/s $\downarrow$. (c) 5.47 m/s $\downarrow$; 6.31 m/s $\downarrow$;
1.195 m/s $\downarrow$.

12.18 (a) 50.2 kg. (b) 357 N.

12.20 (a) $\mathbf{a}_A = 11.15$ ft/s^2 $\leftarrow$;
$\mathbf{a}_B = 7.43$ ft/s^2 $\leftarrow$. (b) 7.69 lb.
SI: (a) $\mathbf{a}_A = 3.40$ m/s^2 $\leftarrow$;
$\mathbf{a}_B = 2.26$ m/s^2 $\leftarrow$. (b) 34.2 N.

12.22 (a) $\mathbf{a}_A = 3.72$ ft/s^2 $\leftarrow$; $\mathbf{a}_B = 2.48$ ft/s^2 $\leftarrow$.
(b) 9.23 lb. SI: (a) $\mathbf{a}_A = 1.134$ m/s^2 $\leftarrow$;
$\mathbf{a}_B = 0.756$ m/s^2 $\leftarrow$. (b) 41.1 N.

12.24 24.6 m.

12.26 19.58 m.

12.28 (a) 14.49 ft/s^2 $\leftarrow$. (b) 6.44 ft/s^2 $\leftarrow$.
(c) 8.05 ft/s^2 $\rightarrow$. SI: (a) 4.42 m/s^2 $\leftarrow$.
(b) 1.962 m/s^2 $\leftarrow$. (c) 2.45 m/s^2 $\rightarrow$.

12.30 $v = \sqrt{2gh(1 - h/l)}$; $v_{\max} = \sqrt{gl/2}$.

12.32 $(v_0\,m/T_0) \ln 2$.

12.34 $\mathbf{a}_A = \mathbf{a}_B = 3.58$ ft/s^2 $\uparrow$; $\mathbf{a}_C = 14.32$ ft/s^2 $\downarrow$.
SI: $\mathbf{a}_A = \mathbf{a}_B = 1.090$ m/s^2 $\uparrow$;
$\mathbf{a}_C = 4.36$ m/s^2 $\downarrow$.

12.36 8 lb. SI: 3.63 kg (mass).

12.38 $\mathbf{a}_A = 4.76$ m/s^2 $\rightarrow$; $\mathbf{a}_B = 3.08$ m/s^2 $\downarrow$;
$\mathbf{a}_C = 1.401$ m/s^2 $\leftarrow$.

12.40 (a) 64.0 N. (b) 3.25 m/s.

12.42 $8.02 \text{ ft/s} < v < 13.90 \text{ ft/s}$.
$2.44 \text{ m/s} < v < 4.24 \text{ m/s}$.

12.44 1.656 m/s; 9.16 m/s^2.

12.46   (a) $T_{CD} = W$. (b) $T_{CD} = \frac{1}{2}W$;
$\mathbf{a} = 0.866g \searrow 60°$.

12.48 $v_A = 67.0 \text{ mi/h}$; $v_B = 104.5 \text{ mi/h}$.
SI: $v_A = 107.8 \text{ km/h}$; $v_B = 168.2 \text{ km/h}$.

12.50 0.974 m/s.

12.52 0.815.

12.54 31.6 mi/h. SI: 50.9 km/h.

12.56 $\delta = (eVlL)/(mv_0^2 d)$.

12.58   The length of the plates must be
doubled.

12.60   (a) $F_r = 4 \text{ N}$, $F_\theta = 0$. (b) $F_r = -21.3 \text{ N}$,
$F_\theta = 21.3 \text{ N}$.

12.62   (a) $F_r = -2.30 \text{ lb}$, $F_\theta = 0$. (b) $F_r = 0$,
$F_\theta = -1.839 \text{ lb}$. SI: (a) $F_r = 10.23 \text{ N}$,
$F_\theta = 0$. (b) $F_r = 0$, $F_\theta = -8.18 \text{ N}$.

12.64   (a) 8.39 lb. (b) 1.118 lb.
SI: (a) 37.3 N. (b) 4.97 N.

12.66 12.96 N.

12.70   (a) 24 in./s. (b) $\rho_A = \frac{2}{3} \text{ in.}$, $\rho_B = 18 \text{ in.}$
SI: (a) 0.610 m/s. (b) $\rho_A = 16.93 \text{ mm}$,
$\rho_B = 457 \text{ mm}$.

12.72 $238,000 \text{ mi}$. SI: $383\,000 \text{ km}$.

12.76   (a) 5880 km/h. (b) 5630 km/h

12.78 2060 mi/h. SI: 3315 km/h.

12.80 $12,160 \text{ mi}$. SI: $19\,570 \text{ km}$.

12.82 $32\,800 \text{ km/h}$; $14\,770 \text{ km/h}$.

12.84   (a) 5590 ft/s. (b) 94 ft/s.
SI: (a) 1704 m/s. (b) 28.7 m/s.

12.86   (a) $35\,200 \text{ km/h}$. (b) 5150 km/h.

12.88 $4 \text{ h } 29 \text{ min}$.

12.90 $3 \text{ h } 45 \text{ min}$.

12.92 4560 m/s.

12.94   Angle $BOC = 44.8°$.

12.96   (a) $\frac{1}{3}v_0$. (b) Vehicle must increase its
speed by $0.1869 v_0$. (c) $75.9°$.

12.98 9.56 ft/s^2. SI: 2.91 m/s^2

12.100 1.141 m.

12.102   (a) 342 lb. (b) Impossible. SI: (a) 1521 N.

12.104 $-62.7° < \theta < 62.7°$.

12.106 10.36 ft/s. SI: 3.16 m/s.

12.108 $r_{max}/r_0 = \alpha^2/(2 - \alpha^2)$.

CHAPTER 13

13.2   (a) 699 ft·lb; 87.3 ft. (b) 699 ft·lb; 530 ft.
SI: (a) 948 J; 26.6 m. (b) 948 J; 161.5 m.

13.4   (a) 15.20 m. (b) 7.74 m/s.

13.6 60.1 ft. SI: 18.32 m.

13.8 14.40 N.

13.10 14.16 ft/s. SI: 4.32 m/s.

13.12 1.981 m/s.

13.14 1.180 m/s.

13.16 2.45 ft. SI: 0.747 m.

13.22   (a) 43 m. (b) 44 m/s^2.

13.24 $\mu_s = 0.268$; $\mu_k = 0.151$.

13.26   (a) 5.67 ft/s. (b) 8.66 in.
SI: (a) 1.728 m/s. (b) 0.220 m.

13.28 $\mathbf{a}_n = 2g \uparrow$, $\mathbf{a}_t = \frac{3}{4}g \leftarrow$.

13.30   (b) 5.25 m/s.

13.32   (a) 39.1 ft/s. (b) 27.1 in.
SI: (a) 11.92 m/s. (b) 0.688 m.

13.34 1718 km.

13.36   (a) $10,980 \text{ ft/s}$. (b) $16,480 \text{ ft/s}$.
SI: (a) 3350 m/s. (b) 5020 m/s.

13.38   (a) $\frac{1}{2}kl^2(\sin \theta + \cos \theta - 1)^2$.
(b) $-mgl \sin \theta$.

13.40 4.83 lb SI: 21.5 N.

13.42   (a) 9.35 m/s. (b) 4.46 m.

13.44 56.7 ft/s. SI: 17.28 m/s.

13.46   (a) 2.64 m/s. (b) 0.

13.48   (a) 2.43 m/s. (b) 2.71 m/s.

13.50   (a) 4.91 ft/s. (b) 0.250 lb.
SI: (a) 1.497 m/s. (b) 1.112 N.

13.52   (a) 22.7 ft/s. (b) 7.75 ft.
SI: (a) 6.92 m/s. (b) 2.36 m.

13.54   (a) 2.67 m. (b) 2.02 m.

13.56   (a) 1.448 in.; 17.52 lb. (b) 2.00 in.;
12.00 lb. SI: (a) 36.8 mm; 77.9 N.
(b) 50.8 mm; 53.4 N.

13.58   (a) $33.5 \times 10^6 \text{ J/kg}$.
(b) $45.0 \times 10^6 \text{ J/kg}$.

13.60   (a) $mgR[1 - (R/r)]$. (b) $\frac{1}{2}mgR^2/r$.
(c) $mgR[1 - (R/2r)]$.

13.62 68.9%.

13.6-   (a) 25.0 kW. (b) 6.13 kW.

13.66 2.86 ft/s^2. SI: 0.872 m/s^2.

13.68   (a) 26.6 kW. (b) 26.6 kW.

13.70   (a) 27.3 s; 1523 ft. (b) 38.3 s; 2970 ft.
SI: (a) 27.3 s; 464 m. (b) 38.3 s; 905 m.

13.72 3.89 km/h.

13.74 9.33 ft. SI: 2.84 m.

13.76   (a) $1.716 \text{ m/s} \leftarrow$. (b) $4.20 \text{ m/s} \nearrow 65.9°$.

13.78 $6 \, mg$.

13.80 $5.29 \text{ m/s} \rightarrow$.

13.82 4.47 in. SI: 113.6 mm.

13.84 233 N/m.

CHAPTER 14

14.2   219 s.
14.4   (a) 2.80 s. (b) 5.61 s.
14.8   (a) 38.9 s. (b) 10.71 kN T.
14.10 (a) 19.75 ft/s; 5.20 s. (b) 9.17 s.
       SI: (a) 6.02 m/s; 5.20 s.
14.12 (a) 9.03 m/s. (b) 0.
14.14 (a) and (b) 111.1 kN.
14.16 50.2 lb ←, 204 lb ↓.
       SI: 223 N ←, 907 N ↓.
14.18 18.46 ft/s. SI: 5.63 m/s.
14.20 (a) 5.2 km/h →. (b) 3.9 km/h →.
14.22 (a) 0.375 mi/h →. (b) 4270 lb.
       SI: (a) 0.603 km/h →. (b) 18.99 kN.
14.24 0.742 m/s →.
14.26 (a) 233 lb · s. (b) 159.1 lb · s.
       SI: (a) 1036 N · s. (b) 708 N · s.
14.28 (a) v_A = 9.45 ft/s →; v_B = 12.25 ft/s →.
       (b) 0.24 ft · lb. SI: v_A = 2.88 m/s →;
       v_B = 3.73 m/s →. (b) 0.32 J.
14.30 0.875.
14.32 v_A = 0.948 m/s →; v_B = 1.092 m/s →;
       v_C = 1.960 m/s →.
14.34 (a) $\tan \theta = \sqrt{e}$. (b) $\sqrt{e} \, v_0$.
14.36 (a) 20.9 in. (b) 10.20 in.
       SI: (a) 0.531 m. (b) 0.259 m.
14.38 (a) 0.913. (b) 33.3 in.; 14.61 in.
       SI: (b) 0.846 m; 0.371 m.
14.40 v_A = 0.721 v_0 ∡16.1°; v_B = 0.693 v_0 ←.
14.42 (a) 76.2°. (b) 37.1 N.
14.44 (a) 0.80. (b) 0.32.
14.46 10,140 lb. SI: 45.1 kN.
14.50 (a) 6670 J. (b) 24 km/h.
14.52 (a) 1.288 lb. (b) 1.479 lb.
       SI: (mass) (a) 0.584 kg. (b) 0.671 kg.
14.54 (a) 1.155 m. (b) 5.20 m/s.
14.56 2.84 ft/s. SI: 0.866 m/s.
14.58 v_r = 9.05 ft/s, v_θ = 9.14 ft/s.
       SI: v_r = 2.76 m/s, v_θ = 2.79 m/s.
14.64 5160 ft/s; 79.9°. SI: 1573 m/s; 79.9°.
14.66 2280 m/s.
14.68 v_{max} = 27,200 ft/s, ϕ_0 = 90°;
       v_{min} = 16,330 ft/s, ϕ_0 = 0.
       SI: v_{max} = 8290 m/s, ϕ_0 = 90°;
       v_{min} = 4980 m/s, ϕ_0 = 0.
14.70 65.7° ⩽ ϕ_0 ⩽ 114.3°.
14.76 (b) $\frac{1}{3}\sqrt{6}\, v_{esc}$, $\frac{1}{2}\sqrt{2}\, v_{esc}$.
14.78 218 lb. SI: 970 N.
14.80 $Q_1 = \frac{1}{2}Q(1 - \sin \theta)$; $Q_2 = \frac{1}{2}Q(1 + \sin \theta)$.

14.82 $A_x = A\rho v_A^2(1 - \cos \theta)$ ←,
       $A_y = A\rho v_A^2 \sin \theta$ ↑,
       $M_A = A\rho v_A^2 R(1 - \cos \theta)$ ↩.
14.84 C = 149.7 N ↓; D_x = 250 N →,
       D_y = 283 N. ↓.
14.86 C_x = 67.1 lb →, C_y = 367 lb ↑;
       D = 433 lb ↑. SI: C_x = 298 N →,
       C_y = 1632 N ↑; D = 1926 N ↑.
14.88 49.7 kg/s.
14.90 461 mi/h. SI: 742 km/h.
14.92 4310 lb. SI: 19.17 kN.
14.94 1143 lb. SI: 5.08 kN.
14.96 (a) 32.4 m/s. (b) 101.9 m³/s.
       (c) 64 600 J/s.
14.98 170.0 rpm.
14.100 (a) $\eta = 2(V/v_A)[1 - (V/v_A)]^2(1 - \cos \theta)$.
       (b) $\frac{16}{27}$.
14.102 (a) qv.
14.104 (a) $mgvt + mv^2$. (b) $(l - vt)mg$.
14.106 qv.
14.108 $v = m_0 v_0/(m_0 + qt)$;
       $a = -m_0 v_0 q/(m_0 + qt)^2$.
14.110 (a) 53.7 lb/s. (b) 13.42 lb/s.
       SI: (a) 24.4 kg/s. (b) 6.09 kg/s.
14.112 (a) 11.19 m/s²; 0. (b) 28.4 m/s²; 814 m/s.
       (c) 200.2 m/s²; 3952 m/s.
14.114 (a) 814 m/s. (b) 5210 m/s.
14.116 522,000 ft. SI: 159.2 km.
14.118 (a) Five. (b) 2 m/s →. (c) Same as
       original.
14.120 (a) 7220 lb, 2.47 ft below B. (b) 5240 lb,
       6.80 ft below B. SI: (a) 32.1 kN, 0.753 m
       below B. (b) 23.3 kN, 2.07 m below B.
14.124 Case 1: (a) $\frac{1}{3}g$. (b) $v = \sqrt{2gl/3}$.
       Case 2: (a) gy/l. (b) $v = \sqrt{gl}$.
14.126 (a) 5.54 ft/s. (b) 0.642 ft from B.
       SI: (a) 1.689 m/s. (b) 0.1957 m from B.
14.128 (a) 135.6 N · s. (b) 108.5 N · s. (c) 368 J;
       294 J.

CHAPTER 15

15.2   (a) 2 s. (b) 2 rad; 8 rad/s².
15.4   3.93 rad/s; 20.6 rad/s².
15.6   (a) 60 rev. (b) 750 rev.
15.8 $\alpha = 8.38 \, t$, $\omega = 4.19 \, t^2$, $\theta = 1.396 \, t^3$;
       16.76 rad/s.

15.10 (a) 1524 ft/s; 0.1111 ft/s². (b) 1168 ft/s; 0.0851 ft/s². (c) 0. SI: (a) 465 m/s; 0.0339 m/s². (b) 356 m/s; 0.0259 m/s². (c) 0.

15.12 1.174 s; 7.05 rad/s.

15.14 (a) 6.93 rad/s ↻. (b) 2.09 s.

15.16 (a) 50 rpm. (b) 3.29 m/s² ↑; 13.16 m/s² ↑.

15.18 (a) 24 rad/s² ↺. (b) 30.6 rev.

15.20 (a) 8.92 s. (b) ω_A = 26.8 rad/s ↻; ω_B = 44.6 rad/s ↺.

15.22 $\alpha = bv^2/2\pi r^3$.

15.24 (a) 2.22 rad/s ↻. (b) 18.77 ft/s ∠40°. SI: (b) 5.72 m/s² ∠40°.

15.26 (a) 3.46 rad/s ↺. (b) 0.693 m/s →.

15.28 (a) 3.93 rad/s ↺. (b) 0.578 m/s →.

15.30 (a) 1.500. (b) $\tfrac{1}{3}\omega_A$ ↻.

15.32 210 rpm ↻.

15.34 (a) ω = 0; v_D = 84 in./s →.
(b) ω = 3.5 rad/s ↻; v_D = 24.5 in./s →.
(c) ω = 0; v_D = 84 in./s ←.
SI: (a) ω = 0; v_D = 2.13 m/s →.
(b) ω = 3.5 rad/s ↻; v_D = 0.622 m/s →.
(c) ω = 0; v_D = 2.13 m/s ←.

15.36 6.52 m/s ↓; ω_{BD} = 20.8 rad/s ↻.

15.38 ω_{BD} = 1.5 rad/s ↺; ω_{DE} = 3 rad/s ↻.

15.40 ω_{BD} = 6 rad/s ↻; ω_{DE} = 2 rad/s ↻.

15.42 (a) $\mathbf{v}_A$ = 10 in./s →; ω_{AC} = 0.
(b) $\mathbf{v}_A$ = 38.1 in./s →; ω_{AC} = 0.968 rad/s ↺.
SI: (a) $\mathbf{v}_A$ = 0.254 m/s →; ω_{AC} = 0.
(b) $\mathbf{v}_A$ = 0.968 m/s →; ω_{AC} = 0.968 rad/s ↺.

15.44 (a) 1.5 rad/s ↺. (b) 0.520 m/s ∠30°.

15.46 Vertical line intersecting zx plane at $x = 0$, $z = 5.25$ ft. SI: $z = 1.600$ m.

15.48 (a) 12 rad/s ↺. (b) $\mathbf{v}_R$ = 2.4 m/s →; v_D = 2.16 m/s ∠56.3°.

15.50 (a) 3.08 rad/s ↺. (b) 83.3 in./s ↖73.9°. SI: (b) 2.12 m/s ↖73.9°.

15.52 (a) 5.33 rad/s ↺. (b) 105.8 in./s ↖49.1°. SI: (b) 2.69 m/s ↖49.1°.

15.54 (a) ω_{AB} = 2.25 rad/s ↻; ω_{BD} = 5.00 rad/s ↻. (b) 0.938 m/s ∠36.9°.

15.56 (a) ω_{AB} = 1.039 rad/s ↻; ω_{BD} = 0.346 rad/s ↺. (b) 69.3 mm/s →.

15.58 (a) 3 rad/s ↻. (b) 5.78 in./s ←. (c) 15.12 in./s ↖52.5°. SI: (a) 3 rad/s ↻. (b) 0.1468 m/s ←. (c) 0.384 m/s ↖52.5°.

15.60 (a) 0.250 m/s ∠36.9°. (b) 0.500 m/s ↑.

15.74 $\mathbf{a}_A$ = 20 ft/s² ↑; $\mathbf{a}_B$ = 4 ft/s² ↓. SI: $\mathbf{a}_A$ = 6.10 m/s² ↑; $\mathbf{a}_B$ = 1.219 m/s² ↓.

15.76 (a) 200 mm from B. (b) 225 mm from A.

15.78 $\mathbf{a}_B$ = 632 m/s² ↓. $\mathbf{a}_E$ = 632 m/s² ←.

15.80 (a) $\boldsymbol{\alpha}_A$ = 40 rad/s² ↺; $\boldsymbol{\alpha}_B$ = 20 rad/s² ↺; $\boldsymbol{\alpha}_C$ = 40 rad/s² ↺. (b) $\mathbf{a}_A$ = 10 in./s² →; $\mathbf{a}_B$ = 5 in./s² →; $\mathbf{a}_C$ = 0. SI: (b) $\mathbf{a}_A$ = 0.254 m/s² →; $\mathbf{a}_B$ = 0.1270 m/s² →; $\mathbf{a}_C$ = 0.

15.82 $\omega_{AB} = \tfrac{1}{2}\omega_0$ ↻, $\boldsymbol{\alpha}_{AB}$ = 0; ω_{BC} = 0, $\boldsymbol{\alpha}_{BC} = \tfrac{3}{2}\omega_0^2$ ↻.

15.84 (a) 296 m/s² ↑. (b) 157.0 m/s² ↑.

15.86 (a) 2.7 rad/s² ↺. (b) 5.4 rad/s² ↻.

15.88 (a) 1.848 rad/s² ↻. (b) 9.24 ft/s² ↘60°. SI: (b) 2.82 m/s² ↘60°.

15.90 (a) 1.848 rad/s² ↻. (b) 8.12 ft/s² ∠80.2°. SI: (b) 2.47 m/s² ∠80.2°.

15.92 (a) 12 rad/s² ↻. (b) 1.875 m/s² ↑.

15.94 $\omega = (v_B \sin \beta)/(l \cos \theta)$

15.96 (a) $v_B = r\omega \cos \theta$. (b) $a_B = r\alpha \cos \theta - r\omega^2 \sin \theta$.

15.98 $a_x = (v^2/r) \sin (vt/r)$; $a_y = (v^2/r) \cos (vt/r)$.

15.100 $\omega = -v_0/(2l \sin \theta)$; $\alpha = -(v_0^2/4l^2)(\cos \theta)/\sin^3 \theta$.

15.102 (a) $\omega = (v/R) \cos^2 \theta$. (b) $\alpha = -2(v/R)^2 \sin \theta \cos^3 \theta$.

15.104 (a) $\omega = (v/r)(\cos \theta)/\tan \theta$. (b) $\alpha = -(v/r)^2(1 + \sin^2 \theta)/\tan^3 \theta$.

15.106 $\boldsymbol{\alpha} = 2(v_A/b)^2 \sin \theta \cos^3 \theta$ ↻.

15.108 (a) 2.73 rad/s ↺. (b) 0.616 m/s ↗70°.

15.110 ω_{BP} = 4.38 rad/s ↻; ω_{AH} = 1.327 rad/s ↻.

15.112 (a) $\omega_{BD} = \omega_{AH}$; $\mathbf{v}_{P/AH} = l\omega$ ↓; $\mathbf{v}_{P/BD}$ = 0. (b) $\mathbf{v}_{P/AH} = l\omega/\sqrt{3}$ ↗60°; $\mathbf{v}_{P/BD} = l\omega/\sqrt{3}$ ↘60°.

15.114 $\mathbf{a}_1 = [r\omega^2 \to] + [2u\omega \downarrow]$, $\mathbf{a}_2 = [r\omega^2 \downarrow] + [2u\omega \to]$, $\mathbf{a}_3 = [r\omega^2 \leftarrow] + [u^2/r \leftarrow] + [2u\omega \leftarrow]$, $\mathbf{a}_4 = [r\omega^2 \uparrow] + [2u\omega \uparrow]$.

15.116 $a_P = \omega^2(r^2 + 4b^2)^{\frac{1}{2}}$

15.118 (a) 1.3 m/s² →. (b) 22.9 m/s² ←. (c) 16.22 m/s² ↘48.2°.

15.120 61.2 mi/h. SI: 98.5 km/h.

15.122 8.08 rad/s² ↻.

15.124 (a) 0. (b) 2.31 $l\omega^2$ ↖30°.

15.126 95.1 m/s² ↗41.9°.

15.128 (a) 6.91 rad/s² ↺. (b) 1.671 m/s² ←.

15.130 0.731 ft/s² ↙. SI: 0.223 m/s² ↙.

15.132 7.85 ft/s $\leftarrow$; 92.7 ft/s^2 $\rightarrow$.
SI: 2.39 m/s $\leftarrow$; 28.2 m/s^2 $\rightarrow$.
15.134 (a) 0.110 m/s $\uparrow$. (b) 0.1667 rad/s $\downarrow$.
15.136 (a) $\alpha_{BC} = 7.5$ rad/s^2 $\downarrow$;
$\alpha_{CD} = 3.75$ rad/s^2 $\uparrow$.
(b) 97.5 in./s^2 $\nearrow 67.4°$.
SI: (b) 2.48 m/s^2 $\nearrow 67.4°$.
15.138 2.25 rad/s $\uparrow$; 23.3 rad/s^2 $\downarrow$.

CHAPTER 16

16.2 (a) 5 m/s^2 $\leftarrow$. (b) $\mathbf{B} = 41.6$ N $\uparrow$;
$\mathbf{C} = 36.9$ N $\uparrow$.
16.4 (a) $\frac{1}{2}W$. (b) $2r/5$.
16.6 (a) $0.612g$. (b) 5.00.
16.8 (a) 2.55 m/s^2 $\rightarrow$. (b) $h \leq 1.047$ m.
16.10 17.36 ft. SI: 5.29 m.
16.12 42.3 ft. SI: 12.89 m.
16.14 5.44 m.
16.16 (a) 5.37 ft/s^2 $\leftarrow$. (b) $\mathbf{B} = 300$ lb $\uparrow$;
$\mathbf{C} = 200$ lb $\uparrow$. SI: (a) 1.637 m/s^2 $\leftarrow$.
(b) $\mathbf{B} = 1334$ N $\uparrow$; $\mathbf{C} = 890$ N $\uparrow$.
16.18 $F_{AD} = 5.39$ N C; $F_{BE} = 45.6$ N C.
16.20 (a) 27.9 ft/s^2 $\nearrow 60°$. (b) 1.443 lb $\uparrow$.
SI: (a) 8.50 m/s^2 $\nearrow 60°$. (b) 6.42 N $\uparrow$.
16.22 (a) $0.500g$ $\searrow 30°$. (b) Platform:
$1.250g$ $\searrow 30°$; block: $0.625g$ $\downarrow$.
16.24 $M_B = 6.37$ N $\cdot$ m.
16.26 89.7 N $\cdot$ m.
16.28 61.8 rad/s^2 $\downarrow$.
16.30 (1) 19.62 rad/s^2 $\uparrow$; 39.2 rad/s $\uparrow$;
19.81 rad/s $\uparrow$. (2) 14.01 rad/s^2 $\uparrow$;
28.0 rad/s $\uparrow$; 16.74 rad/s $\uparrow$.
(3) 6.54 rad/s^2 $\uparrow$; 13.08 rad/s $\uparrow$;
11.44 rad/s $\uparrow$. (4) 10.90 rad/s^2 $\uparrow$;
21.8 rad/s $\uparrow$; 10.44 rad/s $\uparrow$.
16.32 (a) 5.66 ft/s^2 $\downarrow$. (b) 8.24 ft/s $\downarrow$.
SI: (a) 1.725 m/s^2 $\downarrow$. (b) 2.51 m/s $\downarrow$.
16.34 112.2 kg $\cdot$ m^2.
16.36 153.1 lb. SI: 681 N.
16.38 $\alpha = \dfrac{2g}{r}\dfrac{\mu}{1+\mu}$ $\downarrow$.
16.40 (a) 10.44 rad/s^2 $\uparrow$. (b) 4.01 s.
16.42 (a) $\omega_A = 221$ rpm $\downarrow$; $\omega_B = 368$ rpm $\uparrow$.
16.44 $I_R = \left(n + \dfrac{1}{n}\right)^2 I_0 + n^4 I_C$.
16.46 $\alpha = (2g/r)\sin\phi$ $\uparrow$.
16.48 (a) 16.10 rad/s^2 $\uparrow$. (b) 8.05 ft/s^2 $\rightarrow$.
(c) 12 in. from B. SI: (a) 16.10 rad/s^2 $\uparrow$.
(b) 2.45 m/s^2 $\rightarrow$. (c) 0.305 m from B.

16.50 (a) 3.10 m/s^2 $\uparrow$. (b) 0.206 rad/s^2 $\uparrow$.
16.52 0.593 rad/s^2; $a_x = a_y = 0$,
$a_z = 0.200$ m/s^2.
16.54 $\mathbf{a}_A = 21.5$ ft/s^2 $\uparrow$; $\mathbf{a}_B = 10.73$ ft/s^2 $\uparrow$.
SI: $\mathbf{a}_A = 6.55$ m/s^2 $\uparrow$; $\mathbf{a}_B = 3.27$ m/s^2 $\uparrow$.
16.56 $\mathbf{a}_A = 8.46$ m/s^2 $\uparrow$; $\mathbf{a}_B = 1.362$ m/s^2 $\uparrow$.
16.58 (a) 10.00 rad/s^2 $\uparrow$. (b) 1.990 m/s^2 $\uparrow$.
16.60 (a) $1.098g/a$ $\uparrow$. (b) $1.761g$ $\nearrow 36.6°$.
(c) 1.416 g $\searrow 2.0°$.
16.62 (a) $\bar{\mathbf{a}} = \mu g$ $\leftarrow$; $\alpha = 2\dfrac{\mu g}{r}$ $\downarrow$. (b) $t = \dfrac{1}{3}\dfrac{v_0}{\mu g}$.
(c) $x = \dfrac{5}{18}\dfrac{v_0^2}{\mu g}$. (d) $\bar{\mathbf{v}} = \dfrac{2}{3}v_0$ $\rightarrow$;
$\omega = \dfrac{2}{3}\dfrac{v_0}{r}$ $\downarrow$.
16.64 75.5 mm.
16.66 (a) 107.1 rad/s^2 $\downarrow$. (b) 21.4 N $\leftarrow$,
39.2 N $\uparrow$.
16.68 (a) $\alpha = 3Pg/WL$ $\downarrow$. (b) $\mathbf{A}_x = \frac{1}{2}P$ $\leftarrow$,
$\mathbf{A}_y = W$ $\uparrow$.
16.70 13.64 kN $\rightarrow$.
16.72 490 lb $\rightarrow$. SI: 2180 N $\rightarrow$.
16.74 (a) $\frac{1}{4}W$ $\uparrow$. (b) $\frac{3}{2}g$ $\downarrow$.
16.76 (a) 27.8 ft/s^2 $\downarrow$. (b) 32.2 ft/s^2 $\downarrow$.
SI: (a) 8.47 m/s^2 $\downarrow$. (b) 9.81 m/s^2 $\downarrow$.
16.78 (a) 25.3 N $\cdot$ m $\uparrow$. (b) 52.4 N $\leftarrow$, 112.2 N $\uparrow$.
16.80 (a) 35.3 lb $\leftarrow$. (b) 2.46 lb $\rightarrow$, 26.3 lb $\uparrow$.
SI: (a) 157.0 N $\leftarrow$. (b) 10.94 N $\rightarrow$,
117.0 N $\uparrow$.
16.84 4.19 ft. SI: 1.278 m.
16.86 1.266 m.
16.88 (a) 24 rad/s$^2 \downarrow$; 3.84 m/s^2 $\rightarrow$. (b) 0.016.
16.90 (a) 8 rad/s^2 $\uparrow$; 1.280 m/s^2 $\leftarrow$. (b) 0.220.
16.92 (a) Does not slide.
(b) 23.2 rad/s^2 $\downarrow$; 15.46 ft/s^2 $\rightarrow$.
SI: (b) 23.2 rad/s^2 $\downarrow$; 4.71 m/s^2 $\rightarrow$.
16.94 (a) Slides. (b) 12.88 rad/s^2 $\uparrow$;
3.22 ft/s^2 $\leftarrow$. SI: (b) 12.88 rad/s^2 $\uparrow$;
0.981 m/s^2 $\leftarrow$.
16.96 7.16 ft/s^2 $\leftarrow$. SI: 2.18 m/s^2 $\leftarrow$.
16.98 (a) $\mathbf{a}_A = 2g/5$ $\leftarrow$; $\mathbf{a}_B = 2g/5\downarrow$.
(b) $\mathbf{a}_A = 2g/7$ $\leftarrow$; $(\mathbf{a}_B)_x = 2g/7$ $\leftarrow$,
$(\mathbf{a}_B)_y = 2g/7$ $\downarrow$.
16.100 (a) $\alpha = \dfrac{P}{2\,mr}\dfrac{\pi}{(\pi - 2)}$ $\downarrow$. (b) $\frac{1}{2}P/(mg + P)$.
16.102 51.0 rad/s^2 $\uparrow$.
16.104 (a) 23.6 N $\cdot$ m $\uparrow$. (b) $\mathbf{A} = 51.2$ N $\uparrow$;
$\mathbf{B} = 6.93$ N $\leftarrow$.
16.106 (a) 29.9 lb. (b) 76.6 lb.
SI: (a) 133.0 N. (b) 341 N.
16.108 (a) 3.53 rad/s^2 $\downarrow$. (b) 176.6 N. (c) 358 N.

16.110 $T_{BD} = 14.00$ lb; $T_{CF} = 6.36$ lb.
SI: $T_{BD} = 62.3$ N; $T_{CF} = 28.3$ N.

16.112 $A_x = 218$ lb $\leftarrow$, $A_y = 2.5$ lb $\uparrow$;
$B_x = 313$ lb $\rightarrow$; $B_y = 2.5$ lb $\uparrow$.
SI: $A_x = 970$ N $\leftarrow$, $A_y = 11.12$ N $\uparrow$;
$B_x = 1392$ N $\rightarrow$, $B_y = 11.12$ N $\uparrow$.

16.114 38.1 N $\rightarrow$.

16.116 $A_x = 3mr^2\omega_0^2 \leftarrow$, $A_y = mgr \uparrow$.
$B_x = \frac{1}{2}mr^2\omega_0^2 \rightarrow$, $B_y = mgr \uparrow$.

16.118 (a) $\frac{1}{2}a \rightarrow$. (b) $2d \rightarrow$.

16.120 (a) 74.4 rad/s^2 $\downarrow$. (b) 24.8 ft/s^2 $\downarrow$.
SI: (b) 7.56 m/s^2 $\downarrow$.

16.122 (a) 63.5 rad/s^2 $\downarrow$. (b) 20.5 lb.
SI: (b) 91.2 N.

16.124 (a) 12.14 rad/s^2 $\downarrow$. (b) 11.21 m/s^2 $\searrow 30°$.
(c) 14.56 N $\measuredangle 60°$.

16.126 $\alpha_{AB} = 8.28$ rad/s^2 $\downarrow$;
$\alpha_{BC} = 41.4$ rad/s^2 $\uparrow$.

16.128 (a) $\frac{1}{3}g \uparrow$. (b) $\frac{5}{3}g \downarrow$.

16.130 $A_x = \frac{9}{20}W \leftarrow$, $A_y = \frac{13}{40}W \uparrow$; $M_B = \frac{3}{10}WL$.

16.132 (a) $M_{\max} = PL/3\sqrt{3}$ at $L/\sqrt{3}$ below A.

16.134 $V_{\max} = \frac{1}{3}mg$ at A; $M_{\max} = 4\,mgL/81$ at $\frac{1}{3}L$ to right of A.

16.136 101.4 mi/h. SI: 163.2 km/h.

16.138 178.9 km/h.

16.140 (a) 45.0 km/h. (b) $A_n = 3470$ N $\leftarrow$,
$B_n = 1434$ N $\leftarrow$.

16.142 (a) 33.7°. (b) 0.67.

16.144 $C = 630$ N $\uparrow$, $D = 1680$ N $\downarrow$.

16.146 $A_x = -B_x = -46.0$ lb,
$A_y = -B_y = 15.33$ lb.
SI: $A_x = -B_x = -205$ N,
$A_y = -B_y = 68.2$ N.

16.148 (a) 48.3 rad/s^2. (b) $A_x = -B_x = 0.75$ lb,
$A_y = -B_y = 2.25$ lb.
SI: (b) $A_x = -B_x = 3.34$ N,
$A_y = -B_y = 10.01$ N.

16.150 (a) 4.00 N $\cdot$ m.
(b) $A_y = -B_y = -19.49$ N,
$A_z = -B_z = 8.66$ N.

16.152 $x = 2.67$ ft, $y = 4$ ft.
SI: $x = 0.813$ m, $y = 1.219$ m.

16.154 $A_x = A_y = 0$, $A_z = \frac{1}{4}\sqrt{2}\,ma\omega^2$;
$M_x = \frac{1}{6}ma^2\omega^2$, $M_y = M_z = 0$.

16.156 (a) $\cos\beta = 3g/2l\omega^2$. (b) $\sqrt{3g/2l}$.

16.158 (a) $4M_0/ma^2$. (b) $A_x = -\sqrt{2}\,M_0/a$,
$A_y = A_z = 0$; $M_x = M_y = 0$, $M_z = \frac{2}{3}M_0$.

16.160 (a) 21.5 ft/s^2 $\rightarrow$. (b) 15.95 ft/s^2 $\rightarrow$.
SI: (a) 6.55 m/s^2 $\rightarrow$. (b) 4.86 m/s^2 $\rightarrow$.

16.162 (a) $a_x = 0.3g \leftarrow$, $a_y = 0.6g \downarrow$.
(b) $a = 0.630g \downarrow$.

16.164 (a) 22.4 N $\uparrow$. (b) 30.8 N $\uparrow$.

16.166 5.57 rad/s^2 $\downarrow$; 27.2 ft/s^2 $\searrow 65.8°$.
SI: 5.57 rad/s^2 $\downarrow$; 8.29 m/s^2 $\searrow 65.8°$.

16.168 (a) $\bar{r} = \bar{k}$. (b) $\frac{1}{2}W \uparrow$.

16.170 (a) 6.54 m/s^2 $\downarrow$. (b) 5.19 m/s $\downarrow$.

CHAPTER 17

17.2 71.6 N $\cdot$ m.

17.4 14.06 in. SI: 0.357 m.

17.6 (a) 8.00 lb $\cdot$ ft $\cdot$ s^2. (b) 3.98 rev.
SI: (a) 10.85 kg $\cdot$ m^2. (b) 3.98 rev.

17.8 $v_A = 1.293$ m/s $\uparrow$; $v_B = 2.59$ m/s $\downarrow$.

17.12 71.2 rev.

17.14 18.06 N $\cdot$ m.

17.16 200 lb. SI: 890 N.

17.18 9.27 rev.

17.20 9.27 ft/s $\downarrow$. SI: 2.82 m/s $\downarrow$.

17.22 (a) 2.40 rev. (b) 21.4 N $\swarrow$.

17.24 $\sqrt{4gh/3}$.

17.26 (a) $\sqrt{4gh/7}$. (b) $2W/7$.

17.28 (a) $\sqrt{\frac{10}{7}g(R - r)(1 - \cos\beta)}$.
(b) $W(17 - 10\cos\beta)/7$.

17.30 (a) $\omega = \sqrt{3g/l}$; $R = 2.5\,W \uparrow$.
(b) 5.72 rad/s $\downarrow$; 36.8 N $\uparrow$.

17.32 (a) 1.420 m/s $\rightarrow$. (b) 1.529 m/s $\rightarrow$.

17.34 5.75 ft/s $\leftarrow$. SI: 1.753 m/s $\leftarrow$.

17.38 (a) $v_A = 1.332$ m/s $\rightarrow$;
$v_B = 0.769$ m/s $\downarrow$. (b) $v_A = 0$;
$v_B = 4.20$ m/s $\downarrow$.

17.40 3.78 ft/s $\leftarrow$. SI: 1.152 m/s $\leftarrow$.

17.42 14.63 rad/s $\downarrow$.

17.44 $v_B = 0$; $v_D = 34.0$ ft/s $\downarrow$.
SI: $v_B = 0$; $v_D = 10.36$ m/s $\downarrow$.

17.46 4.37 m/s $\downarrow$.

17.48 36.4°.

17.50 (a) Zero. (b) 188.5 W.

17.52 (a) 0.365 lb $\cdot$ ft. (b) 1.824 lb $\cdot$ ft.
SI: (a) 0.495 N $\cdot$ m. (b) 2.47 N $\cdot$ m.

17.54 (a) 3.76 m/s $\nearrow 45°$. (b) 3.18 m/s $\downarrow$.

17.56 $\omega_A = \dfrac{2n}{n^2 + 1}\sqrt{\dfrac{\pi M_0}{I_0}}$.

17.58 (a) 16.61 ft/s $\rightarrow$. (b) $A = C = 418$ lb. $\uparrow$.
SI: (a) 5.06 m/s $\rightarrow$.
(b) $A = C = 1859$ N $\uparrow$.

17.60 $\sqrt{g/3r}$.

17.62 6.47 ft/s $\downarrow$. SI: 1.972 m/s $\downarrow$.

CHAPTER 18

18.2 $128.4 \text{ lb} \cdot \text{ft}$.   SI: $174.1 \text{ N} \cdot \text{m}$.

18.4 $0.904 \, r$.

18.6 $t = (r\omega_0)/(2\mu g)$.

18.10   (a) $3.33 \text{ N} \cdot \text{m}$.   (b) $\omega_A = 23.5 \text{ rad/s} \downdownarrows$; $\omega_B = 39.2 \text{ rad/s} \circlearrowleft$.

18.12 2.72 s.

18.18   (a) $12 \text{ m/s} \rightarrow$.   (b) $100 \text{ N} \leftarrow$.

18.20 $21.5 \text{ ft/s} \leftarrow$.   SI: $6.55 \text{ m/s} \leftarrow$.

18.22   (a) $5\bar{v}_0/7 \rightarrow$.   (b) $2\bar{v}_0/7\mu g$.

18.24   (a) $3.09 \text{ rad/s} \circlearrowleft$.   (b) $-2.40 \text{ ft} \cdot \text{lb}$.   SI: (b) -3.25 J.

18.26   (a) 334 rpm.   (b) -6.51 J.

18.28   (a) and (b) 5.71 rad/s.

18.30 41.1 rpm.

18.32 $\omega_A = \omega_B = 159.1 \text{ rpm} \downdownarrows$; $\omega_P = 20.9 \text{ rpm} \circlearrowleft$.

18.34 $v_r = 3.97 \text{ m/s}$; $v_\theta = 2.86 \text{ m/s}$.

18.36   (a) $1.333 \text{ m/s} \rightarrow$.   (b) $6 \text{ kN} \leftarrow$.

18.38 $\mathbf{v}_A = 4 \text{ ft/s} \leftarrow$; $\mathbf{v}_B = 20 \text{ ft/s} \rightarrow$.   SI: $\mathbf{v}_A = 1.219 \text{ m/s} \leftarrow$; $\mathbf{v}_B = 6.10 \text{ m/s} \rightarrow$.

18.40 $2r/5$.

18.44 $L/\sqrt{3}$.

18.46 $\omega_2 = 6\bar{v}_1/7L \circlearrowleft$; $\bar{v}_2 = 3\sqrt{2}\bar{v}_1/7 \nearrow 45°$.

18.48 $\omega_2 = 3\bar{v}_1/2b \downdownarrows$; $\mathbf{v}_x = \frac{3}{4}\bar{v}_1 \rightarrow$; $\mathbf{v}_y = \frac{1}{4}\bar{v}_1 \uparrow$.

18.50 $\omega_2 = (2 + 5\cos\beta)\omega_1/7 \circlearrowleft$; $\bar{v}_2 = (2 + 5\cos\beta)\bar{v}_1/7 \leftarrow$.

18.52   (a) $\mathbf{v}_A = 0$, $\omega_A = \omega_0 \downdownarrows$; $\mathbf{v}_B = v_0 \rightarrow$, $\omega_B = 0$.   (b) $\mathbf{v}_A = 2v_0/7 \rightarrow$, $\omega_A = 2\omega_0/7 \downdownarrows$; $\mathbf{v}_B = 5v_0/7 \rightarrow$, $\omega_B = 5\omega_0/7 \downdownarrows$.   (c) The motion of part a is the final motion.

18.54 $\omega_2 = \dfrac{\bar{v}_1}{l}\, \dfrac{6\sin\beta}{3\sin^2\beta + 1} \downdownarrows$.

18.56 $59.0° \measuredangle$.

18.58 $\omega_{AB} = \frac{1}{4}\omega_0 \downdownarrows$, $\bar{v}_{AB} = \frac{1}{8}\omega_0 L \uparrow$; $\omega_{CD} = \frac{3}{4}\omega_0 \circlearrowleft$, $\bar{v}_{CD} = \frac{1}{8}\omega_0 L \downarrow$.

18.60   (a) Upward.   (b) To the left.

18.62 $7950 \text{ N} \cdot \text{m}$.

18.64 $22.7 \text{ lb} \cdot \text{ft}$.   SI: $30.7 \text{ N} \cdot \text{m}$.

18.66 $A_y = 0$, $A_z = +813 \text{ lb}$, $B_y = 0$, $B_z = -813 \text{ lb}$.   SI: $A_y = 0$, $A_z = +3620 \text{ N}$, $B_y = 0$, $B_z = -3620 \text{ N}$.

18.68 0.210 N.

18.70 1677 rpm.

18.72 6270 rpm.

18.74   (a) Slides.   (b) *Pipe:* $\bar{\mathbf{v}} = 0.981 \text{ m/s} \rightarrow$, $\boldsymbol{\omega} = 6.54 \text{ rad/s} \circlearrowleft$; *Plate:* $\mathbf{v} = 6.17 \text{ m/s} \rightarrow$.

18.76   (a) 0.522 ft.   (b) 4.47 rad/s.   SI: (a) 0.1591 m.

18.78   (a) $\omega_2 = 3\bar{v}_1/L \downdownarrows$; $\bar{v}_2 = \frac{1}{2}\bar{v}_1 \downarrow$.   (b) $\omega_3 = 3\bar{v}_1/L \circlearrowleft$; $\bar{v}_3 = \frac{1}{2}\bar{v}_1 \uparrow$.   (c) $\omega_4 = 0$; $\bar{v}_4 = \bar{v}_1 \uparrow$.

18.80 $h_A = h\left(\dfrac{3m}{6m + m_P}\right)^2$.

18.82 $\bar{\mathbf{v}}_x = mv_0/M \rightarrow$, $\bar{\mathbf{v}}_y = 3mv_0/5M \uparrow$.

18.84 $\bar{\mathbf{v}} = 6mv_0/5M \rightarrow$.

CHAPTER 19

19.2   (a) 0.1900 m.   (b) 2.39 m/s.

19.4   (a) 2.49 mm; 0.0979 in.   (b) 0.621 mm; 0.0245 in.

19.6   (a) 0.497 s.   (b) 0.632 m/s.   (c) 8.00 m/s^2.

19.8   (a) 0.1348 s.   (b) $2.24 \text{ ft/s} \uparrow$; $20.1 \text{ ft/s}^2 \downarrow$.   SI: (b) $0.683 \text{ m/s} \uparrow$; $6.13 \text{ m/s}^2 \downarrow$.

19.10   (a) 0.679 s; 1.473 Hz.   (b) 0.1852 m/s; 1.714 m/s^2.

19.12 4.55 lb.   SI: 2.06 kg (mass).

19.14   (a) 0.994 m.   (b) $3.67°$.

19.16 0.971 s.

19.18 84.4 min.

19.20 1.904 Hz.

19.22   (a) 1.794 s.   (b) 1.825 s.   (c) 2.12 s.

19.24 1.850 s.

19.26   (a) 0.297 s.   (b) 2.64 ft/s.   SI: (b) 0.805 m/s.

19.28   (a) 1.103 Hz.   (b) $T_B = 79.1 \text{ N}$; $T_C = 61.1 \text{ N}$.

19.30 $\tau = 2\pi\sqrt{2m/3k}$.

19.32   (a) $f = (1/2\pi)\sqrt{4k/m}$.   (b) $f = (1/2\pi)\sqrt{12k/m}$.

19.36   (a) $f = (1/2\pi)\sqrt{3g/2l}$.   (b) $\frac{1}{6}l$.

19.38   (a) $\tau = 2\pi\sqrt{7l/6g}$.   (b) $\tau = 2\pi\sqrt{5l/6g}$.

19.40   (a) 1.737 s.   (b) 150 mm.

19.42 $\bar{k}_x = 6.08 \text{ ft}$; $\bar{k}_z = 6.74 \text{ ft}$.   SI: $\bar{k}_x = 1.853 \text{ m}$; $\bar{k}_z = 2.05 \text{ m}$.

19.44   (a) $\tau = (2\pi l/b)\sqrt{h/3g}$.   (b) $\tau = 2\pi\sqrt{h/g}$.

19.46   (a) 0.288 s.   (b) 0.458 m/s.

19.48 0.456 in.   SI: 11.58 mm.

19.54 $498 \text{ lb} \cdot \text{ft} \cdot \text{s}^2$.   SI: $675 \text{ kg} \cdot \text{m}^2$.

19.56 $\tau = 2\pi\sqrt{0.866l/g}$.

19.60 $\tau = 2\pi\sqrt{2l_1 l_2/g(l_1 + l_2)}$.

19.62 0.926.

19.66 $\tau = 2\pi \sqrt{m/3k \cos^2 \beta}$.

19.68 $\tau = 2\pi \sqrt{m_c/k \cos^2 \beta}$.

19.70 (a) 1.429 s. (b) 1.845 s.

19.72 2.48 s.

19.74 $\sqrt{k/2m} < \omega < \sqrt{3k/2m}$.

19.76 $\sqrt{g/2l} < \omega < \sqrt{3g/2l}$.

19.78 (a) 400 rpm. (b) 0.00167 in.
SI: (b) 42.3 μm.

19.80 0.703 mm.

19.82 0.750 in. or 0.1875 in.
SI: 19.05 mm or 4.76 mm.

19.86 0.794 mm.

19.88 245 rpm.

19.90 0.509 Hz; 0.657 Hz.

19.92 70.1 km/h.

19.100 (a) $x = x_0 e^{-pt}(1 + pt)$. (b) 0.1108 s.

19.102 $\sqrt{1 - 2(c/c_c)^2}$.

19.104 (a) 1.509 mm. (b) 0.583 mm.

19.106 0.1194 in. SI: 3.03 mm.

19.108 (a) 270 rpm. (b) 234 rpm. (c) 8.84 mm;
9.45 mm.

19.110 $m\ddot{x}_A + 5kx_A - 2kx_B = 0$;
$m\ddot{x}_B - 2kx_A + 2kx_B = P_m \sin \omega t$.

19.112 (a) E/R. (b) L/R.

19.118 (a) $m\ddot{x}_m + k_2 x_m + c(\dot{x}_m - \dot{x}_A) = 0$;
$c(\dot{x}_A - \dot{x}_m) + k_1 x_A = 0$

(b) $L\ddot{q}_m + \dfrac{q_m}{C_2} + R(\dot{q}_m - \dot{q}_A) = 0$;

$R(\dot{q}_A - \dot{q}_m) + \dfrac{q_A}{C_1} = 0$.

19.120 (a) $m_1\ddot{x}_1 + c_1\dot{x}_1 + c_2(\dot{x}_1 - \dot{x}_2) + k_1 x_1$
$+ k_2(x_1 - x_2) = 0$;
$m_2\ddot{x}_2 + c_2(\dot{x}_2 - \dot{x}_1) + c_3\dot{x}_2$
$+ k_2(x_2 - x_1) + k_3 x_2 = P_m \sin \omega t$.

(b) $L_1\ddot{q}_1 + R_1\dot{q}_1 + R_2(\dot{q}_1 - \dot{q}_2) + \dfrac{q_1}{C_1}$

$+ \dfrac{q_1 - q_2}{C_2} = 0$;

$L_2\ddot{q}_2 + R_2(\dot{q}_2 - \dot{q}_1) + R_3\dot{q}_2$

$+ \dfrac{(q_2 - q_1)}{C_2} + \dfrac{q_2}{C_3} = E_m \sin \omega t$.

19.122 1.363 s.

19.124 5.98 in. SI: 151.8 mm.

19.126 (a) 5.81 Hz; 4.91 mm; 179.2 mm/s.
(b) 491 N. (c) 159.2 mm/s ↑.

19.128 4.8 mm.

19.132 2.48 s.

APPENDIX

9.72 (a) $I_{AA'} = mb^2/24$; $I_{BB'} = mh^2/18$.
(b) $I_{CC'} = m(3b^2 + 4h^2)/72$.

9.74 (a) $\frac{1}{4}m(r_1^2 + r_2^2)$. (b) $\frac{1}{2}m(r_1^2 + r_2^2)$.

9.76 $ma^2/20$.

9.78 $\frac{1}{3}ma^2$; $a/\sqrt{3}$.

9.80 $\frac{1}{6}m(a^2 + 3h^2)$; $[(a^2 + 3h^2)/6]^{1/2}$.

9.82 $\frac{1}{20}m(a^2 + 12h^2)$.

9.84 $\frac{3}{10}m\dfrac{r_2^5 - r_1^5}{r_2^3 - r_1^3}$.

9.86 1.514 kg·m²; 155.8 mm.

9.88 (a) $m(\frac{2}{5}a^2 + R^2)$.
(b) $(a/R)^2/[(a/R)^2 + 2.5]$. (c) 9.98a.

9.90 $5ml^2/24$, $ml^2/12$, $7ml^2/24$.

9.92 0.000283 lb·ft·s²; 1.031 in.
SI: 0.384 × 10⁻³ kg·m²; 26.2 mm.

9.94 (a) 1.575 lb·ft·s². (b) 2.714 lb·ft·s².
(c) 1.202 lb·ft·s². SI: (a) 2.13 kg·m².
(b) 3.68 kg·m². (c) 1.629 kg·m².

Centroids of Common Shapes of Areas and Lines

Shape		$\bar{x}$	$\bar{y}$	Area
Triangular area			$\dfrac{h}{3}$	$\dfrac{bh}{2}$
Quarter-circular area		$\dfrac{4r}{3\pi}$	$\dfrac{4r}{3\pi}$	$\dfrac{\pi r^2}{4}$
Semicircular area		0	$\dfrac{4r}{3\pi}$	$\dfrac{\pi r^2}{2}$
Semiparabolic area		$\dfrac{3a}{8}$	$\dfrac{3h}{5}$	$\dfrac{2ah}{3}$
Parabolic area		0	$\dfrac{3h}{5}$	$\dfrac{4ah}{3}$
Parabolic spandrel		$\dfrac{3a}{4}$	$\dfrac{3h}{10}$	$\dfrac{ah}{3}$
Circular sector		$\dfrac{2r\sin\alpha}{3\alpha}$	0	αr^2
Quarter-circular arc		$\dfrac{2r}{\pi}$	$\dfrac{2r}{\pi}$	$\dfrac{\pi r}{2}$
Semicircular arc		0	$\dfrac{2r}{\pi}$	πr
Arc of circle		$\dfrac{r\sin\alpha}{\alpha}$	0	$2\alpha r$